GREEN MURDER

A LIFE SENTENCE OF
NET ZERO WITH NO PAROLE

IAN PLIMER

connorcourt
PUBLISHING

This book is dedicated to Honorary Dr Gina Rinehart, one of the very few prominent business leaders in the world who has called human-induced climate change for what it is: nonsense. Other business leaders run with the hare and hunt with the hounds and do not have her courage, perspicacity, moral fibre and leadership qualities.

Mrs Gina Rinehart is the most remarkable woman Australia has ever produced and, despite being generous, highly intelligent, astute, charming and caring for her country, she gets constantly pilloried by the envious green left who have yet to make a contribution to the nation.

This book is endorsed by the environmental group Sleeping Giants™ without reservation

Information, podcasts and comments are on www.greenmurder.com

Published in 2021 by Connor Court Publishing Pty Ltd

Second Edition, Published by Connor Court Publishing Pty Ltd, 2022.

Connor Court Publishing Pty Ltd
PO Box 7257
Redland Bay QLD 4165
sales@connorcourt.com
www.connorcourt.com
Phone 0497-900-685

Printed in Australia

ISBN: 9781922449825

Front cover design: Ian James

Contents

About the author

PROFESSOR IAN PLIMER is Australia's best-known geologist. He is Emeritus Professor of Earth Sciences at the University of Melbourne, where he was Professor and Head of Earth Sciences (1991-2005) after serving at the University of Newcastle (1985-1991) as Professor and Head of Geology. He was Professor of Mining Geology at The University of Adelaide (2006-2012) and in 1991 was also German Research Foundation Professor of Ore Deposits at the Ludwig Maximilians Universität, München (Germany). He was on the staff of the University of New England, the University of New South Wales and Macquarie University. He has published more than 120 scientific papers on geology and was one of the trinity of editors for the five-volume *Encyclopedia of Geology*. This is his eleventh book written for the general public, the best known of which are *Telling lies for God* (Random House, 1994), *Milos-Geologic History* (Koan, 1999), A *Short History of Planet Earth* (ABC Books, 2000), *Heaven and Earth* (Connor Court, 2009), *How to get expelled from school* (Connor Court, 2011), *Not for greens* (Connor Court, 2014), *Heaven and Hell* (Connor Court, 2016) and *Climate change delusion and the great electricity rip-off* (Connor Court 2017). He frequently has newspaper and magazine opinion pieces published as well as chapters in books. This book is that of a scientist which has been a labour of love and hate unfunded by third parties.

He won the Leopold von Buch Plakette (German Geological Society), the Clarke Medal (Royal Society of NSW), the Sir Willis Connolly Medal (Australasian Institute of Mining and Metallurgy). He is a Fellow of the Australian Academy of Technological Sciences and Engineering and an Honorary Fellow of the Geological Society of London. In 1995, he was Australian Humanist of the Year and later was awarded the Centenary Medal. He was Managing Editor of *Mineralium Deposita*, president of the SGA, president of IAGOD, president of the Australian Geoscience Council and sat on the Earth Sciences Committee of the Australian Research Council for many years. He won the Eureka Prize for the promotion of science, the Eureka Prize for *A Short History of Planet Earth* and the Michael Daley Prize (now a Eureka Prize) for science broadcasting. He was an advisor to governments and corporations and was a regular broadcaster.

Professor Plimer spent much of his life in the rough and tumble of the zinc-lead-silver mining town of Broken Hill where an interdisciplinary scientific knowledge intertwined with a healthy dose of scepticism and pragmatism are necessary. He is Patron of Lifeline Broken Hill and the Broken Hill Geocentre. He worked for North Broken Hill Ltd, was a consultant to many major mining companies and has been a director of numerous exploration public companies listed in London, Toronto and Sydney. In his post-university career he is proudly a director of a number various unlisted private Hancock Prospecting companies.

A new Broken Hill mineral, plimerite, was named in recognition of his contribution to Broken Hill geology. Ironically, plimerite is green and soft. It fractures unevenly, is brittle and insoluble in alcohol. A ground-hunting rainforest spider *Austrotengella plimeri* from the Tweed Range (NSW) has been named in his honour because of his "provocative contributions to issues of climate change". The author would like to think that *Austrotengella plimeri* is poisonous.

Ian Plimer identifies as $ZnFe_4(PO_4)_3(OH)_5$ and demands the appropriate pronoun be used.

Acknowledgements

The idea for a book evolved from discussions, reading and thinking about the practical aspects of green solutions to what are non-problems. Some of the ideas came from long discussions with my stepson Bob who thought deeply, looked at the history of innovation, experimented and thought in a different dimension to most people. On some matters we did not agree but the juices got going for both of us.

Ten months of 2020 were spent in Melbourne for treatment for a pretty aggressive cancer and during daily treatment I managed to read metres of background literature, type notes one-handed while a cannula was in the other arm and I thank Dr Peter Eng for providing me with a combined clinic-office so I could focus on and think positively about this book.

While in Adelaide in 2021, work duties were combined with writing and continuing daily cancer treatment. In the last 6 months of intense writing, my wife Maja fed and watered me, endured the long hours of my physical and mental absence, watched the house, garden, studies and libraries become shambolic and gave a huge amount of support for what evolved into an almost never-ending task. Companions during writing were the dog who enjoyed being at my feet for so long and the other was Radio Swiss Classic.

My employer gave me the space, time and encouragement to abandon all but essential work for a few months to complete this book. Regarding temperature and carbon dioxide, it was Margaret Watroba's question I used in this book: *"What came first, temperature or carbon dioxide?"*

The three reviewers, Andrew Drummond, Dr Ø[2] and Jeff Rayner, spent weeks of their own time editing the writings of someone who is a bad typist, somewhat dyslectic, tends to stray off on tangents and had a few live versions of the same document just to make their lives interesting. I owe heartfelt thanks for their massive effort and the opinions in this book are mine, not theirs. Andrew Drummond edited the book *Rocks in our heads: Stories of exploring for mineral deposits in exotic lands* (Connor Court Publishing, 2020) and his editing experience and general knowledge showed. His book shows that geologists work in all sorts of environments around the world, sometimes they are placed in very dangerous situations and those who related their experiences were those who survived. In today's world wokeness, cancellation, censorship and banning of freedom of speech has happened in many organisations around the world. Dr $\varnothing^2$, already an elegant writer, was a reviewer and was meticulous, pedantic, intellectually rigorous, critical and helpful. Dr $\varnothing^2$'s idea ended up as the cover of the book. It is an appalling world when a senior person like Dr $\varnothing^2$ is unable to be named in a robust book for fear of a career growth impediment in years to come. Dr $\varnothing^2$'s contribution has been acknowledged in other ways. Jeff Rayner, a work colleague from the Middle East, Africa, Australia and South America went to a lot of trouble to check on my statements, references and inconsistencies. His literature searches found material that I had overlooked. I could not have had three better peer review editors. Andrew Davis and John Loudon looked at very early drafts and Ross Cameron made useful comments on the last version.

My publisher, Dr Anthony Cappello of Connor Court Publishing, needs to receive a medal struck for dealing with such a tome, a large number of diagrams, constant impossible demands from the author and time pressures that I put on him but would not put on myself.

1

INTRODUCTION

If green activism achieves its aims, the Third World will remain in poverty. Western countries will become impoverished and even more reliant on China which uses climate change as a weapon against the West.

Many green leaders have an unhealthy obsession with death, killing, catastrophes, totalitarianism and restriction of freedoms.

Green leaders are anti-environment, hypocritical and fraudulent and use green policies to make money for woke Western businesses and China.

If green activists were actually concerned about their fellow humans and the environment, they would support cheap reliable 24/7 fossil fuel- and nuclear-generated electricity.

Over the last 50 years, global *per capita* GDP, longevity, health, crop land productivity and forests have increased due to technology, use of fossil fuels and a slight increase in plant food in the air. We are now feeding more people from less land. Global pollution has decreased. The planet is improving. Wealth has solved environmental problems, not greens.

Coal brought people out of grinding poverty and misery in the Industrial Revolution and later hundreds of millions of Chinese out of crippling poverty. Policies of green activists attempt to reverse these gains.

It has never been shown that human emissions of carbon dioxide drive global warming. Notwithstanding, Australia sequesters more carbon dioxide than it emits.

If green activists or school children on strike are to be taken seriously, they should demonstrate in Tiananmen Square against the world's biggest emitter of carbon dioxide and should give up the use of all electronic equipment.

Mining and export of coal, iron, uranium and other minerals keeps Australia solvent. If green activists want to maintain a society that supports social security, subsidies and green schemes, they must be loud public supporters of mining and emission-free nuclear power otherwise they are hypocrites.

Australia has 24 coal-fired power stations compared to the planned and existing plants in China (3,543), India (1,036) and the EU (495). Civilisation advances one coal-fired power station at a time. Closing Australian power stations will have no effect whatsoever on global carbon dioxide emissions.

The long march of the left through schools and universities has produced green activists and a generation that cannot write, read, calculate, think, solve problems and look after themselves. They have no knowledge of the past, Western civilisation, science, critical thinking and the brutalities of previous communist and socialist regimes.

When green activists, politicians and celebrities live in caves as sustainable hunter gatherers, we may seek their advice. While they travel around the world in private jets to chastise us for our emissions of carbon dioxide, they are hypocrites.

Trigger warning

For more than 50 years, green activists made false and failed predictions about the end of the world, famine, health, pollution, sea level rise, ice sheet melting, extinction, global cooling, global warming and population. They still do. If just one prediction was correct, we would not be here.

Green's promotion of renewable energy under the guise of human-induced climate change gives China complete dominance over our energy systems, infrastructure, food production, manufacturing and defence. We are exposed, have learned nothing from the 1930s and will have to rebuild the country. Green activists are essentially Chinese prostitutes.

Many green leaders have an unhealthy obsession with death, killing, poverty, catastrophes, totalitarianism, restriction of freedoms and gaoling or killing those who have an alternative view.[1,2,3,4] This is no surprise considering the green movement arose from eugenics and totalitarianism.

Instead of mounting convincing arguments using knowledge and logic, green activists resort to abuse, threats, cancelling and violence reflecting their totalitarian tendencies.

Virtue signalling by greens attempting to demonstrate their moral superiority has failed. Green activists use fossil fuels for some or most of their energy and some of the 5,000 products from petroleum, including medicines. By their actions, they are hypocrites.

The green movement must be constantly brought to account to avoid lowering the quality of life and standard of living, national bankruptcy resulting in a loss of sovereignty, loss of property and freedoms and an education system that produces idealogues rather than people who can think, create, solve problems, add value to society and look after themselves.

The green policies produce infant mortality, shortened lives, poverty, pollution and starvation and allow fatal curable diseases to persist. Many greens state their wish to greatly reduce population and, although the number to be exterminated is often given, the method of extermination is unspecified.

Good to be alive

By any measure, we live in the best of times and not the worst of times. There should be street demonstrations thanking aspirational past generations for grinding through hardship to give us such a wonderful world. Of course, there are still many wrongs to be righted but if the past is any measure, then we can only have a positive and constructive view of the future.

The world gross product of the last 2,000 years was static until the late 19th Century and then it rose 1,000-fold concurrent with a decrease in the proportion of the global population in abject poverty despite the increase in the global population. Food supply has increased as has the increase in forests. The number of democracies, global happiness, school enrolment worldwide, average years of education, women in politics, global literacy rates, global life expectancy and quality of life, growth of protected areas worldwide, global cereal production and yields, global meat, fish and dairy consumption, global access to electricity, global access to improved potable water and sanitation, global access to mobile phones, oil and gas reserves have risen contrary to 'peak oil' hysteria, and the internet and global international tourism all have increased. Since the time of Jesus, global GDP *per capita* has increased 30,000 times.

Global trend	Direction
World gross product and per capita GDP 1-2000	Up
Global population in absolute poverty 1820-2018	Down
Food supply 1961-2017 (calories per person per day)	Up
Tree cover 1982-2016	Up
Global urban population 1950-2015	Up
Democracies 1800-2017	Up
Autocracies 1800-2017	Down
Worldwide interstate wars 1946-2017	Down
Deaths from natural disasters 1900-2016	Down
UN human development index 1990-2017	Up
Global income inequality 1952-2017	Down
People living in slums 1990-2014	Down
Women in politics 1907-2018	Up
Births per woman 1700-2017	Down
Global literacy rates 1820-2016	Up
Gross school and university enrolment worldwide 1970-2017	Up
Average years of education 1870-2020	Up
Global life expectancy 1500-2016	Up
Global deaths per 1,000 people 1960-2017	Down
Worldwide infant mortality 1950-2016	Down
Maternal deaths per 100,000 births 1990-2014	Down
Global access to potable water and improved sanitation 1990-2015	Up
Global mobile phone subscriptions and internet access 1990-2017	Up
Global homicide rate 1990-2016	Down
Global undernourishment rate 1969-2016	Down
Global cereal production and yields 1961-2016	Up
Global access to electricity 1990-2016	Up
Average hours worked in high income countries 1950-2018	Down
Work-related death rates 1913-2015	Down
Global child labour trends 2000-2016	Down
Wage gap in high-income countries 1995-2016	Down
Countries with legal slavery 1800-2018	Down
Growth of protected areas worldwide 1872-2017	Up
Global carbon dioxide emissions kg/2010 dollar of GDP 1960-2014	Down
Global oil and gas reserves and production 1980-2018	Up

Table 1: Global trends showing that we are lucky to live now rather than 50 or 100 years ago.[5]

Autocracies, wars, deaths from natural disasters, global death rate, infant mortality, maternal mortality, global homicide and battle deaths, military expenditure and nuclear warheads, births per woman, people living in slums, average hours worked in high income countries, work-related deaths, wage gap in high income countries, the number of countries with legal slavery and child labour, global income inequality, carbon dioxide emissions per dollar of GDP, global undernourishment rate, global hours worked to produce one unit of a commodity and average cost of computer storage all have decreased.

Abundant cheap reliable energy has solved global problems. As a result, food has become more abundant, healthier and cheaper. Energy has led to a great increase in health, diet, longevity, finances and leisure over the last five generations of humans using any measure we chose. There was no change over the preceding 20,000 generations of humans and change was brought about by fossil fuels.

The world has changed from a rural subsistence population in brutal poverty to a city-based population employed in service industries. Because of geology, mines are in isolated areas and food growing is mechanised requiring far fewer people. Cheap reliable energy is required for both. Many mines have their own independent electricity source. City populations are now detached from where and how their food, fibre, water, energy and metals are produced.

Green activists ignore the fact that we live in the best time ever to be a human on planet Earth. Rather than trying to bring us back to the horrible times centuries ago, they should be positive and try to make the world an even better place. If green activists disagree, they should give up everything that the last five generations using cheap fossil fuel brought them.

Perspective

It has never been shown that human emissions of carbon dioxide drive global warming.

Over the history of time, six of the six great ice ages started when there was more carbon dioxide in the air than now. If high atmospheric carbon dioxide drives global warming, how could ice ages have occurred?

Who leads the dance? Temperature or carbon dioxide? Which came first? Warming or an increase in carbon dioxide in the air? Ice core measurements going back almost a million years show that temperature rises 650 to1600 years **before** the carbon dioxide in the air rises.

During each ice age, ice cover expanded (glaciation) and retreated (interglacial). We are currently in an interglacial phase of an ice age that started 34 million years ago.[6] For more than 80% of time, there has been no ice on planet Earth.

There are over 150 volcanoes along a volcanic chain stretching 3,000 km under the ice in Antarctica.[7,8,9] The effect of these is not used in any calculation of Antarctic ice melting or climate change models.

There have been warming and cooling events during the current 14,700 year-long interglacial. Previous warmings were warmer than the current times. Why are these previous warmings natural yet the latest warming is claimed to be due to small emissions of a trace gas? Previous Grand Solar Minima have brought bitter cold. We are at the start of another Grand Solar Minimum. Are we going the wrong way along a one-way street?

The Intergovernmental Panel on Climate Change (IPCC) states that 3% of all emissions of carbon dioxide are from humans. Even if human emissions of carbon dioxide did drive global warming, why is it that only human emissions drive global warming whereas the 97% comprising natural carbon dioxide emissions don't?

If 3% of all annual emissions drive global warming, it must also be shown that the 97% of emissions from natural processes don't. This has not been done. Natural carbon dioxide emissions are mainly from ocean degassing.

If the 24 million people in Australia produce 1.3% of the total of 3% of annual global carbon dioxide emissions[10] then no matter what Australia does to reduce emissions it will have no effect whatsoever on global climate, especially as the annual increase in emissions from China far outweighs Australia's total annual emissions.

USA emits 14 times and China emits 26 times more carbon dioxide than Australia. Australia has 0.33% of the global population. Even if it could be proven that human emissions drive[11] global climate change, why do anything unless China greatly reduces emissions?

China has very clearly indicated that they will follow the path of economic growth driven by burning coal rather than handcuffing themselves with the ideology of climate change.[12] Many predicted that COP26 would be FLOP26, in part due to China and India wanting to bring their own people out of poverty using cheap coal-fired electricity.

Australia's emissions derive from domestic activities and from food production for 80 million people and metal and ore production. Australia exports a significant share of the global refined aluminium, zinc, lead, copper, nickel and gold and hence takes a large *per capita* emissions hit for countries that import and use Australia's food and metals because agriculture, smelting and refining in Australia result in carbon dioxide emissions from burning fossil fuels.

In 1997, China was granted an exemption[13] from any obligations to reduce carbon dioxide emissions under the Kyoto Protocol (as was India) on the ground that it was a "developing nation". Despite its posturing over the Paris Agreement, China's carbon dioxide emissions have soared as it has grown more powerful economically and militarily. China claims it is a Third World poverty-stricken backwater yet it has nuclear weapons, the world's largest army, the world's second largest economy and has landed rovers on the Moon and Mars.

India also has a space program, nuclear weapons and a large army. China has also become more totalitarian and intolerant and is crushing dissent within its borders and beyond them. It builds more coal-fired power stations than any other country on Earth and remains exempt from the draconian financially crippling demands of the Paris Agreement. We've been conned.

In 2021 the US coal production was rising because of demand associated with an improving economy and an increase in gas prices. And this is despite President Biden's climate goals and rejoining the Paris Accord scrapped by President Trump.[14] As global economies improve after COVID-19, global carbon dioxide emissions have risen driven by the resurgence of coal use in the power sector.[15] I can hear plants cheering.

International agreements, such as the non-binding Paris Accord, are totally useless. Only three of the 194 countries that signed the Paris Accord met their promised emissions for 2019. They were the Marshall Islands, Suriname and Norway which contribute 0.1% of all global emissions.[16,17] China's Paris Accord commitment is business as usual and it emits more annual emissions than any other country. Why would any country even bother to send taxpayer-funded delegates to such COP charades to sign a piece of toilet paper?

There is no point in getting into a lather about the Paris Agreement. The commitments of Paris signatories vary from country to country, countries are free to determine their own commitments and timetables and there is a huge disparity between promises that politicians make in the global spotlight at talk fests around the world and the policies they implement back home. The IPCC has no powers to enforce compliance and does not even know the costs of compliance.[18]

More than 90% of all new emissions are in nations exempt from the Paris Agreement.[19] Why should a miniscule emitter such as Australia even bother? The UN has got itself into a bit of a lather because most nations ignore the Paris Agreement.[20] How dare sovereign states act in their own interests, look after their economies and not do the same as every other nation!

Past climatic changes have been driven by moving continents and ocean plates (tectonic cycles), supernovae explosions, changes in the Earth's orbit solar cycles, ocean cycles and lunar cycles. We are expected to believe that these monstrous planetary forces changed climate in the past but now don't change climate and that a slight increase in a trace gas in the air drives a major planetary system.

Climate has always changed. Scotland was once a desert; inland desert Australia and North Africa were once covered by tropical vegetation; when Australia was at the South Pole it had temperate forests, grasslands and roaming dinosaurs; the sites of the major cities of Australia were once covered by ice. More recently, sites of New York, London, Paris and Berlin were covered by ice.

Geology textbooks in the 18th, 19th and early 20th Centuries contain much scientific discussion about the most recent and ancient climate changes. Geologists were studying climate "science" more than 250 years before the scary climate "science" of today was invented.

We have already adapted to live in a great diversity of climates and, if there was a change in climate, it would not create any existential problems for the human race as we already live in areas with temperatures as low as -50°C and as high as +50°C.

What is the ideal climate for a human? The climate that Eskimos prefer might not be the same climate that a Bedouin or subsistence hunter in Borneo prefers. Humans adapt. There is no ideal climate for humans and with energy, technology and innovation, we have already shown that we can survive climate extremes.

Grim facts

The world is a far better place than decades or centuries ago. Increased affluence makes the world a better place and affluence enables us to finance and solve any potential environmental problems. Up until the 17th Century, foot power and wood were used to produce energy. In Europe, people and animals died like flies and land was completely deforested a number of times for the making of iron and glass and ship building. In the 18th Century, wood was the main source of energy and forests were destroyed at a very rapid rate. The area of forests now is far greater than it was in the 18th Century.

In the 19th Century, coal became the main form of energy. This was a period of great innovation. It was the high energy density of coal that drove this innovation and the Industrial Revolution. In the 20th Century, it was oil that drove industrialisation, transport and trade and coal was used for smelting and coal-fired power stations for industry and domestic uses. Once there was abundant cheap coal-fired electricity in the 20th Century, economic growth accelerated. Electricity is the multiplier of economic growth.

If it were not for coal, the forests of the UK, US, Canada and Europe would have disappeared during industrialisation. The 21st Century is the century of gas. Industry, domestic heating and cooking, transport, power generation and some smelting uses gas. And what will the future centuries bring?

Resources of gas as methane hydrates are huge, cheaper methods of creating hydrogen are constantly being invented and we may finally enter a nuclear age with fission and perhaps fusion. All of my life, nuclear fusion has been only 20 years away. It's getting close to being a reality.

A city could be powered from a device the size of a shipping container, nuclear fusion has an energy density far greater than any other energy source, it's far safer than conventional nuclear fission and the waste product is inert non-radioactive helium. The environmental footprint is tiny, there is no disposal of toxic waste such as from turbine blades and solar panels and would fulfil the desires of those that lust for Net Zero.[21]

Some 75% of the original European forests have now been cleared for farming, industry, infrastructure and housing yet Europe criticises Brazil for doing the same. Do the European greens want to keep Brazilians in poverty and deny them the living standards of Europe that Western civilisation and capitalism has brought to it?

tnog_

In 1908, fossil fuels were about 85% of total energy consumption. In the nuclear age, it was 81% in 2020. China is still adding two coal-fired power stations a week. In the first half of 2021, China approved the building of 43 new coal-fired power stations and 18 new blast furnaces for steel manufacture. This will add 1.5% to its current annual emissions[22].

China is pressing ahead with building coal-fired electricity generators in the developing world yet Beijing claims it will slash carbon dioxide emissions. In 2020, China opened three quarters of the world's new coal generators, more than 80% of the newly announced coal power plants[23] and burned 53% of the world's coal despite Xi Jinping's net zero by 2060 pledge.[24] A few decades ago, China was at the level of 18th Century Europe. Who can blame China trying to get into the 21st Century in record time?

The Western world is going the opposite direction. In Western countries, there has been a market-driven decarbonisation with a higher proportion of gas burned. Gas burning releases less carbon dioxide than burning coal. The slight decarbonisation was not driven by wind or solar power. In 2003, nuclear contributed to 6% of all energy consumed. Now it is 4% and the US Energy Information Administration[25] forecast it will rise to 6.7% in 2025.

One only has to travel to some African countries to see that environmental degradation results from poverty. As these countries creep out of grinding poverty, the environmental degradation decreases. The International Energy Agency states that more than 75% of people in Africa have no electricity.[26]

In most of the rest of sub-Saharan African countries only 50 to 75% of people have electricity. There are 150 million people in Ethiopia, Congo and Sudan that have absolutely no electricity.[27] And in countries where there may be electricity, the wires may be live for two hours every second day. Wires very often get stolen and sold for their metal value. Hospitals, schools and productive businesses cannot be run with irregular electricity.

Without cheap electricity, prosperity can only be a dream. Nearly 730 million Africans rely on wood, twigs, leaves and dung for cooking and heating and 620 million have no access to electric lights. That's two thirds of the population. In sub-Saharan Africa, 40% of people in urban areas have no electricity and 85% of people in rural areas have no electricity.

A report on global sustainability released by the World Health Organisation (WHO), UNICEF and the scientific journal *The Lancet* and funded by the

Bill and Melinda Gates Foundation put 10 African nations at the top of the list of 180 nations.[28] Australia was 174 on the list, just behind the US and ahead of Saudi Arabia and Bahrain. The authors stated that developed nations *"were threatening the future of children because of high 'carbon' emissions, junk food, alcohol advertising and time spent on-line"*.

I'm sure the children of Africa are tickled pink that they live so sustainably enjoying their short life of child slavery, child soldiering, famine and curable diseases and, without electricity, education opportunities and hope. Failed African states are among the most diseased, cruel, corrupt and violent in the world and are torn apart by wars, coups and poverty.

The WHO director-general Tedros Adhanom Ghebreyesus described the report as a *"wake up call."* He's right. The ideological drivel, set out in bureaucratic speak, should wake up Western politicians to rethink why they send millions of taxpayers dollars to the UN and its affiliates each year.

Globally, 1.3 billion people don't have access to electricity and 2.7 billion people rely on wood, twigs, leaves and dung for cooking and heating which causes harmful indoor air pollution and death. Without cheap electricity, production cannot be increased, goods can't get to market, vaccines cannot be refrigerated and hundreds of millions of young people cannot study after the Sun goes down to escape poverty via education.

If green activists were really concerned about their fellow humans and the environment, they would hold demonstrations and press conferences to support cheap reliable 24/7 electricity for Africa.

At the Millennium Summit of 2000, the UN declared that universal access to energy should be a development goal but only if it is *"sustainable"*. African countries added 8.4 GW of new coal-fired generating capacity between 2006 and 2019 and EU countries added 23 GW over the same period despite having half the population. Germany, with 8% of the area of sub-Saharan Africa and twice the GDP added 10 GW of coal-fired generation.

If the world's wealthiest and most advanced economies cannot dismantle their own energy infrastructure and replace it with renewables, then how can they expect poor African countries to do so starting from scratch? Without cheap reliable electricity, there is no cooking, lighting, heating and cooling in African households.

Without cheap reliable electricity in sub-Saharan Africa, factories, schools, hospitals, businesses and transport cannot function, there is no investment, economies cannot develop and massive unemployment continues. This is the perfect recipe for terrorism. Why should the EU deny Africans what Europeans take for granted? The UN, EU and green activist hypocrites are killing Africans.

The world is now feeding more mouths from less farm land, consuming more calories and spending less income proportionally on food than previously. Fossil fuels have driven greater productivity. More people before COVID-19 were spending their increased wealth with travel to poor countries where they spent money.

In 1800, more than 90% of people in the world were extremely poor with incomes the same as the current poorest countries in Africa. The development of capitalism and economic growth reduced the proportion of extremely poor people in the world to less than 10% despite a sevenfold population increase. Growth has led to a reduction in hunger and poverty. Don't give me the good old days. They weren't.

Effective nutrition in the first two years of a child's life helps develop the brain, improves education results and creates better skilled adults. Nutrition costs $140 per child and boost that person's income by $4,500. Why do greens focus on climate change rather than human problems in the Third World? Are the green movements anti-human?

The world has got better. In the times of Cicero in the first century AD, average life expectancy was 25, a third of all babies died within a month of birth, 50% of the population was under 20 and about 80% were dead by 50. They didn't use coal and didn't have electricity. For decades, the BP Statistical Review of World Energy showed that over 80% of global energy consumed came from fossil fuels for transport (oil), heating (gas) and electricity (coal and gas) and smelting (coal).[29]

Life expectancy at birth has increased more than twice as much in the last century than in the previous 200,000 years. Of the 20,000 generations of *Homo sapiens*, only the last four have experienced massive declines in mortality rates.

In the last 140 years, there have been 106 major famines, each of which has cost millions of lives. The death toll has been particularly high in socialist countries such as the Soviet Union, China, Cambodia, Ethiopia and North Korea. Tens of millions of their people were killed through forced transfer of private means of production to governments and the weaponisation of hunger. Central governments are not the best solution to food production and environmental problems. The biggest socialist experiment in history, Mao's Great Leap Forward in the 1950s, killed more than 50 million Chinese.

Since 1970, the rate of increase in the population has slowed. As the *per capita* GDP in a country increases, the birth rate decreases and demands on the environment also decrease. Poverty and environmental problems are solved by wealth. Why is the wolf population increasing, the lion population decreasing and the tiger holding its own?

Wolves live in rich countries, lions in poor countries and tigers in middle-income countries. We can now see wolves in Europe more easily because we are getting taller due to better diet. Prosperity of the people is the best thing that can happen to a country's wildlife and environment.

As people get richer, they can afford to buy electricity rather than chop down trees, they can buy meat rather than hunt for wild animal bush meat and they can get a job in town rather than scratch a living by flattening a forest and trying to grow crops inefficiently on a pocket handkerchief-sized block of land.

They can stop worrying about whether their children will starve and they can care for the environment. Over the last decades, the area of land needed to produce a given quantity of food was 65% less than in 1961 due to better fertilisers, pesticides and biotechnology.[30]

Where genetically modified crops are grown, there has been a 37% decrease in the use of pesticides with an increase of 22% in crop yields. Much of the land freed up has now been converted into national parks.[31] Over the last 40 years, wildlife that was headed for extinction has reappeared, many suburban areas now have wildlife and the cleaning of rivers and estuaries has resulted in re-colonisation by fish and birds.

There is no scientific consensus on human emissions driving global warming. It is a scientific battle between well-funded mathematicians who create models that attempt to predict the future and those who measure, conduct experiments and integrate all branches of science. Reconciliation and validation are fundamental scientific principles.

One side has politicised a scientific debate with scare stories that are food for a scientifically illiterate media looking for a daily dose of disaster sensationalism. Many of the hysterical environmental over-promoted scare stories that we see in the media prove to be, after closer examination, false. Those earlier incorrect media stories are not subsequently corrected.

Legitimate scientific researchers are passionate about the scientific method, use data and logic and do not do science by social media or politics. The cumulative accumulation of scientific knowledge is not sensational and can even be boring. Real science is not easy to market. Politicians are trapped between a public scare story beat up and those painted as deniers. Politicians pretend to solve non-problems, so gain respect and hence votes and all of this is done with your money.

The combative Australian adaptation of the Westminster system, with its short electoral cycles, has seen society seemingly incapable of initiating and sustaining meaningful policy debate. Discussion of critical issues drifts along without a meaningful comprehensive and non-partisan debate taking place.

Environmental activism, animal rights, sustainability, renewable energy and climate change are the concerns of the rich in First World countries. In the Third World, people lacking food for their family don't care about endangered species. They need food. A cold person does not worry about where the fur to keep them warm comes from or whether the animal was ethically killed. A person with no shelter is not concerned about deforestation. Only rich and selfish people can trade the health of a river for food production.

Is the climate "crisis" really an environmental, scientific or political problem? No. It is a problem of national sovereignty. UN officials state that that the climate crisis has nothing to do with the climate and everything to do with global economic power. Ottmar Edenhofer (2010) said[32] *"One has to free oneself from the illusion that international climate policy is environmental policy. This has almost nothing to do with the environmental policy anymore"* and *"We redistribute de facto the world's wealth by climate policy"*.

Edenhofer was a lead author for the Fourth Assessment Report of the Intergovernmental Panel on Climate Change from 2004 to 2008. Climate change is not about climate or the environment, it is about unelected green socialist activists gaining control over your every action such as the food we eat, house heating and cooling, the number of children we have, the type of car we drive, travel and education. It can only be achieved by destruction of the past and those who have a contrary view.

Green activists claim[33] *".... Most climate change impacts will confer few or no benefits, but may do great harm at considerable cost"*. These is just a collection of rapid-fire deceitful words and such dogmatic statements are the *modus operadi* of greens. This book shows that facts are contrary to this false statement.

UN officials also later stated that that the climate crisis has nothing to do with the climate and everything to do with global economic control by the UN. Christiana Figueres, the Executive Secretary of the United Nations Framework Convention on Climate Change (UNFCCC) stated[34] *"This is probably the most difficult task we have even given ourselves, which is to internationally transform the economic development model, for the first time in human history"*.

The cost to lift everyone on the planet out of extreme poverty is less than $100 billion a year. Western countries are committed spending $1 trillion to $2 trillion a year on the ineffective Paris Accord[35]. This is obscene. Every month the Paris Accord costs are what could be spent every year to lift everyone from extreme poverty. Poverty is real. It has yet to be shown that human emissions allegedy driving climate change is real. Does the UN want to end poverty or enrich elites?

When green activists scream that we are in a climate emergency and governments respond, who benefits?

The gas of life

What's the problem? Carbon dioxide is good for the planet.

We are all environmentalists and want a better world for ourselves and the next generation. All statistics show that the world is becoming a better place. Pollution kills and no one wants pollution of the atmosphere, waterways and soil. We do not deliberately foul our air, water and soils.

However, pollution occurs and it is only wealthy countries that can afford to rectify pollution. The environmental movement was necessary. As a result, we are all concerned about waste, pollution and land degradation but these major problems are not being addressed in favour of concentration upon a scientifically unsupported scam.

Carbon dioxide is plant food. It is not a pollutant. Carbon dioxide is a blessing for the environment. How can it be dangerous if we breathe it in and breathe it out? Without sunlight, water and carbon dioxide there would be no life on Earth. To call carbon dioxide a pollutant is fraud, as is using the term "carbon pollution".

Except for diamond, all other forms of carbon are black. The printing ink in this book is carbon. It is black. If we had carbon pollution, we would either have sparkling diamonds in the sky or a black sky. Forest fires create a haze of carbon pollution. The term "carbon pollution" is used by the green activists and the media to mislead and deceive.

In air, carbon dioxide is a trace gas (0.04%) with other gases being nitrogen (78%), oxygen (21%) and argon (almost 1%). There is far more water vapour in air than carbon dioxide which is a minor greenhouse gas. Up to 95% of the greenhouse effect in the atmosphere is due to water vapour.

The air in rooms containing people has 0.1% carbon dioxide[36], US nuclear submarines contains 0.2 to 0.5% carbon dioxide and sailors have been tested and didn't suffer when carbon dioxide was 1.5%[37], some limestone caves contain 1% carbon dioxide[38], soil carbon dioxide is very variable[39] and can be over 1% and the air we breathe out has more than 4% carbon dioxide. If the air had 80% carbon dioxide we would die because of the lack of oxygen, not because of the effects of carbon dioxide.

There are thousands of studies that show the positive effects in increasing carbon dioxide on plants. Plants like more than twice times the current atmospheric carbon dioxide content to really thrive.

One study showed photosynthesis increased when exposed to increasing carbon dioxide levels for 45 crops that supplied 95% of total world food production over the period 1961 to 2011. The study shows that the annual total monetary value of this benefit grew from $22.7 billion in 1961 to over $170 billion by 2011, amounting to a total sum of $3.9 trillion over the 50-year period 1961-2011.[40]

Whatever the weather throws at farmers, according to the UN's Food and Agriculture statistics of 2nd September 2021, the world's cereal production keeps increasing and the forecast for 2021-2022 is that 40.1 million tonnes more will be produced than in 2020-2021.[41]

CNN Business claimed, contrary to harvest records, that climate-driven crop-driven failures are driving up food prices and 'extreme' weather is here to stay.[42] A 30-second search on a smart phone would have shown CNN journalists that their story had no legs. Are we really in a climate crisis, a credibility crisis or a crisis of media honesty?

Local weather, climate change, insects, tired soils and lack of nutrients have historically affected crop yields. People died of starvation because of bad weather and not climate change. When people die of starvation, it is for a lack of calories not micronutrients. Famine has been reduced, thanks to Mendelian genetics.

In 1865, Gregor Mendel showed that by using recessive and dominant genes, the plants that we grow and harvest could be improved.[43] This is now done experimentally rather than waiting season after season until a mutant appears. If activists choose to be carbon neutral or against genetically-modified foods, then they can only live on air and water.

Advances continue to be made in all branches of science. Increased yields and weights of potato and rice can be gained by manipulation of the RNA in plant cells[44], again showing that more food can be produced from less land. Greens object to anything that has a sniff of genetic engineering, yet queue up for a mRNA Pfizer, Moderna or AstraZeneca COVID-19 vaccine.

We've been eating genetically modified grains with micronutrient and vitamin additives for 80 years. Flour enriched with iron, thiamine, riboflavin and niacin have been part of the American diet since 1941 and have helped eradicate beriberi and pellagra from the US. Folic acid was started to be added in 1998 and its presence in flour is responsible for the decline in the incidence of neural tube defects in babies by 23% in the US and 54% in Nova Scotia (Canada).[45]

I am happy for green activists to live sustainably somewhere in the bush well away from me gathering native grass seeds, living in caves and exposing themselves to curable medical conditions.

Activists also claim that floods and droughts will disrupt agriculture, but the IPCC shows that there has been no increase in floods and droughts with warming and the global drought levels from the Global Integrated Drought Monitoring and Prediction System show a small decline.[46]

Widespread greening[47] over 25 to 50% of the global vegetated area from 1982 to 2009 occurred with the carbon dioxide fertilisation effect explaining 70% of it and warming explaining 8%. Satellite imagery showed an 11 percent increase in foliage when atmospheric carbon dioxide increased.[48]

A paper published in *Nature Climate Change* in 2016 shows a widespread increase in growing season leaf area over 25% to 50% of the global vegetated area with the carbon dioxide fertilisation effect explaining 79% of the observed greening trend. Climate change explains 8% of the greening trend, predominately in the high latitudes and the Tibetan Plateau.[49]

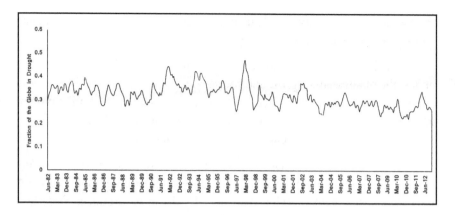

Figure 1: *Proportion of the globe suffering drought over the last 40 years.*

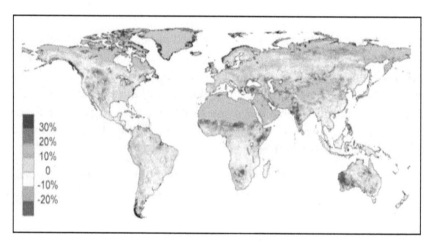

Figure 2: *Satellite data showing the change in foliage cover between 1982 and 2010.*

The world's forests increased their rates of carbon sequestration by 2.8 times in response to a 30 part per million increase in the carbon dioxide content of the air.[50] Elevated levels of atmospheric carbon dioxide causes a long-term biomass increase of 130% for conifers and 49% for deciduous trees.[51]

The Plant Growth Database shows that all plants have benefitted from a slight increase in atmospheric carbon dioxide.[52,53,54,55] This has been validated from crop yields, satellite measurements and experiments. Some cyanobacteria show enhanced antioxidant, antibacterial and anticancer properties when carbon dioxide is higher. Carbon dioxide-induced greening has also led to a surface cooling of the land, contrary to the green activist catastrophist narrative.

Data from the Harvard Forest Long-Term Ecological Research site in New England (USA) shows that the rate of carbon dioxide captured has nearly doubled between 1992 and 2015. The *Eurekalert* news release about the study says *"The scientists attribute much of the increase in storage capacity to the growth of 100-year-old oak trees, still vigorously rebounding from colonial-era land clearing, intensive timber harvest, and the 1938 Hurricane - and bolstered more recently by increasing temperatures and a longer growing season due to climate change. Trees have also been growing faster due to regional increases in precipitation and atmospheric carbon dioxide, while decreases in atmospheric pollutants such as ozone, sulfur, and nitrogen have reduced forest stress. ... The trees show no signs of slowing their growth"*. The faster growth of trees and other plants show how misleading and deceptive it is to label carbon dioxide as a pollutant.

Observations of leaves show that the global rate of carbon dioxide fixation by photosynthesis has risen 31% since 1900.[56] The NASA Global Vegetation Index shows rapid greening of planet Earth.[57] The Index rose 10% over the last 20 years. The Sahara shrank 700,000 square kilometres. Green activists don't realise that the world is getting greener despite their efforts to destroy the planet.

Researchers using Landsat satellite data report the Arctic region has become greener as warmer air, higher soil moisture and carbon dioxide fertilisation led to increase plant growth.[58] The news release about the study says *"When the tundra vegetation changes, it impacts not only the wildlife that depend on certain plants, but also the people who live in the region and depend on local ecosystems for food. While active plants will absorb more carbon from the atmosphere, the warming temperatures are also thawing permafrost, releasing greenhouse gasses. ... Between 1985 and 2016, about 38 percent of the tundra sites across Alaska, Canada and western Eurasia showed greening. Only 3 percent showed the opposite browning effect, which would mean fewer actively growing plants"*.

A paper reviewed in the March 2020 quarterly newsletter of Friends of Science shows that gross primary production (GPP) has increased by 35% since 1900, which is primarily caused by increasing carbon dioxide concentrations. The cumulative biophysical carbon sink is equivalent to 17 years of human emissions of carbon dioxide.

A new study published in *Nature Reviews* shows that since 1980, 29% of human-caused carbon dioxide emissions were offset by carbon dioxide-induced greening of the Earth.[59] The cooling effect of evaporative cooling is nine times greater than the albedo warming effect of global greening. Warming is the major cause of greening in boreal and Arctic regions.[60]

Rising atmospheric carbon dioxide enhances photosynthesis and vegetation greenness by partially closing leaf stomata, leading to enhanced water-use efficiency. The trend in the land carbon sink has further accelerated since the late 1990s. The rate of carbon dioxide uptake during 1998-2012 was three times that of 1980–1988.[61]

For decades, horticulturalists have been burning fossil fuel gas and releasing the exhaust fumes into glasshouses to increase yields. Why? Because the exhaust fumes are warm, rich in carbon dioxide and contain water vapour and the plants respond by growing bigger, better and quicker. What is it about carbon dioxide that green activists don't know about?

To claim that there is a climate emergency due to what is effectively plant food production is the cry of the unbalanced. There is no climate emergency, climate crisis or climate catastrophe. These words flow like honey off the tongues of green activists yet not one journalist stops them and asks "Show me?"

In the past there have been times when the carbon dioxide in the air was hundreds of times higher than now. There were no tipping points, runaway global warming or climate catastrophes. The planet is a self-correcting dynamic system. During these times of high atmospheric carbon dioxide, there were ice ages as well as warm times.

Any person with a bone to pick about human emissions of carbon dioxide is allowed to demonstrate in Western countries. Greens should display their principles and demonstrate against the world's biggest emitter. Economic terrorism is on the statute books in China. It is not in Western countries.

If climate activists want to demonstrate against human emissions of carbon dioxide, why not march around Tiananmen Square, demonstrate outside Chinese embassies, consulates and businesses and try to live life without buying anything made in China? Glue yourself to a major road leading into Tiananmen Square and wait for the response. It will be one that demonstrators won't remember.

The Chinese would not be as kind as the Russians. Greenpeace activists were arrested by Russians for trying to stop Russians drilling for oil in Russian waters.[62] What did Greenpeace expect? To try to climb onto a drilling rig in Arctic waters is against all safety protocols, endangered the lives of others and was a breach of Russian sovereignty.

Does Greenpeace really think that it is so important that it is above the law of a country it invades? Apparently so. There were government travel warnings that the activists ignored and, after arrest, those imprisoned called upon their governments to help them. Again, this cost the taxpayer a huge amount. Greenpeace tried the same stunt in Norwegian waters.[63]

Previous generations feared the cold. They knew that far more people die in cold weather than warm weather. The current generation of school pupils may be the first generation to fear warmth.

A climate activist[64] claimed *"Deaths attributable to heat waves are expected to be approximately five times as great as winter deaths prevented"*. This contradicts a huge literature that shows deaths from Jack Frost are far higher than summer deaths. For example, the death rate in Canada in January is more than 100 deaths per day greater than in August. The medical literature is full of similar studies from Europe, the UK, the US and elsewhere.

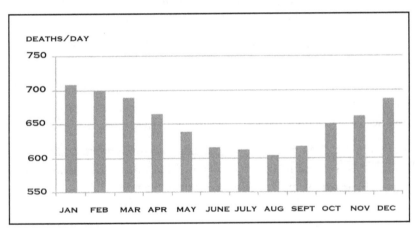

Figure 3: *Monthly deaths in Canada between 2007 and 2011.[65] I wonder why Canadians go to the Florida, the Caribbean, central America and Ecuador for mid-winter holidays?*

Cold weather kills 20 times as many people as warm weather according to an international study analysing more than 74 million deaths in 384 locations in Australia, Brazil, Canada, China, Italy, Japan, South Korea, Spain, Sweden, Taiwan, Thailand UK and USA.[66]

A more recent study by 68 medical scientists used data on 130 million deaths from 43 countries on five continents and showed that cold temperatures contribute to 10 times more deaths than hot temperatures. Between 2000 and 2019, deaths from extremes of cold or heat decreased.[67] How could such a sample size be at odds with green activist ideology? Simple. It's called data.

Assessments of the impacts of global warming give short shrift to the benefits of global warming, whether driven by human activity or whether natural. Similar arguments are given for humans, with claims that higher temperature will increase mortality. This is contrary to history and modern observations.

The daily mortality rate for the last 30 years in southwest Germany shows that cold spells lead to increased mortality as do heat waves with the greatest mortality in cold times.[68] Those that suffered heat stress were old or sick, whereas others adapted to temperature changes far more rapidly.

Mortality data from many countries[69,70], regions[71] and cities[72] with cold, temperate[73], subtropical[74], tropical[75,76] and arid[77] climates shows that mortality is substantially higher in the cold months than in the warm months. Jack Frost and the Grim Reaper are good mates.

The only good thing about cold weather is that it forces politicians to keep their hands in their own pockets.

In many parts of the world, retirees move to warmer climates (e.g. the Sun Belt of USA; south of France or Spain by British retirees). The evidence is there. Warmer weather does not increase mortality rates.

We are told that if human emissions of carbon dioxide are not curbed soon, then there is a threat to humanity. The green left environmental activist group 350.org wants carbon dioxide reduced from the current 410 ppm to the 1988 level of 350 ppm. Since 1988, global GDP has increased 60%, infant mortality has decreased 48%, life expectancy has increased 5.5 years and the poverty headcount has dropped from 43% to 17% despite a population increase of 40%.[78]

The past was not very pretty and the world has never been better as shown by crop yields, food production, hunger, access to clean water, biological productivity, life expectancy and standard of living.[79] It seems that the climate catastrophists ignore the fact that human ingenuity can solve most problems and has made the planet a better place. If 350.org wants to go back to the Dark Ages or Little Ice Age, they are encouraged to lead by example.

We cannot afford to wastefully allocate our sparse resources to address false problems. We cannot be guided by activists who cherry-pick information to suit their ideology, often provided as false press releases, bogus web sites and interviews willingly lapped up by a scientifically-illiterate press starved for its daily sensationalism.

We must be very wary of banks, corporations, traders and lobbyists seeking huge subsidies for schemes to reduce our emissions and quality of life using our money. We must be very wary of those self-appointed saviours who claim that they know what is good for us. They don't. They are trying to lighten your wallet and reduce our freedoms.

What we do know is that climate always changes, carbon dioxide has a very slight warming effect at low concentrations[80] and that humans can have a very slight effect on local climate (e.g. from land clearing or heat generated in cities) if at all and previous climate changes had nothing to do with humans. Why should our current times of change be related to human activity?

The greens want to ban plant food to change a major natural planetary process. It doesn't matter whether they are vegans, vegetarians or omnivores, the end result of this policy is that there will be less food. But then again, since when were green activists concerned about their fellow human beings?

We are part of the carbon cycle. Carbon dioxide is this miracle gas of life that is being demonised by greens as a pollutant.[81] Are the basics of photosynthesis and the carbon cycle taught in our schools? If greens object to our emission of carbon dioxide, then the only honourable thing to do is for them to drop dead.

Climate change ideology is a new idealistic cultic fundamentalist religion that offers certainty, guilt, indulgences, redemption and a green theological reason for living. It is replacing Christianity in Western civilisation and is the ultimate reward for the dumbing down of the education system. Unlike Christianity, the new climate religion has no history, no art, no scholarship, no music and no philosophical grounding.

The young are attracted to environmental anti-capitalist protest movements, possibly because they have achieved little and have nothing to lose. They generally drift away once they have children, a mortgage, employment, experience, travel, read history and gain an awareness of the disastrous environmental, health, employment, economic and political implications of climate change policy.

In earlier decades, communism killed hundreds of millions of people, created huge economic pain and forced great losses of liberty before it was rejected. Only those with little knowledge of history, politics and economics now have a naïve undergraduate attraction to communism and socialism. Communism allowed no discussion, debate or criticism.

To quote Marcus Tullius Cicero "*To be ignorant of what occurred before you were born is to remain always a child. For what it is the worth of human life, unless it is woven into the life of our ancestors by the records of history?*" Nothing has changed since 43 BC and green activism continues in a childish history-free zone.

The same applies to green activists promoting human-induced global warming. It is unrelated to science and is a modern political-religious movement. A large number of politicians, bureaucrats, scientists and institutions have nailed their colours to the human-induced global warming mast and, as with the rejection of communism, it could be a slow death for the West. After that, there will probably be another foolish fraudulent frightening fad which will be used to keep people scared in order to extract hard-earned money and reduce liberties.

Why don't the rich and powerful greens build a net zero emissions town called Utopia that is entirely dependent upon wind and solar for its electricity, is not connected to the grid and neither trades nor uses any products made from fossil fuels? All food, communications, transport, hospitals and schools should be totally self-sufficient.

Australian idealists tried this once before in Paraguay[82]. It failed.

No coal, no dole

Coal is of vegetable origin. It is organic. It is natural. Coal is a form of solar energy because it's highly concentrated solidified photosynthetic energy. It contains far more energy than sunshine or wood. Burning of coal recycles carbon dioxide back to the atmosphere. It therefore should be revered by greens.

It contains organic carbonaceous matter, inorganic material (minerals) and fluids such as coal seam gas (methane) and water. The carbon and hydrogen compounds in coal burn to produce heat, light, gas and a residual solid ash. Many coals contain fragments of fossilised wood, leaves, bark, fungus, pollen and spores and often tree roots pass from coal down into the underlying rocks.

Since the Industrial Revolution, life expectancy for people living in the Western World has more than doubled. Ready access to cheap energy allowed for better living conditions and easier lifestyles. Coal was the fuel of the Industrial Revolution and steam was its power.

There was also a transport revolution to haul coal initially along canals and then later along railway systems from pits to steel and cotton mills. These transport systems allowed workers to travel short distances, enjoy a holiday at the seaside and to broaden horizons.

Towns were lit by coal gas, train timetables meant that workers needed watches and inventors flourished to capitalise on the great technological advances that came with the Industrial Revolution.

Coal allowed a life of desperation to be replaced by a life of aspiration. In northern England, there were community gatherings, sporting competitions[83], brass bands and outdoor activities away from the darkness of an underground coal mine. Workers sought a better education and hundreds of Mechanics' Institutes and similar structures were built to provide education for workers. Women and children working in underground coal mines were replaced by pit ponies that later were replaced by machines.

Coal has advanced hundreds of millions of people over generations from poverty to prosperity, despite population- and economy-destroying wars. The green, white, wealthy Westerners are immoral telling billions of people in Asia, Africa and South America that they can't escape from poverty and develop to our standard of living using the same coal, gas and carbon-based energy that took the West out poverty. This is racial oppression.

In today's world, people are better fed, better sheltered and better protected than in any time in history. Overall prosperity has increased in the last century, as has mine. The average person's standard of living has improved ten-fold over this period of time, as has mine. We have moved from locally sourced sustainable farming to international trade. For thousands of years, international trade was for high unit value commodities. Now we trade everything.

Cheap inexpensive transportation, a consequence of the Industrial Revolution, has revolutionised the spread of trade, services and ideas. Without international trade, people could starve. For example, massive local rains in France in 1694 AD resulted in the third consecutive crop failure. Some 15% of people in France starved although plenty of food existed elsewhere in Europe. At that time, France was "sustainable", its population lived off the land and it had very little trade with other countries.

A convoy of 120 ships left Norway with grain for France. The convoy was captured by the Dutch and was recaptured by the French and escorted to Dunkirk. It was too little too late and there was still not enough grain to feed everyone in France. Starvation continued.[84]

Sustainability works when the Sun is shining and the birds are singing but as soon as there is localised disruption caused by foul weather or trade blockades, people living sustainably die like flies. Like it or not, we live in a global world. Those that don't are North Koreans.

In 1820, more than 80% of people were miserably poor. The Copenhagen Consensus Centre and the World Bank show that the proportion of extremely poor people has more than halved over the last 30 years, from 42% of the global population in 1981 to 17% in 2010.

The World Bank reported that in 1981, 42% of people in the developing world had to live on less than a dollar a day. Some 30 years later, this percentage has been reduced to 14%. This is a huge change in a short period of time.

Such a change is unparalleled in history. The greens' actions attempt to reverse this trend. That is a moral conflict for the greens. Their immoral solution: ignore it.

The World Energy Outlook of the International Energy Agency[85] demonstrates a near perfect correlation between global electricity production from coal and gross domestic product. Consumption of coal grows at 5% *per annum*. Asia leads the consumption race. The world is becoming a better place.

The *per capita* food intake, longevity and wealth have increased whereas child mortality, disease and land area used for food production has decreased. There is still much to do but for a long time the trend has showed that the planet is becoming a better place. The same authority also shows that 3.6 billion people still have no or only partial access to electricity and that the quickest and cheapest way to produce reliable 24/7 electricity is still from coal.

African cities now have spots of lights and mile after mile of darkness. How can an African child do homework at night? Without electricity, there is no hope. With no hope, there is crime and terrorism. The African Development Bank will no longer fund reliable cheap electricity[86], nor now will China, which will just perpetuate Africa's poverty.

African nations are ignoring Western green activists and planning over a thousand new coal and gas power plants. Fossil fuels will account for two thirds of all generated electricity across Africa by 2030[87] and an additional 18% of electricity will be from hydro. Of Africa's 2,500 planned power-generation projects, half are coal and gas plants.[88] Why don't Africans listen to their former colonial masters in the EU whose green wind and solar energy policies will keep Africans permanently impoverished?

There are still 20% of the world's population who are illiterate. Although this sounds dreadful, some 70% of the world's population was illiterate in 1900[89]. China has recorded the biggest improvement.

In 1950, South Korea and Pakistan had the same level of income and education[90]. Today, the average South Korean has had 12 years of education, in Pakistan it has not yet reached 6 years. The South Korean *per capita* income has grown 23-fold whereas Pakistan's growth in the same period was three-fold. Illiteracy costs.

In 1900, global illiteracy cost 12% of global GDP, it is now 7% and the estimates are that in 2050 it will be 3.8%. These are world problems and we cannot ignore them. However, the world is a far better place than it was 100 years ago and global problems are being solved. But not by the greens whose actions push us back in time and stifle progress.

There seems to be a view amongst the greens that we face crises for which there is no solution. If we look at the long history of experimentation, initiative, creativity, inventiveness, science and engineering, we can see that humans have a great ability to innovate and solve problems.

We humans are very good at overcoming hurdles to our existence and this we have shown repeatedly since we began to live in villages more than 10,000 years ago. When we humans are dealt a wild card from the pack, we are pretty good at adapting and improving. We humans live on ice sheets, in the mountains, in deserts, in the tropics, at high latitudes, at the seaside, on boats, in orbit in the international space station and soon on the Moon or Mars. More and more people now live in sterile concrete boxes high above the ground and some in opal-mining settlements live underground. If the need arose, no doubt civilisation would find a way to live under water and on other planets.

We are living at a time in history when there are two nations with more than a billion consumers. They have had thousands of years of gripping poverty and they now have a rapidly emerging large middle class.

The world is in the middle of the biggest industrial revolution in its history. Hundreds of millions of people have moved from rural China to the cities. This is the greatest diaspora the world has ever seen. These immigrants to the city want our Western standard of living. They have had a smell of it and there is no turning back. These aspirants use commodities and the statistics from China and India are mind-boggling.

India is adding more coal-fired electricity generation capacity as demand skyrockets despite the global push to reduce coal use.[91] Demand increase is driven by an expanding economy, increasing population, the shift of millions into the middle class and increased need for goods and services.[92]

At present, the US has 27.6% of the planet's coal reserves, Russia has 18.2% and China 13.3%.[93] Coal resources are unknown but are orders of magnitude higher. Because coal is present in terrestrial sedimentary rocks that drape over most countries, almost all countries have resources and reserves of coal.

Many nations keep coal in their core energy mix as a national security consideration and restrict export of it. Coal also provides more bang for the buck compared with wind, solar, biofuel, wave or tidal energy. Despite green propaganda and wishful thinking, coal is not about to die.

Global coal use in increasing. Coal will continue to thrive because it does what no other energy source can do. It is cheap and can be used to make electricity, metals, chemicals and pharmaceuticals. Coal is cheap and has every possibility of getting cheaper although top quality coking coal is becoming harder to source.

Coal prices have quadrupled since September 2020. Despite green activist and UN aspirations, coal is essential for energy and metals production and the price hike and increased consumption reflects increasing post COVID-19 demand, especially in Asia.[94]

Over the last decade, China has built more coal-fired power capacity than any other country on Earth.[95] In 2020, China built more than three times the amount built elsewhere in the world.[96]

At the same time, China speaks with a forked tongue about its climate aspirations and green activists actually believe them. President Xi Jinping won fawning praise from green groups for pledging in 2019 to make

China carbon neutral by 2060.[97] Is this pledge worth anything? The green activists are China's *"useful idiots"*.

China is recovering from its COVID-19 epidemic and now has a target of 6% growth but provides no goals on climate policy. China claimed that it plans to reach peak emissions by 2030 before net zero emissions in 2060.[98]

While the rest of the world struggled with unemployment and reduction in industrial output due to the COVID-19 epidemic, China's carbon dioxide emissions surged 4% in the second half of 2020 indicating increased industrial activity at the expense of the West that struggles with health and economic recovery.[99]

The average coal-fired power station has a life of at least 50 years and building of coal-fired power stations continues at a fast pace so China pledging to be carbon-neutral in 40-years time is a good game to play.[100] Do we really think that the cheapest form of energy will be abandoned by China because of Western green ideology?

China's steel carbon emissions account for more than 60% of global steel carbon emissions. About 1.7 tonnes of coking coal is needed to produce a tonne of steel. In 2021 China produced 888 million tonnes of pig iron and 1,053 million tonnes of crude steel.[101] If China wants to grow, it needs top quality metallurgical coal. Whatever happens, China will still use huge amounts of coking coal.

European producers have reduced the coking coal consumption and hence the carbon dioxide emissions by increasing the proportion of pellets for the blast furnace. China is trying to do the same.[102] Coking coal is horrendously expensive and China is moving to pellets to cut costs rather than to cut emissions.

UK was once a major producer of steel. UK Steel reported that it is now paying £254 million in additional costs when compared to their French and German counterparts. UK subsidies to renewable electricity generators now amount to about £10 billion a year and the consumer pays. China now accounts for well over half of the world's steel production and UK production is now less than that of Egypt, Belgium and Vietnam.[103]

In the UK, not only is the steel industry in trouble, the energy crisis is such that it is too expensive to make glass and paper. The brilliant solution to this is to go to a 4-day working week resulting in slashing of wages, loss of cash flow and loss of a taxation base from profitable industries. Plans are to greatly reduce road transport. All this will result in a decrease in the standard of living, hunger and deaths.

Turning to India, coal-fired power generation is projected to surge in India as the expanding wave of renewable energy capacity cannot keep up with the electrification growth.[104] According to the CEO of the World Coal Association, it is projected that coal will be the largest single source of electricity in India by 2040. The state-run Coal India Ltd has approved 32 mining projects in the 2021 financial year.[105]

According to the 2019 BP Statistical Review of World Energy, China (51.7%), India (11.8%) and USA (7.2%) are the world's biggest coal users and are gearing up to burn even more.[106] Whatever Australia does as an act of economic suicide in reducing emissions or destroying coal-fired electricity will make no difference to global emissions of carbon dioxide.

In India, some 13% of households still lack electricity. In 2020, fossil fuels accounted for 90% of the country's energy consumption. Coal is the mainstay of India's power sector accounting for just over 72% of total power generation in 2020.[107] For Indians to increase their gross national income *per capita* they will have to burn increasing amounts of coal. Why do green activists want to keep Indians in poverty? Are green activists not aware of Indira Gandhi's statement[108] *"Poverty is the worst form of pollution"*.

By 2030, India claims it will overtake the EU as the world's third largest energy consumer after China and the US.[109] The state-run Coal India Ltd raised production by 180 million tonnes giving a total production of 615 million tonnes in 2016-2017 compared to 436 million tonnes in 2011-2012. By 2021, production will be 596 million tonnes.[110]

Even if China drops off the planet, no matter what Australia does about emissions, it will make not one *iota* of difference to global emissions while countries such as India grow rapidly. Energy use in India has doubled since 2000 with 80% of demand still being met by coal, oil and solid biomass.[111]

India's oil demand is projected to rise by 74% as a result of a five-fold increase in *per capita* car ownership, the electricity system will be bigger than that of the EU and coal and gas consumption will rise by 50%.[112]

No Indian is going to choose eternal poverty because some green activist in Australia wants carbon dioxide emissions to reduce and to have fewer coal-fired power stations. Green activist policy is racist and results in poverty and death in a non-Western country.

India's percentage of carbon dioxide emissions rose more slowly in 2016-2019 than in 2011-2015 but was well above the world average rise of 0.7%. In the same period China increased by 0.4% and the US decreased by 0.7%. China's Five-Year Plan released in March 2021 shows it will invest more in coal to power its economy with only modest increases in renewables.[113,114,115,116]

China needs coal and its growth aspirations clash with its rhetoric about carbon neutrality. If green activists were serious about global climate change, they would spend all their time and money trying to convince the Indians and Chinese to change their ways. Whatever Australia does internally will make no difference globally.

As reported by *Clean Energy Wire* (29th May 2019), of the 21 EU member states that still use coal, only eight of these are committed to phasing out coal by 2030. We'll get back to you on that because there is no pot of gold under the rainbow and 10 years is a long time for reality to set in.

The UK was a significant coal producer. The use of steam in the UK did not become common until late in the 18th Century. As a result, British coal consumption rose from 25 million tonnes a year in 1825 to its peak of 292 million tonnes *per annum* a century ago. It now produces less than 10 million tonnes *per annum*. But for coal, there would have been no forests in the UK resulting in species extinction. There is no shortage of coal reserves in the UK, and the EU has put its green bureaucratic fingers into the internal affairs of the UK energy industry.[117]

The EU is also meddling in Australian affairs. The EU has threatened to make Australia's free trade agreement conditional unless it takes tough action on climate change. However, the EU's own statistics agency Eurostat claimed that the 27-nation EU might not achieve its aim of cutting "pollution" by 40% by 2030.

The EU has sought to position itself as a global leader on the environment as the EU parliament pushes ahead to revise the target from 40% to 55%. Hypocrisy and economic suicide are now European characteristics.

In 1999 the UK was at peak oil production of 2.9 million barrels per day. In 2012, it was 1.9 million barrels per day.[118] The UK now imports almost all the fossil fuel it burns and is greatly exposed to the vicissitudes of the energy market. The UK is phasing out its nuclear power plants which have provided energy to the grid for almost 70 years. The lights could very easily dim in the UK.

Australia is a large coal producer and its costs are also rising. Coal is the biggest export earner yet more than half of Australia's coal mines have production costs above the global average. This is compensated by the outstanding quality of Australian coals that attract a premium price. With capital costs in Australia two thirds above the global average, Indonesia has now overtaken Australia with steaming coal exports to China.

In response to noisy political pressure, we might decide to reduce emissions of carbon dioxide. The only effect will be to make our industries even more uncompetitive, increase unemployment and increase our cost of living. This has already started to happen. Environmental carping about coal and carbon dioxide is unrelated to reality.

If one country for some bizarre reason decides to stop exporting coal to save the world from a speculated carbon dioxide-driven global climate change, another country will fill the gap and steal the markets. Once markets are lost because of unreliable supply, trust is lost and markets cannot be won back. Nothing would happen to global climate by one country alone reducing its carbon dioxide emissions. Export earnings and jobs would be lost. Forever.

China accounted for 90% of the new coal-fired electricity generators in the first half of 2020.[119] China's new coal power plant capacity in 2020 of 38.4 GW grew more than three times than the rest of the world. In 2020, the rest of the world made cuts of 17.2 GW.[120,121]

For Australia to try to change global climate by reducing emissions is like trying to drain the oceans with a bucket. China rules the world of carbon dioxide emissions. In 2020, China had the only economy that grew and China's COVID-19 comeback will increase its emissions even more.

Southern Africa is gearing up to build a dozen or so coal-fired power stations contrary to Western green left environmental pressure. Why did sub-Saharan Africans ignore Western pressure? Because they want a quality of life similar to those who want to deny them electricity.

Proven coal reserves in South Africa are about 32 billion tonnes of economically recoverable coal and the resources are probably an order of magnitude higher. Two massive 4,800 MW coal-fired electricity generators are under construction and there are proposals for building many smaller plants in the order of 300 MW.[122]

Southern Africa needs cheap electricity, needs to profit from its vast coal resources and will prosper from building coal-fired power stations.[123] This is the first step out of widespread poverty.

Landlocked Botswana needs to diversify from its economic dependence on diamonds and tourism and create wealth from its vast coal deposits, although developing coal resources will require capital and infrastructure. Tanzania generates 1,000 MW, requires 2,000 MW and the GDP is rising by 7% *per annum* hence a great increase in the amount of electricity is needed. As with every other place in the world, demand for electricity rises with income. Tanzania, which at present has mainly biomass energy, aspires to have an electrical energy mix of 30% gas, 30% hydro, 30% coal and 10% biomass.

Reports about the death of coal are false. Across Asia, unreliable expensive wind and solar have been snubbed in favour of coal-fired plants with nuclear running a close second. The new high-efficiency low-emission (HELE) ultra super-critical coal-fired power stations are being built in China, Japan, India and SE Asia with older sub-critical plants being decommissioned.[124]

Australia now has only 24 coal-fired power stations and there is green activist pressure to close them because they emit carbon dioxide. Closure of Australian coal-fired power stations will only produce higher costs, more unemployment and poverty. The difference to carbon dioxide emissions will be trivial because of the number and size of coal-fired power stations power elsewhere around the world.

The EU is telling the world to cut coal use. They have 465 existing and 28 planned coal-fired power stations. Of the 21 EU member states that still use coal, only eight of these are committed to phasing out coal by 2030.[125] As soon as the weather turned nasty, they recommissioned coal-fired power stations and tried to import coal from Russia. There is no way eastern European EU countries are going to impoverish themselves by not using coal and going back to the dark days when they were poor and communist. In effect, renewables have made EU products so expensive that they are now attempting to raise the cost of non-EU competitors by moralising, tariffs and carbon taxes.

Civilisation advances one coal-fired power station at a time.

Nation/Union	Existing coal plants	Building	Total
EU	465	28	495
Turkey	56	93	149
South Africa	79	24	103
India	589	446	1,036
Philippines	19	60	79
South Korea	58	26	84
Japan	90	45	135
China	2,363	1,171	3,534
Australia	24	0	24
Total	**3,743**	**1,893**	**5,636**

Table 2. *Australia's reliable cheap 24/7 industrial energy death spiral*

By stopping emissions from Australian industry and not building any more coal-fired power stations, do the green activists really think that this would have any effect whatsoever on global emissions? Why should Australia stop emissions of plant food that has not been shown to drive global warming?

Why not reverse the economic death spiral that anti-industry anti-employment Australian green activists have forced on us? Publicity-seeking climate "scientists" must shoulder much of the blame for Australia's energy disaster because they only presented their pet idea and not the full breadth of available information. This deceit in other parts of the world would be branded economic terrorism.

Australia can't have it both ways and ban nuclear as well as coal-fired electricity. In Australia, coal-fired power stations are the cheapest form of 24/7 electricity and currently cheaper than nuclear. Numerous modular nuclear reactions should be cheaper than any other form of non-subsidised energy.

Electricity is the key economic impact factor underlying economic growth hence cheap and abundant energy is economically beneficial. Australia has thousands of years of coal[126] and uranium resources.[127] Given cheap and abundant coal, it could be argued that nuclear is not necessary. However, diversification of energy sources increases energy security.

The establishment of nuclear power stations accompanied by the necessary enrichment and reprocessing plants would provide the basis for a nuclear industry that would keep generations employed. Nuclear submarines are just the toe in the water. Modular reactors would have great use in isolated towns and mine sites which, after mining ceases, could be transported elsewhere. This currently happens with diesel gen sets.

There is an international market for coal and most other materials. Some greens have an objection to international trade and finance, not for environmental reasons but for political reasons. For thousands of years there has been international trade. For example, there was Neolithic trade of obsidian from the Greek island of Melos all over the Mediterranean and inland into what is now Turkey.[128] Persian bronzes contain copper from Esfahan (Iran) and tin that moved along the Silk Road from southern China.[129]

The Romans had international trade in high unit value commodities and necessities, a large integrated market and a finance system for this market, as shown by the Sulpicii tablets discovered in Pompeii in the 1950s.[130] The Sulpicii accepted deposits, lent money, acted as brokers, transferred funds between customers and changed money.

Roman living standards were extraordinarily high compared to living standards of others at that time and this standard of living was only surpassed in the Industrial Revolution 1,500 years later. It was this participation in a market economy that produced, bought, sold, transported, invested, lent, borrowed and innovated that gave the Romans a high standard of living.

Romans were able to chase their own dreams. A market economy allowed this to happen. Slaves obtained an education and could buy their freedom. Freeing slaves was more common than in any other slave economy. Some 10% of Roman slaves were freed every five years[131], whereas in the south of the US it was 0.2%. Freeing slaves after the American civil war led to their starvation.[132] Slavery still exists, especially in Africa, the Middle East, China and the subcontinent. There are more slaves alive today than at any other time in history.

When the Roman Republic collapsed around 40 BC, an elite of about 2% owned most of the assets, including 25% of the population who were their slaves and the rest of the people struggled.[133] This is what's happening today. We are seeing the end of a few centuries of democracies with the reversion to the previous despotic power systems. Climate change politics and resultant economic disasters are the siren.

The UN Agency for International Development recognises that electricity is important for refrigeration to store food, fluids and vaccines; to allow school children to study at night; to allow businesses to stay open later; to keep food fresh overnight and to make communities safer.

However, the UN African energy financing policies support wind and solar electricity generation and not coal generation thus perpetuating poverty, misery, disease, despair and early death. They justify their anti-people policies by claiming that they are helping the world to adapt to climate change. I'm sure an impoverished African would prefer a good feed rather than a feel-good ideology.

China has 20% of the world's population. It consumes 53% of the world's cement which is made by using coal-fired electricity to cook limestone[134], a rock that contains 44% carbon dioxide that is released to the air during that cooking. Concrete is made from gravel, sand, cement and water and any modern industrial growth is underpinned by steel-reinforced concrete because it is a strong, cheap, long-lasting and durable building material.

The high *per capita* consumption of cement shows that China is building, modernising and growing. It is no wonder that China is the world's biggest carbon dioxide emitter. This is good news as it means more and more people in China are escaping poverty. We should also thank China for putting so much plant food into the atmosphere.

If China reduced its carbon dioxide emissions, then its growth would slow which would produce internal social and political problems and perhaps another famine. In China, there are many areas where mineral exploration is banned because of food security. This is an ongoing problem for China.

Cities are big emitters of carbon dioxide from human activities. In a review of 167 global cities, the top 25 global cities are responsible for 52% of the planet's urban greenhouse gas emissions. China is home to 23 of these top 25 cities.[135]

If Western greens want China to reduce carbon dioxide emissions as a moral imperative, then they are knowingly and hypocritically trying to keep hundreds of millions in poverty as a result of their ideology. China wisely does not listen to Western greens. Why should we?

Over 400 billion kWh per month of electricity is consumed by China. The United States Energy Information Administration projects that China will bring on over 450 GW of new coal-fired capacity by 2040. The demand for coal in China more than doubled from 2011 to the present.[136]

China is reducing domestic steaming coal production and is continuing to close their small and inefficient coal mines which produce high-ash high sulphur coals. Chinese domestic steaming coal adds sulphur gases and particles to the atmosphere thereby providing serious pollution in industrial centres. Many of us Westerners have coughed incessantly and had watery eyes when visiting Chinese cities and industrial areas.

There was a decline in air pollution in China from 2013 to 2017 with a reduction in the critical tiny particulate matter called PM2.5[137] by about a third during which time China's coal-fired power plant capacity increased by 28%. In 2013, Beijing's PM2.5 concentration was 40 times higher than the recommended WHO standards and in 2017 it was 27 times higher.[138]

This reduction resulted from more investment in new and more efficient coal-fired power plants burning higher quality coal, but also in filters, electrostatic precipitators, better boilers, replacement of old factories and new emission controls for vehicles. For generations it has been shown that increased appropriate investment reduces pollution in the Western world, and China is now doing the same.

China is also looking to impose local production limits for the special and scarce resource of coking coal. With reduced access to domestic coking coal, China's demand for imported coking coal will increase. If the greens want Australia to save the planet and not export clean coal to China, then China will be forced to use dirtier coals. Is this the environmentalism the greens want?

In a fit of confected moral outrage about Australia asking the basic question about the origin of the COVID-19 virus in 2020, China greatly reduced Australian coal imports. As a result, there were electricity shortages, rationing and blackouts in China over the 2020-2021 winter. China reverted to using its own filthy polluting coal, Australian coal finds other ways to enter China and there is an increasing amount of dirty low-energy Russian and Mongolian coal travelling along long and expensive railway networks to keep driving the Chinese economy.[139]

China's domestic coking coal reserves are depleted and Chinese coking coal is poor quality thereby needing blending with better quality coking coal for steel making. In late 2020, China banned the importing of Australian coking coal and sourced coking coal from the US. In the fourth quarter of 2020 imports of US coking coal imports increased 748%.[140] It is simpler for Chinese buyers to import US coking coal which is very similar in quality to Australian coking coal rather than adapt steel mills to use lower quality coal from elsewhere in Asia.

China consumes 48% of the world's iron ore production. About two million tonnes of steel is made in China each day from local and imported iron ore and coking coal.[141] China also imports about 60 million tonnes of steel each year and imports the same tonnage of iron ore each month. China produces 11 times as much steel as the USA yet China is only four times as populous.[142] The US grew during the Industrial Revolution in the 19th and 20th Centuries from coal and steel. China is now catching up.

This steel is used for buildings, construction and for the 20 million cars that China makes each year.[143] It is also used in the expanding motorway and railway system. Over the next decade, China will build 40,000 km of railways. It already has the world's fastest train and the largest high-speed rail network in the world.[144] China's third west-east gas pipeline was made of steel and is 7,400 km long.[145]

In order to keep the lights on and to make steel and other products, China imports 47% of the world's coal production. This coal is coking coal for smelting and thermal coal for electricity generation.[146]

China is also the number one producer of wind and solar power generators and lithium batteries. They are made by using coal-fired electricity from imported commodities. They are not silly, they don't use such technology themselves to drive their own industrial revolution.

They sell them to fools in the West whose governments subsidise unreliable electricity generation and live off debt. Not only does the West become more inefficient and internationally uncompetitive with unreliable subsidised electricity, but the Chinese are laughing all the way to the bank.

And the Chinese bank is getting bigger. China purchased 2,600 tonnes of gold in 2013 at a cost of $104 billion. This is more than the world's annual production of gold. China has now opened its borders to billions of dollars of gold which is flooding into state and private hands as bars, coins and jewellery.[147]

China now imports 75 tonnes of gold every month worth $3.5 billion. The Chinese also have the world's largest stockpile of foreign currency reserves. China accumulated $3.2 trillion in foreign currency reserves, the largest currency stockpile on planet Earth.[148] It is ironical that the country that invented paper money using mulberry bark is shifting its paper money and bonds into bullion.

The Chinese also eat. There are more pigs in China than in the next 43 pork-producing nations combined.[149] The list goes on and on and on.

However, there is not only an industrial revolution in China, there is also a knowledge revolution. They have shifted from 14th to 2nd place in the number of published scientific research articles and have the world's fastest supercomputer.

Whatever carbon dioxide emissions savings the UK, Europe, Timbuktu or Betoota manage to achieve, it will have not one *iota* of difference to global carbon dioxide emissions by humans. It also makes not one *iota* of difference to the global emissions of carbon dioxide if Western countries reduce emissions.

China, India and East Asia want a better standard of living and to do this they will emit increasing amounts of carbon dioxide. Why should wealthy Westerners stop their fellow human being from having a better life? No carping moralising green activist in the Western world can stop this increase in the standard of living and, if they try, they are hypocrites.

In China there has been a 45% increase in national income, protein sources are more varied and agricultural productivity has increased.[150] Wine is a measure of consumer affluence. China now owns wineries in France and the New World[151] and, as well as making its own wines, imports wines from many parts of the world. This would have been unheard of 50 years ago. Even though there has been an increase in longevity in China, 50,000 cigarettes are consumed each second.[152]

The area of China's corn harvested over the last half century has doubled, each harvested hectare has become more than 4.5 times more productive than 50 years ago.[153] The 120 million hectares of land thus spared from land clearing is twice the size of France.

No wonder Chinese forests have expanded more than 30% over the last 50 years.[154] As affluence in China increases, the rate of population increase is falling.[155]

The environment is far better than you think. Cars are cleaner and far more efficient. Turbo petrol and diesel cars are more fuel efficient than hybrids. No longer are our cities blanketed in a brown photochemical smog.

Aeroplanes are far more fuel efficient and quieter. Increased farm production has resulted in an increase in wilderness areas, global trade has enriched us, poverty is declining, population growth is no longer a threat, urban living has less impact, storms and droughts are not getting worse, there are more oil reserves in the ground than ever before and more innovation results in a relatively low tonnage of commodities used per unit of production.

The 7.7 billion people on planet Earth today are far better fed than the four billion in 1980. Famine has almost been extinguished. Carbon dioxide is one of the building blocks of life and cannot be a pollutant. We need as much of it as we can get for feeding a growing population.

It is claimed that times were good before the burning of coal. These were apparently the good old days. We apparently sat around arm-in-arm happily singing, picking berries in the forest and living off locally produced organic tofu. We lived sustainably and had a very small carbon footprint.

Since then, sinful modern humans have degenerated into selfish greedy capitalists wilfully emitting dangerous carbon dioxide into the atmosphere. This is rather like the fundamentalist Christian view of the world with original sin, the fall from grace, absolution and redemption.

Nothing could be further from the truth. There was no utopian past just as there will be no utopian future. The journey from the past to now was like climbing a sand dune with three steps forward and two backwards. In previous centuries in what is now the First World and currently in parts of Africa, Asia and South America, forests were clear felled. Anything that could be burned was burned.

European forests were clear-felled for making metal, the regenerated forests were clear-felled for making glass, the forests were yet again clear-felled for making naval ships and again the forests were clear-felled for food production for the post-Little Ice Age population increase. In Europe, a pristine forest is an illusion except in the most remote uninhabited areas.

We killed everything that might give us protein or was a perceived threat and polluted everything we touched. Even in ancient Egypt, the Nile River was so polluted that rather than drink the water, the men drank beer and the women and children drank "small beer" which was what we now call light or low alcohol beer.[156] The process of fermentation to make beer kills bacteria in water.

If we were frightened of an animal or thought a person was a witch, we killed them. We died at an early age, normally from a big bacterial blast or a violent event. Peace, longevity and security were not the natural state of affairs.

In my early years on the land, solar power grew our crops, domesticated animals were vegans and ate grass and crop stubble. Wind power pumped water for stock and irrigation. Hens survived on kitchen scraps and provided eggs. Sometimes they wouldn't. A large vegetable garden was necessary for survival. If we had an excess of one crop, we would share and swap with others in those frugal post-war times.

There were many worse off than us. Sheep and scythes were used as lawnmowers. The vegetable garden and fruit trees gave seasonal foods, gardens were fertilised with waste, farm animals had to be fed whether they were working or not, rabbits were a good source of protein from the bush and a few bob could be made selling the skins. When there were floods, droughts, heatwaves or fires, we didn't blame governments or industry. We adapted and got on with life.

The greens want us to go back to this alleged "sustainable" life and have us using "renewable" energy. I don't. I am old enough and have lived in the "good old days" in a semi-rural environment and it was not that long ago. In every way, the modern world is far better than the "good old days".

We are so lucky to have a choice of good healthy diets. Well, actually this was not luck. This access to good food has derived from thousands of years of agriculture, trial and error with successes and disaster, genetics, innovation, mechanisation, international travel and just plain hard work. We owe it to those who came before us to revel in their life's work and achievements.

We now have so much good food that you can decide not to eat the more digestible meat protein and become a vegetarian or vegan. Most people in the world don't have such a luxury. I didn't. You ate what you were given because there was nothing else.

Greens who play the game of flight shaming to embarrass people into not flying for environmental reasons need a cold shower. They need to show that not one item they use or anything they eat has been delivered as air freight. International freight, as have nuclear reactors, have given us medicines and accessibility to them.

Ask any green activist undergoing cancer treatment whether they are anti-nuclear and they would probably say yes yet it is nuclear medicine that has given us PET scans and radioactive isotopes to target tumours. Flying and medical cures give us freedom and the world is a better place for it.

In the West, no one is shooting at you, there are no such things as armed highway robbers any more, you don't have to travel from town to town in a convoy with a small army, if you are ill then modern medicine can cure you and lengthen your life, you do not have to toil in the fields to grow your own food, you can read and write, you can travel in planes and cars and you have the ability to instantly communicate all over the world. What feature of the modern world don't the greens want?

The internet, social media and the mobile phone are wonderful inventions that previous generations never had. When the internet is down, snowflakes go into meltdown because all of life rotates around it. It is that same internet that consumes about 10% of the world's electricity. Many of these are the same people who have "strikes" against climate change.

In my younger days, there was no plastic. Paper bags were used many times and no school lunch paper bag was thrown out. Most consumer items had no packaging and spare string, cardboard, paper, ribbons, buttons, bottles, jars, tin cans, bolts, screws, rope, chain, cable, nails and wood were all kept just in case there was some future use for them. There was no garbage collection because it was not needed.

This is now called recycling and apparently recyclers have the high moral ground today, despite using more transport and energy, and in the end being dumped as waste landfill.[157] I have had enough recycling in my early years to get me a front row seat in Heaven. Old habits die hard. I am a hoarder or, through different eyes in today's world, a virtuous recycler.

As a child, there was no obesity as people walked everywhere and food was not abundant. There was no junk food. We ate what today would be called health food because there was nothing else available. We walked to a village shop with a basket, to school, the train station, church, and friend's places. It was only later that we had primitive bicycles for faster and more distant travel.

This might sound romantic to some greens who have only ever experienced affluence. It wasn't. We were rescued from this frugality by large amounts of cheap energy generated from coal. Some people had reticulated town gas made from heating coking coal that produced gas and left a coke residue.

In the 1950s, there was an energy revolution. There were fewer coal miners' strikes because conditions in underground coal mines improved. New coal-fired power stations were built.

Electricity generated from coal became cheap, plentiful and reliable. Diesel and petrol became cheaper and we could afford to buy an old second hand car. These were clapped out dreadful English cars like an Austin or Morris.

Electricity could be used for heating, cooking and refrigeration, lawns could be mowed by a machine that did not leave dung everywhere, there was less need for muscle power, there was more free time and foods that had travelled could be sampled. A major event was connection of the house to the town sewerage system.

We very rarely went to a doctor or dentist. It was only if we were seriously ill we saw a doctor and then the doctor came to us in his car. His was one of the few cars on the road. There was no Medicare. If you were sick, it cost a lot of money. If you were so sick you could not go to work, you didn't get paid. Most of the world still lives like this.

It took only half a century during my lifetime to change to the current affluence. This was due to cheap coal-fired electricity. The Australian economy is now underpinned by exporting coal, gas and iron ore. If greens object to mining and the use of fossil fuels, then they should not be beneficiaries of the export income and the products of mining.

Society and economics allow us to spend our free time doing things we enjoy as opposed to being factory fodder or hunting and gathering just to survive. We now have assets and are far wealthier. We can now afford to go to sporting events, concerts and restaurants. We can now even afford to be greens.

Don't give me the "good old days". They weren't. No green is going to tell me that I should go back and live that frugal "sustainable" life again. I've done my bit in my childhood and it did not change climate.

The greens hate the modern world, hate humans and want us to change our lifestyle but yet want to live their modern lifestyle at our expense. No thanks. Those greens that want this allegedly environmentally friendly hypocritical romantic life are quite welcome to do it themselves.

When greens stand at the entrance of their caves after their fifth consecutive unsuccessful day of hunting and gathering in foul weather and give me the benefits of their "sustainable life", I might just listen to them.

Until I see the major movers and shakers of the green movement living in caves without the trappings of the fossil fuel driven world such as medicines produced from fossil fuels or nuclear reactors, then they are evil hypocrites with no credibility.

Green morality of course would not allow them to accept any form of social welfare because the country's coffers derive from immoral mining and hydrocarbon production from oil and gas wells that pierce Gaia until she screams. If ever I need expert advice from a green oracle, I'll visit their cave in the bush. Until then, get out of my life.

The greens are far too keen to say *"Do as I say"* rather than *"Do as I do"* and want to take away my freedoms while they retain all the benefits of a modern life. With no knowledge or experience of such a life, I doubt whether city-based greens would be capable of living the "sustainable" life that I had.

If they want "sustainability" and "renewables", greens should have to pay for it themselves. Where are all the massive wind turbines in parks and on rooftops in inner city green suburbs? Where are 200 metre-high wind turbines along city beaches or waterside suburbs? Why don't green city dwellers stop eating food brought to the cities in fossil fuel-powered trucks?

Many greens like to eat "organic" food, whatever that is. Why do they eat food derived from fossil fuel-driven machines that plough, seed, weed, harvest, sort and transport food to them? I can go on and on about the breathtaking hypocrisy, ignorance and backwardness of greens.

Greens should lead by example from their caves and leave me to enjoy my hard-earned comforts of life.

If greens are on any sort of welfare, and I suspect that a large number of them are, then this welfare derives from coal, iron ore, gas, minerals and agricultural exports.

No coal, no dole.

The long march

After I started school teaching in 1967 and university teaching in 1969, I experienced the long march of the socialists through the education system. My university teaching life ended in 2012 although I still supervise PhD research.

Places that were once outstanding for scholarship have now evolved into institutes for eminent mediocrity. They now even sack staff for asking questions.

For much of my life I oscillated between the academic world and employment in the exploration and mining industry. In my view, a university is a community of scholars comprising students and academic staff. Both are there to increase their knowledge. The role of academic staff is teaching and research and the role of students is learning, everyone else, including vice-chancellors, are there to support these activities.

In the practical vocations of university teaching, staff got on with the job and were aware that graduates worked and made a better world. It was just not possible to teach politically-correct hard science. A few academics I worked with were unusual beasts, scorned the taxpayers who supported them and gave little value to them. Some academics had a passion or skill for teaching and some took a compassionate personal interest in young people and their inevitable problems. There were some notable exceptions. We all remember our great and inspirational teachers.

As one who has been teaching for nearly half a century, in my experience the most disruptive pupils are those who ask questions and want to argue using logic. They are often the best. The worst teachers are those who do not encourage argument, criticism and analysis and just tell you what they want you to believe are facts.

It's easy to teach when there are no questions. The job of a teacher is easy if what might be perceived as facts are trotted out and no alternative views, discussion or analysis are encouraged.

For many years, I joined a group of half a dozen scientists with 16 year-old school children in the outback for a week of practical science. The six staff were outstanding, no job was too much trouble, they communicated well and led by example, were respected by the pupils, and were concerned about the intellectual, emotional and moral growth of their 40 charges. For many pupils, this was a life-changing experience.

After a decade of such field trips, the shiny bum administrative burdens made these trips impossible. It was most enjoyable, I did really enjoy communicating with 16 year-olds and having the support of enthusiastic committed teachers. This is contrary to what we read about what is taught in some schools and the illiteracy of pupils. What is wrong?

Teaching can be a very tough and depressing job. Many of the outstanding school and university teachers I know are totally demoralised because of the mindless administration, asinine school syllabus, lack of school discipline, lack of support and the lack of tangible recognition for inspirational teaching that changes lives. It can also be an easy job if the teacher is lazy, dumb, flogs propaganda and has no interest in pupil growth.

I suspect that unionists have crept into every aspect of teaching, especially on the syllabus committees, and the teachers I know are too busy with their teaching and family life to devote their time to creating an educational teaching nirvana. Control of the system has been lost to idealists.

When an article is published on the poor quality of school education[158], there are hundreds of comments by parents and ex-teachers giving examples of how teachers don't know the language and how the children are taught nothing about reading, writing and our language. I suspect that these comments relate to a few teachers in key areas because that has not been my experience with many teachers I interact with. J. K. Rowling taught children to be voracious readers for the price of a book rather than the billions that are shovelled into the education process.

The most important gift from school is that pupils should have been taught to think critically and analytically. This is how flaws in arguments are easily detected. Most pupils were not given such a school education.[159] In fact, they were taught to stop thinking for themselves and asking questions from a very early age. This is not because they were dumb.

In October 2021, the most prominent ex-pupil who is the greatest woman Australia has ever produced and large benefactor of St Hilda's Anglican School for Girls in Perth was invited to address the pupils. The parts of Mrs Gina Rinehart's video presentation dealt with challenging orthodoxy, climate change, the thinking of Mrs Thatcher, learning to be independent critical thinkers and inspiring the girls to be positive about the future were censored. When Lord Monckton and I addressed the same school on climate change 11 years earlier, the pupils were greatly inspired and about half of them changed their minds about human-induced climate change. We can't have that. Why should the children be taught to critically think using evidence?

Many teachers perceive that pupils who ask questions are disruptive (which they are) and so they pushed their own ideas, propaganda and politics rather than teaching pupils to think and giving them the basics for life. This is not the way to create the next generation.

Academics that strut around with peacock feathers and titles have no predators, have no common purpose, make a huge amount of noise and create costly educational Ponzi schemes that become political dynamite if there are attempts to change the system. They are very serious about their importance. This is especially the case in university departments that don't produce graduates for the employment market.

Less than a few percent of all academics have ever worked outside a university. However, most are totally insular about the ways of the world and have no idea how wealth is generated to provide the taxation base that funds them. Show me a well-dressed academic and I'll show you one who understands the outside world.

After nearly 45 years on the staff of various universities and a Chair for nigh on 30 years, all I can state is that many current university staff have been trained in a system that has been dumbed down. It shows. Some of them, although now middle-aged, tenaciously hang on to their undergraduate young and foolish Marxist view of the world to make their arrested development complete. This is especially prevalent in the humanities.

Much university research is actually by academics following their hobby and almost all published peer reviewed scientific research papers are read by fewer than 20 people around the world and forgotten very quickly. The quality of research is judged by the host university on the dollars won in research grants and the word scholarship disappeared long ago in universities.

I have seen many researchers in totally useless or outdated fields of my science create an army of clones because the university system rewards bums on seats and not outstanding scholarship. Very little research builds on the shoulders of giants and one can count on a sawmiller's hand the research that creates wealth or repays taxpayers.

My experiences are not restricted to Australia. In the US, in major universities, such as Yale, education has been undermined by emphasis on pumping out as many publications as possible, no matter how trivial and by changing the scholarly research ideal to chasing and embracing the latest fad that will attract taxpayer research grant funding.[160]

Anthony Kronman, the former dean of the Yale Law School spent most of his working life there and admits to being an old-fashioned leftie and not some gouty conservative with a bone to pick.

He, like me, defends unabashed elitism, and calls on universities to turn away from politics and reclaim their role as protectors of independent thought and the free search for truth.

My university teachers were returned soldiers, refugees from war-torn Europe and people who had been in the work force many years while they studied part time. They had experienced the horrors of war and wanted to make the world a better place through education. Experiencing a Messerschmitt 109 or Stuka 87 diving at them taught them how to handle stress and gave them survival skills.

Small classes allowed lecturers to take a personal interest in each and every student. Less than 2% of those who started high school were later accepted for university entry. Entry standards were high, degrees involved long hours of lecture, laboratory and field work and graduates gained employment. At those times, one "read" for a degree.

The current COVID-19 crisis has not allowed the best teaching methods to prevail. Face-to-face teaching inspires young people, practical and laboratory experiments provoke argument and realising that a cherished idea is wrong changes young people's lives. This cannot be achieved with the necessary electronic teaching during the COVID-19 crisis. There is a lot of theatre in delivering a lecture and this is used as a memory retention technique. This is lost with electronic teaching.

University entrants now cannot communicate using cursive writing, cannot express simple ideas in English and have no idea of grammar or spelling.[161] They have not read great literature, can't undertake simple mental calculations, have minimal scientific knowledge and have very little knowledge committed to memory.

History has passed them by as has poetry, music, languages and a knowledge of the basics of Western civilisation. Thinking with critical and analytical skills almost don't exist. Information, albeit commonly skewed or incorrect, is found with an instant search on a smart phone only to be forgotten quickly.

People who go to university now expect to be rewarded despite obtaining useless non-rigorous degrees in disciplines that will never see the graduate able to support themselves.

In my various outback travels to mines and exploration areas, I have met many miners and bushies with no degree. Many never finished school. They have a far better grasp of common sense, logic and argument than many city folk with degrees and some I have met were experts in minerals, astronomy, botany, zoology, engineering, history and even French classical literature. If something didn't work, they fixed it using first principles.

Just because someone does not have a degree does not mean that they are unintelligent or unread. I would have difficulty now in advising a school leaver to go to university in today's world unless they chose a very specific area of scholarship in one of the few remaining real universities.

In 2021, I saw Alan Jones ask the NSW Minister for Education to describe what a sentence was.[162] She couldn't do it and yet she's the boss. God help us.

Former Griffith University education lecturer Gregory Martin stated[163] "*A major task for leftist academics is to connect education with community struggles for social justice*". Silly me. I thought education was to give students the skills to look after themselves later in life by being taught literacy, numeracy, critical and analytical thinking, history and science with essential life skills being taught by parents, leaders and mentors.

A former Victorian premier and education minister and Premier was happy to show her colours with this quote[164] "*Education has to be reshaped so it is part of the socialist struggle for equality*". I think she meant that we would all be equally dumb and equally miserable under her education system.

Secondary and tertiary education has become the playground for socialist and green ideologues to push their agenda. Pity about the students. The former head of the Australian Education Union Pat Byrne stated[165] "*We have succeeded in influencing curriculum development...conservatives have a lot of work to do to undo the progressive curriculum*". It seems the union's objective is not to make school children more employable.

Why should unions have anything to do with the school curriculum if they want to use it as a playground for political ideology? Why shouldn't the basics be taught?

An example. Water vapour is the most potent greenhouse gas in the Earth's atmosphere. It accounts for about 60 to 95% of the Earth's greenhouse effect. Children are taught that 84% of greenhouse gases are carbon dioxide, 9% methane, 5% nitrous oxide and 2% fluorinated gases[166] and told nothing about water vapour.

In January 2021, I was at a school car park on the day teachers had pre-class meetings and noted that about half the vehicles were fossil fuel guzzling 4WDs and SUVs only a few years old yet the same teachers indoctrinate pupils about evil carbon dioxide emissions and human-induced global warming.

The other vehicles were old bangers which not only add carbon dioxide to the air but hydrocarbons and particulates. Looks like a rule for one and not the other.

Teachers' unions in Australia[167], the UK[168] and the US[169] oppose national standards testing and merit pay for the best teachers and embrace fads and support pupils' climate change boycotts of classes. There are some outstanding school teachers who are greatly outnumbered by the rest yet they have been disempowered.

One outstanding teacher I know had the bureaucrats try to discipline her for speaking the truth. Her principal had never sat in on one of her classes, I had sat in on some 160 hours of her teaching on outback field excursions and was able to very forcibly support that teacher. Don't blame the teachers for everything, the system is rotten and a fish rots from the head down. Education standards have been declining markedly over a long period of time with the left's long march through our schools despite vast spending increases.[170,171,172]

Less than half of Australian 15-year-olds have reached the minimum standards in maths and reading.[173] Tests of 15-year-old students shows that their maths is 3.5 years behind their peers in China, three years behind their peers in Singapore and six months behind where they were in 2003.[174] Year 9 pupils in Singapore and China are now at the same level as Year 12 pupils in Australia. Why?

When Federal Education Minister asked the Singapore High Commissioner why Singapore has such outstanding success in school education[175], he was told *"It boils down to the human capital you recruit"*. I know a number of dedicated hard working school teachers who have left because of the teaching fads, dogma, bureaucratic social engineering and the lack of discipline. In Australia, we have lost much of the human capital and there is too much dross replacing the good teachers.

According to the Australian Productivity Commission[176], per student funding increased in real terms (above inflation) between 2007-2008 and 2016-2017 by over 14%. Australia now spends more per school student as a dollar amount than the OECD average and above the expenditure of Finland and Japan.[177] School funding has increased from $34.6 billion to $57.8 billion in the last decade.[178] Where are the results? Where is the value for money?

University admission standards are far too low and more than half of university education degree entrants are accepted with an Australian Tertiary Admission Rank of less than 50%.[179] Teaching methods, curriculum, activism, lack of discipline and electronic learning have replaced teaching methods that were tried and proven because of the march of the ugly face of socialism and green politics into our schools.

No wonder Australian school children are slipping down the ratings. Teaching has not been helped by the breakdown of the family unit, the lack of discipline, the growth of child narcissism and the left's destruction of morality and ethics.

The education system has failed generations of children and the teachers' unions, weak governments and a politicised public service must shoulder most of the blame. Teachers' unions lobby for decreased class sizes (i.e. more unionists), increased pay and conditions for teachers but not for increasing literacy, numeracy and education standards.

These unions use the education system to promote their climate catastrophist green ideology rather than teaching the basics. For example, climate change propaganda is now taught in English courses! Maybe instead the unions should have demanded an iron-clad guarantee that pupils don't spit and swear at them, attend school, remain disciplined in class, don't physically attack teachers and respect the teacher's control and authority.

It's bad enough that Johnny can't read, write, do simple mental calculations or know what questions to ask. Johnny can't think, does not even know what thinking is and believes that feelings are the same as thinking. Johnny's alleged education has made him perfect fodder for climate activists, green bureaucrats and politicians.

Because universities are federally funded, the solution is simple and would address many literacy, numeracy and funding problems in schools and universities. Establish a highly competitive national university entry examination whereby those who pass have the right to enter university and the top few percent would win a bursary.

Many universities accept students just because they have a pulse and the tertiary entrance scores for those who want to become teachers is appallingly low. Universities are for the elite and not for every man and his dog. There is no equality of intellect. In my experience, many mature age students and students from the land excelled because of their life experiences.

Those who don't pass a federal university entry examination yet finish the school Year 12 with marks above a certain threshold would be able to attend colleges of advanced education, polytechnics and village glee clubs.

After university graduation, employers continue to educate graduates. There is no way Australia can ever be a clever country when we produce so many graduates who are unemployable and have no skills upon which an employer can build.

There are now so many soft degrees in areas that lack scholarship and rigour that it's little wonder that there is an army of unemployable people paying off a large education debt by doing menial work. We are free to make bad life choices for which we are responsible. We are not victims.

In post-war Australia, school class sizes were far higher than now[180], teachers had authority and respect and as a result we had a generation of baby boomers who received a good education. Teachers then had a real degree in a scholarly discipline with a one-year add-on Diploma of Education. Teachers had a deep understanding of their discipline and actually knew things. They wore a jacket and tie every day.

Teaching was once a calling for those with honours and masters degrees. It is now just a job. There was no "child-centred" learning because the teacher took responsibility for pupil development. Teachers had respect, did not go on strike and did not use a captive class to enforce their political views on trapped impressionable young people.

Once teachers started to go on strike and use schools to promote ideology, respect was lost. In many countries, teachers are held in very high esteem. Not in Australia. They did it to themselves. Many of those entering teaching now are drongos and become unionised teachers more interested in how they can use the system for their own benefit rather than bringing out the maximum potential of their pupils.

Many young people think that socialism is a preferable political system to democracy. This preference is based on living in a democracy rather than in a socialist system. Well, do I have news for you based on personal experience. Many of you may think that socialism is the same as being social or socialising. Now that would be good but it's not true.

We are in the process of creating nations of cretins in the Western world because the education system has failed to give young people an understanding of the history of civilisation, human thought, philosophy, values and the fundamentals of how we know what we know. Such a society is ripe for manipulation, totalitarianism, false worship and sacrifice. We have seen it before. To keep the Sun moving across the sky and to preserve the Aztec way of life, the Sun god Huitzilopochtli was fed with human hearts and blood.

Socialism is more destructive to populations than nuclear bombs have ever been. The estimated number of those killed and injured in 1945 by atomic bombs at Hiroshima is 150,000 and Nagasaki is 75,000.[181] We'll never really know exactly how many were later killed by radiation damage from those bombs but it's hundreds of thousands of people.[182]

In the communist Soviet Union, ancient Soviet nuclear technology, mismanagement, stupidity, lack of care for workers and a lack of adequate modern safety systems[183] killed about 50 people resulting from a nuclear reactor meltdown at Chernobyl (Ukraine) in 1986.[184] We'll also never really know how many people have been killed in laboratories and from observing nuclear explosions but my guess is that it is under 5,000.

There have been 2,056 atomic bomb tests worldwide over the last 75 years.[185] Some of those who died witnessed the 12 British atomic bomb tests in the Montebello Islands (Western Australia), Emu Field (South Australia) and Maralinga (South Australia). These tests took place between 1952 and 1957.[186]

The Black Death (the plague) killed 75 to 100 million people in Europe in the 14th Century, some 50% of all Europeans.[187'188] The world-wide Spanish flu pandemic in 1918 infected 500 million people and killed 50 to 100 million people[189], some 5% of the global population. Socialism and communism killed more people than any disease pandemic over the last 500 years.

It is not possible to know exactly how many people have been killed by communism and its cousin socialism. There were probably 100 to 160 million based on sparse and incomplete data and maybe many more because no country will ever give the names and addresses of the citizens they murdered. These were just normal people, not soldiers in a war.

Ordinary people were killed by execution, labour camps, re-education camps, deliberately engineered famines, incarceration and "treatment" in psychiatric hospitals, ethnic cleansing, flight across no man's land at borders, attempted escapes in leaky boats and long treks over inhospitable terrain. Systematic murder and genocide get little media coverage because journalists are imprisoned and killed and foreign journalists expelled.

People were deliberately murdered because of their religion, race, politics, education and language. They were also murdered because of what they read, said, thought or did. Many were murdered because of false accusations, lies and perceptions. If you were arrested, it was obvious that you were guilty. Arrest meant conviction. Conviction meant death.

Socialism is not a fair redistribution of wealth, it is total control of everything by murderous thieving despots. It is the end result of political correctness enforced by governments. If we do not defend free speech, our democracy, our property rights and fight totalitarianism, then we could go the same way. It's happened before in countries that were once democratic and it does not take long.

In communist countries, people were dobbed into the authorities by their family, children, neighbours, work mates and friends because of malice, envy, neighbourly disputes, settling scores, competition at work and family tensions. If you were arrested, there was a good chance that your friends, work colleagues, relatives and even infant relatives were also arrested and often murdered.

Fear ruled. In public places, no eye contact was possible, people walked with their eyes looking at the footpath and a joke was no laughing matter. I've been there and seen such behaviour.

Do children really get taught about the realities of communism at school? I doubt it. Communism kills its own. Figures of 100 to 160 million people murdered from the start of communism and socialism derived from Russia (from 1917), the Baltic States of Latvia, Lithuania and Estonia (from 1939) and in post-World War eastern Europe (East Germany, Poland, Czechoslovakia, Hungary, Yugoslavia, Albania, Bulgaria and Romania).[190,191] France, Spain, Portugal, Italy and Greece have flirted with communism.

Communism was not restricted to Europe and came to China (1949) and North Korea (1949) and various countries in SE Asia in the 1950s (Vietnam, Laos, Cambodia). There were communist uprisings in other parts of Asia and the sub-continent (Malaysia, Thailand, Singapore, Philippines, Burma, Pakistan, India, Bangladesh, Sri Lanka, Afghanistan) and, since the 1940s, many African countries such as the Congo and Angola have flirted with communism. The same flirtation has occurred in Central America (Nicaragua, El Salvador, Guatemala, Honduras), the Caribbean (Cuba, Granada) and most South American countries.

In 1959, Cuba was wealthier than European countries such as Ireland and Finland.[192] Now, after more than 60 years of communism, the annual average Cuban earnings are $6,000[193], Finland €48,000 and Ireland €52,000.[194] Cuba is a militarised one-party state with no civil liberties. Public gatherings such as demonstrations are banned. At least 300,000 people were murdered and 300,000 left to rot in gulags. Che Guevara was not a freedom fighter on a motorcycle. He killed some 140,000 Cubans and his special target was gays. In mid-July 2021 there was an anti-communist demonstration in Cuba. Two weeks later it was reported[195][196] that 200 people who were at the demonstration are now "missing".

In 1929 in the Soviet Union, party activists went to the farming areas and forced farmers to move from their private farms to modern collective farms where tractors would replace horses. Farmers with private holdings were required to make a contribution to feeding the cities; if the quotas were not met then it was obvious that the farmer was hiding food and in the last half of 1932 more than 100,000 farmers were sentenced to 10 years gaol for refusing to meet their quotas.[197]

Mortality rates suddenly rose and party activists on the collective farms who had no farming experience or related knowledge made matters worse. As always, once a central government starts to interfere with local matters, disaster follows. We also see this all the time in Western countries.

Every time a country becomes communist, people die unnecessarily, people mysteriously disappear, the standard of living decreases, education systems become an agent of propaganda, freedoms are lost, property is stolen, there are forced migrations at gunpoint, communication becomes mortally dangerous, infrastructure is destroyed, economies collapse and the country goes from bad to worse. If you want socialism, then move to a socialist country. Socialism means you can't socialise with anyone.

Why have people risked their lives to move from communist countries to the West yet Westerners do not queue up to emigrate to a communist country?

Some of these countries even have what they call elections. Most commonly one candidate wins more than 99% of the vote and if you didn't vote for this candidate, then you are arrested, gaoled, tortured, killed or "disappear". Sometimes, opposition candidates are arrested, "disappear" or are murdered.

China's communist agricultural and social programs probably killed 50 million people in the time of Mao Zedong.[198] China massacred countless thousands of its own citizens in Tiananmen Square in Beijing in 1989. The BBC estimated that the figure was 10,000.[199] Soldiers shot innocent people and tanks ran over people wanting democracy in China.

After the Tiananmen Square murders, 12% of Chinese newspapers were shut down, 32 million books seized, 150 films banned and many tens of thousands arrested.[200] There are tens of thousands of operators in China who censor the internet every day, remove any reference to the Tiananmen Square murders and China now claims that there was "stability" after the event. It's easy to get stability in a totalitarian system when people who speak out are rubbed out.

China currently has "re-education camps" for at least a million of its Uighurs. God knows what happens in these camps because, as with other socialist and communist regimes, there is absolutely no news flow.[201] It can't be pretty. China struggles with its history. If you live in China, there is no way you could find out about the Tiananmen Square massacre. There is nothing accessible on the internet, nothing in newspapers and local people don't dare speak a word.

This silence about murders, rewriting of history, blanket censorship and rounding up of people is a characteristic of all regimes that are socialist and communist. These countries excel at the knock on the door at midnight. There is no such thing as a little bit of socialism with alleged equality, kindness and compassion. If you are attracted to socialism then you are supporting murder. Is this what you want for Australia?

Communist North Korea executes those who may have a different political view, do not succeed in negotiations with the US or may be perceived a threat. Even family members of the current leader were executed.[202] North Korea starves its people yet spends a fortune on rockets and nuclear weapons.

Early in the 20th Century, Argentina was the second wealthiest country in the world after the US.[203] Wealth was mainly from cattle and crops. And then along came socialism and it was only a few decades before the country couldn't feed itself despite being an agricultural Garden of Eden. It then went bankrupt and had to be bailed out by the International Monetary Fund.

More than a century later, Argentina has regressed to being a developing country. Australia could go the same way with debt, red and green tape, poor education, the loss of the work ethic and the wasting of huge amounts of money on climate change capers.

Socialist Venezuela has collapsed in front of our eyes over the last few years.[204] It was the wealthiest country in South America from the 1970s because it has oil reserves greater than any other country in the world, ahead of Saudi Arabia, Canada and Iran.[205] It still has the world's largest oil reserves, oil accounts for 99% of the country's export revenue and is close to the US and European markets. The Venezuelan oil industry is owned and run by the state.

Venezuela also had a huge agricultural industry with products exported to the US and Europe. Venezuela had food, water, money, health care, electricity, stability, property rights and jobs. Inflation and debt were low and manageable.

Then along came the socialist leader Hugo Chavez and it only took 20 years to totally destroy the country. There is now massive inflation and huge debt, no food, no tap water, no electricity, no sewage system, no jobs, no medical assistance, no property rights and starvation resulting in a shrinking of children's body size, weight and IQ. As for going to school, forget it. The income-earning state-run oil fields have collapsed because of lack of equipment, no maintenance and incompetence.

The same is now taking place in South Africa and took place in Zimbabwe, which is also a basket case. Any country that prints a 100 trillion-dollar note is ruined. This amount of money buys very little and many of us have this Zimbabwean bank note as a tragic reminder of how a great agricultural and mining country was quickly destroyed by sticky-fingered socialist despots.

As soon as a government starts to talk about or actually removes property rights, then this is the road to Venezuela or Zimbabwe. Each time an Australian political party has flirted with socialist ideas, it gets thrown out yet such ideas do not die easily in Australia. Thanks to green activists.

The National Socialist Workers' Party (which in English we call Nazi) started in 1920 in Germany.[206] It was elected to government and did not gain power by storming the barricades in a revolution. Many members of the National Socialist Workers' Party were animal lovers, vegetarian, cultured and had children. Does that make them good virtuous people? In slanging matches in Australia, those with a different view or considered right wing are often called Nazis when their arguments cannot be answered.

However, the Nazis in Germany were socialists and not a right-wing party[207] and quickly became totalitarian after Hitler was elected Chancellor in 1933. Actions by the National Socialist Workers' Party led to the calculated extermination of at least six million people and the deaths of 70 to 85 million soldiers and civilians during World War II. That was 3% of the world population at that time.

Before you claim to be socialist or argue that socialism may be a better political system than democracy, talk to an older person who escaped European, Russian, Chinese or African socialism. If you are still attracted to socialism or communism, go and live in Zimbabwe, Cuba, Venezuela, North Korea or China for a year or so. If you survive, you will change your mind in a flash.

Do as I did. Travel to totalitarian countries and get out of the normal tourist traps. Read about the history of revolutions and socialism. You will find that there were very few revolutions that resulted in a better life for the average person. Socialism means death, poverty, starvation, inequality and slavery and the transfer of huge amounts of people's money to the bank accounts of dictatorial demonic despots.

If a teacher told you about joys of socialism or communism, beware. All you were told was theoretical and not the brutal reality. Your teacher was wasting class time to teach the political propaganda of death. Such teachers have never lived under the murderous brutality of socialism or communism, they would not survive or would be carted off to a "re-education" camp before you could say boo. These teachers probably know very little about totalitarian systems yet use such ideology to show their envy of hard-working successful people in our capitalist world.

Greens evolved out of socialist idealism and hence need more lies to sustain them to the point where honesty is just not possible. Green idealism has an attraction to anyone under the age of 40, especially if they retain threads of youthful socialism. Older greens are total hypocrites.

Capitalism may not be perfect but it is the best economic system we've devised. Democratic countries are capitalist, free and have property rights. Under capitalism, we have equality of opportunity and we can vote. In capitalist countries, people born on the wrong side of the tracks can become phenomenally successful and wealthy. Those born with a silver spoon can bomb it. And they do.

Once in the embrace of socialism, you have a terminal disease. Why must our children be exposed to socialism through a degraded education system?

If our children are taught at school that climate change results from evil capitalists raping the planet, then they should do something about it. They should give up their smart phone, iPad, television, Xbox, computers, internet, social media, health care, hot water, electricity, heating, cooling, lighting, hot meals, refrigeration, batteries, new shoes, new clothes, school bags, jewellery, watch, soap, shampoo, make up, paper, pens, cutlery, crockery, glass windows or glass ware, car or aeroplane travel and holidays and make sure they walk to school. If children are in favour of banning coal and oil, then have them live in a cave. Once they have experienced this life as children, they are on the road to understanding the consequences of climate change policies. It's called reality.

In a capitalist society we have choices. We can be lazy, choose not to work, spend our life on drugs and be a whingeing slob and yet feel entitled to be supported by those who choose to get out of bed, work, pay tax, make sacrifices and acquire skills and an education in order to look after themselves and others.

That is the unfairness of capitalism.

Do as I say, not as I do

In the 18th and 19th Centuries, we used whale oil for heating and lighting.[208] This was replaced by fossil fuels in the late 19th Century. What would you prefer for your heating and lighting? Energy from coal and oil or energy from whale oil? Australia only stopped whaling in late 1978. If you are against the use of liquid fossil fuels, will you support a return to whaling?

In 2017 Senator Sarah Hanson-Young[209] took her sick daughter whale watching at taxpayer's expense in the Great Australian Bight. It was claimed as "electorate business". Do whales vote? She is Senator for South Australia and represents all people in South Australia. Did the electorate get value for money for almost $4,000 of taxpayer's money? Why can a Senator spend time whale watching when so many of those who pay her more than $200,000 a year are on struggle street? How can such a Senator attack big business, perks and capitalism with a straight face?

After the trip, the Senator posted photographs of the Head of Bight Whale Centre, plates of fresh oysters and her 10 year-old daughter admiring sunsets and whales. She claims she took her daughter because she was sick and claimed *"I didn't have anyone that could look after her at home"*. Most parents would have stayed at home with a sick child and almost everyone in the workforce would not be able to claim a whale-watching holiday as an expense.

Did she explain to her daughter that whales were slaughtered almost to extinction for lamp oil and it was fossil fuels that saved the whales because it was cheap and technology made it easy to extract and refine? Fossil fuels and the free market saved the whales, not the green policies.

Did she also explain to her daughter that the planes and cars they travelled in to go whale watching were made by and powered by fossil fuels? Maybe a window seat allowed her to explain to her daughter that if there were net zero carbon emissions, there would be neither the agriculture nor fishing that could be seen below.

These are the same planes that take Hanson-Young business class on the short sectors to and from Canberra. She gets picked up and dropped off by fossil fuel consuming cars and survives the heated Senate in freezing Canberra winters by using coal-fired electricity.

In 2018, she was at it again. Before politics, Senator Hanson-Young was a bank teller. She often ranted about company tax cuts yet was late to pay back her taxpayer-funded bills with invoices unpaid for over 120 days on numerous occasions and overspent on staff travel by $20,460.[210] Is something not quite right when the Senator cannot pay her bills on time and overspends yet votes on complex financial legislation in the Senate and travels at her own cost to the World Economic Forum in Davos (Switzerland)?

I guess being a former bank teller trained her to make financial decisions as a politician. How did she get to Davos without using huge amounts of fossil fuels in aeroplanes and ground transport?

Politicians must declare their financial interests. Greens Party Richard Di Natale failed to declare his 20-hectare family farm *Twin Gums* in Victoria's Otway Ranges for 15 months. He declared a North Melbourne investment property on July 28th 2011 but not *Twin Gums*, his primary residence. I suppose we can forgive Richard Di Natale for forgetting about his $2.3 million 50 acres of land that, if owned by someone else, he would claim was "stolen land". I guess it's like the small change we squirrel away and just forget about. A $2.3 million property here or there is just so easy to forget. On October 2012, he transferred the jointly-owned farm to his partner Lucy Quartermaine. In parliament, the very same Di Natale attacked Labor MP David Feeney for failing to declare his $2.3 million property.[211]

What sector of society do the Greens represent? I guess that all those hypocrites in society need parliamentary representation and they have their own party to do just that.

In his maiden speech in 2011, Australian Greens politician Richard Di Natale championed the Greens as *"a party that represents the best traditions of liberalism, expressed through its support for individuals to make decisions without interference from government"*.

In 2019, the Australian Greens leader Richard Di Natale proposed to criminalise free speech on subjects he does not like and to stop conservative commentators. Has he ever heard about hypocrisy or the boot being on the other foot?

On 19th December 2019, the leader of the Australian Greens Senator Di Natale said *"What kind of message does it send when a multi-billion dollar corporation like Whitehaven Coal pays less tax than a nurse on $54,000 a year?"* and was supported by Greens Senator Larissa Waters who said *"Our democracy is sold out. We need to stop the rorts that let big corporations get away with tax avoidance and remove the toxic influence of donations from our parliament"*.

Because the Annual Report of Whitehaven Coal shows that $235 million was paid in tax in 2018, it is clear that Senator Di Natale is trying to attract anti-business socialist voters.

Company directors need to demonstrate financial competence before they are allowed to make financial decisions. Why not the same for Senators who are dealing with our money and far larger amounts than any company deals with?

If an Australian Greens Party politician tells a lie, their followers cheer and they still collect annually more than twice the Australian annual average household income. Was this attention seeking for the envious left or were they trying to raise matters of concern that should have discussed with Treasury and the Australian Taxation Office before making a public statement?

The Greens Party seems to object to political donations. However, the Greens web site bleats for donations from the naïve and some of the biggest political donations in Australian history have been from businessmen to the Greens and the Electrical Trades Union of Australia (Victoria Branch) has made a substantial donation to the Greens.

Don't let the Greens anywhere near money. They only thing they understand is that public money is for pillaging for whale watching or jaunts with an au pair, wasting on pet schemes promoted by green activists and creating unemployment. Greens belong in salads, not parliament. In salads they are good for you; in parliament they are toxic, cost huge amounts of money and kill forests, wildlife and humans.

Leading green activist Bob Brown was a former Greens Party leader in Australia. He was also the chairman of Sea Shepherd Australia. This is the same Bob Brown who fulminated when a grounded Chinese coal ship released some oil on the Great Barrier Reef: "*Studies of previous accidents shows damage to the Reef occurs through physical damage to the coral substructures and toxic pollution from marine anti-fouling paint, as well as impacts from oil spills...Because of the sway the industry has over the government, the Great Barrier Reef has been turned into a coal highway... The Greens are calling for a Royal Commission into how this situation could occur. Certainly, the coal industry should be held to account*".

However, in early 2014, the anti-whaling ship *Sea Shepherd* pleaded guilty in the Cairns Magistrates Court (Australia) for polluting Barrier Reef waters with 500 litres of oil.

Apart from the *Cairns Post* and a few blogs, there has been no mainstream media mention of *Sea Shepherd* polluting the Great Barrier Reef. By contrast, the news outlets got themselves into a huge lather about the grounded Chinese ship releasing oil onto the Reef a few years earlier.

So, it's OK for a green's anti-whaling ship to pollute the Great Barrier Reef but not OK for a coal carrier to do the same thing. It seems that there is a pecking order of environmental concerns and that saving whales from Japanese fishing boats is higher priority than dropping oil in Barrier Reef waters.

Bob Brown had an extremely well-conceived thought bubble. In 2019 just before a Federal election, he organised a group of green activists to raid the coalfields of Central Queensland. They drove thousands of kilometres from southern Australia in fossil fuel burning cars made of steel. Steel is made by burning coal and emitting carbon dioxide. Gravel roads are bound with a fossil fuel residue to make a sealed road. The convoy created massive roadkill.

Brown condescendingly advised Queensland coal miners that they should give up their high paid coal-mining jobs in order to stop human-induced global warming. He suggested that there was work in the green industry. What miners knew is that for every green job created, three productive jobs were lost and green workers get paid a third of the salary of a miner.

Brown did not tell the miners who were to become unemployed and impoverished that such a fate was green policy.

As a result, the green activist Australian Labor Party (who used to represent miners and other workers) and the Greens Party in Queensland were wiped out in the election.[212]

The end result of this hypocritical green activism is that the Queenslanders voted for jobs. They voted for the coal mining industry. And they voted for Adani's Carmichael Mine which at maximum capacity will produce only 0.16% of the world's coal output. Queenslanders voted for common sense, something not taught at school.

Coal mine workers and businesses in Clermont (Queensland) should erect at statue of Bob Brown to remind them that the great southern enemy can be defeated by common sense. Southern greens greatly outnumber Queensland workers but in many battles, numbers don't count.[213]

Environmental movements have a shocking track record. Even the co-founder of Greenpeace, Patrick Moore, has washed his hands of them because he claims they are now no longer an environmental group and are a thuggish communist political pressure group with no regard for the truth.[214] They were originally concerned about a green planet and peace. They are now a business of hypocritical environmental thugs who rely on donations from the naive.

For a while in 2020 I was living in the Melbourne suburb of Fitzroy, the home of the hard left loopy greens and home of the Greens Party MP Adam Bandt. Yes, you know him. The politician who didn't disown the strong anti-Semitic statements from a former Greens Party candidate.[215]

The Spectator Australia article asked whether the Green Party has hundreds with such views and that Bandt maybe did not dare to cross his supporter base? A deleted 30th May 2019 tweet suggests he has anti-Semitic *Der Sturmer*-style form. A closer look at the benefactors, expenditure and every Greens Party tweet is an obvious theme for investigative journalism or another book.

Back to Fitzroy. Despite having lived in industrial centres at times in my life, Fitzroy was one of the most polluted areas I have inhabited. Piles of rubbish, constant noise at all hours, bumper-to-bumper cars, meaningless graffiti, soup kitchens tucked in behind swanky bars and restaurants, multi-million dollar houses next to rented dumps and a constant smell of decomposing discarded food characterised the area. Homelessness was everywhere and I noticed that there were no Greens Party charity soup kitchens; only those run by the established churches.

House gardens, flowers and attempts to beautify residences were not common. The green activists in Fitzroy can obviously see the world clearly because the trees have gone. This is the environment that inner city green activists prefer to inhabit. Green activists need to get their own house in order before they start to preach to others.

The hypocritical spleen venting urban green activist lives in a place that was once a forest. It was built from mine products and forest timber. A massive amount of coal was burned to construct the residence. Forests had to be cleared, wildlife killed and soils sterilised to allow the green's accommodation to be constructed. Only hot air is now generated from this land rather than food, fibre, energy or metals.

There's no need to be a scientist to understand what is good for the environment. Fitzroy might be perfect for cockroaches and feral rats but it is not good for us. It is not good for plants and animals. It is disgusting. We can all make daily and long-term decisions about how much of our own crap we want to live with. You can do it and your actions are the most effective way you can change the environment.

The cities need to be cleaned up before city-based environmentalists can pontificate about the environment of rural Australia. Cities are built on land that was once pristine forest yet city environmentalists are too keen to stop foresters harvesting timber, despite it being a renewable resource, and farmers who might want to clear land for food production find it nearly impossible to gain regulatory approvals. When city-based environmentalists get out of their concrete bunkers in cities and live sustainably in caves, then they may have some credibility.

There are other simple decisions we can make, even as a city person. Don't sit on your bum and complain about environmental degradation. Do something useful. Why not eradicate every introduced feral animal and plant using the tried and proven methods of guns, traps, poisons, genetic engineering, organism specific viruses and pesticides? Put a pack on your back, go bush, destroy feral animals and plants and forget killing the economy or your fellow humans with hare-brained environmental policies.

If you see a feral animal on the road, run it over. I do. Rabbits have almost destroyed this country a number of times and I have pretty good aim when a bunny is in the headlights. Cane toads are really easy and give a very satisfying pop when the wheels flatten them. Foxes, feral rats, feral cats and feral dogs are much harder to run over as they are far more crafty. I don't suggest you take on a feral pig, goat, deer, donkey, buffalo, horse or camel unless you want to have massive car damage. Don't try any of this on a motor bike unless you want to be a Darwin Award-winning organ donor.

In Australia, feral cats kill zillions of rare native birds and many rare mammals, feral goats eat everything they can, feral dogs form marauding killer packs, feral horses destroy alpine upland areas in national parks, feral foxes and rats plunder wildlife, cane toads wipe out native life, feral pigs and buffalo pollute and destroy waterways, feral camels kill vegetation in sensitive desert areas, introduced fish take over from native fish and introduced plants destroy ecosystems. It is a very long list of introduced plants and animals that have gone feral.

You can have a great impact on the planet. Organise groups to go bush and kill introduced plants and animals and to clean up our waterways and beaches. Camping out with a group is fun and you can do something important for the environment: kill ferals. If you can't find anything to kill, collect plastic.

Why don't the greens show the world that they are really concerned about the environment and actually do something? Move away from the city pulpit, get outdoors into nature and do something to make the world a better place. Go hunting and kill feral animals.

Humans exhale about three billion tonnes of carbon dioxide each year[216], presumably promoting deep and meaningful discussions about whether human breathing heats the planet. Why don't Adam Bandt and his fellow hypocritical green activists who are so concerned about carbon dioxide emissions completely stop adding carbon dioxide to the atmosphere during Earth Hour?

Just remind me again how Greens Party parliamentarians travel to and from Canberra. By carbon dioxide-emitting jets.

The world's airlines emit 2.5% of the human emissions of carbon dioxide by burning fossil fuels[217] and these emissions were rising 70% more quickly than expected.[218] If activists and green politicians were not hypocritical, they would travel using a yet-to-be-invented machine from which there were no carbon dioxide emission during construction, use, maintenance and decommissioning.

Do Greens Party parliamentarians insist that the air conditioning, lighting, central heating and electricity be cut off at Parliament House? No. Electricity for Parliament House derives mainly from coal. Greens Party politicians consume food and drink at Parliament House that has been heated, refrigerated, transported or fermented using fossil fuel-generated electricity.

How do they use a mobile phone or computer without electricity? Green politicians are hypocrites and should be constantly called out. Green activists broadcast their messages by using coal-fired electricity for radio, television, internet and social media.

The only thing of interest with Earth Hour is the number of children conceived. April 22nd, Earth Day, March for Science, Lenin's birthday. Appropriate. The annual Earth Hour, is when people in rich Western countries are supposed to turn off their lights for 60 minutes.[219] This is to repent for using coal-fired electricity for our houses, schools, hospitals and businesses and the first to suffer would be those children on "strike".

Of course, computers, refrigerators, heating, traffic lights and air conditioning are not turned off and toxin emitting candles are used for pious gathering in the semi-darkness. For most of the young, it is fun. For all participants it is total hypocrisy and makes life for the poor tougher.[220]

While the Earth Hour celebrants claim to be concerned about the poorest in the world, their policies produce soaring electricity prices, fewer jobs and lower living standards in the Western world and, in developing countries, perpetual poverty, disease, malnutrition and premature death.

We are paying more and more for minimal improvements in the environment, this is combined with ever-expanding government and activist control of our lives and with stonewall opposition to reliable, cheap energy for the Third World.

Earth Hour should be renamed Green Energy Poverty Hour as recognition of how green schemes damage the poor, the elderly, the working class and families in the developing world.

The school "strike" on the Ides of March makes kids useful idiots. A strike is when worker's labour is withdrawn. The school "strike" was a fun social day skipping school and meeting new people. It was backed by teachers' unions.[221] Why not strike about the poor educational standards that place Australian school kids at the same academic level as those in impoverished Kazakhstan?

Why not strike about the poor people in Africa and India who don't have access to cheap reliable coal-fired electricity thereby allowing them to also study at night? Why not strike about the national debt which must be paid back? Guess who will pay it back?

Listen to some of the interviews with the children. They are passionate, dogmatic and traumatised as a result of long-term child abuse about a forthcoming climate catastrophe and they jettison rigour for jargon and have no knowledge.

They clearly have not been taught to think critically, would know nothing about the Roman and Medieval Warmings and would not know that every positive aspect of their lives has come from cheap fossil fuels. They use smart phones, clothing and transport systems that are dependent upon fossil fuels.

The world's biggest carbon dioxide emissions are from China. Unless climate activists have their Earth Day or school "strikes" outside against Chinese embassies, consulates, Chinese businesses and in China, then they are hypocrites.

Children are a chaotic mass of undeveloped cells which is why they can't vote, can't sign contracts, fight for the nation, drive cars and legally drink alcohol or smoke tobacco. They need time, discipline in emotional regulation, a rigorous education, the ability to think and life experience before there is cellular organisation evolved from chaos. Children need to pick their parents wisely.

Children are full of potential but need to harness the chaos and won't peak until they reach their 40s after they have dealt with life's setbacks and mastered resilience. This is not the type of fluffy resilience taught in schools or woke businesses but the type that been earned and experienced.

These same children use electricity to broadcast their opposition to fossil fuels and their banners, clothing, electronic equipment, travel, smart phones and instant media communication happen because of fossil fuels. They are hypocrites.

Green activists hate nuclear power for the same reason I love it. Nuclear power works 24/7, no matter what the weather is like. By having nuclear power, unelected green activists lose the power to control the everyday aspects of each person's life because nuclear fission does not emit carbon dioxide. Anyone who claims that carbon dioxide is destroying the planet and is not simultaneously promoting nuclear power cannot be taken seriously.

No country serious about industry, growth and employment relies on solar and wind power. There is not one single industrialised modern country that survives on sea breezes and sunbeams. The weather cannot run an industrial economy. China is building coal-fired power stations as quickly as possible. China's nuclear capacity is also increasing rapidly and, at the current rate, will become the world's top nuclear power producer.[222]

Doomsday climate activists in the US are telling people to give up their selfish demand for constant electricity.[223] Maybe the doomsayers should do a poll of those who like to have electricity such that their children can have hot meals, hot drinks and do school assignments on computers. Maybe they should do a poll of people in hospitals on electricity-driven life-support machines.

Having a few hours of electricity each day is a feature of the Third World and this is where these doomsday climate activists want to take us. These killers should lead by example and immediately stop using electricity. A practice run would be to be stuck between floors in an elevator for a day or two with no lights, air conditioning, communication or access to a bathroom.

Environmental activism, animal rights, sustainability, renewable energy and climate change are the concerns of the rich in First World countries. In the Third World, people lacking food for themselves and their family are not able to be concerned about endangered species. They need food.

The NZ Prime Minister Jacinda Ardern demanded that Prime Minister Morrison increase Australia's action on climate change.[224] She was not critical of China's emissions that are growing at the fastest pace in seven years and are more than the US and EU combined. New Zealand's *per capita* emissions rate increase over the past 27 years has outstripped Australia's by five times.[225]

It is a phenomenon extending through all levels of hierarchy, whether it be dictators, prime ministers or tinpot popinjays. Speaking of the latter, the mung beans of Melbourne's Moreland City Council announced a "Meat-free Monday" in the name of combating climate change.[226] Henceforth, its caterers will serve only vegetarian dishes at events and meetings falling on that day.

Roman Catholics have had hundreds of years of meatless Fridays and meat-free Lent. Climate was not affected. Why have a meatless Monday in a local government area of Melbourne because a few crazies think they can change the global climate by eating grass clippings one day a week? One enterprising meat producer, managing director Steven Castle of Koallah Farm, responded by offering discounts and free delivery to Moreland's 45,271 constituents.[227] Not all heroes wear capes.

Not to be outdone the ACT Legislative Assembly, another grandstanding town council, considered a proposal by Greens MLA Caroline Le Couteur to force school canteens to provide vegan options.[228] Additionally, she wants hospital meals to be plant-based: *"It's easy to feel overwhelmed by climate change"* she stated. *"A simple adjustment in diet ... can help in our collective efforts to deal with the unfolding climate emergency"*.

Hospitals are miserable places at the best of times where one maybe can get some joy by devouring a hearty meal after fasting. Imagine, however, waking up ravenous after an operation only to be presented with a tofu salad drizzled with green pea puree. For the love of God, even a condemned man on death row is allowed a decent meal.

The UK's Committee on Climate Change urges the public transitions to a plant-based diet and reduce consumption of meat and dairy by at least 20% by 2050 to achieve "carbon neutrality". It will also create far more climate sceptics.

Dame Emma Thompson flew 8,870 km from London to Los Angeles for her 60th birthday bash and then flew the 8,780 km back to her Hampstead (north London) home to play her part in the Extinction Rebellion demonstrations against climate change.[229] She was photographed in seat 2A in First Class drinking champagne on the return flight from the USA.

Just remind me again, what are the bubbles in champagne? Carbon dioxide. The good news is that Emma Thompson could have taken a private jet (which she has done in the past) rather than British Air. She is a passionate advocate of carbon dioxide emissions reduction and claims that we mere mortals should fly less. However, living the high life as a left-wing actress makes it hard for her to be anything else but be a first class hypocrite. Her London agent failed to say whether she took part in any carbon offsetting schemes, such as planting trees.

At the Extinction Rebellion demonstrations, a tanned Emma in dungarees made victory V symbols with both hands while being filmed at the protestors base at Marble Arch. In the short clip, she said *"Hello fellow humans. I wasn't here on the 15th of April, I was with my husband because I just turned 60...I absolutely wanted to be arrested on my 60th birthday but I haven't quite managed that, I bet that will happen in the future".* Something is wrong with Emma. She can only have one 60th birthday and cannot possibly have another one for her arrest.

She added *"It's hard, it's inconvenient for people sometimes but it's much more inconvenient to leave a planet that's so destroyed that our grandchildren will be up against things that we cannot even imagine".*

An Extinction Rebellion spokesman said *"If Emma Thompson wants to come and help out, that's great, she's using her platform which is incredibly valuable to anyone. If she has to fly around the world like a climate lawyer might have to fly around the world, it seems counter productive in the short term but we are looking at the bigger picture".*

Extinction Rebellion insists that no flights be taken, even in Economy Class, unless in an *"extreme emergency"* and insisted that any flight Emma had taken was an *"unfortunate cost in our bigger battle to save the planet".* What vomit-inducing hypocritical bilge.

Dear Emma helped buy land near London's Heathrow Airport in a failed attempt to stop the building of a third runway. She took off and landed at Heathrow for her 60th birthday bash. So let me get this right. It's OK if Emma flies First Class and takes off and lands at Heathrow because she is on planet-saving business but if a worker wants to take a short flight to enjoy Spanish warmth on annual holidays, then Emma's Extinction Rebellion wants to prevent those flights.

Why should we take any notice of so-called celebrities lecturing us about our lifestyle? These people only have to remember a few lines at a time after doing multiple retakes. Who are they?

Politicians should lead by example even though they sometimes may have urgent business on the other side of the world. *"If you offset your carbon, it's the only choice for someone like me who is travelling the world to win this battle…I can't sail across the ocean, I have to fly to meet with people and get things done".* This quote is from John Kerry, President Biden's special climate envoy, explaining to reporters why he took his private jet to accept the Arctic Circle award for "climate leadership" in Iceland.[230,231] If he was not hypocritical, he would have rowed a boat to Iceland.

John Kerry also fought to keep wind turbines out of sight from his summer house in Nantucket.[232] No incessant humming and health-damaging buzzing and bird and bat slicing by turbines for the great hypocrite. Wind turbines, we are told, reduce carbon dioxide emissions and I thought that Kerry would be happy to be able to see these wonders of the modern emissions reduction age.

Google's Camp Conference held at a luxury resort in Sicily was for world "leaders" to address global climate problems. The last week in July 2019 saw the A-listers jetting into Sicily for a party for 300 guests to groan about a concocted climate change crisis. There were only 144 gas-guzzling jets and many huge cruisers worth up to $400 million to carry the glitterati who lectured us about our excessive emissions of carbon dioxide and the crisis of climate change.

The Google summer camp sufferers had a modest $100,000 supper at an ancient Greek Temple of Hera. No plastic straws were used at the supper because they wanted to reduce their carbon footprint. Guests were carted around in 200 gas-guzzling Maserati SUVs that had to be shipped to the island.[233] This is unlike any camp I have attended.

Former president Barack Obama, actor Leonardo DiCaprio and Prince Harry added scientific credibility to the breast beating at the Verdura Resort which, although it won a "sustainability award", boasts of three water guzzling golf courses, a 60-metre heated swimming pool, four outdoor pools, steam baths, saunas, purge pools and a private beach covered with sand flown in from elsewhere. The energy used by the resort would keep a modest-sized town going for a year.

Although Prince Harry wore no sackcloth, after a visit to a podiatrist he lectured barefoot about the need to "*save the environment*" and stated that he would confine himself to two children.[234]

Prince Harry also lectured about climate change being a "humanitarian issue". He agrees with me. It is a humanitarian issue for those denied reliable cheap coal-fired electricity in Third World countries by Western hypocrites. I will believe that there is a climate crisis when the glitterati behave as if there is a climate crisis. Until then, I don't want to be lectured to by hypocritical narcissistic ignorant drongos about climate myths.

Later Prince Harry and his lady were both made to look like right royal dills when it was revealed the duke and duchess of sustainability had passed up commercial flights and taken four private jet journeys in 11 days, including a trip to Nice, France, to stay at Sir Elton John's home.[235]

A petulant John responded[236] "*To support Prince Harry's commitment to the environment, we ensured their flight was carbon neutral, by making the appropriate contribution to Carbon Footprint TM*". We're all not in this together. The rich, famous and powerful buy indulgences from carbon scamsters. These people are appalling hypocrites and expect us to take notice of their ravings.

Scientists who specialise in climate change fly in carbon dioxide-emitting aeroplanes more often than other researchers, according to a study by Cardiff University.[237,238] Are these the same scientists who shriek hysteria in the media about human emissions of carbon dioxide driving global warming and that there is a climate "emergency"? These hypocrites use our money for jaunts around the world for climate conferences with their mates and then attack we mere mortals for flying.

Hypocritical oxygen thieves leave a large carbon footprint. It's a case of thee and me.

2

LOOK BACKWARDS TO SEE FORWARDS

Climate "science" ignores the scientific method as does climate "science" peer (or pal) review. Consensus and settled science are terms to enforce green activist and political certainty. These are not words of science.

Climate "science" is not in accord with what is known from the geological past, archaeology, history and experiments.

Climate "science" is one method of promoting government policy, the UN and beneficiaries such as China, Russia and many Western businesses. Global climate will not change by gifting more money to the UN.

In the past, climate changed due to tectonic, galactic, orbital, solar, ocean and lunar cycles. Past climate changes were not driven by carbon dioxide. Nothing has changed.

Ice has been on the planet for less than 20% of time. Each of the six great ice ages started when atmospheric carbon dioxide was higher than now. Ice sheets wax and wane, sea ice comes and goes, sea levels rise and fall by more than 600 metres and life diversifies after ice ages.

We are currently in the interglacial warm phase of an ice age that started 34 million years ago. There have been hundreds of warming and cooling cycles in our current ice age with temperature changes of up to 15°C. Why should a total 0.8°C increase during a period of rising and falling temperatures over the last century drive us to oblivion?

The Earth has been warming since the cold Maunder Minimum in the Little Ice Age 350 years ago. Which part of this warming is natural and which part is due to human emissions of carbon dioxide?

If human emissions of carbon dioxide drive global warming, why wasn't there cooling as a result of the massive decrease in human emissions in the Global Financial Crisis (2007-2008) and the COVID-19 crisis (2020-2021)?

Past climate changes were greater and more rapid than those measured today, carbon dioxide in air increased after natural warming and humans and other life thrived when the temperature and sea level were higher.

There have been at least 26 periods of significant global warming over the last 2.6 million years. The Greenland ice sheet shows 40 warmings over the last 500 years. Why is the current slight warming any different?

Reefs have appeared and disappeared for 3,700 million years. The Great Barrier Reef has appeared and disappeared 60 times over the last three million years.

There have been five major mass extinctions of complex life, over 20 minor mass extinctions and the number of species on the planet is increasing despite species turnover. Cold climates kill life.

The oceans are heated from below and above, the amount of submarine volcanic heat and volcanic carbon dioxide added to the oceans is monstrous and is ignored by the Intergovernmental Panel on Climate Change (IPCC).

The under estimation of carbon dioxide by the IPCC from unseen sources means that the IPPC's estimate of the human contribution of 3% of total annual carbon dioxide emissions is far too high. Why does this purported 3% drive global warming yet the 97% of natural emissions not drive warming?

How do we know what we know?

This chapter begins treatment of that which constitutes real science and contrasts that with the deficiencies and errors of climate "science". Science is one way of knowing things because science is married to evidence that is repeatable. Science also bathes in modest uncertainty. To have consensus, make definitive statements or make predictions without declaring the uncertainties is not science. This is why real scientists often appear so boring.

The nature of science is scepticism. Science encourages argument and dissent. Scientific evidence is derived from repeatable and reproducible observation, measurement and experiment and must be in accord with previous validated evidence.

Computer models are not evidence. Evidence in geology is interdisciplinary, terrestrial and extra-terrestrial and shows the complex and fascinating intertwining of evolving natural processes on a dynamic planet. Much geology is validated with experiments that duplicate processes that took place deep in the Earth.

Scientists engage in healthy argument about the veracity of evidence. The debate is whether the processes of evidence collection were valid, what the errors were, what was the order of accuracy, what assumptions were made, whether such assumptions were valid and whether the data is reproducible.

Much scientific data is collected by assistants and students and then given to a scientist to analyse. What happens if the primary data is collected incorrectly by an inexperienced person?

Much climate "science" research comprises a mathematical and computer model analysis of other people's data. In most cases, this data cannot be independently validated by the modellers. Data for modelling has been "homogenised" hence it is no surprise that models and measurements don't agree.

Much of what has been predicted in climate "science" is within the order of accuracy of older measurements and hence is meaningless. Once we tamper with data as happens in the climate "science" industry, then the question "How do we know what we know?" means that a method of reaching the truth has been abandoned.[239]

Exploration geologists learn this very quickly. Testing of ideas for targets at depth with a drill hole based on the latest mix of geophysics, geochemistry and geology is a very humbling and expensive experience. There are always surprises. Models based on geophysical measurements are most commonly shown to be wrong by drilling. The new subsurface evidence from drilling and measurements on the diamond drill core are then used to refine the geophysical model.

The model may still be wrong but it is better than the first iteration and can be tested again. With testing in exploration geology, failure often results in loss of employment. This is not the case in the climate "science" industry. More grants are just around the corner.

Scientific evidence must be reproducible. Science transcends countries, cultures, religion, gender and race because any scientific evidence can be checked by another person. By this method, it was shown that a peer-reviewed paper on cold fusion was based on instrumental errors and the theory of cold fusion was then discarded.

Post-modernists, Marxists and the cancel culture comrades try to deconstruct science because it is underpinned by evidence, reproducible and is a way of understanding a truth. Whether we drop an apple a hundred times or a million times, we conclude from this repeatable experiment that gravity is a truth.

When it is judged that there is a suitable body of evidence, it needs to be interpreted and explained. Primary or raw data is not adjusted, amended or "homogenised". This is fraud. Misrepresenting exploration data to investors can land a geologist in gaol. If climate data is misrepresented to governments who invest billions on false data, the taxpayer bleeds through the eyes and there are no consequences for the climate "scientist".

Sometimes, data is judged unreliable and must be recollected. Some material just cannot be recollected (e.g. lunar, Martian, asteroid and space samples) although it can be re-examined many times by different techniques because scientific protocol is that such material must be preserved.

Not so in climate "science". A scientific theory is the best available explanation of all the evidence from a diversity of disciplines at the time, it may change with new evidence and must be in accord with the existing body of validated knowledge. If it is claimed that the science is settled, then we are dealing with propaganda and not science and no longer have use for science. Research grants should therefore cease.

Science is anarchistic, has no consensus, bows to no authority and it does not matter what a scientific society, government or culture might decide. It is only reproducible validated evidence that is important. Great science is often undertaken by those without a big name. You don't need to be top of the pile to make a great discovery.

There have been exceptions, such as Lysenko and climate "science", that persist for a few decades. Like any other area, science suffers from fads, fashions, fools and frauds; has short-term leaders and can be cultish. Science advances with each funeral of a scientific leader.

The media views science as a popularity contest in which those with the greatest numbers win. Consensus is politics, not science. If there is a hypothesis that human emissions create global warming, then in science only one item of evidence is needed to show that this hypothesis is wrong. Dozens of items of evidence are listed here and in other writings to show that the human-induced global warming hypothesis is wrong. If a hypothesis is wrong, it must be rejected.

Climate "science", an infant scientific discipline, is built from a synthesis of other well-established disciplines. Mechanisms of climate "science" are complicated because they involve complex, non-linear, chaotically interacting systems which in turn involve poorly understood fluid dynamics.

It may take many decades before climate "science" can come to a better understanding of climate. In the interim, we have a "consensus" and "settled science" yet a large number of physicists, astronomers, geologists and scientists in non-Western countries oppose the popular view that human emissions of carbon dioxide drive climate change. Consensus is just mob rule.

With climate "science", not only are there huge amounts of research funds sloshing around but the economic implications of a carbon trading scheme, new taxes, renewable energy subsidies and the transfer of wealth have brought together all sorts of disparate and desperate groups.

Science funding opportunities, the rise of green activism and green political parties and financial opportunities in the carbon marketplace have combined to produce an almost perfect storm of quasi-religious hysteria with all the hypocrisy that one associates with a fundamentalist religion. Funding, fame, power and self-interest drive a consensus.

Fear of the environmental consequences of ever-expanding economies combined with the deterioration in science education encourages people to think of simple solutions to problems (or non-problems). This is amplified by a lazy media that exploits fears to increase sales and can't really be bothered to get to the bottom of a story.

There has been a great decrease in science ethics. We now have senior scientists seeking to suppress publication of research that undermines their beliefs. We have senior scientists making statements that the "science is settled" on human-induced climate change regardless of contrary findings of others. We now have previously well-regarded institutions discarding scientists whose work is against popular opinion. The age of open transparent science in some areas may be coming to an end.

The history of planet Earth has been ignored in the current popular catastrophist story of human-induced climate change. If large bodies of knowledge from the past are ignored, then this provides an unbalanced, misleading and deceptive view of global climate. If scientists ignore integrated interdisciplinary empirical evidence, then they have politicised science to gain government favours and research grants and they are operating fraudulently.

Consensus views in the past were that the Sun rotated around Earth, that burning material was due to a substance called phlogiston, that health was driven by humours, that leeches could cure most diseases, that malaria was caused by bad air, that earthquakes were an act of God, that heavier-than-air machines could never fly, that the continents could not move and that duodenal ulcers were caused by stress. All these views strongly supported by science at the time were the consensus and settled science. They were wrong.

The normal procedure in science is a new idea is discussed with colleagues. It is then aired at conferences in front of peers. After critical analysis, it is then written up and submitted to a scientific journal for peer review. Publication pressure now is such that much work is submitted for publication in haste.

At any scientific conference, one can see the various putsches prancing around narcissistically competing for attention, fame and research grant fortune while espousing their competing theories and demonising opponents. This is normal human behaviour. Scientists do not hold the high moral ground and have the same weaknesses and foibles as others.

Much is made by green activists about peer review. They proclaim that certain scientific papers supporting human-induced global warming have been peer reviewed, so must be right, and the science is thus settled.

Peer review depends on the discretion of the editor and integrity of the referees. The editor can decide to send a paper to peer reviewers who will give the desired expert opinions. A scientific paper or grant proposal can easily be buried by sending the work to a referee in another camp. As an editor, member of a research grant committee or member of an appointment committee, it is useful to know who the players are in competitive tribalism because an opposing clique often buries good work and outstanding people.

Peer review is only an editorial aid. It involves rejecting or accepting a paper and making suggestions and changes to the submitted paper. Much of peer review is about style and language rather than science. The number of reviewers that can be called on for a submitted paper is generally two or three people.

The peer review process is very susceptible to the influence and bias of small groups promoting and protecting their own interests. Most reviewers are anonymous. This only serves to strengthen the influences, biases and cowardice via anonymity. Pre-publication review is important but the most important process occurs after review. This is when the scientific community at large can refute, check, confirm or expand on the ideas published.

The climate clan is peer reviewing and publishing their own grant-earning dogma and some of the scientific community is showing how weak this science is post-publication. Peer-review, far from being the gold standard, is used by the climate industry to approve or censor science according to its own agenda.

In the climate industry, it is not peer review, it is peer bullying and peer pal pressure to conform to the orthodoxy. The work of Newton, Darwin and some of that of Einstein was not peer-reviewed and it is wrong to impute that only good science is peer reviewed. Most industry funded scientific research is not peer reviewed.

Over a lifetime of science, I have seen shoddy published science and failings in the peer-review system. As an editor of international scientific journals I had the sneaking suspicion that many reviewers just ticked every box and did not review the paper conscientiously because it was not a high priority and they possibly did the chore because they wanted to curry favour with the editor for their next publication in that journal.

I also had peer reviewers and thesis examiners who must have spent days reviewing documents as a service to their science and helping young scientists who will eventually push them off their perch.

Scientific journals should take no position on any issue; they should be indifferent to politics, should publish the spectrum of competing hypotheses and leave it to the market (i.e. the scientific community) to distil, validate or replicate over time.

A lot of what is published is incorrect. These are not my words. In medicine, maybe half of what is published is untrue. Studies with small sample sizes, tiny effects, invalid analyses, flagrant conflicts of interest, or an obsession with pursuing fashionable trends of dubious importance seem to be the norm.

Poor scientific methodology gets results and produces compelling stories. This is until they are checked or an attempt at replication is undertaken. There is so much concern about published incorrect medical sciences that the Academy of Medical Sciences, Medical Research Council and the Biotechnology and Biological Sciences Research Council have now put their reputational weight behind such concerns.

Statistical fairy tales abound, journals have poor peer review and editing processes and processes of granting research monies all contribute to bad scientific practices. If this is the case for the medical sciences[240], what about other fields? I doubt it is any different.

There have been some high-profile errors in physics (e.g. cold fusion), that resulted in changing procedures such that intensive checking and rechecking of data takes place prior to publication. What about climate "science"?

I suspect that with Climategate, failed climate models, failed predictions, exaggerations, omission of contrary data and unfounded statistical studies, this figure would be far higher because I suspect that the fields of physics and medicine are more rigorous than climate "science". Your guess is as good as mine but it's not a pretty sight. Maybe with climate "science" it's a case of never let the science get in the way of a hefty research grant underpinned by a scare campaign promoted in the media.

In the climate industry there are no parallels. There appears to have been a profession-wide decision that there can be nothing published that might threaten their fame- and fortune-giving ideology that global warming is driven by human emissions of carbon dioxide. This lack of scientific objectivity is such that one has to question whether any literature from the climate industry can even be treated seriously for the next 30 years.

The climate industry has openly and flagrantly violated every aspect of the core ethos of science. Climate "science" is advocacy in the service of one arm of politics, is not dispassionate and is no more valid than the tobacco industry publishing science about the harmlessness of smoking.

Very often green activists have too much to say about the peer review system which only demonstrates that they have never had their own work peer reviewed, have never been a reviewer and have never been an editor.

Many of us have written papers with our critics and rarely do scientific disputes become personal. At times our papers are rejected by peer reviewers who are our personal friends. This is irritating but it's not the end of the world.

Scientific theories are testable and once the scientific theory has been tested over time, it becomes accepted into the body of knowledge. Just one contrary validated body of evidence can destroy a scientific theory. This has happened many times in all fields of science. The scientific idea examined in this book is that human emissions of carbon dioxide drive global warming and that this warming may be irreversible and catastrophic.

This is easily tested by comparing with the reproducible validated evidence from past climate changes. This is the coherence criterion of science. Any new idea needs to be in accord with validated evidence from other areas of science.

Human-induced global warming is not supported by measurable geological evidence. There are thousands of examples that show past events of global warming were not driven by carbon dioxide, that the planet has been far colder and warmer in past times and that the rates of temperature change have been far faster than any changes measured today. The past is a story of constant climate change and our present climate is the result of the past.

The idea that human emissions of carbon dioxide will lead to catastrophic global warming is therefore invalid and to continue to promote such an idea is ignorance, fraud or perhaps both. No wonder those who call themselves climate "scientists" don't want to debate geologists.

The climate "scientists" solution to this spot of bother is to ignore geology, astronomy and history; demonise dissent, and yet claim that they are involved in science. All this shows is that they are green activists funded from the public purse. Most science today is in search of research grants and continuing employment rather than a search for new knowledge. Research is not cherry-picking through an existing body of knowledge. It is discovering something new.

There may be one hundred sets of evidence in support of a scientific idea. It only takes one piece of validated evidence to the contrary and that idea must be rejected. That is Karl Popper's concept of falsification or refutation.[241] Adolf Hitler sought to discredit the work of Albert Einstein who was Jewish by commissioning a paper[242] called *"100 Authors Against Einstein"*. Einstein responded *"…it doesn't take 100 scientists to prove me wrong, just one fact will do…"*

This is rather like having a list of scientists and professional societies who support human-induced global warming. The Royal Society comes to mind, which is contrary to their motto[243] *"Nullius in verba"*. The human-induced global warming farce has evolved in a similar fashion.

Creationism has the same characteristics, long lists of allegedly eminent titled folk with impressive looking post-nominals. Some post nominals are easily acquired. I once spent $25 to purchase a Doctorate of Divinity for a neighbour's slobbering grinning cattle dog.

My criticism of the climate industry is that much of the data (especially temperature and carbon dioxide measurements and corrections) is contentious, that computers have to be tortured to confess the pre-ordained result, that computer codes are not freely available, that neither the data nor the conclusions are in accord with what we know from the present and the past, that publication is within a closed system that challenges the veracity of the peer review system, that financial interests are not declared, that the financial rewards for publishing a scary scenario are tempting and that the process of refutation seems to have been overlooked.

The climate industry needs to show it is dispassionate, has no pre-ordained conclusions, is unrelated to green activism, is driven by intellectual curiosity and can accept and critically analyse competing theories.

In areas of science where much of the debate by experts is obscure to the layman and relies on truthfulness (including uncertainties), the climate "science" clique has let down the side and operated as green activists for certain policies by singing from the same hymn sheet as the funder of their "science" (i.e. governments).

As Judith Curry[244] wrote *"In the climate change problem, it seems that often one's sense of social justice trumps a realistic characterisation of the problems and uncertainties surrounding the science and the proposed solutions"*.

For scientists to declare certainty, to make predictions and to ignore large bodies of contrary science shows their personality weaknesses overwhelm the scientific method. In my experience, academics have a huge chip on their shoulder and feel undervalued hence an opportunity to be in the spotlight and not be interrogated is tempting.

There is a view in society that a belief is evidence and a strongly held belief must be correct. This is emotion, not evidence. Most public discussions about climate change are underpinned by emotion and normally the intensity of emotion is inversely proportional to the amount and veracity of evidence.

The word belief is not used in science because belief is untestable. Belief is a word of politics and religion. Science is unable to make judgments about what is good or bad. These are judgments which vary with time and are based on contemporary politics, religion, aesthetics and culture.

That science should face crises in the late 20th and early 21st Centuries is inevitable. Science is primarily funded by governments who are unable to appreciate that great scientific ideas and discoveries are made by individuals and not structured teams or committees[245] following government policy.

These individuals are often dysfunctional loners who are difficult to manage, fiercely independent, don't accept the popular paradigms and don't work well with others. Their discoveries are often serendipitous. These scientists are pilloried by peers, have difficulty in winning research grants, their work is often ignored for decades and they are often grudgingly given posthumous accolades when recognition eventually comes.

Not one great scientific discovery has ever been made by consensus. They have been made by researchers going out on a limb and challenging the popular contemporary knowledge. The grant-awarding political policies of the Department of Climate Change and Energy Efficiency and the Australian Research Council guarantee that there will be no great discoveries in Australia about climate.

That Department only appears interested in those who support the current political dogma and is not genuinely interested in either the environment or energy efficiency. This does not appear to be a wise way to create policy or prepare for the future.

The normal run of the mill scientist adds to the body of knowledge, does not solve great problems and gives little return for a large taxpayer investment. Most innovation is meandering and incremental, not disruptive. Very little of today's scientific research is of use now yet there is the faint chance that some of it may be useful in decades time. This is why many scientific organisations have media units who loudly promote interim findings as a mechanism for winning future funding and keeping the gravy train rolling.

Most scientists are paid to pursue their hobby and can end up in fields so narrow that only a few people in the world know their work. Maybe these scientists should fund their own hobby as the rest of the community does. In the 19th Century, science was funded by individuals who had a burning curiosity. Corruption of science started when governments started to fund science.

Science is easily corrupted because governments fund science and want metrics such as publications, scientific citations, honours and awards, all of which operate in a closed community. Most applied science is funded by industry but almost all natural science is government funded.

We were warned in President Eisenhower's farewell speech in 1961[246] that government funding would change the nature of the *"free university, historically the fountainhead of free ideas and scientific discovery"*. He also warned that Federal funding of science might come to dominate science and a *"science-technological elite"* might dominate science-based public policy. Eisenhower's concerns became a reality.

Governments have difficulty with dispassionate science which may solve one problem, create another two and give no definitive answer. Elected governments respond to publicity, media pressure and lobbying and hence fund science to avoid political embarrassment.

Power corrupts and science and much of the education system today is as the Catholic church was around the start of the 16th Century. The church was used to having its own way and dealing with heretics by excommunication and not argument.[247]

So too with modern science where individuals, heretics and non-conformists are frozen out of the research grant process and have difficulty in having work published in literature controlled by those in the club.

When some of us were senior in our field, we became editors of major scientific journals, sat on editorial boards, refereed too many scientific papers, sat as experts on research council panels, supervised and examined doctoral theses, were called as expert witnesses in Court, advised industry and other scientific institutions and continued as scientific authors. One also sat on review committees of university departments and on chair selection and promotions committees all around the world.

As these activities involved the judgement of the scientific works of others, the power had to be used very responsibly as such duties present a wonderful opportunity to bury the careers of critics. It happens all the time.

Modern scientists have been polluted by the current education system. Kindergarten children are taught about the evils of human-induced global warming and that it is heresy to question such "facts". This process continues in high school and undergraduate studies. By the time a person has finished postgraduate studies and is poised for a research career, the ability to think critically, analytically and individually has been well and truly diminished. Pedestrian run of the mill science on a particular useless non-competitive hobby subject becomes an easy option. Only outstanding people can rise above their formal education.

Many scientists promote that science is pure, truthful, free from bias and a noble quest for knowledge, and that fraud is the preserve of business people, lawyers and crooks. How wrong can one be? Scientists are funded by the public to selfishly pursue their passion, their quest is to keep research funds rolling in, they become narcissistic, uncommunicative and out of contact with those that fund them.

Because science is a closed shop, it is far easier for fraud to flourish. Most scientific fraud involves omission and "homogenisation" of data, poor statistical treatment of data, claiming credit for the work of others, ignoring contrary findings and producing pre-ordained conclusions. There is certainty in the mob rule of consensus science and that keeps the gravy train moving.

Governments may deem that a scientific area such as human-induced climate change is a national priority and should be preferentially funded. Immediately, all sorts of phenomena are claimed to be related to climate change. There is a very real danger that charlatans with self-interest persuade governments to follow a certain theme in science.

We see this today with governments, self-interested businesses and government-funded scientists claiming that a conference in Paris can twiddle the planetary dials and lower global temperature by 2°C.

This is why government funding for research into human-induced climate change gives the result that political policy requires. Everyone wins. Governments can continue a scare campaign using the great force of fear; can control society and increase taxation revenue; scientists are funded if they continue to produce the desired results; and the public is seduced into feeling that their taxpayer monies are creating a better society and protecting them.

Government ideology and an uncritical media have led to the promotion of false scientific concepts such as human-induced global warming. This has led to billions of wasted dollars, poverty and an electricity crisis. It's happened before.

One example was Trofim Denisovich Lysenko, a self-promoting Russian peasant who invented a process called vernalisation based on a trial of growing crops on half a hectare for just one season.[248] Seeds were moistened and chilled to enhance later growth of crops without fertiliser and minerals. It was claimed that the seeds passed on their characteristics to the next generation.

Stalin wanted to increase agricultural production, vernalisation never had scientific scrutiny. Lysenko became the darling of the Soviet media and was portrayed as a genius and any opposition to his theories was destroyed. Lysenko's theories dominated Soviet biology.[249] They did not work.

There was famine, hundreds of dissenting scientists were sent to the gulags or the firing squads and genetics was called the "bourgeois pseudoscience". The evidence-based Mendelian genetics was rejected in favour of an ideology (Michurin's hybridisation).

The plummeting of grain production resulted in the deaths of more than 14 million people between 1929 and 1933 and was a direct result of Lysenko. More Soviet people were killed by Stalin's Lysenkoism and collectivisation of agriculture than were killed in World War I or Hitler's genocide.[250] History has a habit of repeating itself.

Lysenkoism and climate "science" are very similar. They both have what they call "facts". I have been met with the argument by climate activists that they have their facts and I that have my facts! The repeatable validated interdisciplinary evidence that critics raise with activist climate "scientists" is dismissed because they have their facts; other facts are inconvenient hence can be ignored and there is consensus. The method of knowing what we know is abandoned.

Underpinning the global warming and climate change mantra is the imputation that humans live on a non-dynamic planet. Change is normal on all scales and is driven by a large number of natural forces. Change can be slow or very fast. However, we see political slogans such as *Stop Climate Change* or government publications such as *Living with Climate Change* demonstrating that both the community and government think that climate variability and change are not normal hence are the result of human activity.

Those using such slogans think that climate can be changed by just twiddling one dial that controls human emissions of carbon dioxide. Sorry to disappoint you folks but the planet is just a little more complicated than having just one dial that is twiddled to change climate.

We have always lived with climate change and, whether we have elected governments, kings, dictators or emperors, climate will change. By using the past as the key to the present, we are facing the next inevitable glaciation at some unknown time. Yet the climate, economic, political and social models of today assess only the impact of a very slight warming and do not evaluate the far greater effect of another glaciation.

Geology, archaeology and history show that during glaciation, famine, war, depopulation and extinction were the norm.

The global warming scare is the biggest scientific fraud in the 2,500-year history of science.

It will result in the biggest waste of money since the beginning of time.

Who's who in the zoo?

So why is an understanding of geology and its development as a fundamental science so relevant in understanding climate?

In previous centuries, geology was called natural philosophy. It still is. There is no such scholarly discipline as climate "science". Those trained in sociology, economics, politics, mathematics, statistics, physics, chemistry, botany, zoology, environmentalism, activism, legal studies, agriculture, government, marketing, oceanography and meteorology now call themselves climate "scientists". Even yoga teachers sign climate petitions, presumably because they know a huge amount about climate "science".

At times one is criticised because, as a geologist, it appears that I am not a climate "scientist" hence I know nothing about climate. Well, have I news for you.

There have been hundreds of years of study of climate change by geologists. In studies of modern and fossilised plants, Georges-Louis Leclerc, Comte de Buffon (1707-1788) realised that climate had changed in Siberia and Europe. Buffon concluded that past Northern Hemisphere climates were warmer and that the Earth was cooling from its original molten state. As with all healthy science, this idea was criticised by Joseph Fourier (1768-1830).

Studies of modern and extinct Siberian animals led Georges Cuvier (1769-1832) to conclude that there had been a past freezing event and he argued that catastrophe was an important component of Earth history. This was the first time the concept of an ice age appeared in the earth sciences. Catastrophism was criticised by James Hutton (1726-1797) and Jean-Baptiste de Monet, Chevalier de la Marck, commonly known as Lemarck (1744-1829).

In the Paris Basin, Lemarck recognised that many fossils were tropical species. Both Lemarck and Antoine-Laurent Lavoisier (1743-1794) recognized shallow- and deep-water sedimentary rocks in the Paris Basin and concluded that sea level rose and fell. The later mapping of the Paris Basin by Alexandre Brongniart (1770-1847) validated earlier work of climate and sea level changes.

Baron Alexander von Humboldt (1769-1859) used integrated interdisciplinary science, as geologists do now, to show that the planet has climate zones. The seminal works of Charles Lyell (1797-1875) showed that the planet is dynamic, that there were past warmings and coolings and that coral atolls increased in size when sea level rose. Giovanni Battista Brocchi (1762-1826) also showed that there were tropical fossils in ancient rocks, this time in Italy suggesting that climate changes were not local.

Danish natural philosopher Henrick Beck (1799-1863) and Frenchman Gérard Deshayes (1795-1875) independently validated that in the past that Europe was once tropical. After discussions between Lyell and von Humboldt, Lyell realised that climate changes can be latitudinal and global and was getting close to thinking about continental drift in order to explain tropical fossils at high latitudes and polar fossils at low latitudes.

Lyell realised that there had been ice sheets on Earth and the Swiss geologist Jean Louis Rodolphe Agassiz (1807-1873) concluded that a vast ice sheet had covered Europe in the recent past and had left large blocks of rocks upon melting. He concluded these rocks had been carried long distances from elsewhere by ice. Agassiz showed that Swiss glaciers were retreating and concluded that Europe had been covered by an ice sheet from the North Pole to central Europe.

At that time, little was known about Greenland and its ice sheet. Greenlanders had died out in the Little Ice Age. It was only when James Clark Ross (1800-1862) visited the Ross Sea in 1841 and 1842, that the frozen southern continent of Antarctica supported the view of giant ice sheets and zoned climate on Earth.

There was great controversy about large boulders of different rock types spread across Scotland resting on different basement rocks. Lyell's former tutor, William Buckland (1784-1856), concluded that these boulders were relics from Noah's Flood. On a visit to Scotland by Agassiz, Buckland and Roderick Murchison (1792-1871), Murchison concluded that the boulders were left behind by retreating ice sheets during global warming. This resulted in decades of controversy about ice sheets. Murchison later worked in the Perm Basin of Russia and made advances on the understanding of coal-forming conditions, fossilised tropical animals, deserts and glaciers that occurred in the long ago.

The comprehensive work on Scottish glaciation by Archibald Geikie (1835-1924) led to a recognition that sea level rises and falls, land level rises and falls and ice sheets retreat and advance as a result of climate change. Archibald Geikie's younger brother (James Geikie 1839-1915) published the definitive work on climate change (*The Great Ice Age*).

Later work by James Croll (1821-1890) showed that calved icebergs carry stones out to sea and these are later dropped on the sea floor when the iceberg melts. Even in the 19th Century, geologists knew that ice sheets wax and wane, that the retreat of an ice sheet was a normal event unrelated to human activities, that global warming had occurred long before humans were on Earth and that sea level rises and falls.

Geological studies on climate for more than 250 years have involved discussion, argument, criticism, replication, validation and refutation. This does not take place with modern climate "science".

There is a fundamental difference between geology and climate "science". Geology creates wealth whereas politicised climate "science" uses funding created by geology to destroy wealth and the fabric of society.

Geology has no need to be politicised because it is useful.

Did you know?

Climate cycles

The current climate is not unprecedented. It is driven by the same cyclical natural processes that have operated for billions of years.

If you don't want to read the following summarised science, there are only a few take-home messages. Climate is driven by that great ball of heat in the sky. No one has ever shown that the gas of life, carbon dioxide, drives global warming. What has been shown is that the atmospheric carbon dioxide content increases after temperature rises. Past and present climates are cyclical and, over the past 2,000 years, two cyclical peaks were warmer than now. Planetary systems do not work like a kitchen oven where temperature can be changed by twiddling the dials.

The Earth's climate has constantly changed for 4,567 million years. There are long-term climate cycles, medium-term climate cycles, short-term climate cycles and sporadic events that suddenly change climate.

Long-term climate cycles are in the order of 400 million years when continents pull apart and then stitch back together. At those times carbon and oxygen cycles operate in tandem and may drive evolution and extinction.[251]

Living in the wrong galactic address every 143 million years results in an ice age. Galactic climate cycles derive from increased bombardment of the Solar System with cosmic rays. These cosmic rays induce the formation of low-level clouds that reflect heat. The Earth then cools.

The six major ice ages that planet Earth has endured occurred when Earth was in the Sagittarius-Carina (twice), Perseus, Norma, Scutum and Orion Arms when there was increased cosmic radiation. These have been measured in ancient sediments[252] and calculated from astronomy. No legislation or Conference of Parties (COP) jaunt such as Glasgow COP26 can change where our Solar System lies in the galaxy or how continents move. I'm predicting it will be termed FLOP26 in due course.

There are medium-term orbital climate cycles that position the Earth closer or more distant from the Sun. For some odd reason, we have global warming when we are closer to the Sun and cooling when we are further away. These are Milankovic Cycles and have been known for more than a century. No green activist can change the Earth's orbit.

An elliptical orbit cycle of 100,000 years[253] has our planet closer to the Sun for about 20,000 years and distant from the Sun for about 80,000 years giving us warm interglacials and cold glaciations.

Our planet tilts and wobbles in its elliptical orbit. This is a characteristic of the smaller rocky planets in our Solar System. Every 19,000 to 23,000 years the orbital plane of the Earth changes.[254] Every 41,000 years the tilt of the axis of rotation changes[255] from 22° to 25°.

Peak warming is about 12,000 years into the interglacial. In the current interglacial which we enjoy, peak warming was some 6,000-4,000 years ago when temperature was about 5°C warmer than now and sea level was some three metres higher. Orbital wobbles can be measured and are well documented in ice cores, sea floor and lake sediments and the distribution of glacial debris. During the many interglacials that have been measured, plants and animals migrate towards the poles, alpine tree lines rise, new cities appear, coral reefs expand, populations of all animals increases and ice sheets get smaller.

During the last interglacial 116,000 to 128,000 years ago, warm climate animals such as water buffalo, hippos and elephants inhabited northern Germany. During glaciations, ice sheets and sea ice expands, biodiversity, rainfall and crop yields decrease, dust storms and desertification increase and plants and animals migrate towards the equator. During the last glaciation, Scandinavia was covered with 5 km of ice, the UK was covered by an ice sheet 3 km thick as was Canada and the northern states of the USA; Tasmania and the highlands of eastern Australia were covered by ice as was much of South America and South Africa, forests were destroyed and coral reefs died.

Inland Australia was even drier, had howling sandstorms and sea level was 130 metres lower. People could walk from Papua New Guinea to Tasmania. And they did. Geology and history show us that this is the time of extinction. We are past the peak of our current interglacial and, unless the UN and governments can legislate to change the earth's orbit, we face the next glaciation.

We are not prepared for greatly enlarged ice sheets, increased sea ice, desertification, low rainfall and the loss of food production. The larger land masses in the Northern Hemisphere will be hit harder than the Southern Hemisphere as was the case in the many previous glaciations. Once Northern Hemisphere summer temperatures cannot melt all the winter snow, the ice sheets quickly expand and join each other as snow reflects solar energy back into space. It's happened many times before.

The Sun has a number of regular cycles and outbursts of energy. The same occurs with other stars. These influence climate because they result in changes in the solar magnetic field that, in turn, protects the Earth from cosmic ray bombardment. Cycles result in more or less heat striking Earth from the Sun.

The cyclical dimming and brightening of the Sun has a profound effect on climate and can occur very quickly. The temperature effect of carbon dioxide on the atmosphere is insignificant compared to that big ball of energy in the sky. Not surprisingly, the Sun has a bigger effect on climate in cooler times than in warm times.

There are variations in radiation emission, magnetic field intensity, magnetic polarity, particle emissions, and surface convection to consider. These changes affect the Earth in several ways that manifest through auroras, magnetic storms, changes in galactic and solar cosmic rays and climate change. Solar activity can be random and cyclical.

Changes in sunspots produce solar cycles of 11 years (Schwabe Cycle) which is expressed as a 22.2 year cycle (Hale Cycle). This has been known for 400 years. For example, Adam Smith noted a correlation between sunspot numbers and grain prices. A low sunspot number led to cold weather, low yields and high prices. Modulations of this cycle gives other cycles such as the Gleißberg Cycle (87 years), DeVries-Suess Cycle (210 years) and the Dansgaard-Oeschger Cycle (1,500 years).

The daily rotation of the Earth causes continual changes in the incoming solar energy and outgoing back radiation. This gives us the cool dawns and warm afternoons. The overlap of the daily solar cycles and the monthly lunar cycle gives us variations in short term weather driven by air pressure and tides and currents in the oceans and atmosphere. The tilting of the earth's rotational axis gives us the seasons as a result of the variation in the amount of incoming solar radiation. This variation can be extreme near the poles.

These cycles are a surrogate measure of solar radiation and are shown on Earth by cycles of sediment chemical fingerprints that were initially generated in the upper atmosphere by decreased solar activity; lake water levels; droughts and floods; sea surface temperature and temperature-sensitive floating micro-organisms; dust in ice sheets and ocean sediments; vegetation, pollen and spores, bogs and swamps; and historical records. The 22-year and 1500-year cycles dominate.

The 1,500-year solar cycle has produced cycles of warming, ice and ocean sediments. For example, the Ancient Egypt New Kingdom, Roman Warming, Medieval Warming and Modern Warming are all 1,500 years apart. The Modern Warming is up to 5°C cooler than the others. Yet we are told to believe that it is only the Modern Warming that results from human emissions of carbon dioxide, is unprecedented and the other warming events are natural.

Sometimes several solar cycles exhibit greater activity for decades or centuries. We've had our Grand Solar Maximum and have thrived in the Modern Warming. Population, wealth and food productivity increased. Four solar cycles were out of phase in 2020 and some solar physicists are suggesting that we face another cold period like the Maunder Minimum (1645-1715 AD). God help us if this happens. We are totally unprepared.

We've seen Solar Minima before with the Wolf Minimum (1300 to 1320 AD) that occurred immediately after the Medieval Warming (900-1300 AD). Rivers froze and there was crop failure resulting in the Great Famine (1310-1322 AD). As per usual, the poor were the most affected. The weakened population was wiped out by the plague in 1347 AD. The Spörer Minimum (1410-1540 AD) had the same effect as the Wolf Minimum.

In the extremely bitter and long Maunder Minimum (1645-1715 AD), the growing season was reduced, harvests failed, rivers froze, the number of snowy days increased, alpine glaciers advanced and buried many villages, tree lines in the Alps dropped, ports were blocked by ice, there was mass famine and a huge depopulation. The milder Dalton Minimum (1796-1820 AD) was similar to the Wolf and Spörer Minima.

There are cycles in the oceans[256] that produce 60-year-long climate cycles and change currents. These have been known for thousands of years and calendars and planting times were based on these climate cycles.

Sea level, tides and polar sea ice are affected by lunar orbit cycles every 8.85 and 18.61 years and these occur as cycles every 18.61, 9.305, 8.85 and 4.25 years. Sea level changes of more than 4 metres and 6 metres occur in some places with the 18.61- and 8.85-year cycles. The 18.61-year cycles push warm water into the Arctic Ocean.

The Pacific Ocean has *El Niño-La Niña* events[257] which greatly influence rainfall and drought on both sides of the Pacific. Combination of solar and ocean cycles with *El Niño-La Niña* events can produce catastrophic weather with long droughts or huge floods.

Surface Pacific Ocean waters are warmer in *El Niño* events. There is some evidence that this warming derives from sea floor volcanoes and/or increased solar activity.

Sudden climate changes occur after asteroid impacts and large volcanic eruptions which pump out ash and gases. These are currently unpredictable. Terrestrial supervolcanoes (e.g. Yellowstone, USA; Taupo, NZ; Kamchatka, Russia) eject thousands of cubic kilometres of aerosols into the atmosphere during an explosive eruption.

These can have a profound effect on cooling over decades and may trigger or exacerbate longer cooling events (e.g. Toba 73,000 years ago). Filling of the atmosphere with aerosols from smaller explosive volcanic eruptions that blast out in the order of 30 cubic kilometres of aerosols induce short term cooling such as the five years of cool climate after Krakatoa (1883), Tarawera (1886) and Pinatubo (1991). Asteroid impacts and upper atmosphere comet explosions add dust to the atmosphere and extinction-producing large asteroidal impacts are almost instantaneous catastrophic events where one has little time to kiss body parts goodbye.

There have been thousands of warmings and coolings over the history of time due to a variety of reasons. There is no evidence that shows global warming is driven by human emissions of carbon dioxide.

Earth went into an ice age 34 million years ago. We are currently in a warm interglacial phase of the current ice age. The interglacial has lasted almost 12,000 years because we are slightly closer to the Sun. When the orbital eccentricity inevitably takes us further from the Sun, the interglacial will end. We don't know when this will occur but I'm sure we will be suffering the next glaciation on a Thursday.

Geological processes that took place long ago are still taking place today. This is why a geological perspective on climate cannot be dismissed.

For the purposes of understanding climate over decades, centuries and millennia, the Sun, the Earth's orbit and ocean oscillations have been the most important climate drivers while humans have been on Earth.

The dim distant past

An overview of the Earth, and carbon dioxide's place in it, now follows.

Climate "science" mainly deals with the atmosphere and is not an integrated view of the planet, its Sun and the Solar System. By cherry-picking one small aspect of the planetary system, climate "science" is poor science.

The Earth has experienced a complex and fascinating evolution, which continues to this day. Carbon dioxide has had a key role in it and the atmospheric content of it has risen and fallen for many reasons. Humans arrived exceedingly late in the Earth's evolution and their role needs to be understood in the context of Earth history and processes.

Our planet is a wet warm volcanic planet. It formed 4,567 million years ago by the condensation and recycling of stardust associated with the formation of an exceptionally stable star in a good galactic address. We live in the best house in the best street.

Earth materials have been constantly recycled and the Earth and all associated systems have been dynamically evolving. The Earth has not stopped being an evolving dynamic system just because humans now live on the planet.

Early in the history of the Solar System, Earth was bombarded by massive asteroids. The intensity and frequency of impacting has decreased over time and now about 40,000 tonnes of extraterrestrial material is added to the Earth each year.

One asteroid some 4,500 million years ago broke off a bit of the Earth. This we now call the Moon. Every time a primitive sea formed by condensation of volcanic steam, it was vapourised by impacting. This was the limitation to the formation of life on Earth.

Volcanic activity was degassing Earth. It still is. The main gases released from volcanoes then and now are the greenhouse gases water vapour, carbon dioxide and methane together with minor helium, hydrogen, nitrogen compounds, sulphur compounds, acids and rare gases.

As soon as the surface and atmosphere of the Earth had cooled and asteroid bombardment decreased, rainwater accumulated and life formed. The evolution of life is inextricably linked to a very weird molecule. It is called water. Without water, there would be no volcanoes, no recycling of crustal rocks, no oceans, no climate change and no life.

The Earth has had three atmospheres in its history. The earliest atmosphere had a high ammonia and sulphur dioxide composition and these gases were quickly scrubbed out of the atmosphere by rainfall. These volcanic gases accumulated to form a primitive oxygen-deficient greenhouse gas atmosphere.

First life on Earth formed at least 3,800 million years ago and by 3,500 million years ago, this bacterial life had colonised and there were reefs. Some bacteria fed on methane. There have been reefs on Earth since that time; they come and go due to sea level fall, cold water and inundation by sediment and volcanic ash.

These bacterial colonies are still with us. So are viruses that appeared at least 3,000 million years ago. Bacteria were the first life on Earth, bacteria is still the dominant life on Earth, bacteria have survived all the natural catastrophes on Earth and have always been the largest biomass on Earth.

Every now and then thin landmasses would flip and lavas far hotter than those of today flooded the land surface. There is some evidence to suggest that there were minor periods of local glaciation. The Earth's second atmosphere was warm, rich in carbon dioxide and methane and lacking oxygen. Atmospheres of this composition still exist on other planets.

Around 2,500 million years ago a number of irreversible events took place. The continents became thicker and did not flip over. They started to drift, pull apart and stitch back together. At that time the first great global ice age occurred.

There have been six great global ice ages over the history of time. Ice has been on Earth for less than 20% of time. The rest of the time planet Earth has been a warm wet volcanic greenhouse planet. Continents were clustered around the equator when the first ice age took place, the Earth was covered with ice and thick ice sheets occurred at sea level at the equator. This was the first of two events of snowball Earth.

The Earth then had its second atmosphere which contained at least 10% carbon dioxide and probably more than 20%. Tell me that scary bedtime story again. Because we are adding traces of carbon dioxide today to the atmosphere we will fry-and-die and I will toss and turn all night wondering how it was possible to have a past ice age when there was more than 100 times the current amount of carbon dioxide in the atmosphere and life continued to evolve.

Moving ice picks up soil and rocks and grinds them to rock flour. After the ice age during warming, meltwaters washed rock flour nutrients into the oceans. Bacteria loved their new diet. Bacteria then had all their cellular material surrounded by a single cell wall. With a nutrient-rich diet, these bacteria evolved into another group of bacteria that had a cell nucleus that protected all cell machinery. This type of cell was surrounded by fluid which was then surrounded by another cell wall.

These cells emitted large quantities of oxygen which accumulated in the air and chemically reacted with soils and the seas. Bacteria with a nucleus produced oxygen and evolved whereas the more primitive cells without a nucleus suffered mass genocide. Survivors of this genocide live inside you and many other animals to assist digestion. Many survivors live in bogs, swamps and in other places with no oxygen (e.g. the bottom of the Black Sea).

Around 2,500 million years ago, limey rocks rich in carbon dioxide and black carbon-rich sediments started to become abundant. Carbon dioxide started to drawdown from the air and get sequestered in rocks.

Around the same time, soils rusted from green to red and excess oxygen was trapped in soils. Iron was dissolved in the oxygen-poor oceans. As the ocean iron content increased the dissolved iron in the oceans was oxidised and precipitated with silica on the sea floor as banded iron formations. This event is called the Great Oxidation Event.

Leaching of these iron oxide-rich rocks billions of years later in tropical climates removed much of the silica and rocks became even more enriched in iron. The great iron ore fields of planet Earth form from these 2,500-year-old banded iron formations (e.g. Pilbara, WA).

The slight addition of oxygen to the atmosphere shows that life, the atmosphere, the oceans and the rocks interacted, through processes that are still occurring 2,500 million years later on our dynamic evolving planet. Change is constant. Climate "science" mainly deals with the atmosphere but not the total planetary interaction, hence it is very poor science.

For at least the last 2,500 million years, the continents have been pulled apart and stitched back together as a result of convection currents deep within the Earth. Every time the continents are pulled apart, huge quantities of volcanic water vapour, carbon dioxide and methane are released from deep in the planet into the atmosphere. This process is still adding carbon dioxide mainly from submarine volcanoes and gas leaks.

When continents collide and stitch together, mountain ranges form. Mountains are stripped of soils, new soils form and remove carbon dioxide from the atmosphere. These newer soils are in turn stripped from the land and the carbon dioxide becomes locked in sediments. These are later compressed to rock. Mountain range building also releases carbon dioxide into the air from chemical reactions at depth, from melting of rocks and from hot springs.

Because of the inverse solubility of carbon dioxide in water, glaciation removes carbon dioxide from the air. When the oceans warm, they later release carbon dioxide. Carbon dioxide does not create warming of the oceans. Warming oceans release carbon dioxide over long periods of time during after a warming event.

Try the experiment. Sit and watch a cold carbonated soft drink or champagne keep bubbling as it warms. As the liquid warms, it can hold less carbon dioxide which is then released as a bubble. When the experiment is complete, drink the warm liquid which will have no fizz. Yuk.

Try another experiment. Fill a bath with hot water. The air in the bathroom becomes warm. Try the inverse. Run a cold bath, point a radiator at the bath water and the bath water doesn't heat up. It is the oceans that hold the planet's surface heat and not the air. Water has a high heat capacity. Air doesn't. If we have global warming of the atmosphere, the oceans will not warm. If the oceans warm from the Sun and submarine volcanoes, then the atmosphere will warm.

Some 2,000 million years ago, cyanobacteria appeared. They may have appeared earlier but the dim distant past is hazy. Cyanobacteria use carbon dioxide and sunlight to obtain energy and release oxygen as waste in a process known as photosynthesis. Eucaryotic cells, those with a nucleus, thrived and later divided into the ancestors of modern plants, fungi and animals.

About 1,800 million years ago when the population of oxygen-producing bacteria had greatly increased, oxidation of seawater again resulted in the precipitation of iron ores on the sea floor (e.g. Michigan, USA).

A little later between 1500 and 1700 million years ago, thinning and then pulling apart of the Earth's crust led to the formation of giant lead-zinc deposits (Broken Hill, Mount Isa, McArthur River, Australia; Gamsberg, Aggeneys, South Africa) and enormous copper deposits (Mount Isa, Olympic Dam, Australia). A large amount of carbon dioxide, sulphur gases, methane and hydrogen leaked into the atmosphere when some of these metal deposits formed.

This was followed by a period of time called the boring billion when a warm wet greenhouse Earth with an oxygen-bearing carbon dioxide-rich atmosphere had a giant supercontinent (Rodinia) that was stitching itself together. It received the normal dose of raining space dust, meteorites and asteroids from somewhere out there. This giant supercontinent was at the equator and mid-latitudes and not at or near the poles.

Rodinia started to break into pieces and drift in all directions some 830 million years ago releasing massive quantities of water vapour, carbon dioxide and methane to the atmosphere.

Fragments of Rodinia are found on all continents on Earth and have been put back together to show what Earth looked like in the dim distant past. Parts of Rodinia exist today in the geological records of India, West Africa, East Antarctica, Australia, China, Siberia, Scandinavia, Brazil and northern North America.

Snowball Earth

The first of a few snowball Earth events took place 2,500 years ago. This was when the Earth's surface was entirely frozen. The far better-known Snowball Earth event was later and described later. If the whole ocean had been frozen and not just the surface, the planet would have remained an icy planet.

Green activists claim that the slight increase in atmospheric carbon dioxide today will lead to runaway global warming yet all previous glaciations started when the atmospheric carbon dioxide content was up to hundreds of time higher than now.

The origin of the greatest climate change on Earth is an enigma. The second Snowball Earth event occurred between 800 and 600 million years ago in what is called the Cryogenian. There were two major glacial events and numerous smaller events. These glacial rocks are still very well displayed in much of inland Australia, southern Africa, China, Russia, Canada and the US showing that this was a global event.

Before, during and after this time, massive volumes of the mineral dolomite (a magnesium-rich limestone) were precipitated in shallow warm seas. Dolomite contains 48% carbon dioxide. Back calculations from dolomite experiments show that dolomite can only precipitate when the carbon dioxide content of air is at least 100 times more than now.

The Earth's second atmosphere was very rich in carbon dioxide which was being drawn down into limey sediments. This sequestration of carbon dioxide from the air was the start of the planet's third atmosphere, today's oxygen rich one.

These two glacial events and all the numerous smaller events were at a time when the air had far more carbon dioxide than now. According to the global warming green activists, the planet should have had a run-away global warming.

Sea level in the Cryogenian changed by at least 600 metres and interglacial sea temperatures were +40°C. Kilometre-thick ice sheets were at the sea level and at the equator which experienced temperatures -40°C. The surface of the oceans was covered with sea ice.[258] In many places, glacial debris is directly covered by limey sediments precipitated from a shallow warm sea showing that climate change was rapid and extreme.

If the oceans had not contained deeper liquid water under sea ice, then bacterial life may not have survived. Similar conditions exist today in Antarctica's Lake Vostok where bacteria live under ice in cold, dark, nutrient-poor water.

The survival of bacterial life under sea ice in cold, dark, salty water is no surprise. Bacteria currently live in oxygen-poor or oxygen-rich conditions in acid and alkaline environments, hot or cold springs, in highly radioactive water, deep in the Earth in darkness in cracks in hot rocks, in volcanic rocks in mid ocean ridges, in clouds, in ice and on every environment on Earth. Bacteria some 400,000 years old have been resurrected from the Greenland ice sheet and from salty liquid inclusions of water in the 250 million year-old salt beds of Texas.

Some 90% of cells in humans are bacteria and 15% of human body weight is bacteria. You are a creeping colony of critters. Bacteria keep you alive and if one type of bacteria wins the constant biological warfare inside your body, then you shuffle off. Thank God for antibiotics and modern medicine which did not exist when we lived in the "sustainable" world that green activists want for humanity.

Bacteria always have been the dominant biomass on Earth and yet their role in interaction with the atmosphere is unknown. The greatest biomass on Earth lives in the top 4 km of the Earth's crust and not forests or oceans.

A massive amount of carbon dioxide and methane is stored in and released to the air from bacteria in soils, sediments and rocks. A slight change in land level, sea level or temperature can release large volumes of bacterial carbon dioxide and methane. Bacteria rule the world and always have.

About 650 million years ago, ice sheets started to retreat and leave behind debris and rock flour. The seas were again quickly filled with nutrients from meltwaters, bacteria thrived in warm nutrient rich water and evolved into complex multicellular life. This was the first appearance of reefs[259] with coral and other complex life.

This interglacial did not last long and the Earth again was covered by ice in the second Cryogenian snowball Earth event.[260] Sea level dropped, the experiment with complex life failed and reefs died, ice sheets scoured the surface, sea ice covered the oceans, bacteria took refuge in the water under the sea ice where there was no oxygen or circulation in the dark cold water.

Global warming took place again. Ice retreated; sea ice disintegrated and melted, sea level rose by at least 600 metres; the sea again had contact with the air and currents started to circulate oxygen. Dissolved iron precipitated as iron ores[261] on the sea floor and melting sea ice dropped boulders that had been picked up by moving ice sheets on the continent. Limey rocks precipitated in warm water lie directly upon glacial debris showing that global warming was very rapid and that the atmosphere was carbon dioxide rich.

Meltwaters flooded the oceans with more nutrients, bacteria evolved again into multicellular animals at about 583 million years ago[262], called the

Ediacaran Fauna, and the global carbon dioxide content started to decrease with sequestration into sediments and micro-organisms started to pump much more oxygen into the air.

These multicellular animals were soft-bodied, some had back bones and some had a primitive alimentary canal. These are your distant relatives. They grazed on sea floor algal mats and, at times, were rolled into a death mask during storms.

Evolution continued; some soft-bodied animals grew shells, scales and skeletons. This process removed dissolved carbon dioxide from sea water. Marine multicellular animals developed shells, skeletons and protective coatings because there were enough nutrients in the oceans for muscle functions. This explosion of life created a massive drawdown of carbon dioxide from the atmosphere. It took 20 million years for most of the major animal groups to evolve. Since that time, no further animal phyla have evolved.

These armoured animals were predators, the Ediacaran fauna were on the menu and there was an explosion of predation and life 543 to 520 million years ago when the Ediacaran fauna became extinct. The original consensus was that the Ediacaran fauna was the oldest multicellular life on Earth. New work about a decade ago showed that the Arkaroola Reef was made up of older multicellular life and that the consensus was wrong. That consensus had lasted for 60 years.

Life and death

Life on Earth is because of a good address in the Solar System, water, plate tectonics and numerous other parameters. Water and air temperature were not as benign as today's climate and past cooler and warmer conditions allowed life to exist.

The explosion of life from about 535 to 489 million years ago gave us all of the major life forms currently present on Earth. It was in fact an explosion of predation. Rapid diversification via evolution took place. Despite five major mass extinctions and more than 20 minor mass extinctions since that time, the number of species on Earth continues to increase at a rate far faster than species extinction turnover.

Some 50 million years later after the explosion of marine animal life, land plants appeared. Amphibians also appeared as meals were also now available on the land. Land plants have only been on Earth for 10% of time. The first of the five major mass extinctions of complex life took place 440 to 420 million years ago, life quickly recovered, vacated ecologies were quickly filled and life continued diversifying.

There was a minor ice age after which life thrived again. More minor mass extinctions followed. About 365 million years ago, there was another major mass extinction caused by asteroid impacting what we know today as Sweden or due to a loss of ocean oxygen.

Life recovered and plants, amphibians and land animals bounced back. Between 365 and 251 million years ago, the world's major coal deposits formed from plant material that had used carbon dioxide as plant food. This caused a huge drawdown of carbon dioxide from the air.

Rotting vegetation greatly increased the methane content of air and the oxygen content increased to the point where it was common for the Earth's atmosphere to spontaneously ignite. Monstrous forest fires led to increased soil erosion and sediments from that time contain charcoal and fossil insects with 1 metre-wide wing spans which exploited the high oxygen content of the dense atmosphere.

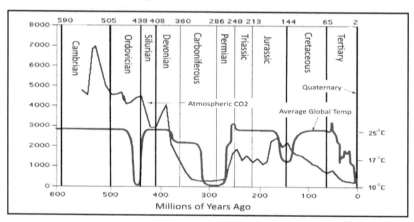

Figure 4: *Plot of atmospheric carbon dioxide and temperature since the explosion of life 520 million years ago overlain by the geological time periods.*[263] *Four of the four ice ages at this time started when the global carbon dioxide content was far higher than now. Note the massive drawdown of carbon dioxide during the major coal forming times (late Carboniferous-early Permian) which were also cold times. Atmospheric carbon dioxide is currently 0.04%.*

This explosion in vegetation came to a sudden end when the giant southern continent Gondwana drifted across the South Pole. There was a major 50 million-year-long ice age yet the atmosphere was blessed with a very high carbon dioxide and oxygen content. Rocks which are now in northern Europe formed in an environment like the modern Red Sea and its surrounding deserts and salt pans.

Life continued to diversify. Minor mass extinctions continued and the biggest major mass extinction on Earth took place 251 million years ago. Some 96% of species suddenly became extinct and some, like trilobites and rugose corals, disappeared forever.

Obvious evidence for as asteroid impact is weak, but the smoking gun is volcanicity in Siberia. The shock rebound from an impact triggers melting deep in the Earth and lavas rise to form volcanoes. The impact crater may have been completely filled with lava, and so is not so apparent to us.

Massive volcanic activity over a very short period of time exhaled huge volumes of carbon dioxide into the atmosphere as well as a white haze of sulphur gases that reflected heat and light. The resultant climate was cooler, acid rain from sulphur-rich gases may have destroyed large tracts of vegetation thereby creating a collapse of terrestrial environments for plants and animals. The surface of seas probably became acid for a short time. Burrowing and deep-sea animals survived. Subsequently, life diversified even more and filled the vacated ecologies. These types of rare events we should be fearing, but you can't make money from or morally shame an unforeseen 'act of God'.

Another major mass extinction took place 217 million years ago. A swarm of asteroids hit the Northern Hemisphere, as a continental mass was being fragmented to form the Atlantic Ocean. Large volumes of lava erupted and massive amounts of carbon dioxide were again released to the atmosphere.

Plant life again thrived with the extra carbon dioxide in the air and some coal formed. Earth again recovered from the mass extinction; life continued to diversify, the number of species continued to increase and the continents continued to drift. The planet was then still a warm wet greenhouse planet with the normal cycles of rising and falling sea levels, rising and falling land levels and changing climates.

Flowering plants appeared about 130 million years ago and underwent rapid evolution. Some 120 million years ago, Australia was at the South Pole enjoying a temperate climate. There were minor glaciers in the highlands, volcanoes were active and dinosaurs adapted to the long periods of darkness by evolving enlarged eyes.

There are numerous Large Igneous Provinces both on land and in the oceans, and this is where extraordinary volumes of basalt lava were erupted onto the ocean floor and land 120 million years ago. The volume of basalt was so large that sea level rose.

Warm sea temperatures at that time may have been due to these basalts heating sea water as lava cooled to solid rock and the oxygen-poor conditions of the ocean at that time may be a result of sulphur gases released into the sea from solidifying lava.

The atmosphere contained far more carbon dioxide than now and, because basalt can dissolve up to 15% by weight of carbon dioxide which is released as the lava rises from deep in the Earth, the smoking gun for the high carbon dioxide content of the atmosphere at that time are the submarine Large Igneous Provinces.

Global sea level was more than 100 metres higher than at present, the sea surface temperature was 10 to 15°C warmer than now, currents flowed latitudinally and many continents were covered by shallow tropical seas. Coral did not die in these warm oceans. It flourished. Planet Earth was a warm wet greenhouse paradise and thick vegetation covered the land masses.

From 251 to 120 million years ago, the global carbon dioxide content varied greatly and increased to a peak 120 million years ago. This derived from intense volcanic activity associated with ocean formation as the continents drifted apart. Thick vegetation covered the land masses.

The atmospheric oxygen content greatly increased to 35% 300 million years ago, decreased and then increased to 27% 150 million years ago. It is currently 21%. During times of high atmospheric oxygen and methane, there was spontaneous combustion of the atmosphere, global bushfires and increased erosion.

Australia started to pull away from Antarctica at about 100 million years ago. It drifted northwards at 7 cm per year, the Tasman Sea opened and the Indian Ocean expanded as India starting to drift away from Western Australia. India eventually collided with China and the Tibetan Plateau started to be pushed up 50 million years ago. It still is.

If you are ever on a quiz show and asked how high is Mt Everest, the only correct answer is: when? The measured average rise of Mt Everest is 2 cm per year. Limey rocks that now sit on the peak of Mt Everest sequestered the carbon dioxide from the carbon dioxide-rich atmosphere 120 million years ago.

The rocks on the peak of Mt Everest are limestones formed in a shallow warm sea. Creationists want such rocks to have been deposited in Noah's flood 4,000 years ago. If this was the case, the 8,849 metre-high Mt Everest must have risen at 2.2 metres each and every year for the last 4,000 years without anybody noticing it or the massive earthquakes that are always associated with the breaking of rocks to produce uplift. There are always nutters around the fringes of science.

Australia got rid of New Zealand 100 million years and pushed New Zealand into the Pacific Ocean where they now enjoy their well-deserved volcanoes and earthquakes. Volcanoes such as Lord Howe Island, Balls Pyramid and Norfolk Island in the Tasman Sea formed because cracks went deep into the Earth as the rift zone between the countries opened up.

The opening of the Tasman Sea produced the rise of the Great Dividing Range, the inland diversion of the major river systems and changes to the climate of eastern Australia. Topography is another dynamic influence on climate and is always slowly changing. A minor mass extinction of life 90 million years ago was the result of volcanoes in the Indian and Pacific Oceans belching out gases into the oceans and atmosphere. The top of the oceans became acidic for a very short time, some 26% of advanced life became extinct and there was a short warming until volcanism waned.

An extraterrestrial visitor 65 million years ago smashed into the Earth and there was a major mass extinction of life at that time. There was also a huge volcanic event in India with the release of much carbon dioxide, methane and sulphur gases into the atmosphere. Ancient soils, vegetation and rock chemistry show that conditions were tropical. An asteroid impact closely followed by a dirty big volcano would have been a killer punch for much life on Earth. Goodbye dinosaurs.

A minor mass extinction at 55 million years ago was caused by a Caribbean volcano. There was a rise in sea temperatures by up to 8°C for 100,000 years, atmospheric carbon dioxide was 10 times that of today. The surface of the oceans became acid for a geological instant and the oceans lost dissolved oxygen while the ocean floors released methane into the atmosphere.

During these warmer times, plankton sucked up the atmospheric carbon dioxide, mammals thrived and life filled the vacated ecologies. We often hear the throw-away line that the Amazon is the lungs of the Earth. Wrong. It is phytoplankton. The Amazon rainforests are far more resilient to climate change than previously thought.[264]

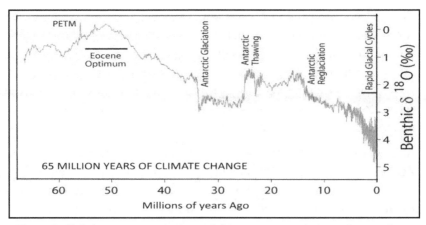

Figure 5. *Global temperature constructed from temperature proxies showing the short warm Palaeocene-Eocene Thermal Maximum (PETM), the warm Eocene optimum followed by 50 million years of cooling, ice sheet formation with the Antarctic glaciation with thawing and reglaciation and extensive cooling over the last 15 million years.*[265]

One of the myths put out by the media and certain world celebrities, is that the Amazon rainforests are the "Lungs of the World" and major source of the Earth's oxygen supply plus a large carbon sink absorbing a quarter of the carbon dioxide taken up by forests around the world every year.

Amazon plants consume about 60% of the oxygen they produce in respiration. Microbes, which break down the forest biomass, consume the other 40%. The net contribution of the Amazon ecosystem to the world's oxygen is effectively zero. Plant life stores carbon but only 5% and not 25% as touted by activists. This is just one of the many myths, distortions and exaggerations promoted by green activists.

Atmospheric carbon dioxide decreased from 0.35% to 0.07% within a million years, stayed low until 47 million years ago and then went up and down to about the present level (0.04%) 40 million years ago.

The oceans were very warm from 55 to 33 million years ago, and huge volumes of sediment were deposited on the continental shelves. During this time, tropical conditions were widespread. For example, tropical soils such as bauxite formed in Ireland during warm monsoonal times. Ireland now has peat soils formed in cold wet times.

India collided with Asia 50 million years ago and formed the Tibetan Plateau. During these warm times, the bare rocks reflected much heat back into space and the high mountain mass disrupted the air flow and the

Northern Hemisphere climate.

Continental uplift produced the Tibetan Plateau that started to scrub carbon dioxide out of the atmosphere. Fresh rocks were pushed up and were quickly weathered to soils which sequestered carbon dioxide. Soils were removed downslope by monsoonal weather and deposited as sediments to the south by the great rivers.

The Tibetan Plateau is still rising and carbon dioxide is still being scrubbed out of the atmosphere by these processes. During this time the solar radiation-warmed surface water of an ancient ocean was replaced by the Alps of Europe and the Tibetan Plateau as Africa and India collided with land masses to the north. The planet cooled.

In the dim distant past, we are not able to determine every slight climate change in great detail. Time is against us. However, in younger rocks, we can use the chemical fingerprint of fossilised plankton shells to determine ancient surface ocean temperature. Sediment type and sediment chemistry assists in determining global climate. Once ice sheets formed, chemical fingerprints of ice can give us the polar temperature and trapped air provides an estimate of past atmospheric chemistry.

Today's ice age

The current ice age we are enjoying did not happen overnight. There has been cooling since peak warming 50 million years ago[266], driven by continental drift and collision, orbital cycles and an ever-changing Sun. There is much scientific discussion about the rapid falls and rises in temperature.

The Drake Passage opened as South America pulled away from Antarctica that resulted in a circum-polar current and Antarctica refrigerated. That was the start of an ice age 34 million years ago. We are currently in a short orbitally-driven interglacial of this ice age. We don't know when the ice age will end and when we will return to the planet's normal warm wet greenhouse climate. As planet Earth started to cool 34 million years ago, the Earth's orbit started to have a profound influence on climate and climate changes drove human evolution over the last 5 million years.

The closing of the Mediterranean Sea at the Strait of Gibraltar resulted in the Mediterranean drying up and leaving salt deposits that reflected solar energy back into space. The Antarctic ice sheet formed and the northward flow of polar bottom waters formed climate zones, a feature that had not previously existed during the previous long periods of warm wet tropical climates.

There were a number of minor mass extinctions, comet and meteorite impacts and sea level changes. For example, in south-eastern Australia the Murray Basin became a large inland sea, retreated and then advanced again only to start its final retreat five million years ago. Over 200 old beaches lines in the Murray Basin sit up to 500 km inland from the present seashore.

Warm currents in the Indian Ocean, deflected by landmass, drifted through the Great Australian Bight and up the Pacific coast of Australia. From 17 to 14.5 million years ago, southern Australia was again tropical with mid-latitude temperatures 6°C warmer than today. Atmospheric carbon dioxide was 180 to 290 parts per million compared to the current 410 parts per million (0.041%). This warming occurred when atmospheric carbon dioxide was 30 to 50% lower than today!

Land changes periodically closed the Strait of Gibraltar 7 million years ago and thick salt deposits again formed in the Mediterranean Basin. This effectively removed salt from the oceans. Because there was less salt in the oceans, parts of the surface of the oceans froze. Both the ice and salt reflected sunlight and the planet cooled further. By 5 million years ago, Earth was so cool that the orbital wobbles now had a significant effect on climate.

Climate changes drove human evolution in Africa over the last 5 million years. These climate changes were global and recorded outside Africa. For example, in south-eastern USA, between 5.2 and 2.6 million years ago, atmospheric carbon dioxide content was more than now as were global temperatures (up 2 to 3°C) and sea levels rose 10 to 25 metres.

Cooling changed forests to grasslands and primate extinction and diversification to upright bipeds took place. By 2.67 million years ago, central American volcanoes had closed the seaway between the Pacific and Atlantic Oceans and ocean circulation was disrupted. The planet started to cool. Explosive volcanoes in Kamchatka (Russia) added dust to the atmosphere, dust reflected sunlight and the planet cooled further. This cooling coincided with a supernoval eruption that increased cloud cover further adding to cooling.

It was so cold that the Arctic polar ice cap and sea ice formed. The weight of ice in both Greenland and Antarctica pushed down the land, deep basins filled with thick ice and ice had first to be pushed uphill before flowing as surges down to the sea as glaciers.

For more than a century, we have known that Mt Erebus in Antarctica is an active volcano. Over the last two decades, the scientific literature has recorded sub-glacial volcanoes and hot zones beneath the Antarctic ice.[267]

More than 150 active volcanoes and geothermal areas have been identified underneath the Antarctic ice sheet. For example, the Thwaites and Pope Glaciers in West Antarctica are retreating and are underlain by geothermal areas of elevated heat flow.[268]

The waxing and waning of ice sheets and glaciers is extraordinarily complicated and air temperature is only a minor factor. Sub-glacial volcanoes and hot spots emit heat and are not considered in sea ice and ice sheet models, sea level models and the IPCC climate predictions which all suggest that only human emissions of carbon dioxide are melting the Antarctic ice cap.

Why is this critical piece of data omitted by the IPCC?

The last glaciation

Glaciations are cyclic and another will come but, meanwhile, we have nothing to fear from warming in the interim.

Climate fluctuated between warm and cold periods on 41,000-year orbital cycles. About one million years ago, the climate also started to fluctuate between cold and warm periods on 100,000-year cycles. We are currently in the warmer interglacial phase of an ice age that has been in progress for 34 million years during which time sea level rises and falls were about 130 metres. We do not know when the current ice age will end. However, we cannot escape the fact that the current interglacial will end and we will suffer another 100,000 years of glaciation. Previous glaciations resulted in ice sheets many kilometres thick that covered Canada, northern USA, Europe north of the Alps, most of Russia and elevated areas in both hemispheres.

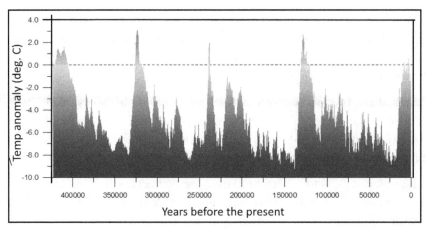

Figure 6: *Orbitally-driven cycles of glaciation over the last 400,000 years showing that the last four interglacials were warmer than the current interglacial and that during glaciation there were rapid events of cooling and warming.[269] Why is it that a slight warming during the current interglacial is due to human emissions of carbon dioxide and yet all other and larger warmings are not?*

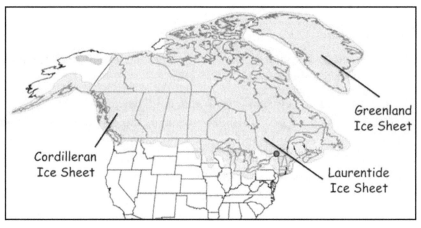

Figure 7: *Distribution and thickness of the North American and Greenland ice sheets during the zenith of the last glaciation 20,000 years ago.[270]*

The penultimate interglacial was 120,000 years ago. *Homo erectus, Homo neanderthalensis. Homo floresiensis* and *Homo sapiens* coexisted, sea level was at least six metres higher than at present and the planet was far warmer and wetter than now.

During that interglacial, there was more vegetation than today and atmospheric carbon dioxide was 78% of today's concentration. After warming, the atmospheric carbon dioxide and methane content increased suggesting that atmospheric temperature rise drives an increase in atmospheric carbon dioxide and methane contents.

Orbital-driven cooling commenced 116,000 years ago. There were very abrupt changes in temperature and the ice sheets changed in size.[271,272] There were abrupt millennial scale climate oscillations.[273] The eruption of the supervolcano Toba (Indonesia) 73,000 years ago blasted 3,000 cubic kilometres of rock into dust that reflected sunlight and accelerated cooling.

Tropical vegetation died, sea level dropped, humans migrated out of the devegetated tropical areas and glaciation continued. *Homo sapiens* had an existential crisis, only about 8,000 humans survived although some molecular biology estimates claim 1,000[274] or even 40 breeding pairs survived.[275] We very nearly joined our relatives *Homo erectus*, *Homo neanderthalensis* and *Homo floresiensis* in extinction. We are not quite sure when our relatives became extinct. We live in unusual times when only one species of the *Homo* genus exists on Earth. Although…when I look around in a crowd I'm not so sure.

During the history of the latest glaciation, great armadas of ice were released into the sea every 7,000 years resulting from the physical failure of thick ice sheets. These had a profound effect on climate. Small cool periods occurred every 1,100 to 1,300 years.

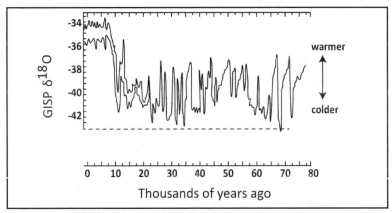

Figure 8: *The latest glaciation showing millennial scale temperature variations and the modern interglacial with much more frequent temperature oscillations. Sea level was lowest after the Toba supervolcano and at the zenith of the glaciation 20,000 years ago. The millennial scale variations are due to solar variability or surges of ice from thick ice sheets into the oceans.*[276]

The zenith of the last glaciation was 20,000 years ago. Sea level was 130 metres lower than it is today, temperature was 10 to 15°C lower than today and there were very strong cold winds. The Northern Hemisphere was covered by ice down to latitude 38°N[277] with more northern areas such as Scandinavia being covered by more than 5 km of ice. The loading of the polar areas with ice changed the shape of the planet, the planet's rotation changed and as a result ocean currents distributing heat across the Earth changed. Humans lived very short lives around the edge of ice sheets.

Tasmania and parts of the south-eastern highlands of Australia were covered in ice and sea level was so low that Aboriginals walked to Tasmania across what is now Bass Strait from mainland Australia. Rainforests disappeared and the Amazon Basin consisted of grasslands and copses of trees. Coral reefs were stressed and disappeared in colder conditions when sea level was lower.

Upland areas, even in the tropics, had glaciers. In areas with no ice sheets, strong cold dry winds shifted sand and devegetation occurred. Australia was buffeted by anti-cyclonic winds that deposited sand dunes and carried sea salt spray that was trapped in the inland basins such as Lake Frome and Lake Eyre. Dunes in North Africa, the Middle East and North America again moved and great wind-deposited loess deposits covered Mongolia, China and northern USA.

Immediately after the peak of the last glaciation, there was global warming of at least 15°C. Now that is global warming! Was this warming due to smokestacks emitting carbon dioxide? Of course not.

Yet today as a result of green activist pressures, governments are wanting to completely and totally destroy economies based on a 0.8°C measured temperature rise over the last century and a speculated 2°C temperature rise over the next century despite the fact that we are at the end of an interglacial period. Bear in mind, the planet has spent most of its time warmer and wetter than now and that was the best environment for life.

Shortly after this sudden warming 20,000 years ago, temperatures dropped suddenly by about 7°C. Temperatures remained cold for several thousand years and fluctuated between 3°C warmer and 3°C colder than now.

As with all science, the scale of measurement is critical. On one scale, there is a correlation between the orbitally-driven cycles of glaciation and interglacials. On a more detailed scale, this is not the case.

What came first? Global warming or an increase in atmospheric carbon dioxide? Ice core measurements of the current interglacial show us what we know from chemistry. As the temperature increased, the atmospheric carbon dioxide increased 650 to 1,600 years later. This is the opposite of what we are being told by green activists who claim that an increase in atmospheric carbon dioxide will lead to an increase in temperature.

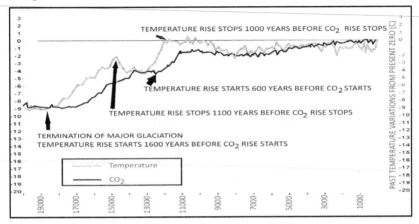

Figure 9: *Antarctic Epica Dome C ice core analysis of temperature showing historic air temperatures and atmospheric carbon dioxide content over time.*[278,279]

At present, the global carbon dioxide content is increasing and the temperature is going nowhere again showing that carbon dioxide does not drive global warming.

Ice core measurements showing temperature rose before carbon dioxide increased is supported by ocean carbon dioxide studies. Because of the lag in carbon dioxide emissions from the oceans as temperature increases, the increase in carbon dioxide in the atmosphere over the last century is due to the higher ocean temperature, maybe in the Medieval or Roman Warmings.[280]

Warming and the rise of humans

The development of civilisations has been made possible by global warming.

Only in the present interglacial have humans had a good and varied diet, travelled and colonised the Earth. There are massive scales of natural cycles in temperature and ice volumes and, rather than a futile attempt to change a planetary scale natural cycle, we humans must adapt to change as we've done in the past. Technology, energy and infrastructure will allow most humans to adapt to the inevitable forthcoming glaciation far better than they could during previous depopulating glaciations.

About 15,000 years ago, there was another sudden rapid warming of about 12°C. This is far greater than anything measured today or predicted by the scariest catastrophist for the future. Was this warming due to human's sinful heavy industry emitting carbon dioxide? No. The large ice sheets that covered Canada and the northern USA, all of Scandinavia and much of northern Europe and Russia started to melt quickly and glaciers retreated.

The northern polar ice sheet started to melt 14,700 years ago. There were very rapid and major temperature fluctuations, sea level rose and fell and the total sea level rise over the last 14,700 years has been at least 130 metres. Land masses previously weighed down by thick ice started to rise. Scandinavia is still rising in elevation and has risen more than 340 metres over the last 14,700 years. As a counterbalance, the Netherlands, south-eastern England, Schleswig-Holstein and Denmark are sinking.

The breaching of dams from melt waters filled the oceans and generated cold surface waters 12,000 to 11,000 and 8,500 to 8,000 years ago. The result was climate change, an increase in sea level and changes to ocean currents. After these intensely cool periods, temperatures rose by 5 to 10°C in the space of a few decades. Sea level rise resulted in the breaching of the Mediterranean into the Black Sea Basin some 7,600 years ago and is probably the origin of the Sumerian, Babylonian and biblical stories of a great flood. Climate changes induced by changes in ocean currents cooled North Africa, grasslands changed to a desert, humans migrated and the great Mesopotamian cities were established.

One of the consequences of the massive sea level rise over the last 14,700 years is that the West Antarctic Ice Sheet was no longer underpinned by the land. Two thirds of the West Antarctic Ice Sheet collapsed into the oceans and sea level rose 12 metres. The final third of the West Antarctic Ice Sheet has yet to collapse to produce a six metre sea level rise as part of the dynamic post-glacial climate on Earth.

There is constant exaggeration by green activists and the media about the fate of the Antarctic Ice Sheet, yet measurements show the ice sheet waxed and waned and a decade of ice loss between 1992 and 2016 represented 0.00045% of measured sea level rise.[281] Over the past four decades, East Antarctica, which covers two thirds of the South Pole, has cooled 2.8°C; West Antarctica has cooled 1.6°C and only the tiny Antarctic Peninsula saw statistically insignificant warming.[282] As per usual, normally the media, climate activist and IPCC hype is the opposite of what has been measured or what the past shows.

A few centuries later, temperatures again suddenly fell about 11°C and glaciers re-advanced. Was this due to a sudden loss of carbon dioxide from the atmosphere? No. About 14,000 years ago, global temperatures again rose rapidly by about 4.5°C and glaciers receded. Was this due to human emissions of carbon dioxide? No chance.

About 13,400 years ago, global temperatures dropped, this time by 8°C and glaciers re-advanced. Was this due to a loss of carbon dioxide from the atmosphere or other factors such as the Sun, ocean heat changes and ice sheet collapse? About 13,200 years ago, global temperatures again increased rapidly by 5°C and glaciers again retreated.

Was this due to coal-fired power stations releasing carbon dioxide? No. They have only been with us for just over 100 years. You get the picture. The lunatics are running the asylum.

Some 12,700 years ago temperatures fell quickly by 8°C and a 1,300-year cold period followed. The Younger Dryas period had begun. We humans survived the Younger Dryas by congregating in fortified villages, developing animal husbandry by using docile wild animals and growing grains rather than gathering wild grass seeds. Humans were still wiping out easily hunted mega-fauna.

A solar outburst 11,700 years ago may have accelerated ice sheet melting at the end of the last glaciation, a shift from warming[283] to cooling[284] of the Younger Dryas cold period and the final mega-fauna kill. Solar outbursts caused glazing on the surface of moon rocks.

After this 1,300-year period of intense cold, global temperatures rose very rapidly by about 12°C marking the end of the Younger Dryas and the end of the latest glaciation. Did this glaciation end rapidly because we sinful humans suddenly put carbon dioxide into the atmosphere? Of course not, there was no fossil fuel-burning industry.

Between the end of the Younger Dryas and 6,000 years ago, there was a prolonged period of warming. Some 8,200 years ago, the interglacial warming was suddenly interrupted by a 500-year period of global cooling. During this time, alpine glaciers advanced. The warming that followed the cool period was also abrupt. Neither the abrupt climatic cooling nor the warming that followed was preceded by atmospheric carbon dioxide changes.

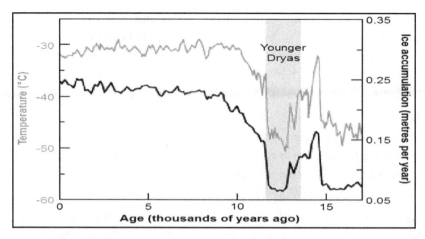

Figure 10: *Temperature and ice accumulation rates over time for Greenland ice. Rapid post-glacial climate change as represented by the Younger Dryas period. Both the magnitude and rate of climate change were far faster than any modern changes.*[285]

After this short sharp cold snap, warming continued until 6,000 years ago which was the peak of the current interglacial.[286] It was up to 5°C warmer then than now and sea level was up to six metres higher than now. What is now the Sahara Desert was grassland with patches of trees. As there was no industry then pumping out carbon dioxide into the atmosphere, it can only be concluded that the maximum warmth in post-glacial times was natural.

As continental ice sheets melted and added water to the oceans, this water also expanded as it warmed. Land that was pushed down during loading by ice now started to rise and, as the sea level rose by 130 metres, the sea floor sank due to loading by water. Although sea level rise was gradual, there were wild fluctuations in temperature and sea level did not respond to such rapid temperature rises and falls. Why should it now?

Areas covered with ice sheets during the last glaciation (116,000-14,700 years ago) sank under the weight of the ice. With the collapse of the ice sheets in the current interglacial, some lands are rising (e.g. Scandinavia and Scotland) and others are sinking (south-east England and The Netherlands). Since the zenith of the last glaciation 20,000 years ago, sea level rose 130 metres. Which part of this sea level rise is due to increased human emissions of carbon dioxide in 1950? Sea level curves suggest none.

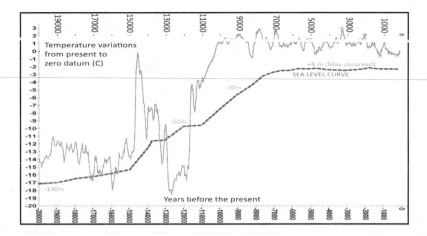

Figure 11: *Sea level and temperature from the Greenland Ice Sheet Project (GISP) ice core since the zenith of the last glaciation 20,000 years ago showing a lack of correlation between them and extremely rapid temperature rises and falls.*[287,288] *Some temperature changes were over 12°C and some changes took place in a decade.*

History shows us that some port cities (e.g. Ephesus, Turkey) are now inland whereas other cities (e.g. Lydia, Turkey) are submerged. In both the Maldives and eastern Australia, relative sea level has fallen. The Maldives is 70 cm higher now than in the 1970s and eastern Australia is two metres higher than 6,000 years ago.

Without a detailed knowledge of local land rises and falls, subsidence, fluid extraction, ocean eddies[289] and sedimentation, sea level predictions about future sea level are, at best, speculation. Just ocean eddies alone show that global mean sea level rise models need to be discounted by 25%.

Since the last interglacial began, there have been alternating warming and cooling phases. Within each of the phases there are numerous smaller warming and cooling events. These are well recorded from ocean sediment, lake and ice core drilling.

The rise and fall of empires

History shows that past natural warmings created economic booms and population increase. This has been validated[290] in a recent paper that showed increased atmospheric carbon dioxide and a 3°C warming is beneficial for the economy. Previous great empires grew in warm times.

Bölling	14,700 to 13,900 years ago
Older Dryas	13,900 to 13,600 years ago
Allerod	13,600 to 12,900 years ago
Younger Dryas	12,900 to 11,600 years ago
Holocene warming	11,600 to 8,500 years ago
Egyptian cooling	8,500 to 8,000 years ago
Holocene Warming	8,000 to 5,600 years ago
Akkadian cooling	5,600 to 3,500 years ago
Minoan Warming	3,500 to 3,200 years ago
Bronze Age Cooling	3,200 to 2,500 years ago
Roman Warming	500 BC to 535 AD
Dark Ages	535 AD to 900 AD
Medieval Warming	900 AD to 1300 AD
Little Ice Age	1300 AD to 1850 AD
Modern Warming	1850 AD to ....

Figure 12: *Major warming and cooling cycles over the last 14,700 years from the end of the last glaciation until now. Time periods are developed from compilation of history and ocean floor, lake and ice core drilling data.*[291] *Unless you are colour blind, (black = warm periods, white = cool) I think you can work out what the next climate cycle brings.*

On a high-resolution scale, there have been 10 warm periods over the last 10,000 years in the current interglacial. All were warmer than now and humans did not fry-and-die. They adapted. The last few thousand years shows that great empires and economies existed during warm times when it was warmer that at present.

Because there was no fossil fuel burning during warmings over the last 10,000 years, humans cannot be blamed for the Egyptian, Sumerian, Minoan, Roman and Medieval Warmings.

Why is it that only the last few decades of the current warming that started when the Maunder Minimum ended just over 300 years ago are due to human emissions of carbon activity? In colder times, there was famine, death, disease and an increased frequency of wars. Since the Minoan Empire 3,500 years ago, there has been a general cooling trend with spikes of warm and cold.[292]

Again, history shows that the hypothesis of human-induced global warming is wrong. How many times does a hypothesis have to be shown to be wrong before it is rejected? Only once! The climate industry still clings to the carbon dioxide hypothesis and the only way this can be done is to ignore the history of the planet and pretend that the planet had a stable benign climate until the Industrial Revolution. This is not science.

Sea levels were three metres higher in the Holocene Optimum 6,000 years ago. It was a few degrees warmer and there was 20% more rainfall. Cold dry periods, glacier expansion and crop failures between 5,800 and 4,900 years ago resulted in deforestation, flooding, silting of irrigation channels, salinisation and the collapse of the Sumerian city states.

Long periods of *El Niño*-induced drought resulted in the abandonment of Middle Eastern, Indian and North American towns. About 1470 BC, Thira (Santorini, Greece) exploded and threw 30 cubic kilometres of dust into the atmosphere. The tsunami, ash blanket and destruction of Thira greatly weakened the dominant Minoans. This led to the rise of the Greeks. One volcano changed the course of Western history.

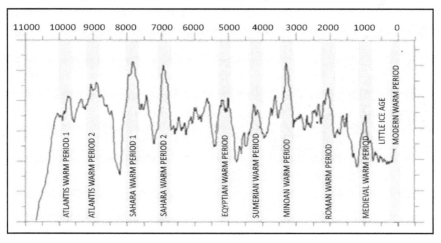

Figure 13: *Interglacial temperatures from GISP-2 core (Greenland) showing a correlation of temperature rise with the expansions of empires and cooling over the last 7,000 years.*[293]

Global cooling from 1,300 to 500 BC gave rise to the advance of glaciers, migration, invasion and famine. Egyptian records show a cool climatic period from about 750 to 450 BC. Global warming commenced again at about 500 BC, there was an excess of food and great empires such as the Ashoka, Ch'hin and the Romans grew.

The last 2,000 years

During the last 2,000 years, the Earth remained in an orbitally-driven interglaciation. However, slight variations in the Sun's output of energy changed the climate from cold to warm times. The colder times were colder than now; but not as cold as a full-blown glaciation and the warmer times, such as the Medieval Warming, were warmer than at present.

When the Sun's magnetic field is intense, it allows few cosmic particles to enter the atmosphere. When the Sun's magnetic field is weak, the Earth's atmosphere is flooded with cosmic particles which react and form short-lived chemicals that can be measured.[294] Cosmic particles are one of the ways clouds form hence cloud formation increases when the Sun has few or no sunspots. This results in cooling.

These short-lived chemicals are a temperature proxy. This change in the Sun's magnetic field has been known from the 22-year sunspot cycles that have been counted for 400 years. The chemical fingerprint measurements used as a temperature proxy show the Modern and Medieval Warmings as well as the exceptionally cold times in the Little Ice Age (Dalton, Maunder, Spörer, Wolf and Oert Minima).

These chemical fingerprints are in accord with what was historically recorded. Other chemical fingerprints showing a cyclical galactic influence on the Earth have now been traced back 32 million years.[295]

If science is not your thing, then history has some good examples of how climate has changed quickly and naturally. Hannibal was able to take his army and elephants across the Alps in the winter of 200 BC. This would not be possible now. Julius Caesar conquered Gaul in about 50 BC. He had to build a bridge across the Rhine River (which now separates France from Germany) in order to mount an attack.

The Rhine acted as a natural protective barrier for nearly 500 years but, as the Roman Empire was disintegrating in Gaul in the 5th century AD, the Vandals were able to walk across the frozen Rhine River to wage war. The Rhine has not been frozen in modern times.

There were sudden short periods of cold in the Roman Warming. The Romans wrote that the Tiber River froze and snow remained on the ground for longer periods. This now does not happen in Rome. During the Roman Warming after 100 BC, the Romans wrote of grapes and olives growing much farther north in Italy than had been previously possible because there was little snow and ice. Wine grapes were grown in northern England where it is now too cold to grow them.

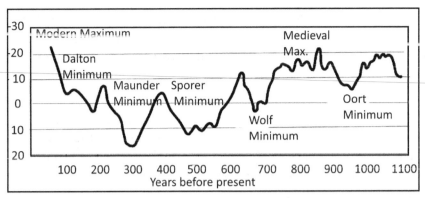

Figure 14: *Temperature reconstruction from proxies showing the coincidence of low sunspot activity with temperature minima.*[296]

Was this warming due to carbon dioxide released from fermentation of grapes to produce wine for the Romans? Hardly. Roman clothing showed times were warm.

In 29 AD, the Nile River froze in a short cooling event in the Roman Warming. It has not frozen since. Sea level did not rise during the 600-year long Roman Warming. Although the Romans did emit some carbon dioxide from smelting and agriculture, the amounts emitted were tiny compared to now. In Roman times it was warmer than now but this could not be due to human emissions of carbon dioxide and again the hypothesis that human emissions of carbon dioxide drive global warming is wrong.

During the Dark Ages, it was very cold. In 535 and 536 AD during a period of solar cooling, there were a number of large volcanoes that filled the atmosphere with white sulphur compounds and dust. These events produced a dusty atmosphere that reflected heat and darkness prevailed. As a result, the climate cooled and there was famine and warfare.

There was a coincidental decrease in sunspot activity and solar-induced cooling. Trees grew very slowly, the Sun appeared dimmed for more than a year after the volcanic eruptions and temperatures dropped in Ireland, Great Britain, Siberia, and North and South America. The plague wiped out some 60% of the weakened population in Constantinople. In 800 AD, the Black Sea froze. It has not frozen since.

Changes in solar output resulted in the Medieval Warm Period from 900 to 1300 AD In Europe, grain crops flourished, the tree line in the Alps rose, the population more than doubled and many new cities arose. Great wealth was generated, as always happens in warmer times.

The warmer climate allowed the Vikings to colonise Greenland in 985 AD. Ice-free oceans enabled Viking sea travel as far as Canada and the Middle East. The depths of the graves show that there was no permafrost. Fishing grounds were enlarged. On Greenland, they raised cattle and sheep, grew barley and wheat in places that are now covered by snow and ice.

This was not the first time that Greenland had been warm and habitable. Fragments of plants in ice from a bore hole in Greenland ice show that expanding glaciers pulverised forests that had thrived during two ice-free warm periods in Greenland in the past few million years.[297]

There are neither crops nor livestock today on Greenland showing that the Medieval Warming was warmer than the Modern Warming that we now enjoy.

In France and Germany, grapes were grown some 500 km north of the present vineyards, again showing that it was far warmer than now. In Germany, grapes are now grown at a maximum altitude of 560 metres. In the Medieval vineyards grapes were grown up to 780 metres altitude showing that temperature was warmer by about 1.0 to 1.4°C. Wheat and oats were grown at Trondheim (northern Norway), again suggesting climates 1°C warmer than now.

The warmer wetter climate of Europe produced excess crops and wealth that resulted in the building of castles, cathedrals and monasteries. As with previous warming events, there was great prosperity.

Prolonged droughts affected southwestern USA. Alaska warmed. The Medieval Warming had a considerable effect on South America. Lake sediments in central Japan record warmer temperatures, sea surface temperatures in the Sargasso Sea were approximately 1°C warmer than today and the climate in equatorial east Africa was drier from 1000 to 1270 AD. Ice core from the eastern Antarctic Peninsula shows warmer temperatures during this period. The Medieval Warming was not a localised warming event in Europe. It was global.

The Little Ice Age (1300 to 1850 AD) started with a rapid decrease in temperature of 4°C. Solar activity had decreased. In 1280 AD, volcanic eruptions on Iceland and a change in ocean currents started a triple whammy and it is no surprise that it took only two decades to change from the Medieval Warming to the Little Ice Age.

The Gulf of Bothnia froze in 1303 AD and from 1306 to 1307 AD. Three years of torrential cold rain led to The Great Famine from 1315 to 1317

AD. The famine was followed by the plague pandemic that attacked the weakened population in 1347 to 1349 AD. That was a real pandemic killing 50% of the European population. There was massive depopulation and it was not until 1550 AD that it reached the population level it had in 1280 AD.

During the Little Ice Age, there were warmer periods associated with sunspot activity. During minimum sunspot activity (1440 to 1460, 1687 to 1703 and 1808 to 1821 AD), the intensely cold conditions were recorded by the Dutch masters and King Henry VIII was able to roast oxen on the frozen Thames River. The Greenland ice sheet expanded as did sea ice. Glaciers expanded worldwide. This was not due to a sudden decrease in atmospheric carbon dioxide.

There were food shortages. The grain-dependent population of Europe suddenly faced a cold and variable climate with early snows, violent storms, catastrophic flooding and massive soil erosion. There were constant crop failures, livestock died and famine re-joined the plague to become became a great killer. Short cold periods occurred after the eruptions of Tambora (1815 AD) and Krakatoa (1883 AD) respectively.

In fact, 1816 AD was known as the "year without a summer". This was the time when English artist William Turner painted stormy oceans and skies full of volcanic dust, Mary Shelley wrote *Frankenstein* and Byron wrote *Darkness*.

In the Little Ice Age, the population of Iceland halved and the Viking colonies in Greenland died out in the 1400s because they could no longer grow enough food or get through the ice for fishing. In parts of China, warm weather crops that had been grown for centuries were abandoned. In North America, early European settlers experienced exceptionally severe winters.

The growing season for all crops was shortened, wine production decreased, rivers such as the Thames in London along with canals in the Netherlands regularly froze. Ice fairs and cooking of oxen took place on the ice. In February 2021, the Thames River froze again for the first time since 1963.[298] Is this due to global warming? Are we seeing hints of a forthcoming (Little) Ice Age?

All travel became hazardous. Glaciers advanced in the Alps, Papua New Guinea, the Andes and New Zealand and they buried villages in the Swiss and Austrian Alps. When New York Harbour froze in the winter of 1780, people could walk from Manhattan to Staten Island.

In June 1644 AD, the Bishop of Geneva, who had a reputation for exorcism, took his flock to pray at a Mont Blanc glacier that was advancing rapidly. The Glacier des Bossons retreated and advanced many times with advances in 1610 to 1643, 1685, 1712, 1777, 1818, 1854, 1892, 1921, 1941 and 1983 AD. The glacier was at its maximum in 1818, towards the end of the Little Ice Age. Since then, the glacier has retreated by 1.5 km.[299] Glacial advance and retreat is complex and often related to local phenomena.

The Glacier des Bossons has retreated 10 times over the last 400 years and just because a glacier retreats does not mean that retreat is due to human emissions of carbon dioxide.

Yesterday and today

Since the beginning of the last interglacial, there has been a warming of at least 8°C. In the Modern Warming since the beginning of the mid 19th Century, temperature has risen by 0.8°C. Why is this catastrophic when previous and far greater temperature rises have stimulated a thriving of humans?

Why didn't humans, polar bears, emperor penguins and other poster animals of the green activists die in the previous warmings when temperature was 5°C warmer than now?

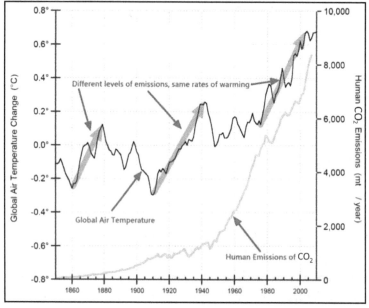

Figure 15: *Plot of global air temperature measurements and human emissions of carbon dioxide since 1850[300] (the end of the Little Ice Age).*

There has been a general rise in temperature since the end of the Little Ice Age. What would you expect temperature to do after a Little Ice Age? Rise or fall?

The coldest period in the Little Ice Age, the Maunder Minimum, was over 300 years ago. Since then, planet Earth has been warming and there is no answer to the key question. Which part of the post-Maunder Minimum warming is natural and which part is of human origin? Until this question can be answered quantitatively, then there is no measured evidence for human-induced global warming. Only computer speculations.

Since thermometer measurements were recorded from about 1850 AD (and earlier from central England), temperature decreases (1880-1910, 1940-1976 AD), no temperature change (1998-present) and increases (1860-1880, 1910-1940, 1976-1998 AD) show that there is no correlation of temperature with increasing atmospheric carbon dioxide.

The planet has not been warming rapidly, changes measured today are well within historic variability and even in recent times of industrialisation, there has been both cooling and warming. With no correlation between global warming and atmospheric carbon dioxide on geological, ice core and historical time scales, there can be no justifiable causation.

If human emissions of carbon dioxide drive global warming, why have there been both rises and falls in temperature since 1850 if temperature is driven by human emissions of carbon dioxide? This we are told is due to human activities.

The obvious question arises. Which part of the Modern Warming is natural and which part is due to humans? Furthermore, the rate of the three temperature rises is the same yet we would expect the rate of the latest temperature rise to greatly increase in accord with the post-World War II increase in carbon dioxide emissions. The story that human emissions of carbon dioxide cause temperature changes just does not stand scrutiny.

Ice cores from Greenland show the position of Northern Hemisphere glaciers changed and temperature oscillated about 40 times with changes on average every 27 years over the last 500 years. All of this has occurred within that last 500 years of the current interglacial as we emerge from the Little Ice to the Modern Warming. None of these changes could possibly have been driven by changes in atmospheric carbon dioxide or human emissions of carbon dioxide.

Why is it that only the modern warming since World War II is due to human emissions and yet all previous warmings were natural?

In Australia, the Great Barrier Reef, the poster child of the green left environmental activists, disappeared during glacial events more than 60 times over the last three million years. Each time the Great Barrier Reef disappeared, it reappeared a little later. It has survived past massive rain and storm events that deposited sediment on the Reef well before humans were around.

The sea level fall and lower temperature during glacial events kills higher latitude coral reefs and they continue to thrive at lower latitudes. The geological record and the occurrence of modern coral reefs shows that coral reefs love it warm. When there is more carbon dioxide in the atmosphere, coral has more building materials.

The Earth is approaching the end of its 52nd warm interglacial period over the last 2.6 million years. Why were the past interglacial warmings natural yet the fading warmth at the end of the current interglacial is now due to human emissions of carbon dioxide?

Global warmings and coolings have taken place over the space of decades. We have seen it before. As the Earth starts to cool, not all winter snow on the land masses melts. The extra snow left in summer reflects more solar radiation and cooling is hastened. The tree line drops, glaciers expand, mountain roads are covered and alpine villages are covered with ice.

A good example of this is the Habachtal emerald deposits in the Alps of Austria that were exposed and mined during the Medieval Warming, covered by ice in the Little Ice Age and are now free of ice again.

During glaciation, sea ice expands, ice sheets grow, rivers and lakes are frozen, sea level falls and coral reefs are left high and dry to die. Because more carbon dioxide dissolves in cold water than in warm water, there is a drawdown of carbon dioxide from the atmosphere into the oceans.

Since 1979, the Japanese Meteorological Agency has been measuring Antarctic sea ice area. It appears that global warming is yet to reach the Antarctica because the mean and maximum sea ice extents have risen and the sea ice area has increased between 1979 and 2020[301]. On 9th September 2021, the Arctic sea ice extent was 4.88 million square kilometres and greater than on the same day every year from 2015 to 2020 and 1.4 million square kilometres greater than in 2012.[302] Yet another case of data trumping propaganda.

During these times of unstoppable global warming, Antarctica leads the way, It had the coldest winter since 1957 when records began of minus 61.1°C. The previous record was minus 60.6°C in 1976.[303] The 1970s was a time of a solar-induced cooling cycle and we have just started a Grand Solar Minimum cycle.

During glaciation events, existing data shows tropical vegetation is reduced from rainforest to grasslands with copses of trees, somewhat similar to the modern dry tropics inland from the Great Barrier Reef. There was no Amazon rainforest during the last glaciation, just grasslands and a few trees.

During glaciation, sea level falls, farming lands are greatly reduced, forests get covered with ice and cold dry winds predominate. Crops fail and there is famine, weakened populations get hit by pandemics, deserts expand, settlements are abandoned, empires fall, some species become extinct and the human population and economies fall. It is no wonder that before we had central heating and double glazing and air conditioning, populations feared the cold.

History and observations shows us that during interglacials sea level rises, ports become flooded, agriculture moves to higher latitudes, tree lines increase in altitude, rainfall increases and population increases as does wealth.

Climate activists assume that the planet was at equilibrium until we dreadful humans came along and gave ourselves a far better life by burning fossil fuels and emitting beastly carbon dioxide. A brief look at the past shows that this assumption is unrelated to the body of knowledge accumulated over centuries.

The past shows us that the role of carbon dioxide in the atmosphere is not simply one of a thermostat for which high is hot and low is cold, as with a kitchen oven. It is not that simple.

For green activists to twiddle the dial to lower human emissions of carbon dioxide and global temperature is not the simple answer.[304] History shows that such a simple answer to a complex natural phenomenon is not science but a long-term green activist propaganda program aimed at destroying Western industrial society.

The last new geological time period internationally agreed was named after the Australian type locality of the Ediacaran fauna. It was called the Ediacaran Period and spans the 94 million years from 635 to 541 million years ago. Climate activists claim that we are living in extraordinary times when major planetary processes are now driven by humans. Their arrogance is breathtaking.

They want to create a new geological time period to reflect what activists consider horrible times and call it the Anthropocene. Geology does not respond to mad hatter green activists trying to change Earth history by inventing a term to describe the time when humans started to emit carbon dioxide from industrialisation and accordingly to enjoy a better life.

As usual, facts tell a different story. The International Commission on Stratigraphy used data and not ideology to divide the post-glacial time when humans were on Earth[305] into three parts, each marked by a significant cooling. They are: Greenlandian (11,700 to 8,200 years ago), Northgrippian (8,200 to 4,200 years ago) and Meghalayan (4,200 years ago to the present).[306]

Geological data tells us that we have been cooling for 6,000 years since a period known as the Holocene Optimum when it was warmer than now and sea levels were higher. Activist greens, bonkers Boris, gormless Gore, flim Flannery, ersatz "scientists", those with vested interests and second-rate performers in the sporting, entertainment and cacophony industries tell us, based on populist ideology, that the planet is heating due to human activity.

It's your choice: data or doomsday ideology based on ignoring unwanted truths.

There has never been a better time to be a human on Earth than today.

Making rocks

Granite makes up a significant portion of the Earth's continental crust. After centuries of dispute about how granite forms, scientists attempted to make granite in the laboratory. Mixtures of the most common minerals in granite were combined and heated until melting. Metallurgists have been melting metal mixtures for centuries and knew about minimum melt mixtures and fluxes. Experiments with rocks gave similar results.

When a cocktail of minerals in a specific proportion was cooked, the mixture melted at a lower temperature than the minerals themselves. More experiments showed granite melt at a lower temperature under high pressure. If a flux like water was added to the mixture, the melting temperature became even lower (~700°C).

Granite is more easily made at high temperature and pressure by cooking a mineral cocktail of wet sedimentary rocks with or without pre-existing granite. Molten granite at depth separates into a number of granite types that can mingle with other molten rocks deep in the crust.

Granite cooling from a melt to a solid traps small amounts of the flux as salty watery fluid inclusions in minerals. At times, there is some carbon dioxide in these inclusions.

Similar experiments with basalt, the most common volcanic rock on Earth, were tried. Basalt also melted at a lower temperature under pressure. In the presence of fluxes, such as steam and carbon dioxide, the melting temperature was again lowered (~1100°C).

Experiments showed that granite could dissolve large amounts of water and very little carbon dioxide whereas basalt could dissolve large amounts of carbon dioxide and little water.[307] The less water dissolved in a molten rock, the better it flows as lava.

The study of meteorites showed that there are three main types: iron, stony iron and stony meteorites. It was concluded that meteorites derived from fragments of a protoplanet between Mars and Jupiter that was impacted and broken up. The iron was the core of the planet and the stony material was the mantle. The planet was still separating into layers and a crust had not formed.

During its early history, the Earth separated into an iron core, a stony mantle and a thin skin of crust on top of the mantle. Rocks of stony meteorite composition have been found on Earth. These were in areas that had been pushed up from great depths. If this were the case, then perhaps laboratory experiments on the melting of rocks of stony meteorite composition could help to determine how basalt formed at great depth.

Mantle rocks were melted and did not produce basalt or granite. However, if mantle rocks were partially melted under different pressures in the presence of fluxes then all the types of basalt known on Earth could be reproduced by partial melting of the mantle under different pressures. From these experiments and modern geochemistry, we can now measure the chemistry of basalt and calculate its depth of formation.

Experiments also showed that almost 15% by weight of carbon dioxide can be dissolved in basalt.[308,309] At depths of 175 km or more, there can be 8% carbon dioxide by weight in mantle rocks which fluxes melting[310] yet some erupted basalts contain only 0.01-0.001% carbon dioxide. Where does all the carbon dioxide go? Up. With rising basalt and into ocean waters.

Experiments also showed that if a solid mantle rock was at high pressure and temperature, a sudden decrease in pressure induces partial melting. This we now know happens with recoil after an asteroid smashes into the planet and when the Earth is pulled apart in the centre of oceans.

The surface of the Moon and Mars are covered with basalt that filled asteroid craters. We conclude that these basalts derive from impacting. Many craters are filled with basalt only to be cratered again with later impacting. Because the Moon and Mars have no plate tectonics, running water, sediments and life, the surface of these bodies has preserved the 4,500 million year history of impacting.

Stony meteorites and mantle rocks contain carbon as carbonates, carbides, carbon dioxide, methane and other hydrocarbons and as native carbon (diamond). The Earth's upper mantle and crust contains 1.9 billion billion tonnes of carbon. The atmosphere contains 5,100 billion tonnes of carbon as carbon dioxide. The planet's carbon is beneath your feet and not in the sky above.

Laboratory experiments to make synthetic diamonds show that diamond comes from depths of at least 150 km. They are plucked off by molten rocks rising at about 20 km/hr through colder rock otherwise diamond would change to its cousin graphite. Experiments and the mineral inclusions found within diamonds show that the fast-rising molten rock is pushed up by carbon dioxide which expands and explodes when close to the surface to give a kimberlite pipe.

A rare volcanic rock type[311] composed of carbonates, and called carbonatite, contains 40 to 48% carbon dioxide. Monstrous amounts of carbon dioxide are released into the air before, during and after eruptions of carbonatite. It is also held as carbonate minerals in the solid rock. Many such eruptions have been observed in the East African Rift (e.g. Ol Doinyo Lengai, Tanzania, 2008).

Some volcanoes explode only carbon dioxide or steam and no lava. These gas volcanoes form distinctive craters called a maar. These are now often preserved today as crater lakes that are in areas where there are no lavas.[312] These emissions are not counted in estimates of natural carbon dioxide emissions. Carbon dioxide still bubbles out of these crater lakes and carbonate minerals in the surrounding broken up rock are common.

In 1984 and 1986, burps of carbon dioxide from the volcanic crater lakes of Monoun and Nyos in Cameroon killed thousands of people and their livestock because heavier-than-air carbon dioxide flowed downhill and filled valleys leaving no air for breathing.

Virtually every hole drilled for hydrocarbons finds carbon dioxide in the ancient sedimentary rocks. Oil and gas separation plants extract carbon dioxide, helium, hydrogen, sulphur gases and other non-flammable gases from the hydrocarbons and normally release most of these gases to the atmosphere.

Every hot spring releases carbon dioxide as carbon dioxide gas or bicarbonate in solution. All spa centres have carbon dioxide either from rocks being pushed together very deep down, from cooling volcanic rocks just beneath the surface or parts of the crust being pulled apart. We see hot springs on the land but don't see the millions of submarine hot springs.

In continental areas where the crust is thinning and being pulled apart, there are basalt volcanoes and hot springs (e.g. Basin and Range, USA; western Anatolia, Turkey; Red Sea, Saudi Arabia and Egypt; East African Rift). Spa centres such as Pamukkale (Turkey) are draped with carbonate minerals precipitated from bicarbonates in solution.

This is from carbon dioxide leaking out after an eruption and, from what we know from experiments, monstrous amounts of carbon dioxide would have leaked into the atmosphere before eruptions. One small hot spring on Melos (Greece) contributes up to 1% of the planet's volcanic carbon dioxide derived from land-based volcanoes.[313,314]

Molten rocks rise from depth because they are lighter than the surrounding rocks. This is partly because they are molten and partly because of the amount of dissolved gas. As molten granite rises, the overlying pressure decreases and dissolved gases boil off. These rising hot gases release energy by fracturing rocks and often deposit minerals such as quartz in the veins. Fluid inclusions in quartz are salty water. Carbon dioxide is a minor component of fluid inclusions.

At depths of anything from 4 to 15 km so much gas has been boiled off that the molten granite now no longer has a flux and solidifies. At times, a molten granite that initially had a small amount of flux can rise to the surface to form a volcano.

At shallow depths, ground water and seawater can mix and join the flux in the molten rock. As molten rock rises to within a kilometre or so from the surface, there is a catastrophic loss of gas because of the added water and decreased overlying pressure. This leads to a massive explosive volcanic eruption (e.g. Santorini and Krakatoa).

Basalt ocean floors are pushed under continents to produce mountain ranges, earthquakes and volcanoes such as in the Pacific Ring of Fire and mid-ocean ridges formed where the ocean floor is being pulled apart.

These volcanoes are composed of andesite, named after the Andes where they are very common. Laboratory work showed that if a basalt is partially melted in the presence of water, then the partial melt is the chemistry of an andesite. A thin skin of sea floor sediments and ocean water assist the melting.

This basalt was again cooked and a small fraction rises to the surface as molten andesite. Sometimes molten andesite separates into a number of different molten rock types. Andesite volcanoes also suddenly release dissolved water. Shallow surface water acquired from groundwaters and seawater can make the andesite even more explosive.

What we now know from experiments and measurements of volcanic rocks and gases is that most of the dissolved fluxes such as steam and carbon dioxide are released as molten rocks rise from kilometres depth to the surface. A little is released during eruption, even if it is explosive, and old volcanoes leak gases for a long time after eruption as hot solid rocks cool. Experimental studies on the solution of carbon dioxide in molten rocks shows that far more carbon dioxide dissolves in basalt than in andesite, the most common rock type in land-based volcanoes.

There are 1,511 active land-based volcanoes, most of which are andesites. Measurements of gas chemistry and temperature are a reliable method of predicting eruptions. Use of geophysics and volcanic gas chemistry to predict eruptions and associated tsunamis has saved tens of thousands of lives.[315] There is a sneaking suspicion amongst some volcanologists that planetary alignments and solar and lunar gravity may trigger eruptions.

There are measuring stations on most active land-based volcanoes, mainly andesites, and the main gases emitted are steam and sulphur gases. Very little carbon dioxide is emitted which is in accord with the experiments on making rocks.

Most of the world's volcanoes are deep submarine basalts where only a few gas measurements have been made. We know from experiments, measurements of volcanic gases and fluid inclusions that carbon dioxide is a major gas component of molten basalt. Seafloor basalt covered by a thin skin of sediment occupies about 70% of the Earth's crust. Basalt volcanism rules the world and the rocky planets.

How can the pre-life carbon dioxide rich atmosphere of the Earth be explained? How can the carbon dioxide rich atmospheres of the inner rocky planets be explained? The original carbon dioxide in the Earth's atmosphere was from planetary degassing, a process that is still operating and adding carbon dioxide to the atmosphere.

What's all this got to do with climate?

Our carbon-blessed world

Without carbon dioxide, there would be no life on Earth. About 12% of the human body contains carbon, we eat carbon-based foods and we expel carbon compounds.

Planet Earth has been cooling down and leaking liquids and gases for 4,567 million years. The hot mantle is covered by a thin skin called the crust. The crust has been stretched, pushed, flexed, broken, melted and recycled and these processes allow mantle heat to be released to the surface of the Earth. This heat is released to the crust, oceans and atmosphere.

Carbon dioxide leaking from rocky planets in the Solar System has been going on since the year dot. For the first 80% of the Earth's history, carbon dioxide was a major component of the atmosphere. Because the surface of the Earth contains running water, sediments and life, there has been a drawdown of atmospheric carbon dioxide to the current very low level. Other planets without water, sediments and life still have a high carbon dioxide content in their atmospheres.

Not only are there climate cycles based on tectonic, galactic, orbital, solar, ocean and lunar cycles as previously presented, there are other interacting cycles on the Earth.

The tectonic cycle changes from pulling apart to pushing together and the construction of a supercontinent has happened four times in the last 2,500 million years.

There is also a rock cycle where sediments are cooked under pressure to metamorphic rocks, melted and later to solidify as igneous rocks. After uplift, weathering and erosion remove, oxidise and hydrate materials that were once metamorphic, igneous and sedimentary rocks to deposit them as new sediments which later lithify to sedimentary rocks. The process then continues. This process is very broadly related to the tectonic cycles and it takes place right in front of our eyes.

Sea level cycles were quantified and used by the oil industry to find when the land was submerged and draped with sediment which could contain oil and gas and where land was once exposed to be and weathered and eroded.[316]

There have been six major sea level cycles over the last 600 million years and one minor sea level cycle every 32 million years occurred over the last 545 million years.[317] These 32 million-year cycles coincide with episodes of flood basalt volcanism and minor mass extinctions. The science is not settled. More work needs to be done on cyclical activities deep in the Earth and on astronomical and solar cycles.

The carbon cycles involve the release of carbon dioxide into the atmosphere from ocean degassing, volcanoes, cooking of rocks at depth, respiration, soils and industry. Uptake is as carbon in sediments and as carbon dioxide into the oceans by dissolution and marine life, as limey rocks, and as chemical reaction with seafloor rocks.

There have been times in the past when volcanism was far more active than today. During times of increased volcanic activity, the atmospheric carbon dioxide content was high and when volcanoes were quiet, there was less carbon dioxide in the atmosphere. The correlation between volcanic activity and atmospheric carbon is poor, perhaps because it uses all volcanism and is not specific to basalt volcanism. There is clearly much more work to be done on the volcanic cycles, the deep carbon cycle, understanding carbon dioxide over time and the relationship between time, carbon dioxide, basalt volcanism and planetary[318] evolution. Maybe basalt volcanoes are far more important than we are led to believe?

Land plants uptake carbon dioxide as plant food. As part of the carbon cycle, methane is released from rocks, springs, soils and bacteria and digestion by animals and is oxidised in the air into carbon dioxide and water vapour.

Back calculation from oxidised rocks and measurements of trace metals in minerals that are easily oxidised[319] have shown that minimum oxygen levels (2-10%) correlate with mass extinctions and the formation of lead-zinc ores in sediments whereas very high oxygen levels (20-30%) correlate with major evolutionary events, global periods of salt and red dune formation and the formation of copper ores in sediments.

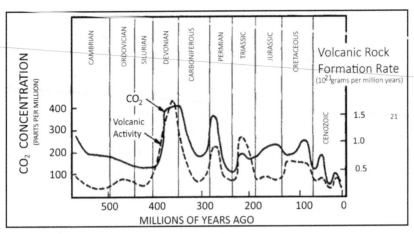

Figure 16: *Estimate of the variations in atmospheric carbon dioxide and the mass of volcanic rocks erupted over time.*[320]

In ancient rocks oxygen cycles were 400 to 600 million years in length. Over the last 500 million years cycles have been 60 to 150 million years in length. These cycles are probably driven by the tectonic cycles.

Science is always full of surprises and there is no such thing as settled science. About 400 million years ago, carbon dioxide levels were high, perhaps 10 times greater than today. In an attempt to establish a relationship between the carbon-silicon cycle and temperature over three billion years, it was shown that there was a sudden abundance of life 400 million years ago.[321] This is contrary to the greens activists' claim that increasing atmospheric carbon dioxide will harm marine life. It's the opposite, as can be shown with so many catastrophist climate activist claims.

There are also sulphur, nitrogen and phosphorus cycles which, although important for life on Earth, do not appear to be critical for driving climate change.

Without carbon, there would be no life on Earth. It is obviously neither a poison nor a pollutant.

Twenty thousand leagues under the sea

Just because we can't see something doesn't mean it does not exist.

Experiments, measurements and observations show that a huge amount of heat is released from submarine volcanoes on the ocean floor into water and that venting of carbon dioxide into deep ocean water occurs before, during and after eruptions.

Air temperature is driven by ocean heat due to the high heat capacity of water. Cooling submarine basalt is a massive source of heat to the oceans. This does not even get a look in with climate models and, when I raise it with climate activists, all I get is a blank look. They have not even heard of heat from below. It is an overlooked mechanism for atmospheric heating.

Ships and buoys have been measuring the temperature of the top 10 metres of the water column for centuries. ARGO buoys which can sample to depths of 2,200 metres have shown that parts of the top 700 metres of the oceans have warmed by 0.03°C per decade[322] and cold-water ocean circulation does not move as expected.[323] There is much to do as the deep oceans are poorly understood.

About 15 years ago, there was an estimate that there were more than one million seafloor volcanoes.[324] The most active submarine volcanoes are well hidden at depths of more than 2.6 km. There are 1,511 potentially active land volcanoes and about 500 have erupted in recorded history.

Most of the planet's volcanoes are submarine basalts and account for 75% of the heat transferred to the surface via molten rocks from great depth.[325] The oceans are heated from above by the Sun and are also heated from below by submarine basalt volcanoes which have a very high content of dissolved carbon dioxide[326] and most are supersaturated in carbon dioxide.[327] The material for these volcanoes, including carbon dioxide derives from the mantle of the Earth.

When molten rocks rise and cool, they release monstrous amounts of hot gases, mainly carbon dioxide.[328] This gas release provided the initial high carbon dioxide content of the Earth's early atmospheres before plant life was abundant.[329]

Land-based volcanoes explode because they suddenly release gas. Volcanic gases such as carbon dioxide escape from the molten rock prior to, during and after eruptions.[330] Most gas is released before eruptions and it is this gas that helps lava rise. After an eruption, volcanoes leak gases for a very long time. Unless gas measurements can be made before, during and after a submarine volcanic eruption with instruments awaiting an eruption, then the amount of carbon dioxide released into seawater cannot be directly measured.

However, if submarine volcanic rocks suddenly freeze to a glass, then they trap carbon dioxide[331,332] although the molten rock may have lost most of its carbon dioxide as the liquid basalt rose. Trapped carbon dioxide in glass gives a minimum figure for the amount of carbon dioxide released into bottom ocean water.[333]

The weight of at least 3 km of seawater on top of rising molten basalt and the lack of abundant dissolved water in it stop submarine volcanoes exploding. The most abundant submarine volcanic fluxes in basalt are carbon dioxide and dissolved water. Basalt can explode if erupted in shallow water.

There is at least 64,000 km of mid ocean ridges at water depths of 3 to 6 km.[334] In the centre of ridges are deep submarine rifts where the ocean crust is being pulled apart and new crust is being formed. To make space, the old crust at the edge of the oceans is pushed away. Ocean floor basalt crust is no older than 200 million years. Pulling apart of today's oceans occurs at centimetres to tens of centimetres a year and produces new ocean crust at the mid ocean ridges and old crust at the edge of oceans is pushed under continents and island arcs.

Basalt volcanoes occur along the mid ocean rifts. Beneath mid ocean ridges are huge volumes of molten basalt and an accumulation of gas, mainly carbon dioxide.[335] Methane is also released at mid ocean ridges. Seafloor and sub-seafloor basalts at 1100°C are cooled by deep frigid water at 2°C. Basalt is cooled, becomes solid and eventually gives up all its heat to the bottom ocean waters. The heat added to the ocean floor has to go somewhere.

These warmer bottom waters are light, rise and may form the "blobs" of warm water detected in the North Atlantic (2012), North Pacific Ocean (2013-2016), Southwest Indian Ocean (2018-2019) and the South Pacific Ocean (2019-2020) associated with submarine volcanism.[336,337,338] The Humboldt Current and most deep ocean currents flow over hot rocks and may distribute heat widely. These blobs may be from submarine volcanicity whereas the decadal cyclical *El Niño-La Niña* events may be from cyclical solar events.[339]

Each year, 10,000 cubic kilometres of cold dense seawater circulates through the mid ocean ridges as a coolant. In places where old ocean crust has been pushed up over the land (e.g. Cyprus), we can measure that ocean water circulated to a depth of 5 km and, during chemical reactions, chemicals were added and subtracted from the ocean floor basalts and seawater. Some of these chemical reactions give out heat.

It takes 200,000 years for the whole of the planet's oceans to do one complete lap through the top 5 km of ocean floor basalts. This means that heat is constantly added to the oceans and the oceans are continually buffered from becoming acid. Green activists frighten the gullible that rising atmospheric carbon dioxide levels will lead to ocean acidification. This is fraud.

In submarine basalt mid ocean ridges volcanoes, the dissolved water does not boil and is released as liquid that can be over 300°C in vents and warm springs that also release carbon dioxide.[340] At the sites of hot springs, there is a slight localised increase in acidity. This is caused by the venting of both carbon dioxide and sulphuric acid.[341]

Submarine hot springs have a highly variable carbon dioxide content[342] and carbon dioxide can be removed from hot springs by precipitation of carbonates in fluid-rock chemical reactions below the sea floor. Mid ocean ridge hot springs release an estimated 0.3 to 1.2% of the annual input of carbon dioxide into the oceans[343] whereas mid ocean ridge volcanoes release a far larger and unknown amount of carbon dioxide into ocean water via gas vents.

Many mid ocean ridge lavas are supersaturated in carbon dioxide[344], the amount of carbon dioxide released is huge[345] and is part of the normal sea floor spreading process.[346,347,348] The large amount of dissolved carbon dioxide in the molten basalt does not solidify to carbonate minerals. It also does not bubble up from the mid ocean ridges and enter the atmosphere because it dissolves readily in the cool, salty, high pressure, deep ocean water.[349,350]

The high-pressure cool bottom waters dissolve all the volcanic carbon dioxide and this abundant source of carbon dioxide never enters the calculations of those attempting to calculate the amount of volcanic carbon dioxide released into the atmosphere.

Upwelling hundreds to thousands of years later releases carbon dioxide to the atmosphere. The present increase in atmospheric carbon dioxide also may be due to the release of carbon dioxide from past warmings such as the Medieval or Roman Warmings.

Some submarine volcanoes and associated springs have pools of liquid carbon dioxide.[351]

In Iceland on the Mid Atlantic Ridge volcanoes result from the pulling apart of the Atlantic Ocean. This process of rifting depressurises hot high-pressure rock beneath Iceland. Water, carbon dioxide and other gases migrate to these depressurised zones, act as a flux and the mantle of the Earth beneath Iceland partially melts. These gases dissolve in the molten rock which buoyantly rises as a basalt melt.

It is rare for basalt volcanoes to be explosive because of the low amount of dissolved water as a flux. Iceland is an exception because water is added to near surface molten rock from ice, ground water and seawater. As the molten rock rises, the weight of the overlying rocks decreases and a point is reached where gas dissolved in near surface molten rock is explosively released. Iceland has a history of explosive basaltic volcanoes.

The eruption of Iceland's Eyjafjallajökull was a very small eruption and took place for most of 2010. Just a modest belch from that small unpronounceable Icelandic volcano caused chaos to international air transport, due to the normal human weakness of panic and over reaction. There had been a bigger eruption from the same vent in 1821-1823.

The Icelandic volcano of Laki had an even larger eruption from June 1783 to February 1784 which produced "ash" and a massive haze of dry sulphuric acid clouds over Europe. The air was acrid, people had difficulty in breathing and those with respiratory problems died. There was deposition of white sulphate films, destruction of plants, crop failure and famine. Mortality rates increased and the following summer and winter were gloomy and cold.

In the Arctic Ocean, huge deep water explosive submarine basalt volcanoes along the Gakkel Ridge have formed large craters that released massive amounts of heat and carbon dioxide into Arctic waters.[352] In other areas of the Arctic Ocean, there is submarine basaltic volcanic and hot spring activity.[353] In 1999, that slow spreading mid-ocean ridge experienced an explosive submarine basaltic eruption. For basalt to explode at such a great water depth, at least 13.5% by weight of the molten rock was carbon dioxide.[354,355] In other areas of the Arctic Ocean, there is further submarine basaltic volcanic and hot spring activity.

Slow-spreading ridges and relatively unexplored ridges such as the 1,000 km-long Gakkel Ridge have surprises such as numerous hot springs and volcanoes along their lengths.[356] The new volcanic rocks were cooled by circulating seawater and the Arctic Ocean warmed for a short time. This warming was coincidental with a lunar tidal cycle of 18.6 years that pushed warmer surface North Atlantic Ocean water into the Arctic. If Arctic sea ice retreats, it may be from heat below and climate cycles rather than atmospheric warming.

The mid ocean ridges are not the only places where heat and carbon dioxide are added from below. There are hundreds of off-axis volcanoes[357,358,359,360] that add heat and carbon dioxide in addition to that from the mid ocean ridges. These off-axis volcanoes are tens to thousands of kilometres from mid ocean ridges.

Some of them even vent liquid carbon dioxide at less than 4°C which is mixed with traces of rotten egg gas, methane and hydrogen and other fluids which are vented from the same field at 103°C.[361] Weird things happen in the oceans and we probably know more about the surface of the Moon than we do about the ocean deeps.

Parts of the ocean floor move over hot spots where plumes of mantle heat come close to the base of the basalt crust on the sea floor. Chains of basalt volcanoes form by partial melting of the hotter mantle and sea floor basalt volcanoes grow so much that they rise from the ocean floor to above sea level (e.g. Canary, Hawaiian, Galapagos and Réunion Islands; Gulf of Alaska).

These hot spot plumes are hundreds to thousands of kilometres wide and long[362] and exhale gases such as water vapour, carbon dioxide, methane, sulphur and helium.[363,364,365] They also add heat, carbon dioxide and nutrients to the oceans and carbon dioxide to the atmosphere. Mid-plate seamounts are estimated to number between 55,000 and 22,000 of which 2,000 are active volcanoes.

The main global measuring station for atmospheric carbon dioxide is actually located on one of these carbon dioxide-emitting volcanoes above a hot spot along the Hawaii-Emperor seamount chain (Mauna Loa, Hawaii). None of the heat emitted or carbon dioxide exhaled from it is acknowledged in any climate calculations.

Ancient submarine basalt volcanoes occur as volcanic islands (e.g. Lord Howe Island) and have grown a pile of basalt more than 3 km high above the sea floor to be above sea level. The submarine slope of basalt volcanoes is gentle because basalt flows easily hence millions of cubic kilometres of basalt cooled and released carbon dioxide to form basalt volcanic island chains. Some of these submarine basalt volcanoes originally rose above sea level and then subsided when all the lava was vented to form seamounts.[366,367,368]

In warm water, atolls grow on these submarine basalt volcanoes. As the volcano sinks, the coral atoll keeps growing upwards. Some of these atolls have been drilled (e.g. Bikini Atoll) and hundreds of metres of coral have been intersected before basalt is reached.

Not only does this provide a story about submarine heat and carbon dioxide release from submarine volcanoes but it shows that coral atoll growth can keep up with a very rapid sea level rise (or volcano subsidence). In places, they have been eroded to flat-topped submarine volcanoes.[369] There are many such submarine volcanoes and very little is known about them except that they are basalt.

An attempt to count the number of off-axis basalt volcanoes in part of the oceans suggested that there are at least 3,477,403 of them. Each has heated the oceans and vented carbon dioxide and are still venting carbon dioxide.[370] It is estimated that 140,000 of these off-axis volcanoes are active[371] and there are up to 10 million hot springs belching out carbon dioxide and heat associated with these off-axis submarine volcanoes.[372] This compares with the 1,511 land volcanoes that emit very little carbon dioxide, yet form the basis of official calculations of global emissions for volcanic carbon dioxide. Why do climate activists ignore such freely-available information?

The ocean floors are criss-crossed with fractures resulting from the sea floor shuffling for space as the oceans open up and the ocean floor is pushed under continents. Many volcanoes occur along these transform fractures[373] and, as with the land, these fractures leak carbon dioxide.

On a far smaller scale, many terrestrial volcanic belts extend from the land to under water[374], resulting in local heating and carbon dioxide release into shallow water.

In their models for the release of natural carbon dioxide and model estimates of future climate change, the IPCC doesn't consider natural carbon dioxide emissions from submarine mid ocean ridge basalts, off-axis basalts, hot spot basalts, submarine basalt volcanoes, hot springs, liquid carbon dioxide or fracture leakage. Nor do they consider heating of the oceans from below.

The IPCC carbon dioxide emissions calculations are based on measurements at land-based measuring centres which have been established to forewarn inhabitants of an impending eruption. Very little carbon dioxide is released from volcanicity on the land. This means that the IPCC has grossly underestimated the volume of natural carbon dioxide emissions. This underestimation results in an overestimation of the human emissions of carbon dioxide.

There is not one deep submarine measuring station hence the emissions of heat, carbon dioxide and methane from submarine basaltic volcanism can only be deduced or ignored. What we do know from experiments, measurements and observations is that the IPCC models have not used most of the planet's volcanic emissions. These come from unseen processes associated with deep submarine basalt volcanoes.

The emissions from mountain springs, hot springs, volcanic gas leaks, carbonatite eruptions, and maar gas leaks let alone leaks of carbon dioxide from soils and bacteria are not factored into the IPCC calculations either.

The IPCC use a unique carbon chemical fingerprint of carbon dioxide in the air to determine the amount of human emissions from burning fossil fuels, grass, stubble, timber, forests and peat. However, there is just one slight problem. The carbon chemical fingerprint of carbon dioxide dissolved in the oceans from past volcanic activity and later released to the air by ocean degassing is the same as that released from the burning of fossil fuels and vegetation.

If the amount of carbon dioxide released from submarine basalt volcanism is massively underestimated, then this makes the human emissions estimates far higher than reality. The IPCC's figure of 3% total annual emissions by humans is probably far lower and I suspect is the order of 1% or less of total annual emissions.

When the carbon fingerprints are analysed, components from the ocean and plants can be detected, a seasonal Northern Hemisphere variation is detected and there is a time delay between the Northern and Southern Hemisphere variations in carbon dioxide.[375] This is inconsistent with a steady accumulation of atmospheric carbon dioxide due to human activity.

Analysis of carbon dioxide measurements at Law Dome (CSIRO, Antarctica) and Mauna Loa (Scripps Institution of Oceanography, Hawaii) shows that for over the last 100 years, some 50% of the annual increase in atmospheric carbon dioxide comes from ocean degassing.[376] Furthermore, using chemical fingerprints of carbon dioxide in the air is not in accord with laboratory measurements.[377]

The Global Financial Crisis and the shutdown of industry did not result in a decrease in carbon dioxide in the air. If the green activists are right, it should have. There was a significant reduction in carbon dioxide emissions by humans during the global COVID-19 epidemic which created production shutdowns and an almost complete cessation of travel yet the rate of rise and concentration of atmospheric carbon dioxide did not decrease.[378]

This was validated by the International Energy Agency (IEA) which reported a record drop of 8% in human emissions of carbon dioxide during the COVID-19 crisis. Global energy demand fell by 6%, seven times greater than the 2007-2008 Global Financial Crisis collapse.

The COVID-19 crisis caused the biggest fall in global energy investment in history. Consumption of aviation fuel and petrol plummeted with the consequent drop in human carbon dioxide emissions but with no drop in global temperature. The carbon dioxide content of the atmosphere kept rising at the same rate shwoing humans have little effect.

There was no decrease in global temperature showing that there is no relationship between alleged global warming and human emissions of carbon dioxide. Why then is there a push for 20%, 50% or 100% emissions reduction when the two global experiments (2007-2008; 2020-2021) showing that nothing happens when emissions are reduced?

During World War II when there were massive emissions of carbon dioxide in the war effort, the consequent global carbon dioxide fingerprint change in the air could not be measured.[379] These three planetary experiments all show that human emissions are much lower than the 3% contribution to annual global emissions claimed by the IPCC.

Something is seriously wrong with the IPCC's estimates of carbon dioxide emissions from submarine volcanoes, the oceans and from industry. As with most science, it is unfinished business and there is no consensus.

Human emissions of carbon dioxide need to be placed in perspective. If the IPCC's annual emissions (3%) of carbon dioxide comprise 33 molecules, only one is from human emissions and the rest from natural processes. This one molecule of human-derived carbon dioxide is mixed with 85,000 molecules of other gases in the air. If human emissions of carbon dioxide drive climate change, then it has to be demonstrated that this one molecule in 85,000 drives climate change and that the 32 molecules derived from natural processes do not.

Furthermore, if annual human emissions are only 1% and not the IPCC's 3% estimate of total emissions, then green activists are attempting to change the Western world economies on the basis that one molecule of carbon dioxide in air mixed with 255,000 other molecules in air drives global warming. If you believe this, then I can sell you London Bridge at lower than market price.

It has yet to be shown that human emission of carbon dioxide drive climate change. In fact, there is only evidence to the contrary.

I think that there need to be more robust reasons to totally restructure the economy.

When green activists, lobby groups and politicians can turn volcanic eruptions on and off whenever they please, then they can have a go at trying to twiddle the dials and change global climate.

Don't wait up.

3

THE END IS NIGH

Not one prediction about the end of the world has happened. If only one prediction were correct, we would not be here.

Not one green activist prediction about the world being wiped out by environmental pollution, famine or over-population has occurred. In fact, the opposite has occurred.

Climate "scientist" predictions in the 1970s about an impending ice age in a few decades were all wrong. Many of the same people who made predictions about a forthcoming ice age are now making predictions that the world will end by global warming.

Green activists' predictions about drought were hosed down shortly after by massive rains and flooding.

A number of green activists are advocating show trials followed by the death sentence for scientists who don't agree with the global warming hysteria. They also argue that global warming is so great that democracy will have to be placed on hold.

No climate "scientists" make public statements against such opinions or against those who shot up the office of an eminent colleague who does not share their view.

Time has shown that all predictions made just before the Conference of Parties (COP) meetings about how a climate catastrophe is upon us, how we have days, months or years left and if we don't act immediately then we are doomed have all been wrong. Not one prediction has eventuated.

Ships packed with climate "scientists", a compliant press and tourists have sailed into polar waters to show that polar ice is melting because of global warming. They have all been stuck in thick sea ice and people had to be rescued by fossil fuel-burning aircraft and ships.

Green activist Al Gore uses nearly 40 times as much energy as the average American, flies around the world in carbon dioxide emitting

jets to tell us that we are all doomed and must reduce our emissions. While warning us that sea level will rise by six metres, he buys waterfront mansions.

A UK court has ruled that Gore's film *An Inconvenient Truth* was a crusade to make a political statement and contained fundamental errors of fact.

Publication of poor quality, hoax and scam papers in the peer-reviewed literature characterises some areas of scholarship. In the physical sciences, the alleged gold standard of peer review may not be a measure of quality and is greatly influenced by the bias of editors and reviewers.

There is a pretty dismal history of self-appointed experts making predictions about the end of the world and other such frightening catastrophes. Time has shown that all such predictions were wrong. Pessimistic predictions attract interest, the media goes into overdrive and there is always a crowd ready to listen to dire apocalyptic predictions.

The education system in most Western countries no longer gives the young the ability to think, reason, criticise and analyse. It used to be based on committing a large body of knowledge to memory. Questioning is taboo yet essential for every human endeavour. No one is going to provide the education that's needed to overthrow a teacher's treasured ideology. The education system in the West is now the playground of idealogues.

We may be the first civilisation destroyed, not by the power of our enemies, but by the ideology and ignorance of our teachers and the nonsense that they are teaching our children. Young people now no longer want to hear the truth, engage in debate or read because they don't want their cherished illusions destroyed.

Some 500 years ago, the mainstream establishment said the Sun rotated around the Earth, 150 years ago the mainstream scientific bodies said that manned flight was impossible, 100 years ago the mainstream scientific opinion was that flights across the great oceans was impossible, 90 years ago the mainstream opinion was that space flight was impossible and 80 years ago the mainstream opinion was that the continents did not move. In all cases, the mainstream was wrong.

For centuries there have been scientific successes as well as blunders, mistakes, wishful thinking and fraud.[380] Examples are the steady state solar system, the calendar, the age of the Earth, Lamarckism, creationism, Piltdown man, perpetual motion, phlogiston, n-rays, aether, cold fusion, the four elements, oxygen, alchemy (which was supported by Newton), Lysenkoism, simian virus 40, rotten fish as the source of leprosy, phrenology, Rorschach ink blot, false memories and more recently peptic ulcers. All were "consensus" theories supported by the experts of the day and all were wrong.

In all segments of society there is fraud, lies, self-interest, narcissism, ignorance, deliberate omissions, exaggeration and narrowness such that one can't see the wood for the trees. Scientists are no exception and don't occupy the high ground in society. For hundreds of years, reputable scientists with an alternative opinion have been denigrated. Over time, this alternative opinion has often been shown to be correct.

The same is happening with the climate catastrophe cacophony comrades today who also ignore Popper's basic methodology of science.[381]

At times there is something in the water, doomsday predictions flow thick and fast and people panic in response. These times were especially in the mid 1500s, the 1970s and in the 2000s.

The end of the world

There is a history of at least 2,000 years of end-of-the-world predictions.[382,383] It's a boringly long list and only some predictions are listed here. If just one had come true, we would not be here. It's very hard to be 100% wrong, but those predicting the end of the world succeeded.

Most predictions, including those of climate zealots, have religious, moral, authoritarian, mathematical and scientific overtones. There are only three certainties in life: death, tax and failed apocalyptic predictions.

The New Testament tells us (Matthew 16:28) the world will end before the end of the last Apostle. It didn't.

In 992 AD, the scholar Bernard of Thüringen announced that the world had 32 years left. The world ended for Bernard. He died before the 32 years had elapsed.

The Last Judgement was due to take place 1,000 years after the birth of Christ. Many subsistence farmers in 999 AD didn't bother to plant crops because they were going to die anyway. They did die. From starvation.

John of Toledo publicised the end of the world would be at 4.15 pm on

23rd September 1179 AD. Clocks were not that accurate in the 12th Century but don't let that spoil a good scare story. The Byzantine Emperor of Constantinople walled up his windows and the Archbishop of Canterbury called for a day of atonement. These actions succeeded. The world did not end.

Catastrophic floods are a scare story welded into the human brain (e.g. The Epic of Gilgamesh, Noah's Flood).[384] A prediction of a global flood in 1523 AD exploited this fear and created panic. Some 20,000 Londoners left the city for higher ground as they preferred to perish outdoors in the hills rather than in the comfort of their own homes. There was no flood.

The 1523 AD prediction was revised to 1524 AD by the astrologer Nicolaus Peranzonus de Monte Sante Marie. Who could possibly doubt someone with such an impressive name? Mercury, Venus, Mars, Earth, Jupiter and Saturn would be aligned in Pisces. Neptune (unknown then), Uranus, Pluto (unknown then) and the Moon were not in Pisces. Of course, the astrological sign of a fish meant there would be a global flood.

Mathematician Georg Tannstetter argued that the world would not end but the hysteria was so great that no one listened to him. Familiar? There was frantic boat building in England and on the continent. In Germany, Count von Iggleheim built a three-storey ark into which he retreated on deluge day (20th February 1524 AD). An angry crowd gathered outside the Count's ark, it rained gently, hundreds were killed in a stampede and the Count was stoned to death. Records show 1524 AD was a drought year in Europe.

The delightfully named Frederik Nausea, Bishop of Vienna, predicted the end would come in 1532 AD because he had seen all sorts of strange things such as black bread falling from the sky. The world continued to do what the world has always done.

An even more precise mathematical calculation was made by Stifelius of Lochau who predicted the world would end at 8 am, 3rd October, 1533 AD. He was precisely wrong. The folk of Lochau gave Stifelius a good flogging for his scaremongering and ran him out of town. He was lucky his world did not end then and there.

In Strasbourg (France), the anabaptist Melchior Hoffman announced the world would be destroyed in a ball of flame in 1533 AD and only 144,000 people would live. The world didn't end and hundreds of millions of people stayed alive. Recalculations showed the world would end in 1542 AD. It didn't. Sounds like many of the predictions coming out of Strasbourg and Brussels today.

Pierre Turrell (Dijon, France) calculated the world would end in 1537, 1544, 1801 or 1814. It only had to end once and that was it. Turrell's predictions look like the climate computer model predictions of today which can't be checked until well after the death of the scare monger. It is now 2021 and I think it would be safe to say Turrell got it wrong.

Cyprian Leowitz calculated the world would end in 1584. He was so confident of his prediction that he published astronomical tables showing planetary movements up to the year 1614. It appears the world didn't end and the planets kept orbiting according to Leowitz's tables.

The sage Johann Müller called himself Regiomontanus. This was a far better name if his predictions were to be taken seriously. In 1588 he predicted the end of the world for well after his expected lifetime. It all went to plan. He kept himself fed and watered all his life by scaring people witless and then shuffled off. Sounds familiar?

Rabbi Sabbatai Zevi of Smyrna (now Izmir, Turkey) told those who would listen that the world would end in 1648, he was the new Messiah and the citizens of Smyrna should give up work and prepare for their return to Jerusalem. Zevi was arrested for sedition by the Sultan and while in prison converted to Islam. Maybe his world ended in prison.

In 1578, Helisaeus Roeslin of Alsace calculated the world would end with a solar eclipse on 12ᵗʰ August 1654. This was a pretty safe bet as the physician Roeslin would have been expected to be pushing up daisies 76 years after a prediction made in his adult years.

The eclipse occurred a day earlier because of a mathematical error and it appears that the world did not end. However, on the appointed day people stayed indoors and the churches were filled. There is no record on whether preferred places of death such as inns or brothels were packed.

It was business as usual in the 18ᵗʰ Century. Cardinal Nicolas de Cusa declared the world would end in 1704. Neither the Vatican nor the Heavenly Father agreed. They had far greater supernatural influence than the Cardinal because the world did not end.

The Bernoulli family of Switzerland produced eight outstanding mathematicians over three generations. Jacques Bernoulli predicted the world would end on 19ᵗʰ May 1719 from a comet impact. The world dodged a bullet and survived. Every family has its black sheep.

The Continent was not the only place for end-of-the-world predictions.

Englishman William Whiston predicted the end of the world for 13th October 1736. That day was a fizzer and the world is still here although some parts of England look like the end of the world.

The Swede, Emmanuel Swedenborg, is famous for his encyclopaedic religious concordance. The angels told him that the world would end in 1757. The angels were wrong.

English sect leader Joanna Southcott claimed that she was pregnant with the New Messiah and that the world would end in 1774. The world did not end which is probably why the New Messiah was never born. I suspect our Joanna was covering her bases after a romp in the hay.

Earthquakes in England on 8th February 1761 and 28 days later led William Bell to claim that on 5th April there would be a third earthquake. Many Londoners left town, mainly by boat. They came to their senses on 6th April and threw William Bell into Bedlam. Bell's predictions were akin to climate predictions and based on a straight line between two points.

Englishman John Turner, prophet and follower of Joanna Southcott, predicted that the end was on 14th October 1820. Nothing happened on what was just another boring day in cold, wet England.

William Miller was absolutely certain of the date the world would end and predicted it would be on 3rd April 1843, 7th July 1843, 21st March 1844 or 22nd October 1844. The end was to be preceded by a midnight cry in 1831. A spectacular meteor shower in 1833 gave Miller an *iota* of credibility. On each appointed date, Millerites would gather on hilltops awaiting the midnight cry. Miller's end of the world was 1849 when he died.

It appears that measurements of pyramids contain ancient knowledge about everything, including the end of the world. The Great Pyramid of Giza told us the end was 1881. I'm at a loss to understand how measurement of a pyramid can determine the end of the world. It was remeasured and the new date was computed as 1936. Further remeasurement gave the date as 1953. We breathlessly await new measurements to give us the next exact date. This is a bit like climate predictions telling us about a speculated climate catastrophe in 10, 20, 30, 50 or 100 years time.

A gentleman with the name of Richard Head republished the 1677 book *The Life and Death of Mother Shipton*. In the reprint, with forged rhymes attributed to Mother Shipton, it was predicted the end was 1881. It wasn't. Mr Richard Head was well named. His end came in 1886.

Failure is getting boringly repetitive so we'll skip the first part of the 20th Century. You get the drift. The conclusion is that we are still here because every end of the world prediction failed.

Despite the horrors of two world wars, John Ballou Newbrough predicted that 1947 was the year. The US and other governments were going to be crushed and Europe again would have another massive depopulation from war. It didn't.

The doctors who predicted in 1953 that Stephen Hawking had only two years to live would have died well before Hawking's death in 2018.

In 1956, models were used to predict that oil production would rapidly decline[385] from around 1965-1970.[386] We now have more oil resources than any time in history due to exploration, better technology, risk takers and entrepreneurs. Just remind me again, what predictions are currently being made based on models? Climate? Mass deaths from COVID-19?

Some creationist religious cults claimed that one day of God represent 1,000 years of man. God toiled creatively for six days and then took a day of rest. On that basis, the planet should take a permanent rest after 6,000 years. This was 1996. Some of us are still toiling and we're still waiting although we are well aware that the creationist's god makes some pretty simple errors.

Quatrain 1072 of Nostradamus tells us that July 1999 was D-Day. It wasn't. Even with such a cool name, no prediction of Nostradamus can be taken seriously.

The Millennium cults and those selling computer software had a field day on December 31st, 1999. Computers did not fail with the Y2K bug, aeroplanes did not fall out of the sky, life went on and some people made bucketloads from frightening folk about a date change.

When someone knocks on your door and tells you the end of the world is coming, sool the dog onto them. History is on your side.

Environmental catastrophes

There have been numerous late 20th Century and earlier predictions of population and environmental catastrophes[387] in the style of Thomas Malthus (1766–1834)[388], all of which have been spectacularly wrong because they omitted to consider advances in science and technology and human ingenuity.

In 1893 Thomas Huxley wrote about ethical evolution.[389] Fairfield Osborn can be credited with starting the modern environmental scare movement and reviving Malthusianism.[390] He was a eugenics advocate, wanted selective mating within humans to breed desirable hereditary traits to improve the species and a prominent Aryan enthusiast. The 20th Century green movement grew out of eugenics, Malthusian doomsday scenarios, Nazi movements and totalitarianism[391] and still retains many of these characteristics.

There was once a consensus about eugenics. The plan was to identify those that were feeble minded (and that variously included Jews, blacks and foreigners) and stop them from breeding by isolation in institutions or by sterilisation. George Bernard Shaw claimed that it was only eugenics that could save mankind.[392] H. G. Wells spoke against[393] "*ill-trained swarms of inferior citizens*" and Theodore Roosevelt wrote in a letter[394] that "*Society has no business to permit degenerates to reproduce their kind*". Other supporters of eugenics cited by the National Catholic Register were Francis Crick, Adolph Hitler, Winston Churchill, Helen Keller and Alexander Graham Bell.[395]

Eugenics research was funded by the Carnegie Foundation and the Rockefeller Foundation, both of whom now fund climate activism. With such eminent citizens and respectable foundations supporting eugenics, how could it not be correct consensus science? The same logic is used for climate "science". We now know eugenics was a racist, murderous, anti-immigration social program masquerading as science.

Paul Ehrlich[396] gained great recognition by attempting to scare us witless by suggesting that "*the battle to feed humanity is over*" and that "*hundreds of millions of people would starve*". Ehrlich was a follower of eugenics enthusiast William Vogt[397] and his solution to perceived population problems was to add chemicals to staple foods to create sterility.

Books by Paul Ehrlich derive from Malthusian ideas, the eugenics movement and communism (all of which underpin the modern green left environmental movement) and espouse the centralised control of every aspect of life. The greens still have an unhealthy obsession with death and killing people.

Some 12% of atoms in the human body are carbon.[398] The history of the green movement suggests that when greens talk about carbon reduction or net zero carbon, they are really talking about getting rid of you.

During World War II in the Pacific and south-east Asia, spraying DDT over swampy areas reduced loss of Allied troops to malaria thereby allowing more troops able to fight. Rachel Carson[399] was primarily responsible for the 1972 banning of DDT as a control on malaria on the grounds that it killed birds and caused a cancer epidemic. Both of these unsupported claims were later shown to be untrue. By 1970, DDT had helped wipe out malaria in 99 countries, including the US. The World Health Organisation reversed the ban on DDT in 2006.

Between 1972 and 2006, at least 50 million people died of malaria unnecessarily. Most of the deaths were in Third World countries and most of the deaths were children[400] and Offit quoted Michael Crichton[401] *"Banning DDT is one of the most disgraceful episodes in twentieth century America. We knew better, and we did it anyway, and we let people around the world die, and we didn't give a damn"*. This is the track record of the greens. Carson has blood on her hands and the order of magnitude of killing puts her on the same pedestal as Stalin, Mao Tse Tung and Pol Pot. Carson's book made her millions, she killed millions and she remains the poster girl of the greens.

Carson also claimed that acid rain was devastating German forests and this was repeated *ad nauseum* by the German Greens in the 1980s. For decades I have had a close association with Germany and very often have engaged in *spazieren gehen* in their forests. The German forests have been so devastated that they have expanded![402]

Al Gore wrote a *Washington Post* article telling the world that the ozone hole was making rabbits and salmon blind.[403] I'm not so sure why rabbits would want to look at salmon or salmon need to look at rabbits but neither animal went blind. In Australia, farmers would welcome a reduction in rabbits, an introduced species.

In 1967 it was predicted that there would be major global famine by 1975.[404] Then the global population was 3.48 billion, it is now 7.7 billion yet there has been no general famine and proportionally fewer people are malnourished. What went wrong? We used science, engineering, capitalism and human ingenuity to make the world a better place.

The Salt Lake Tribune of 17th November 1967 wrote that Paul Ehrlich had predicted a dire famine in 1975. Wrong. Each year since 1967, people have been eating more and more food and the calorie intake has increased. In some countries now, the food intake is so great that there is an obesity epidemic. Famine was once common. It is now rare. However, famine of reason is becoming very common.

In 1968, it was predicted that overpopulation will spread worldwide.[405] There has been a worldwide population increase, people are eating better, producing more food from smaller acreages and living longer. Some countries have experienced population increase whereas others such as Russia are experiencing a population decline.[406]

Paul Ehrlich was at the height of his predicting abilities when in 1969[407] he said *"The trouble with almost all environmental problems is that by the time we have enough evidence to convince people, you're dead"* and *"We must realise that unless we are extremely lucky, everybody will disappear in a cloud of blue steam in 20 years"*. Maybe I was not paying attention or perhaps I am colour blind but I saw no cloud of blue steam in 1989 and I'm still here. Maybe it was because I was at university in 1969 and not experimenting with psychedelic drugs.

In 1970, Peter Gunter from North Texas University stated[408] *"Demographers agree almost unanimously on the following grim timetable: by 1975 widespread famines will begin in India; these will spread by 1990 to include all of India, Pakistan, China and the Near East, and Africa. By the year 2000, or conceivably sooner, South and Central America will exist under famine conditions...By the year 2000, thirty years from now, the entire world, with the exception of Western Europe, North America, and Australia, will be in famine"*. Wrong. This was consensus science.

By 1980, wheat yields in India and Pakistan had doubled. Gunter did not even know what was happening next door. In 1940, Mexico was importing half the grain it needed to feed its people. By 1963, Mexico went from an importer of wheat to an exporter. One can be wrong, spectacularly wrong or environmentally wrong.

In 1970, *Life* magazine (January 1970) reported *"Scientists have solid experimental and theoretical evidence to support...the following predictions: In a decade, urban dwellers will have to wear gas masks to survive air pollution...by 1985 air pollution will have reduced the amount of sunlight reaching the earth by half..."*. Really!

If you are looking for informed opinion, then *Life* magazine is not for you. The only places where gas masks might have assisted breathing in 1985 were in the urban industrial areas of communist countries. They still are. I know. I went there many times over the last 50 years.

Paul Ehrlich was excitedly preaching death again by predicting in 1970[409] that "...*air pollution ...is certainly going to take hundreds of thousands of lives in the next few years alone*". Ehrlich fantasised about 200,000 Americans dying during "*smog disasters*" in New York and Los Angeles. He was wrong. Again. Capitalism and economic growth greatly reduced air pollution.

In 1970, Ehrlich warned that DDT and chlorinated hydrocarbons[410] "... *may have substantially reduced the life expectancy of people born since 1945*". He also warned that Americans born after 1946 would have a life expectancy of only 49 years and predicted that if current patterns continued this expectancy would reach 42 years by 1980 when it might level out. The exact opposite happened, life expectancy reached over 80 and now junk food gluttony, drugs and lack of exercise not carbon dioxide reducing American's longevity

In 1970, Washington University biologist Barry Commoner wrote[411] "*We are in an environmental crisis which threatens the survival of this nation, and of the world as a suitable place of habitation*". Change the date and this is no different from what we hear today about climate. The US survived as a nation and the world is a far better place than 50 years ago.

The day after Earth Day in 1970, the *New York Times* tried to frighten us by writing "*Man must stop pollution and conserve his resources, not merely to enhance existence but to save the race from intolerable deterioration and possible extinction.*" This is a fail. These are the same words used by the death cult Extinction Rebellion freaks today.

Paul Ehrlich wrote in 1970[412] "*Population will inevitably and completely outstrip whatever small increases in food supplies we make*" and "*The death rate will increase until at least 100-200 million people per year will be starving to death during the next ten years*". Wrong. There is currently a PhD research topic on why environmentalists are so obsessed with death, doom and gloom and get great glee from terrifying children.

Ehrlich's death cult blackness appears again[413] "*Most of the people who are going to die in the greatest cataclysm in the history of man have already been born.*" and "*By 1975, some experts feel that food shortages will have escalated the present world hunger and starvation into famines of unbelievable proportions. Other experts, more optimistic, think the ultimate food-population collision will not occur until the decade of the 1980s*". Wrong. I get the feeling that Ehrlich is disappointed that hundreds of millions of people didn't die as he missed his opportunity to say *"I told you so"*.

On Earth Day 1970, Ehrlich gave his view of the future assuring readers of *The Progressive* that between 1980 and 1989, some four billion people, including 65 million Americans, would perish in the *"Great Die-Off"*. They didn't but Ehrlich's obsession with mass death continued. Why didn't he join a death cult, go to Jonestown a few years later in 1976 and lead by example?

Death cult Paul Ehrlich was at it again in 1970, the peak year of his failed predictions. This time the US would have water rationing by 1974, the oceans were going to be dead by 1980 and there would be food rationing by 1980.[414] How can such people hold down a job at an American university? Both water and food are still abundant in the US and shortages are due to human stupidity. There are times over history of collective madness. We live in one of these times and 1970 was another.

In 1970, ecologist Kenneth Watt predicted that the world would use up its crude oil by 2000.[415] He speculated that we would drive up to the bowser and say *"Fill 'er up buddy"* and be told *"I am very sorry, there isn't any"*. This was hyperventilating scare mongering. More crude oil has been found from fracking and exploration continues to find new deposits.

In 1970, the Earth Day organiser Denis Hayes set the scene and claimed[416] *"It is already too late to avoid mass starvation"*. Mass starvation was avoided by human ingenuity and not by moaning from the couch.

In 1970, ecologist Kenneth Watt predicted the end was nigh and that[417] *"We have five more years at the outside to do something"*. This sounds very much like the present scary predictions about climate, especially just before Conference of Parties (COP) climate meetings.

Also in 1970, Harvard biologist and Nobel Laureate George Wald wrote[418] *"Civilisation will end within 15 or 20 years unless immediate action is taken against problems facing mankind"*. The problems were unspecified, no action was taken and civilisation did not end. Sometimes it's wise to thoughtfully do nothing.

In 1970, Harrison Brown of the National Academy of Sciences claimed that humanity would totally run out of copper shortly after 2000 and lead, zinc, tin, gold and silver would be exhausted before 1990.[419] Exploration is becoming deeper and more expensive, but new metal deposits continue to be found and there are numerous unmined deposits with lower metal contents awaiting price rises.

In 1970, Sidney Dillon Ripley of the Smithsonian Institution argued that in 25 years, somewhere between 75 and 80 percent of all the species of living animals will be extinct.[420] It's now 50 years since Dillon's scare campaign and the species count keeps growing.

It was a bad year in 1970 for predictions by the unbalanced apocalyptics. It was predicted that nitrogen buildup would make all land unusable.[421] Because of better farming, manufactured fertilisers, genetic engineering and a higher atmospheric carbon dioxide content, a population double that of 1970 can now be fed. Farmed acreage is decreasing and forest areas are increasing.

Barry Commoner predicted in 1970 that pollution would kill all of America's freshwater fish.[422] In 2021, we still have fish, barramundi is on the menu tonight and nothing beats Shane Warne's travelling diet in India of sardines and baked beans.

Fish have survived asteroidal impacts that produced major mass extinctions, have survived for hundreds of millions of years and they are not likely to shuffle off because of an ill-informed prediction by an unbalanced wretch. The irresponsible uncritical media sensationalise such predictions of disaster. Nothing has changed.

In 1970, it was predicted that killer bees (Africanised honey bees) would wipe us out.[423] They didn't. We still have bees, still eat honey and still get stung by bees be they wild, African or Italian.

The First Earth Day year was 1970. Most of the original predictions have been removed from the records thereby allowing the same old catastrophists to continue their trade and, without diligent journalists fact checking, it's doomsday business as usual.

Among the top global-cooling theorists were President Obama's "science czar" John Holdren. In the 1971 textbook *Global Ecology* by John Holdren and Paul Ehrlich, the duo warned that overpopulation and pollution would produce a new ice age, claiming that human activities are *"said to be responsible for the present world cooling trend"*. The pair claimed that *"jet exhausts"* and *"man-made changes in the reflectivity of the earth's surface through urbanization, deforestation, and the enlargement of deserts"* were potential triggers for the new ice age. They claimed that the man-made cooling might produce an *"outward slumping in the Antarctic ice cap"* and *"generate a tidal wave of proportions unprecedented in recorded history"*.

Holdren predicted that a billion people would die in *"carbon-dioxide induced famines"* as part of a new ice age by the year 2020. Later John Holdren was advising President Obama when he was in office about the dangers of global warming. In contradiction of the IPCC 2001 report which stated that global warming would bring *"Warmer winters and fewer cold spells"*, Holdren stated *"A growing body of evidence suggests that the kind of extreme cold being experienced by much of the United States as we speak is a pattern we can expect to see with increasing frequency, as global warming continues"*. Using that thinking, I guess even in-grown toenails are due to global warming.

The Club of Rome, a global doomsday green left think-tank, commissioned a book[424] which predicted that the world would run out of various non-renewable resources in the 1980s and 1990s and that environmental economic and societal collapse would follow. The computer predictions were based on exponential economic and population growth using only five variables.

The world didn't run out of resources but may have run out of common sense. Nothing has changed with computer predictions. The Club of Rome was wrong. Some 30 years later, the authors produced an updated version.[425] Surprise, surprise, the updated doomsday predictions have also been shown to be wrong.

In 1972, it was predicted that oil would be depleted in 20 years.[426] This prediction was based on proven reserves and did not account for upgrading of oil fields with more drilling. The exact opposite has happened.

In 1974, it was predicted that ozone depletion high in the atmosphere was a great peril to life.[427] This was a total fizzer; huge funds were expended on this exaggerated risk, carpet baggers got rich and life goes on.

A 1974 book[428] *The Jupiter Effect* predicted that when the planets were aligned the world would end with massive earthquakes. There have been thousands of previous planetary alignments without the world ending and this was no different. However, there is a sneaking suspicion among some geologists that planetary alignments may influence volcanicity, earthquakes, earth tides and even climate.

In 1975, Paul Ehrlich was suffering an attention deficit disorder after his 1970 star billing and predicted that[429] *"since more than nine-tenths of the original tropical rainforests will be removed in most areas within the next 30 years or so, it is expected that half of the organisms in these areas will vanish with it"*. Sounds scary. Who do I contact to give money? The prediction failed. Again, total bunkum from Ehrlich. If Ehrlich bets on black, then bet on red. You'll win.

In 1977, the US Department of Energy predicted that oil production would peak in the 1990s.[430] Beware of the bluster of bureaucrats. We've heard all this before. We did not have peak oil in 1990 and we now have more oil resources and production than ever before and the US is now no longer dependent upon oil from the Middle East.

In 1980 after 10 years of making dire predictions to a swooning unquestioning media, Paul Ehrlich was challenged to put his money where his mouth is by Yale business professor Julian Simon.[431] Ehrlich stated *"If I were a gambler, I would take even money that England would not exist in the year 2000"*. Simon offered to take that bet. Ehrlich refused. England still exists and very occasionally wins international sporting matches of games it invented.

Eventually the two agreed to bet on resource scarcity between 1980 and 1990 and had a $10,000 wager on the inflation-adjusted prices. Ehrlich chose copper, chromium, nickel, tin and tungsten. Prices fell, Ehrlich lost the bet and paid up. Ehrlich's prediction ignored variables and did not understand the beauty of human ingenuity, mineral exploration and markets.

In the 1980s, we had many choices of dates.[432] For an extra-terrestrial end of the world in the 1980s, astrologer Jeanne Dixon predicted a comet would destroy the Earth. It didn't. It may have destroyed her rationality.

When Saturn and Jupiter were almost in conjunction in Libra on 31st December 1980, the world was to end.[433] Such conjunctions are very common. I don't remember what happened that day leading up to midnight so maybe the world did end.

In 1980, it was predicted yet again that peak oil will be reached in 2000.[434] We now have more oil than any time in history due to exploration, better technology, capitalism, risk takers and entrepreneurs.

In 1980, it was also claimed that acid rain wiped out fish in 107 of New York States lakes in the Adirondack Mountains.[435] Ten years later, a US government research program on acid rain concluded that acid rain was no environmental risk[436] and fish in the New York State lakes continued to do what fish do.

On June 30, 1989, the *Associated Press* ran an article headlined *"UN Official Predicts Disaster, Says Greenhouse Effect Could Wipe Some Nations Off Map"*. In the piece, the director of the UN Environment Programme's (UNEP) New York office was quoted as claiming that *"entire nations could be wiped off the face of the earth by rising sea levels if global warming is not reversed by the year 2000"*.

He also predicted *"coastal flooding and crop failures"* that *"would create an exodus of 'eco-refugees,' threatening political chaos"*. Of course, 2000 came and went, and none of those disasters actually happened.

In 2005 the UNEP were at it again. They warned that imminent sea-level rises, increased hurricanes, and desertification caused by global warming would lead to massive population disruptions. If they were correct, migration would be slow because sea level rise is not instantaneous. In a handy map, the organisation highlighted areas that were supposed to be producing the most *"climate refugees."* Especially at risk were regions such as the Caribbean and low-lying Pacific islands, along with coastal areas.

The 2005 UNEP predictions claimed that, by 2010, some 50 million *"climate refugees"* would be fleeing those areas. However, not only did the areas in question fail to produce a single *"climate refugee"* by 2010, population levels for those regions were still soaring and most Pacific island nations are growing in land area. I am at a loss to understand why hard-working taxpayers fund the UN Environmental Programme.

Oh no, not peak oil again. In 1996, it was predicted that we would reach peak oil in 2020.[437] Why are green activists so obsessed with peak oil when they are against the use of fossil fuels? I suspect they wishfully hope the world will run out of oil so people can suffer because more than 5,000 products in common use derive from petroleum.

Poor delusional George Monbiot predicted in 2002 that famine can only be avoided in 10 years time if the rich give up eating meat, fish and dairy.[438] He stated that vegans were right all along. A decade after the deadline, people are eating more and more meat, fish and dairy, enjoying better health and living longer. The only conversation one can have with a vegan is about veganism. Why has *The Guardian* employed such a prat for two decades?

Oh no, not peak oil predictions again. Not one has been correct. In 2002, it was yet again predicted that we would reach peak oil by 2010.[439] Why should yet another prediction be correct, even if it is made from someone in a prestigious institute? We now have more oil than any time in history.

The media loves scary reports, especially health and environmental ones. For every valid scare, there are hundreds of false ones. In 1981, it was claimed that coffee causes 50% of pancreatic cancers. The scientists who made this claim retracted it in 1986. The International Agency for Research on Cancer took until 2016 to reverse its claim that coffee is a possible cause of cancer. It will take far longer for this scare to leave the mind of the public.

The precautionary principle is a loophole to avoid rigorous assessment of the costs and benefits. Mandated application of the precautionary principle in the specific case of climate change is just an integral component of the whole fraudulent climate marketing when all scientific arguments have been shown to be false and climate change becomes a legal exercise. Coffee drinkers of the world unite. Time for me to zip off and have my third cup of the day. Or is it my seventh? Who cares?

Nina Fascione, Vice President for Field Programs at Defenders of Wildlife stated in 2003[440] *"Frankly, it looks like we're on a crash course towards massive species extinctions in the next 20 years...We could lose one-fifth or 20% of our species within the next two decades. That's a very short amount of time"*. I'm drinking my wine cellar dry as quickly as possible because I have little more than a year before I'm gone.

I hereby declare that the end of the world is cancelled. You heard it here first while you were enjoying a coffee.

Population

Dave Forman, founder of Earth First, stated[441] *"My three main goals would be to reduce human population to about 100 million worldwide, destroy the industrial infrastructure and see wilderness, with its full complement of species, returning throughout the world"*. Eugenics is alive and well and forms the foundation of the green movement. If Dave Forman wants to go back to the lives we humans had thousands of years ago he should lead by example.

Why doesn't Dave, born 1946, shuffle off first in a blaze of narcissistic glory as he is a good allegedly honest socialist. He could get preferred front row seating in his Nirvana ahead of all of us. I prefer to enjoy the triumphs of man's ingenuity, hard work, risk taking and entrepreneurship that has given me longevity, health, democracy and freedom in our industrial capitalist society.

The UN is guilty of almost every …ism invented. Try this from the UN Population Division Policy in 2009[442] *"What would it take to accelerate fertility decline in the least developed countries?"* Whoever wrote this 63-page report does not know the difference between fertility and birth rate. If you are living in poverty in Central America or Africa, the UN is not coming to feed, clothe, house and water you. It's coming to get you and your children. The UN is promoting a children's book called *Pachamama* about an Andean goddess to whom children were sacrificed. Because of a low birth rate, Australia's population growth is due to immigration and in a century it will be a foreign country for native born Australians. The same applies to the US.

John P. Holdren, senior scientific advisor to Barack Obama opined[443] *"Indeed, it has been concluded that compulsory population-control laws, even laws requiring compulsory abortion, could be sustained under the existing Constitution if the population crisis became sufficiently severe to endanger society"*. The Obama administration did its bit by killing people in foreign lands and was being advised to turn on its own people. No thanks. Why didn't Holdren lead by example and abort his two children thereby avoiding having five grandchildren?

Like other Soviet leaders, Mikhail Gorbachev did a pretty good job of reducing the longevity and population of the Soviet Union and stated[444] *"We must speak more clearly about sexuality, contraception, about abortion, about values that control population, because the ecological crisis, in short, is the population crisis. Cut the population by 90% and there aren't enough people left to do a great deal of ecological damage"*. Some of us went to the Soviet Union and the ecological damage inflicted by the communists led the world. The Russians have had some pretty cool ways of bumping people off. What novel method of killing was Gorby thinking of?

Ted Turner, CNN founder, stated in 2016[445] *"A total population of 250-300 million people, a 95% decline from present levels, would be ideal"*. What is Turner's favoured method of genocide? Will he put up his hand to shuffle off first?

Chase Manhattan's David Rockefeller stated[446] "*The negative impact of population growth on all our planetary ecosystems is becoming appallingly evident*". What did the Rockefeller family or Chase Manhattan do about it? David Rockefeller did a little bit to help by departing this mortal coil in 2017.

David Wallace-Wells frightened me witless with an article entitled "*The uninhabitable Earth*" in a New York magazine[447] and claimed that the apocalypse would come very quickly, that there would be rolling death smogs and "*at 11 or 12 degrees of warming, more than half the world's population, as distributed today, would die of direct heat. Things almost certainly won't get that hot this century, though models of unabated emissions do bring us that far eventually*". I was frightened that such nonsense abounds. In order to get such global heating, humidity would have to decrease to almost zero. This, like all the other apocalyptic zombie predictions above, is bilge.

The Georgia Guidestones in the US apparently give wisdom about the future.[448] We are advised to "*Maintain humanity under 500,000,000 in perpetual balance with nature*". Thanks for the advice. Humans are part of nature and there is no guidance about how to have perpetual balance in a constantly changing dynamic system. If we are to have perpetual balance with nature, how many bacteria should we have on Earth? Better advice is given on beer coasters.

Explorer, conservationist and film maker Jacques Cousteau stated[449] "*In order to stabilise world population, we must eliminate 350,000 people per day*". Why didn't someone cut off his air supply when he was diving? There would have been only 349,999 left to murder that day. If the other six in his immediate family were in the queue for that day, Cousteau's genocide program would have only to score another 349,993 killings that day. Like so many green activists, Cousteau was obsessed with mass death of humans yet does not lead by example.

His Royal Highness the late Duke of Edinburgh stated[450] "*If I were reincarnated I would wish to be returned to earth as a killer virus to lower human population levels*". He was certainly no greenie.[451] The past shows us that population crashes are common, especially when there is a viral or bacterial attack on populations weakened by global cooling.[452]

We are all environmentalists. No one wants to see the atmosphere, soils and waters polluted. We can all do something about pollution and the wealthy

Western world has cleaned itself up over the last 50 years. This is because of wealth and culture. A few decades ago, some of the Mediterranean countries were cesspits of garbage on the roadside, vacant land blocks and forests (e.g. Spain, Italy, Greece). They are more environmentally aware today and are still changing the throw away littering culture.

Most developing countries have a culture of throwaway littering and are polluted. They will need to be wealthier and have a change in culture before they are environmentally comparable with the West.

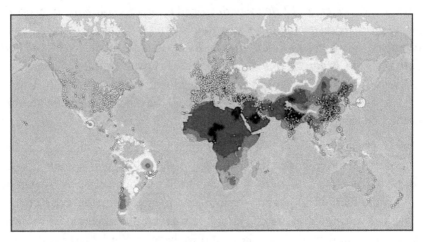

Figure 17. *WHO pollution map of the world*[453] *showing the developed world is far less polluted than the developing world. The darker the shade, the more pollution.*

Predictions and claims by apocalyptic activist greens can be tested with evidence and time. This is what has happened in the West over the last few decades and has shown that the greens never stop crying wolf. Greens are full of negativity and doom and gloom. Put it to the test. Try to get a green activist to be positive or laugh at themselves.

Meanwhile, the mountain of plastics in the oceans continues to grow.

Starvation

In 1800 AD, although there were only one billion humans on Earth, global life expectancy was about 25 years. The world's population had doubled by 1927 and people could expect to live twice as long. At present, there are more than 7.7 billion people on Earth and average longevity is 69 years. Life expectancy in the poorest nations now is far better than in the richest nations 200 years ago.[454] The average life expectancy of a child born today in the Western world is more than twice as long as in my grandparents' day, a little more than century ago.[455]

The UN's Food and Agriculture Organisation (FAO) concluded that crop yields will increase under a wide range of climate scenarios. Humans now produce food for 10 billion people, a 25 percent surplus. Food production will depend upon access to hydrocarbon-driven tractors, trucks, processing plants and transport to markets. Global wheat production was expected to reach a new record of 780 million tonnes in 2021 according to a forecast issued on 4th March 2021 by the FAO.[456]

In June 2020, the International Grains Council said that the world soybean crop for 2020-2021 will be a record 364 million tonnes, up 8 percent from the previous year. Production has more than doubled over the last 20 years. Corn production in 2020-2021 is also forecast to increase. The FAO forecast a 4.4 percent increase in corn production, a 2.6 percent increase in cereal production, a 5 percent increase in maize production and a 1.6 percent increase in rice production.

Despite a 2 percent shrinking of the economy in South Africa, compared to 2019, agricultural production rose by 28 percent with grain production up by 38 percent and citrus production up by 13 percent. If the alleged global warming and increased atmospheric carbon dioxide are to blame, then the catastrophically warm future planet is not really a bad thing after all. We were told by activist greens that the climate change would result in crop failures and starvation. The opposite has happened.

Due to better medicine, health care, nutrition and increasing wealth, infant mortalities have fallen. In the late 19th Century, mortality globally of children under five was 40 percent. It is now 6 percent in the Third World and falling. Mortality rates in Western countries are extremely low. The decreased mortality rates have played a major role in population increase and longevity. So too has food availability.[457] For example, in 1906 life expectancy of a Bangladeshi was 25 years. The average Bangladeshi child born today can be expected to live to 75 years old.[458] This has been achieved by doing the exact opposite of what the current Pope has advocated yet he promoted the irrational fear that we are on the eve of destruction.

The current Pope wrote *"Doomsday predictions can no longer be met with irony or disdain[459]"* and *"There are regions now at high risk and, aside from all doomsday predictions, the present world system is unsustainable from a number of points of view…[460]"* The Pope should stick to theology.

The current Pope also wants us to go back to simple agrarian times with small community plots of land[461] using no manufactured fertilisers[462] and to avoid genetically modified crops and cotton.[463] The Papal Encyclical on the environment is suggesting that we go back to the times when curable fatal diseases were rampant. This can only lead to mass social disruption, starvation, depopulation and the pushing of billions of people into poverty. We have been there before and it is not a pretty sight. Starvation still occurs because of wars, drought, political upheavals, incompetence, unreliable transport systems and despotism. The Pope is fallible.

Crop yields per hectare and food consumption *per capita* have been increasing for a century and have never been higher, despite the population increase. Enough food is now produced for every person on Earth to consume 3,500 calories per day and there is no need now for anyone to starve[464]. Furthermore, this food is produced from less land than decades ago due to manufactured fertilisers, herbicides, insecticides, a slight increase in atmospheric plant food and genetically modified crops. This has led to an increase in forest lands on planet Earth.[465]

In 2013, the EU imposed a ban on using neonicotinoid insecticides because the honey-bee population was declining. The decision was based on faulty science, the honey-bee population was not declining, there were an additional 900,000 hives in Europe and wild bees who are most likely to come in contact with neonicotinoids were thriving. As a result of the ban, oilseed rape (canola) production decreased by up to 20%. This was not the precautionary principle in operation. It was hysteria based on flawed science.[466]

Green left activists have successfully prevented the farming of golden rice, genetically modified rice with vitamin A precursor traits introduced from maize. There are 200,000 to 700,000 children who die prematurely each year because of vitamin A deficiency. One study in 2008 calculated that in India alone 1.4 million person-years of healthy life had been lost for every year the crop has been delayed. This rice was specifically designed to help 250 million children, predominantly in Asia, who subsist on rice and suffer from vitamin A deficiency.[467]

Every step to use genetically-modified (GM) golden rice to save the lives of poor malnourished children lives has been opposed by Greenpeace for years. Why does Greenpeace knowingly kill people? Green groups claim that science supports their position on climate change but have a hostility to the science that saves lives. Green murder.

In Bangladesh, farmers risk their own health and spray egg plants with insecticide up to 140 times a season because the insect-resistant GM variety of the plant is fiercely opposed by greens. After 20 years and billions of meals, there is no evidence that GM foods damage human health. Only the opposite has been seen in this 20-year period. GM foods saved lives and had environmental benefits.

The green movement has repeatedly denied people access to safer and cheaper technologies and forced them to rely on dirtier, riskier or more harmful technologies. This results in death. The movement exploits human's fears of the unknown such as the exaggerated dangers of the unseen like GM crops and climate change. These claims have been shown to be wrong.

Rather than having the green's death cult view of the global population, there is a positive solution. Time and time again has been shown that the wealthier people become and the greater access to technology, the fewer children they have.

A good start to making people wealthier would be avoid wasting trillions on trying to prevent climate change, a natural process, and maybe adapt if and when the time comes.

By every measure, life is better now than it has ever been.[468] Life expectancy is higher, infant mortality is lower, daily calorific intake is higher[469] and more people have access to education, clean water, electricity, housing and travel.[470] Research has shown that a more animal-intensive diet, especially fish, has decreased infant mortality.[471] Probably the best measurement of the current good times is life expectancy. Compared to the Third World, the Western world is very well off and now suffers the curse of affluence.

Not only have longevity, infant survival and health increased, the pocket has become heavier.[472] The rate of *per capita* income has outstripped population growth. Over the past two centuries population has increased by a factor of seven whereas *per capita* annual income has increased from $100 to $9,000.

Although there are currently a few short-term exceptions due to socialism (e.g. Venezuela), economies are expanding, productivity is increasing, poverty is decreasing and pollution is declining. The human race has never been so well off and we are blessed to live on Earth today rather than hundreds or thousands of years ago.

In an amusing capture of the language, those that call themselves progressives want to push us back to the times when we died like flies from starvation and disease.

The evidence is there. Green policies are regressive and kill humans.

Brass monkeys

In 1970, it was predicted that there would be an ice age before 2000.[473] There wasn't. We still have polar ice sheets and mountain glaciers and the changing world is still doing what it always has without the help of humans.

James P. Lodge Jr claimed in 1970[474] that in the first third of the 21st Century, air pollution would trigger a new ice age and that increased electricity generation would boil dry the entire flow of rivers and streams in the US. I know that the 1960s and 1970s were times of drug experimentation and flower power but something boiled people's brains dry such that they made scary predictions. James P. Lodge Jr died in 2001 when the 21st Century had barely begun, the first quarter of the 21st Century is almost over and we are still OK. There's no ice age and rivers and streams still have cool running water.

In 1970, ecologist Kenneth Watt was probably trying to outdo Paul Ehrlich with failed predictions. He stated[475] *"The world has been chilling for about twenty years. If present trends continue, the world will be about four degrees colder in the year 2000. This is about what it would take to put us into an ice age"*. We had no ice age in 2000, the world had been chilling for a while and Watt ignored the 100 years of literature on climate cycles. Change a couple of words and this prediction is no different from that we hear today from global warming green activists.

In a letter to the US President on 3rd December 1970, Kulka[476] and Matthews of Brown University told the President that a group of 42 experts from the US and Europe predicted a great ice age was coming because *"global deterioration of climate was an order of magnitude larger than any experienced by civilized mankind, and indeed a very real possibility and indeed may be due soon"*. Make up your mind. Do I pack my suitcase as a climate migrant for warming or for cooling?

Rasool of NASA[477] claimed that the *"fine dust man puts into the atmosphere by fossil-fuel burning could screen out so much sunlight that the average temperature could drop by six degrees"* and *"If sustained over 'several years'* – *'five to ten,' he estimated* – *such a temperature decrease could be sufficient to trigger an ice age".* Burning coal was going to create an ice age in the 1970s or 1980s and now we are being told it will create global warming. There was no ice age in 2020 due to dust from coal burning.

In 1971, it was predicted[478] that there will be a new ice age by 2020 or 2030. Last time I looked out the window, there was no ice. There is some evidence of cooling with the onset of a Grand Solar Minimum.[479]

In 1972, it was predicted that there will be a new ice age by 2070.[480] This is a great prediction because the alarmist activists would all be dead by 2070. Does this sound familiar? There are those who now predict that we will fry-and-die in 50- or 100 years time well after they'll be dead.

In 1974, it was predicted that satellites showed that an ice age was coming fast.[481] A new ice age has not come. I am a patient man but after 47 years my patience has run out.

The CIA got into the act and argued that global cooling would cause conflict and terrorism.[482] The Center for Strategic and International Studies supported this view and stated *"Since the dawn of civilization, warmer eras have meant fewer wars.*[483] *"*

The left-wing *Guardian* breathlessly provided balance with an uncritical half-truth half-lies story about the work of climatologists who predicted a new ice age was coming.[484] It must have come and gone very quickly because we never felt it.

In an infamous prediction by *Time*[485], we were warned that there was another ice age coming to get us. *"Telltale signs are everywhere – from the unexpected persistence and thickness of pack ice in the waters surrounding Iceland to the southward-migration of a warmth-loving creature like the Armadillo from the Midwest"*. We were all going to freeze and die because human-produced aerosols would block sunlight and heat reaching the Earth's surface. They were wrong.

The same prominent figures in the human-induced global warming industry were just as prominent in the human-induced global cooling industry of the 1970s. They were also wrong. We now get scientific predictions from eminently scientifically unqualified souls like Ross Garnaut telling us that the science is beyond reasonable doubt and that we are going to fry-and-die. How would he be able to evaluate the science presented to him by those with a self-interest?

The only thing beyond reasonable doubt is that these predictions, like all catastrophic predictions in the past, will be wrong. We fragile humans seem to learn nothing from the repeated failures of the prophets of doom.

In 1976, there was a scientific consensus that predicted the planet was cooling and there would be famine. The planet did not cool and there was no climate-induced famine. So much for consensus. Many of those who were promoting global cooling are now promoting global warming. There is now a scientific consensus that the planet is warming and that there will be a climate-induced famine. Make up your mind.

Stephen Schneider claimed that the planet was cooling and the present world food reserves are an insufficient hedge against future famine.[486] It appears that in 1978 there was no end in sight to the 30-year cooling trend according to *The New York Times*. A few years before, Schneider was telling us that the planet was both cooling and warming.[487]

Just a little more than a decade later, it was the same Stephen Schneider who was predicting that we'd all fry-and-die. He did not predict that there would be more than a decade when there was neither significant cooling nor warming.[488] As a failed futurologist, he could have a second career selling clapped out used cars.

It was clear that there was an ice age coming. There was a 'consensus' and *The Washington Post, The Guardian* and *Time* were all running stories about an icy end. In 1979, articles were starting to appear about the Arctic meltdown and global warming due to carbon dioxide emissions. The *Chicago Tribune* was still reporting an ice age scare in 1981 when other media networks were predicting we would fry-and-die.

In 2004, it was predicted that Britain will be like Siberia in 2024.[489] We'll just have to wait this prediction out but, considering that not one environmental prediction has come true, I think we know the result already.

UK newspapers occupy a specially reserved place in the gutter. In a Sunday edition, *The Observer* screamed[490] "*A secret report, suppressed by US defence chiefs and obtained by The Observer, warns that major European cities will be sunk beneath rising seas as Britain is plunged into a 'Siberian' climate by 2020. Nuclear conflict, mega-droughts, famine and widespread rioting will erupt across the world*". This great exclusive scare story was obviously composed by a journalist during a long session in a bar trying to fill the Sunday paper with junk journalism. It has all the elements of a great Sunday paper story: US defence, conspiracy, death, war and climate change. We did see communists rioting in the UK in 2020 but this was nothing to do with a Siberian climate. If a sea level rise were to inundate European cities, then one would have expected the climate to be tropical rather than Siberian.

Humans have adapted to live on ice, in mountains, in the desert, in the tropics and at sea level, in space and will adapt to future changes. History shows that during interglacials, humans create wealth which allows populations to grow whereas glaciation is associated with famine, starvation, disease and depopulation.[491] The cycles of climate change suggest that the next inevitable glaciation will be little different from previous glaciations or Little Ice Ages. No COP agreements or government mandates can change the rotation of the Earth and the energy output of the Sun.

Imagine ice sheets covering the same area as in the last glaciation. Most of Europe, Canada and northern USA would be covered by ice. Sea ice would reduce international trade. Other areas such as China, Mongolia and Australia would have howling cold winds and shifting inland desert sands. Alpine areas would be covered with ice and ice sheets would greatly expand. The food supplies for nearly eight billion people would be really under pressure.

The best we humans can do is prepare for change (which we have never done in the past) and adapt (which we have done in the past). Because of modern technology, any adaptation to modern climate change would be far easier this time.

You can make up your own mind. You don't buy a used car from a shifty unctuous salesman.

Fry-and-die

There are forecasts by scientists and scientific forecasts. Green and Armstrong[492,493] state that there are 140 forecasting principles. From the IPCC reports, they were able to evaluate 89 of these and 72 violated the standard forecasting principles. They conclude that the forecasts were not the outcome of scientific procedures and they were opinions of scientists disguised by mathematics and obscured by complex writing. Green and Armstrong were too polite to state that the IPCC is a propaganda organisation wherein many of the scientific principals have been shown to be fraudulent.

Climate zealots warn us of a future catastrophe and that we must pay penance and change our ways. They use a narrow body of science and some mathematics and the message is given with religious vigour. There is no reasoned argument presented, hence reason cannot be used to evaluate contrary data and to challenge conclusions. They argue that our planet is overpopulated but don't volunteer to shuffle off or offer a solution besides genocide of other people. The method of genocide is not stated.

Politicians keep power by keeping the population frightened. They fear debate, logic and knowledge as it gives power to the people who will show that fears are unfounded. There is no public debate on the hypothesis that human activity causes global warming because it is easily shown that this hypothesis has poor foundations.

Attempts to restrict free speech and calls for censorship of alternative views are made by climate zealots. Such actions have characterised authoritarian salvationist cults down through the ages.

During the 1970s when the world was in a frenzy about a new ice age, it was reported[494] "*Arctic specialist Bernt Balchen says a general warming trend over the North Pole is melting the polar ice cap and may produce and ice-free Arctic Ocean by the year 2000*". This prediction was wrong, as were those by other scientists in the 1970s predicting a new ice age by 2020.

In 1988, it was predicted that regional droughts would occur in the 1990s.[495] They didn't. The cycles of drought and flood continue as they did in the past.

In 1988, it was predicted that temperatures in Washington DC would hit record highs.[496] They didn't although there was an increased amount of hot air emanating from Washington. There was no mention of the inconvenient urban heat island effect where concrete, roads, buildings, air conditioning and motor vehicles add heat to cities. In business, such predictions could land you in court for misleading and deceptive conduct.

The National Aeronautics and Space Administration's (NASA) James Hansen predicted[497] that 1988 was going to be the hottest year ever and there would be massive droughts. Not so. Surely Hansen knew that the 1930s were very hot with a decade of dustbowls. Maybe Hansen was too busy cooking the books rather than reading American literature classics that described these times.[498]

The UN was at it again spending our money trying to frighten us to pay more. The UN's Neil Brown, director of the UN Environment Program, claimed that rising seas could obliterate island nations if the global warming trend was not reversed by 2000[499] "*Coastal flooding and crop failures would create an exodus of "eco-refugees" threatening political chaos ...".*

The year 2000 passed as just another boring year with no climate emergency, eco-refugees or coastal inundation. We should tie our contributions to the UN on the success of their predictions. Now that's a key performance indicator!

It was predicted[500] in 1988 that the Maldives would be under water by 2018. They are not. Some of the Maldives cabinet had an underwater meeting as a publicity stunt to sensationalise their demand for other people's money to save them from inundation. Unlike most cabinet meetings, it was short.

In fact, massive new foreshore developments, hotels, luxury resorts and new airports were being built there in 2018-2021 giving tangible proof that idealists are out of step with investors who normally do a comprehensive due diligence before investing.[501] Climate activist "scientists" lose nothing if they are wrong. Investors lose their shirt.

Hyperventilation came from *The Canberra Times*[502] servicing the isolated bubble that is The Australian Capital Territory when it claimed *"sea level is threatening to completely cover this Indian Ocean nation [Maldives] of 1196 small islands within the next 30 years"* and *"But the end of the Maldives and its 200,000 people could come sooner if drinking water supplies dry up by 1992, as predicted".*

The 30 years has been and gone, the Maldives is thriving and has potable water. More than 30 years later, the Maldives population has doubled and there has been a building boom of waterside tourist facilities. A global scale analysis of 221 islands in the tropical Pacific and Indian Oceans reveals *"a predominantly stable or accretionary trend in an area of atoll islands worldwide"* throughout the 21st Century. Land area for the 221 studied islands had increased by 6% between 2000 and 2017. The Maldives alone expanded 37.5 square kilometres from 2000-2017.[503]

This is in accord with a 2019-global scale analysis of 709 islands in the Pacific and Indian Oceans that revealed 89% were either stable or growing in size and only a few small islands had slightly decreased in size.[504]

It was the UN that warned in 1989 *"entire nations could be wiped off the face of the Earth by rising sea levels if the global warming trend is not reversed"*. The UN was wrong. We were told lies. Did they correct their mistake? Did they apologise? Why not? The land of climate activism is littered with lies, false claims, cooked data and mistakes and yet no corrections are made.

In 1989, it was predicted that rising sea levels would obliterate whole island nations by 2000.[505] Island atoll nations have become larger since 1989.[506]. If any sea level catastrophist had actually looked up Charles Darwin's 1842 book on coral atolls[507] which drew on the work of Lyell[508], they would have seen that knowledge of atoll expansion has been around for a long time. It is a characteristic of green activists that they don't read the breadth of scientific literature and ignore the interdisciplinary validated past. This is fraud.

In 1989, the global warming scare was in full swing after a couple of decades of the ice age scare. A senior environmental official from the UN, Noel Brown, said[509] *"entire nations could be wiped off the face of the Earth by rising sea levels if global warming is not reversed by 2000"* and *"Coastal flooding and crop failures would create an exodus of 'eco-refugees', threatening political chaos"*.

In 1989, NASA climate activist scientist James Hansen predicted that New York's West Side Highway would be underwater by 2019.[510,511] It is still well above water. I wonder if the climate activists will be charged with uttering falsehoods and misleading and deceptive conduct as such predictions from a prominent "scientist" would have influenced investment decisions.

From a *New York Times* quote[512] *"At the most likely rate of rise, some experts say, most of the beaches on the East Coast of the United States would be gone in 25 years. They are already disappearing at an average of 2 to 3 feet a year"*. Nothing has changed 25 years later and the beaches are still there.

George Monbiot, affectionately known in the UK as Moonbat, wrote[513] *"The global meltdown has begun. Long predicted and long denied, the effects of climate change are arriving faster than even the gloomiest prophets expected. This week we learnt that the Arctic ecosystem is collapsing. The ice is melting wiping out the feeding grounds of whales and walruses. Polar bear and seal populations appear to have halved. Three weeks ago, marine biologists reported that almost all the world's coral reefs could be dead by the end of the coming century. Last year scientists found that between 70 and 90 percent of the reefs they surveyed in the Indian Ocean had already expired, largely as a result of increasing water temperature. One month ago the Red Cross reported that natural disasters uprooted more people in 1998 than all the wars and conflicts on earth combined. Climate change, it warned, is about to precipitate a series of 'superdisasters', a 'new scale of catastrophe'. The demographer*

Dr Norman Myers calculates that 25 million people have already been displaced by environmental change, and this will rise to 200 million within 50 years. The London School of Hygiene and Tropical Medicine reports that nine of the ten most dangerous diseases carried by insects and other vectors are likely to spread as a result of global warming. The British Government's chief scientist has warned that climate change could cause the Gulf Stream to grind to a halt..."

Moonbat should be nominated for the Paul Ehrlich Medal for Dud Environmental Predictions. He would have won it many times.

More than 20 years later, every single prediction in Moonbat's July 1999 scary article has been proved wrong. Polar bear and walrus populations have increased, we have no uprooting of people due to climate change, the Arctic sea ice continues to wax and wane, coral reefs have not disappeared from the Indian Ocean, superdisasters have not occurred, hundreds of millions have not been displaced, tropical diseases are not spreading because of global warming and the Gulf Stream has not ground to a halt. Unbalanced green activists like Moonbat have an unhealthy obsession with death, disaster, disease and human suffering.

No one had ever heard of David Viner from the infamous Climate Research Unit of the University of East Anglia[514] until he predicted in 2000 that snow is beginning to disappear from our lives, that winter snowfall will become *"a rare and exciting events"* and *"Children just aren't going to know what snow is"*.

He was front page news in the UK yet no journalist asked him to produce repeatable validated evidence for such a prediction or asked whether Viner's predictions were in accord with the past. Viner and his catastrophist climate colleagues at the University of East Anglia established a global reputation for fraud.[515]

Some 20 years later, the BBC[516] was at it again reporting the UK Met Office's prediction that by the 2040s most of southern England could no longer see sub-zero days and by 2060 only high ground in northern Scotland would experience cold days. These predictions were made after two very cold winters and unusual cold summer weather. Both organisations have an appalling track record for their previous predictions.

I am sceptical when postulated temperature increases of 6.4°C (IPCC, 2001) and 4.5°C (IPCC, 2007) from the doubling of atmospheric carbon dioxide ignore the past. If the same methods are used, then at the times in the past when there was a very high carbon dioxide content in the atmosphere and an explosion of life, the atmospheric temperature should have been between 78 and 108°C. This was not the case.

Computer models used for predictions of a future fry-and-die scenario have been shown time after time to be wrong. A new study by a team of Chinese scientists again showed that "the pause" from 1998 to 2013 could not be reproduced by climate models.[517]

I have difficulty in believing that the IPCC has anything to do with science when it admits that it used a Greenpeace campaigner to help write an "*impartial*" report on green energy suggesting that 77 percent of the world's energy in 2050 could come from sea breezes and sunbeams. The report was not science. It was recycled Greenpeace campaign material. Greenpeace has a pathological hatred of coal-fired power stations, hydroelectric power and nuclear power so the books had to be cooked to show that unreliable uneconomic forms of power will save the world.

If one is to trust the IPCC, they need to have no campaigners, advocates or activists as lead authors of IPCC reports. They should not use sources such as newspaper reports, the grey literature or campaign material as references for their reports. No authors should be assessing their own work and press releases and Summaries for Policymakers should be released at the same time as the full scientific reports to enable scientists to evaluate the basis for conclusions.

As with any other organisation, there should be a transparent policy on conflict of interest. I am very sceptical when Climategate star Phil Jones cites his own work in the IPCC 2007 report "*Studies that have looked at hemispheric and global scales conclude that any urban-related trend is an order of magnitude smaller than decadal and longer time-scale trends evident in the series (e.g. Jones et al., 1990; Peterson et al., 1999)*".

Jones simply ignores the abundant literature that shows an urban heat island effect of up to 11°C in mega-cities like New York. As the Climategate emails show in Chapter 8, Jones does everything in his power to silence critics or those that try to publish contrary views.

Then there are the claims about drought. Some UN alarmists have even predicted that Americans would become "climate refugees" using imagery that may be familiar to those who suffered through the infamous (and natural) 'Dust Bowl' drought of the 1930s. Prominent Princeton professor and lead UN IPCC author Michael Oppenheimer, for instance, made some dramatic predictions in 1990.

By 1995, he said, the *"greenhouse effect"* w ould be *"desolating the heartlands of North America and Eurasia with horrific drought, causing crop failures and food riots"*. By 1996, he added, the Platte River of Nebraska *"would be dry, while a continent-wide black blizzard of prairie topsoil will stop traffic on interstates, strip paint from houses and shut down computers"*. The situation would get so bad that *"Mexican police will round up illegal American migrants surging into Mexico seeking work as field hands"*.

When confronted on his predictions, Oppenheimer, who also served as Gore's advisor, refused to apologise. *"On the whole I would stand by these predictions — not predictions, sorry, scenarios — as having at least in a general way actually come true,"* he claimed. *"There's been extensive drought, devastating drought, in significant parts of the world. The fraction of the world that's in drought has increased over that period"*.

Unfortunately for Oppenheimer, even his fellow alarmists debunked that claim in a 2012 study for *Nature*, pointing out that there has been *"little change in global drought over the past 60 years"*.

George Monbiot predicted in 2002[518] that there would be famine in 10 years. There wasn't. Moonbat also claimed that *"Famine can only be avoided if the rich give up meat, fish and dairy"* in an article e ntitled *"Why vegans were right all along"*. There was no famine, the vegans were wrong, Moonbat has earned his nickname and only *The Guardian* would pay for such tosh.

It was the same ever-reliable *Guardian* that informed spell-bound readers a little earlier[519] that *"Worldwide and rapid trends towards a mini Ice Age are emerging from the first long term analyses of satellite weather pictures"*. If *The Guardian* makes a prediction, the odds are the opposite will happen.

English graduate and tree kangaroo specialist turned to climate catastrophist Tim Flannery said in 2004[520] *"I think there is a fair chance Perth will be the 21st Century's first ghost metropolis. It's whole primary production is in dire straits and the eastern states are only 30 years behind"*. We're well into the 21st Century and, last time I was in Perth, it was thriving. Flannery has such spectacular form at being perpetually wrong, he even has a dud prediction site dedicated to him.[521] Now that's an honour!

On 22nd February, 2004, *The Guardian* stated[522] *"Climate change over the next 20 years could result in a global catastrophe costing millions of deaths and natural disasters.*" Wrong. What is it about the left's obsession with death?

In 2004, environmental hypocrite Tim Flannery stated[523] that Australians are "*one of the most physically vulnerable people on Earth*" and "*southern Australia is going to be impacted very seriously and very detrimentally by global climate change. We are going to experience conditions not seen for 40 million years*". We are still waiting, patiently, for the faintest clue that Flim Flam may have jagged a correct prediction.

In 2005, it was predicted by the ever-reliable ABC that New York, London, Shanghai, Brisbane, Sydney and Melbourne will be under water by 2015.[524] They are still well above the high tide mark. Such announcements just before the Paris COP21 meeting are of course purely coincidental and are uncritically lapped up by the ABC and other media. This was a prediction of a self-interested dope because it predicted a catastrophe in 10 years time rather than in 100 years time. Did any of those making predictions own property near the current sea level?

The 2005 book *The Weather Makers* by Tim Flannery[525] the author breathlessly told us "*Australia's east coast is no stranger to drought, but the dry spell that began in 1998 is different from anything that has gone before....The cause of the decline in rainfall on Australia's east coast is thought to be a climate change doubly whammy – loss of winter rainfall and prolongation of El-Niño like conditions. The resulting water crisis here is potentially even more damaging than the one in the west....As of mid-2005 the situation remains critical...very little time to arrange alternative water sources such as large scale desalination plants*". The reader can work out whether Flannery was wrong, hopelessly wrong or gilding the lily.

One of Flannery's 2005 dire predictions was that Sydney's dams could be dry in as little as two years because global warming was drying up the rains. The dams did not dry and in 2011 were at more than 70% capacity. In 2020 and 2021, Sydney's dams were full. One overflowed and created widespread flooding downstream.

In order to really frighten people, in 2007 he claimed just before heavy rains and great floods that many Australian cities will run out of water and that "*Perth will be the 21st century's first ghost metropolis*" and that Adelaide, Sydney and Brisbane would "*need desalinated water urgently, possibly in as little as 18 months*". Brisbane dams spent much of 2011 at overcapacity.

In an October 2006 opinion piece[526] entitled *Climate's last chance*, Tim Flannery tried to scare readers about sea level change with *"Picture an eight storey building by the beach, then imagine waves lapping its roof"*.

Tim Flannery then lived at sea level in a waterside house. He purchased the adjacent property, also at sea level. Apart from the blatant hypocrisy, he omitted to mention that the most catastrophic sea level predictions were only millimetres per year hence it would take thousands of years to reach his alarmist sea levels.

He clearly does not believe his own sea level change predictions and clearly does not believe that there is a problem with global temperature, as he stated on Melbourne Talk Radio (March 2011)[527] *"If the world as a whole cut all emissions tomorrow the average temperature of the planet is not going to drop in several hundred years, perhaps as much as a thousand years"*.

In 2007, Flannery announced in *New Scientist* and gushingly reported in *The Sydney Morning Herald*[528] that *"Australia is likely to lose its northern rainfall"* and we should forget about moving people and agriculture north. When last I was in the Top End, pastoral and agriculture activities were booming. Check it out.[529]

It was predicted in 2007 that there would be an increase in super hurricanes.[530] As with most catastrophic environmental predictions, the exact opposite has occurred. Both the number and intensity of hurricanes has decreased with time. However, the property damage costs have increased because we are far more wealthy, more expensive houses have been constructed for far more people and many of these houses are at the waterfront.

The national UK broadcaster[531] was true to form in 2007 in uncritically promoting ideology *"Our projection of 2013 for the removal of ice in summer is not accounting for the last two minima, in 2005 and 2007, ... So given that fact, you can argue that may be our projection of 2013 is already too conservative"*. It wasn't too conservative. It was wrong.

In 2007, failed recidivist soothsayer Tim Flannery stated[532] *"...that's because the soil is warmer because of global warming and the plants are under more stress and therefore using more moisture. So even the rain that falls isn't actually going to fill our dams and our river systems, and that's a real worry for the people in the bush. If that trend continues then I think we're going to have serious problems, particularly for irrigation"*. I know, I know, we all had to read the great Tim Flannery's words twice to try to work out what he was really trying to say. Since this mumbo jumbo, Australia has had floods. Many times.

The bad prediction year for Flannery was 2007. He stated[533] *"The one-in-1000-years drought is, in fact, Australia's manifestation of the global fingerprint of drought caused by climate change"*. Never let the facts spoil a good story. This is contrary to the well-documented history of drought in Australia and the largest drought suffered in recent history in Australia was during the Medieval Warming.[534]

More from Flannery's horrible year of drought predictions in 2007.[535] *"Brisbane and Adelaide – home to a combined total of three million people – could run out of water by year's end"*.

Oh no, not Flannery again making more 2007 predictions. He stated that Australia was facing[536] *"the most extreme and the most dangerous situation arising from climate change facing any country in the world right now."* The year 2007 has been and gone. It was boring except for the humour provided by Flannery foolishly flapping his lips about the end of the world.

As a result of not reading history or looking at Bureau of Meteorology records, in 2007 Flannery again was preaching drought catastrophism[537]. Farmers love Flannery making predictions about drought because it is a sure sign that heavy rain and floods are coming soon. *"Over the past 50 years southern Australia has lost about 20% of its rainfall, and one cause is almost certainly global warming. Similar losses have been experienced in eastern Australia, and although the science is less certain, it is probable that global warming is behind these losses too. But the far more dangerous trend is the decline in the flow of Australian rivers: it has fallen to around 70% in recent decades, so dams no longer fill even when it does rain..."*.

Since this breathtaking prediction, eastern and southern Australia have had floods. Many times.

During the drought in 2007, our Tim was cashing in while farmers were suffering. Flannery stated[538] *"In Adelaide, Sydney and Brisbane, water supplies are so low they need desalinated water urgently, possibly in as little as 18 months"*. Some $12 billion of hard-earned taxpayer's money was spent building desalination plants in South Australia, Queensland, New South Wales, Victoria and Western Australia.

The Queensland Labor Premier spent more than $1.2 million of taxpayers' money on a desalination plant on the basis that the lower than usual rainfall would continue. It didn't. The expensive desalination plant has been mothballed and Brisbane has since been flooded twice.

Not to be outdone, the Victorian Labor government spent $2 billion on a desalination plant and, at the time of its opening, catchment dams were more than 80 percent full. The plant was mothballed. Sydney also spent horrendous amounts of money on a desalination plant. It has produced no water since 2012. The cities did not run out of water and the plants have been mothballed before pumping a squirt of drinking water.

In Perth, it was different. There had been a slight decline in rainfall (especially in winter) over the last 40 years in south-western Western Australia.[539] Such cycles are common all over Australia. Furthermore, the slightly lower runoff increased salinity of waters captured in dams (e.g. Wellington Dam) and much dam water became unsuitable for drinking.

Perth was a rapidly expanding metropolis, much domestic and industrial water was from a shallow aquifer and new water supplies were needed. The 2021 flooding winter rains filled the aquifers and dams and there was more rainfall in July 2021 than for the preceding 20 years.

A desalination plant was built near Kwinana (Perth Seawater Desalination Plant) and supplied Perth with 17% of its water needs. The second and bigger plant was built further south at Binningup (Southern Seawater Desalination Plant).[540] Some 50% of Perth's water is now from desalination plants and the energy for these plants derives from fossil fuel (natural gas). Water Corporation WA also purchases the entire output from the 10 MW solar plant at Greenough and from the 55 MW wind plant at Mumbida and claims that the Binningup plant is supposedly "carbon neutral" (except of course at night when the facility runs on gas). A third plant is planned at a cost of $1.5 billion and, with a compulsory purpose-built renewable energy plant, the total cost will be around $5 billion.

Perth residents did not suffer from climate change. They had a population explosion. Residents also did their bit and changed from well-watered European green lawns to native gardens that needed less water. The additional costs have been absorbed into the Water Corporation WA costs and sandgropers are just happy to have water.

The normal suspects were up in arms about building a desalination plant at Kwinana. They claimed that the sea grass meadows off Kwinana were fragile, and that there would be an increase in salinity in the Indian Ocean. The sea grass was going to die, this would create an ecological collapse and that everything off the coast of Western Australia would die.

The sea grass did not die. A simple calculation would show that there would be no measurable increase in salinity in the Indian Ocean off Kwinana as dilution is effectively infinite. As usual, science and engineering solved a problem. As usual, the green left activists were hopelessly wrong and just moved onto the next perceived disaster.

The costs to run each mothballed desalination plant is between $500,000 to $1,000,000 per day every day until contracts run out around 2030. By 2009, the dams for Brisbane, Canberra and Sydney were overflowing. How many politicians and bureaucrats lost their jobs for such woeful decisions? Why do journalists stay silent? Flannery was later sacked after a change in the Federal government.[541]

In 1974, the Wivenhoe Dam was built to protect Brisbane from floods. In 2007, Tim Flannery and others argued that dams would never fill again.[542] Huge rains in January 2011 in Queensland filled the Wivenhoe Dam rapidly, water was released, 28,000 houses were flooded and the damage bill was more than $100 billion. Twelve people and a large number of livestock died. Despite the January 2011 fatal catastrophe, Flannery did not have the strength of character in 2011 to say he was wrong[543] or stop making predictions.

Governments listened to Flannery's predictions despite the fact he was a failure with past predictions. Why did no bureaucrats stop this government folly? Is the public service now filled with green activists? Whatever the reason, scare predictions by Flannery cost the Australian taxpayer billions because it fitted the green activist narrative of catastrophes.

In 2007, Tim Flannery was giving financial advice and inviting all and sundry to invest in green geothermal power. The fawning ABC asked no searching questions. Flannery claimed[544] that hot rocks in South Australia *"potentially have enough embedded energy in them to run Australia's economy for the best part of a century"* and *"the technology to extract that energy....is relatively straightforward"*.

The Rudd Labor government gave $90 million of taxpayer's money to Geodynamics Ltd for a Cooper Basin geothermal project in South Australia.[545] Why should taxpayers' money be gifted to companies listed on the stock exchange? Flannery was a shareholder in this public-listed geothermal company which had a high risk speculative project. The project failed, shareholders lost money, the technology was not straightforward and the project was abandoned. One wonders who bought shares on the advice of Flannery and the legal implications.

Models underpin modern scary predictions by climate activists. The models also claimed that modest warming by carbon dioxide would be amplified by increased water vapour in the atmosphere. This can be tested by checking actual rainfall against the predicted rainfall.

Out of 18 regions tested in USA, eight show that rainfall would either increase or decrease depending upon the model used. One model predicted an 80 percent decrease in rainfall and another predicted an 80 percent increase in rainfall. Neither deserts nor swamps appeared. The models were wrong.

Modellers would only have to spend a few hours looking at the scientific publications in astronomy to see that there is climate change on other planets and their moons like Mars and Saturn's largest moon Titan respectively. This extra-terrestrial climate change could not possibly be related to human emissions of carbon dioxide and clearly the factors that drive extraterrestrial climate change are also important on Earth.

Only one model has been able to reproduce the past temperature and climate fluctuations over the past century. This is the model from Nicola Scafetta of Duke University and this is based on solar, lunar and planetary cycles and does not use carbon dioxide as the controlling variable.[546,547] The latest data from the European Council for Nuclear Research (CERN) particle physics laboratory has also produced a model and foresees no runaway global warming. Instead, it sees an impending cold solar minimum.[548]

In 1990, the IPCC predicted that the rate of global warming would be twice what occurred. In 2007, the IPCC predicted that in the decade following 2005, there would be significant warming. There wasn't even any warming in this decade. In 2013, the IPCC were at it again and predicted short-term warming. This was wrong.

Both were shown to be wrong as a result of measurement. The IPCC is embarrassed. It has now had to admit that the climate models built by their so-called climate scientists cannot predict the temperature that is measured. They meekly state *"For the period from 1998 to 2012, 111 of the 114 available climate-model simulations show a surface warming trend larger than the observations"*. This is the coward's way of saying that they were hopelessly wrong.[549]

Some green extremists suggested that such sceptical scientists should be removed from employment, imprisoned and even assassinated. For example[550] *"On September 20th the British newspaper, the Guardian, published an article on a letter from the Royal Society…calling for the silencing of groups, organizations, and individuals who do not conform to their views on climate change and policy"* and *"The British Greens have called for a purge of officialdom to get rid of anyone who doesn't accept 'scientific consensus on climate change".*[551]

Other scribes have had their two bob's worth on the same theme.[552,553,554] *"I wonder what sentences judges might hand down at future international criminal tribunals on those who will be partially but directly responsible for millions of deaths from starvation, famine, and disease in the decades ahead. I put [their climate change denial] in a similar moral category to Holocaust denial – except that this time the holocaust has yet to come and we still have time to avoid it. Those who try to ensure we don't will one day have to answer for their crimes"*(Lynas 2006) and *"Stopping runaway climate change must take precedence over every other aim. Everyone in this movement knows that there is little time: the window of opportunity in which we can prevent two degrees of warming is closing fast. We have to use all the resources we can lay out hands on, and these must include both governments and corporations"* (Monbiot, 2008) and *"I feel that climate change may be an issue as severe as war. It may be necessary to put democracy on hold for a while"* (Lovelock, 2010).

Richard Parncutt, a professor of systematic musicology at Karl-Franzens-Universität Graz (Austria), has clearly critically evaluated the science of human-induced global warming and gave us his expert opinion[555] *"As a result of that process, some global warming deniers will never admit their mistake and as a result they will be executed. Perhaps it would be the only way to stop the rest of them. The death penalty would have to be justified in terms of the enormous numbers of saved future lives".*

I wonder if these murderous anti-democratic scribes would be prepared to take the chop if they are wrong and the planet cools. They are the deniers who try to tell us that there was no Medieval Warming, that there has been atmospheric temperature rise during the last 20 years and that there is a linear correlation between atmospheric carbon dioxide and temperature. Green left environmentalist activism is the new Lysenkoism and such movements are a honey pot for all sorts of nutters. Many of them also proudly call themselves socialists.

It is safe to conclude that pretty well everything you hear, read or see in the popular media about climate change is wrong, exaggerated or made up. And if models are used, you can be certain that it is wrong.

And if contrarian scientists are threatened with violence without a squeak of opposition coming from climate "scientists", then climate "science" is unrelated to science and is the preserve of thugs.

In testimony in 2015 to the US House Science Committee hearing on the Paris climate treaty, John Christy from the University of Alabama at Huntsville (UAH) showed a massive divergence between models and measurement for atmospheric temperature.[556] Christy argued that he would not trust models and concluded that UAH satellite data that validate the millions of balloon measurements gave the most reliable measurement of the atmosphere's temperature.[557]

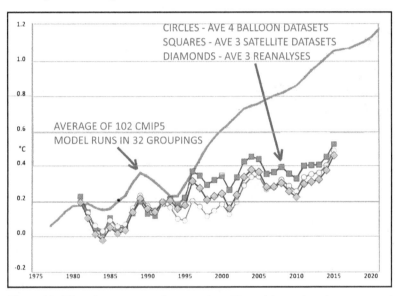

Figure 18: *Plot of temperature forecasts from 32 models against time compared with the averages of balloon and satellite measurements showing consistent overestimation of temperature and the predicted rate of change.*

Christy also showed that there is a close correlation between satellite and balloon measurements which do not show the modelled global warming but show cycles of warming and cooling.[558]

The key pillar of the global warming theory is that carbon dioxide emissions should trap heat in the tropics at an altitude of 10 km in the Earth's atmosphere preventing it from escaping into space.

Despite the release of about 30 million weather balloons since 1950, the modelled hot spot cunningly hid from every single one of these balloons. Balloon measurements of temperature have been validated by NASA satellite data over the past decade that show that our atmosphere releases much more heat into space than the computer models show. The missing heat is lost to space, not to the oceans and atmosphere as the models predicted.

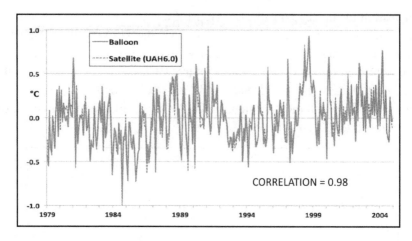

Figure 19: *Plot of balloon and satellite temperature measurements showing an almost perfect correlation and cyclical temperature change.*

It appears that climate activists were not happy with Christy's House Science Committee testimony and seven high velocity bullets were fired at his office.[559] Dr Roy Spencer, another high-profile critic of climate "science" who has an office in the same building stated[560] *"Given that this was Earth Day weekend, with a 'March for Science' passing right past our building on Saturday afternoon, I think this is more than coincidence When some people cannot argue facts, they resort to violence to get their way. It doesn't matter that we don't 'deny global warming'; the fact we disagree with its seriousness and the level of human involvement in warming is enough to send some radicals into a tizzy. Our street is fairly quiet, so I doubt the shots were fired during Saturday's march here. It was probably late night Saturday or Sunday for the shooter to have a chance of being unnoticed. Maybe the 'March For Science' should have been called the 'March To Silence' ".*

The silence from climate "scientists" was deafening. This is commensurate with silence from climate "scientists" when green activists threaten contrarians with silencing or murder. Does this mean that climate "scientists" agree with the violence against eminent people who hold a contrary view based on data? Green activists, climate "scientists" and journalists need to explain themselves.

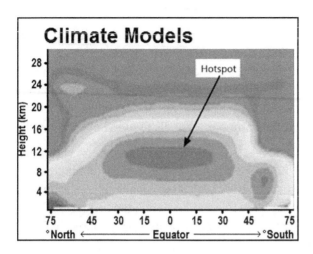

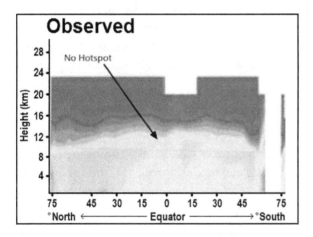

Figure 20: The upper diagram shows the computer model-predicted hot spot in the atmosphere at the equator at 10-12 km altitude in the atmosphere. The lower diagram shows the measured atmospheric temperature from radiosonde balloons. The models of the atmosphere clearly do not correlate with the atmospheric measurements. The models are wrong.

Climate, atmospheric temperature and ocean temperature models over the last few decades have all been checked with measurements. All models were wrong. The models all told us incorrectly that we will fry-and-die and the measurements are telling us that we won't.

Models can't easily be checked because the programs, data and assumptions are not freely available. In this book I show that the data is, at best, highly contestable. It is appallingly arrogant and preposterous to suggest that a computer can predict the future. Recent papers showed that models were overcooked, overegged and overdone[561,562] yet this received little press coverage[563] as it did not fit the narrative. Models run too hot.

One of the biggest weaknesses in computer models, the very models whose predictions underlie proposed political action on human carbon dioxide predictions, is cloud behaviour.[564] Clouds can both cool and warm the planet. Low-level clouds such as cumulus and stratus clouds are thick enough to reflect 30-60% of the Sun's radiation that strikes them back into space and cool the planet.

High-level clouds such as cirrus are thin and allow most of the Sun's radiation to penetrate and act as a blanket preventing the escape of the re-radiated heat to space.[565] High-level clouds near the equator open like the iris of an eye to release extra heat and marine low clouds cool the planet.[566]

The IPCC has spent decades trying to scare us to death with projections well into the future. However, a recent study[567] shows that the IPCC's worst-case climate scenarios are already off track.

Efforts to attribute specific weather events to global warming are in vogue but rife with errors, as are the flawed models forecasts of rain and storms. To make matters worse forecasts of regional climate change crucial for policymaking are close to meaningless.[568]

A new study documents the dominance of internal variability in decadal-scale global temperature changes and suggests we may experience a global cooling trend during the next 15 or even 30 years despite rising greenhouse gas emissions.[569] After 30 years of being wrong, can climate modellers still claim to be right about the alleged impending disaster of warming?

The IPCC's CMIP5 and CMIP6 models versus weather balloon observations have such large differences between models and measurements that there is only one conclusion: the models are invalid.[570]

Peer-reviewed papers from independent teams[571,572] confirm that the latest generation climate models overstate atmospheric warming and the exaggerations have got worse with time. One study looked at 38 models, the other looked at 48. One wonders why so many models are needed, especially as they are inaccurate.

The story that global warming is mainly from human emissions resulting in the need for drastic action hinges entirely on computer models.[573,574] People eventually get tired of failed apocalyptic computer models because there are only a certain number of times the boy can cry wolf.[575,576]

Models forecast that we'll fry-and-die and the solution offered by green activists is that we reduce our carbon dioxide emissions. Forget the fact that it has never been shown that human emissions of carbon dioxide drive global warming. Forget the past where there were numerous ice ages and glaciations that commenced when atmospheric carbon dioxide was far higher than at present.

The models could not even confirm past climates by running backwards without substantial retuning. The real test is whether these models can predict future climate in say 100 years time.

There are over 100 versions of climate models yet they did not even predict a period of no warming for the last 20 years. If they couldn't even get this right, then no matter how much tampering and adjustment, we can safely conclude that they can't predict what will happen a century from now. As soon as a green activist or media person uses the word model to justify their position, just laugh loudly. Very loudly. Models used to predict COVID-19 deaths were also hopelessly wrong.

In 2008, the ABC just accepted Flannery's prediction without critical questioning when he said[577] "*The water problem is so severe for Adelaide that it may run out of water by early 2009*". Flannery, a former Adelaide resident, appears to have a special kind of hatred for the city and for years was wanting it to die of thirst. It didn't because the rains came.

Adelaide did not run out of water. The dams are at more than 90% capacity and the desalination plant is an expensive white elephant that produces no water for pipelines because idealogues in government uncritically listened to Flannery. Why are such decision-making public servants still employed?

Tim Flannery also stated "*Even the rain that falls isn't actually going to fill our dams and our river systems*". On 8th November, 2020, the main dam supplying Sydney with potable water (Warragamba) was 98.1% full. At other times it has been lower, other times it has been higher and later in 2020 and early 2021 it was overflowing.

Maybe Flannery, who has a degree in English, should have read a 100-year-old poem about the land of droughts and flooding rains. I write this while it is raining cats and dogs in Adelaide, the dams are full and the coastal aquifer is saturated.

James Hansen claimed that *"We see tipping points right before our eyes"* and *"The Arctic is the first tipping point and it's occurring exactly the way we said it would"*. Hansen said that in five or ten years the Arctic will be free of sea ice in the summer.[578] It wasn't.

In early 2008, it was predicted that the Arctic would be ice free by late 2008.[579] It wasn't and the sea ice and Greenland ice cap still wax and wane due to a number of interacting complex processes but not due to a trace gas in the atmosphere.

The Guardian opines[580] *"We found that, given all of the above, 100 months from today we will reach a concentration of greenhouse gases at which it is no longer 'likely' that we will stay below the 2°C threshold...So what can our own government do to turn things around today? Over the next 100 months, they could launch a Green New Deal?"*

The planet did not obey, the 100 months has been and gone, temperature has not exceeded the predicted 2°C global increase and this recidivist scaremonger rag ignored climate cycles, the fact that carbon dioxide does not drive climate and the fact that the planet's motion and the Sun would not obey a communist green new deal where freedoms, finances, food and fun all will be lost.

The Australian national broadcaster takes world honours by probably being the worst Western public network. Try this for size from *Earth 2100* their heavily promoted 2009 TV special *"2015 is only six years away, but many experts say that if the world has not reached an agreement to massively reduce greenhouse gases by then, we could pass a point of no return"*. Some 80 percent of the energy for ABC broadcasts comes from burning fossil fuels and only a few inner city zealots would notice if the ABC cut broadcasts by 80 percent as part of their effort to reduce emissions.

The UK Met. Office consistently gets its seasonal forecasts hopelessly wrong. The barbecue summer forecast of 2009 was a washout[581], the October 2010 forecast that December would be warmer than average preceded the coldest December ever[582] and the March 2012 prediction of a dry April was followed by the wettest April on record.[583] The November 2013 prediction was for a drier than average winter.[584] It was not. It was very wet.[585] The same computer procedures that predict we are going to fry-and-die in 2100 are used to make seasonal forecasts.

It appears that every time Flannery makes a prediction, the opposite happens yet he seems uneasy about predictions when he states (*Lateline*, November 2009) "*So when the computer modelling and the real world data disagree, you've got a very interesting problem*".

The obvious way to solve the "*interesting problem*" is to abandon computer models with their alarmist predictions and stick to real observations and measurements. By not abandoning computer models, Flannery allowed ideology to overrule real-world data that underpins science. Why doesn't he have the spine to say he was wrong?

Lateline foamed at the mouth earlier in 2009 when I showed that the planet was cooling yet carbon dioxide emissions were increasing but they did not question the same science in November 2009 when their darling climate commissioner said the same thing "*Sure, for the last ten years we've gone through a slight cooling trend*".

Australia's then chief scientist was getting hysterical with the COP15 catastrophism and stated in December 2009[586] that "*We've got five years to save the world*" from global warming. She would have been sacked in the private sector for being out of contact with her market. The private sector takes risks and pays for correct advice.

On December 13th and 14th, 2009, Al Gore big-noted himself at the Copenhagen COP15 conference and predicted an ice-free Arctic in five years.[587] He had already made this prediction in 2007 and 2008. In 2007, Gore stated "*It could be completely gone in summer in as little as seven years*". Even Gore's hot air could not melt the Arctic ice. Gore's statement derives from him misquoting a Polish climate scientist and is reminiscent of Gore claiming "*During my time in the U.S. Congress, I took the initiative in creating the internet.*"[588]

Thanks to the Internet I was able to show that Gore was lying. There was still Arctic ice in 2020. He was a total fool to make a prediction that could be checked in his lifetime. Gore is, yet again, hopelessly wrong. Then again, if Gore had taken geological advice he would have known that there has been polar ice for some 20 percent of time.

His Royal Highness Prince Charles stated in 2009 that we had 96 months to save the planet.[589] Saved from what? The 96 months have been and gone and we live better lives than we did in 2009. The Prince should stick to shaking hands, kissing babies, keeping out of politics and small talk.

In 2009, the UK Prime Minister stated that there are 50 days to save the planet from catastrophe.[590] The 50 days have been and gone and Gordon Brown is no longer Prime Minister. This was his personal catastrophe.

On 14th December 2009 to a German TV audience at the COP15 Climate Conference, Al Gore predicted the North Polar ice cap would be completely ice free in five years.[591] Gore was wrong. Yet again.

Al Gore continued to provide hilarious hypocritical hype. He predicted in 2009 an ice-free Arctic by 2014.[592,593] This was a slight change from his 2008 prediction. The Arctic would be ice free by the summer of 2014.[594,595] The Arctic is still a frozen wasteland.

In the 2009 MSNBC documentary *Future Earth 2025*, we were meant to quiver in fear in our seats when it was claimed that *"As water levels drop, by 2017 Hoover Dam will no longer provide drinking water to Las Vegas, Tucson and San Diego. And it stops generating electricity to Los Angeles. And if nothing is done, the reservoir will be a dry hole by 2021"*.

Not one of these predictions came true.

The same 2009 MSNBC documentary *Future Earth 2025* claimed that *"Scientists predict the Himalayan glaciers could disappear by 2035, as a warming planet leads to both evaporation and melting of ice"*. Based on the film's track record, I've planned my ascent of Mt Everest for 2036.

The HuffPost quoted the Democrat Senator John Kerry in 2009[596] *"Scientists tell us we have a 10-year window – if even that – before catastrophic climate change becomes inevitable and irreversible. The threat is real, and time is not on our side…Scientists project that the Arctic will be ice-free in the summer of 2013. Not in 2050, but four years from now…Make no mistake: catastrophic climate change represents a threat to human security, global stability and – yes – even to American national security"*. He was totally wrong and delusional. Kerry is now President Biden's US Special Presidential Envoy for Climate. God help us.

"More than 85% of the ice that covered the three peaks of Africa's highest mountains has disappeared in the last 100 years and the rest is melting at such a rate that it will be gone by 2030…Their research shows that the melting is the worst in more than 11,000 years, when the ice was formed…". So says the UK's *Daily Telegraph*.[597] They omitted to state that 14,700 years ago the current warmer wetter interglacial had begun,

water-saturated air precipitated snow on high African peaks and the past and present precipitation is related to humidity.

As climate commissioner at the National Climate Change Forum, Flannery warned Australian families that their summer beach trips would be a thing of the past[598] "*It's hardly surprising that beaches are going to disappear with climate change*". There has been climate change since the beginning of time, beaches only disappear when there is sea ice. Beaches move inland or seawards with rising and falling sea levels.

Flannery should have known this because he then lived at sea level on a drowned river system and along all eastern Australia are submerged beaches, raised beaches, inland beaches, rock platforms, drowned river systems and back dune lakes showing that sea level has been up and down with no catastrophic consequences.

In June 2011, Flannery was at it again "*There are islands in the Torres Strait that are already being evacuated and are feeling the impacts*". This is nonsense. People moved from an island but for other reasons. Furthermore, if sea level is forcing people to leave islands in the Torres Strait, then why are people still living on other islands in Torres Strait and around the coast of Australia?

A quote from Australian Climate Commissioner's "*The critical decade: Climate change and health*" dated 2011 "*We need to act now. Decisions we make from now to 2020 will determine the severity of climate change health risks that our children and grandchildren will experience*". We suffering taxpayers financed these delusional predictions. Enough. Get rid of all such organisations.

The Australian (3ʳᵈ July 2012)[599] had an article entitled "*Enjoy snow now… by 2020, it'll be gone*" reporting a prediction from a Catherine Pickering of Griffith University. In 2019, 2020 and 2021, Australia had recorded summer snows and three consecutive winter snow seasons opened early because of heavy falls.

Later that year, the still eminently unknown Catherine Pickering[600] again tried to raise her shock jock profile by claiming "*We've predicted by 2020 to lose something like 60% of the snow cover of the Australian Alps*". The 2019, 2020 and 2021 snow seasons were long with thick snow. Why should the Australian taxpayer fund fanatical fools making farcical predictions?

It was predicted by alarmists John Abraham and Dana Nuccitelli in 2013 that the Arctic would be ice-free in 2015.[601] *The Guardian, The Times* and *New York Times* went into apocalyptic overdrive with scare stories. Nuccitelli is an American activist who writes regular alarmist sensationalist columns for *The Guardian* and writes blogs.

He does not appear to understand the scientific method and his writings show that he is a radical activist and not a journalist. In 2013, the Arctic was losing ice, it has now recovered and the normal cycles of waxing and waning of ice continue as they have done for millennia. Only someone obsessed with their own self-importance would make such a short-term prediction. If you are in the prediction business, make sure that your end of the world is well after your expected lifetime.

Peter Wadhams and others predicted in a *Nature*[602] article that an ice-free Arctic in two years time heralds a methane catastrophe and this was given sensationalist coverage by the uncritical media.[603,604,605] And just to prove that this was credible, computer models showed that the Arctic would be ice free from 2054-2058.[606]

The Arctic is still not ice free, there has been no methane catastrophe and Wadhams still lives off taxpayers' money by scaring people with fake science. A later *Nature* paper that did not receive star billing in the sensationalist media showed that a rise in the sea floor has triggered methane emissions rather than global climate.[607] This is the monocular essence of climate "science". Every natural variation is due to climate change and no other possibilities are even considered, let along investigated.

A 2013 US Department of Energy-backed research project led by a US Navy scientist predicted that the Arctic would lose its summer ice cover in 2016[608], some 84 years ahead of modelled predictions. This catastrophic prediction was repeated uncritically by *The Guardian*.

Our local soothsayer Tim Flannery pondered in 2013[609] "*Just imagine yourself in a world five years from now, when there is no more ice over the Arctic*". Why didn't Flannery do a 30 second search to validate this prediction? Does he deliberately ignore contrary evidence? Why didn't the taxpayer-funded media fact check and ask him searching questions about the past? Flannery has no shame and, after his 100 percent track record of failure, still pops up on the taxpayer-funded media to promote his next doom-and-gloom prediction.

The New York Congresswoman Alexandria Ocasio-Cortez gets huge media coverage by telling the gullible press that the world will end in 12 years.[610] Ocasio-Cortez received so much ridicule that she disowned what she said before, claimed that she was joking and if you thought that she was speaking literally then you would have had *"the social intelligence of a sea sponge"*. Either she is a Senator who can be trusted with her thoughtful words or she cannot be trusted.

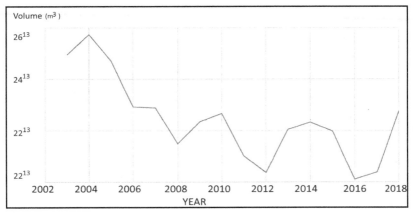

Figure 21: *Arctic sea ice volume in cubic metres on 3rd June over the last decade (from Danish Meteorological Institute).*

There are three ways of doing things. The right way, the wrong way and the French way. In May 2014 just before the COP21 Paris meeting, the French Foreign Minister Laurent Fabius jointly appeared with the US Secretary of State John Kerry and claimed that there were only 500 days before climate chaos.[611,612] The 500 days have been and gone.

The only climate chaos has been that of climate activists changing predictions each time they are shown to be hopelessly wrong. We are getting used to scary predictions made just before every COP conference by climate comrades in order to make such jaunts look relevant yet the media breathlessly and uncritically reports predictions doomed to failure.

In 2015, Tim Flannery became obsessed with cyclones and said[613] *"Sadly we're more likely to see them more frequently in the future"*. If our Tim had looked at the records, he would have seen that cyclone numbers and intensity have been decreasing with time. It took me less than 30 seconds to find the diagram below. Why was Flannery so brazenly sensation seeking and careless? Cyclones (hurricanes or tornados) are great news and the media seeking sensation seek out Flannery after a cyclone to yet again make a fool of himself. In 2016, there was not one severe cyclone in Australia. Flannery was silent. Every village has one.

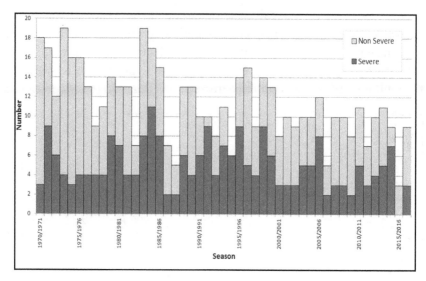

Figure 22: *The Australian Bureau of Meteorology record of severe and non-severe cyclones 1970-2017. Severe cyclones have a minimum central pressure of less than 970 hPa.*

A quote from a University of Arizona professor in 2016 didn't frighten me to death[614] *"In 10 years humans will cease to exist. Abrupt rises in temperature have us on course for the sixth mass extinction".* Blimey, I'd better hurry up as I have such a long bucket list. Even though there is no abrupt rise in temperature and the global species count is increasing, one can't be too careful.

A Glacier National Park (Montana) sign announced[615] *"The small alpine glaciers today started forming about 7,000 years ago and reached their maximum size and number about 1850, at the end of the Little Ice Age. They are now rapidly shrinking due to human-caused climate change. Computer models indicate the glaciers will all be gone by the year 2020".* Never put a time limit on your doomsday prediction. You may get caught out. The signs have since been removed. Quietly.

A recent report[616] claimed that with increased global warming, fewer aeroplanes would be able to fly. This is an unfounded scare campaign, especially as even the most unbalanced environmental activist has never suggested that the mythical global warming will create a planet where temperatures are greater than 45°C. The origin of this scare story is from Phoenix, Arizona where American Airlines cancelled many flights on 21st June 2017 because one particular plane[617] was designed to take off at temperatures less than 45°C. The wings could not generate enough lift in the thinner air to take off safely. However, many other types could take off.

It wasn't that long ago that the World Health Organisation (WHO)[618] called climate change *"the greatest threat to global health in the 21ˢᵗ Century"*. We now have another apocalypse to panic about and the effects of the COVID-19 are already having an effect on health, employment, nutrition, jobs and debt.

The economic depression from the COVID-19 has not affected the public sector but is terminal for much of the small business sector. The national economy is being destroyed by those who suck off the public teat and who have no mandate to destroy people's lives. This has not been helped by the WHO's failure to act quickly, to determine the source of the COVID-19 virus and to escape the controlling grip of Chinese communism. The WHO is discrediting itself and one has to be wary of the self-interest of an organisation that gains 75 percent of its funds from Big Pharma. Who needs climate change when government actions can create far greater damage?

A quote from our activist friends[619] *"The world's leading climate scientists warned that there is only a dozen years for global warming to be kept to a maximum of 1.5°C, beyond which even half a degree will significantly worsen the risk of drought, floods, extreme heat and poverty for hundreds of millions of people…according to the 1.5°C study, which was launched for approval at a final plenary of all 195 countries in Incheon in South Korea that saw delegates hugging one another, with some in tears"*.

Based on past predictions, I will save my tears and hugs and thoughtfully do nothing knowing that the history of predictions has my back covered.

At the COP24 meeting in Katowice (Poland) in 2018, the International Executive Director of Greenpeace (Jennifer Morgan) stated[620] *"Humanity is at stake. Countries needed to act immediately"*. Why was there no one at COP24 giving the contrary view? In the light of all previous predicted disasters, why should we even listen to such bilge? If Greenpeace were a transparent organisation, it would be possible to calculate how their coffers were topped up every time they promoted a scare story.

Prince Charles was true to form with more dud predictions[621], this time to a reception for Commonwealth foreign ministers *"I am firmly of the view that the next 18 months will decide our ability to keep climate change to survivable levels and to restore nature to the equilibrium we need for our survival"*. What on Earth does this mean? People survive at sea level, in high mountains, in deserts and in polar, temperate and tropical areas. What is restoring nature to equilibrium? The planet is dynamic hence never at equilibrium. Does His Royal Highness mean population reduction and, if so, by what technique? The 18 months have been and gone during which time human emissions of carbon dioxide decreased and yet global emissions kept increasing at the same rate. Maybe he should sack his advisors.

Presidential candidate Bernie Sanders in the 2019 MSNBC and *Washington Post* debate tried to win votes by telling Democrats[622] *"We don't have decades. What the scientists are telling us, if we don't get our act together within the next eight or nine, we're talking about cities all over the world, major cities going underwater. We're talking about increased drought. We're talking about increased extreme weather disturbances"*. Those advising him were green activists and did not know that with warmer temperatures, more water is dissolved in air and rainfall increases.

The UN Environmental Program warns that *"Unless global greenhouse gas emissions fall by 7.6 percent between 2020 and 2030, the world will miss the opportunity to get on track towards the 1.5°C temperature goal of the Paris Agreement…Failure to heed these warnings and take drastic action to reverse emissions means we will continue to witness deadly and catastrophic heatwaves, storms and pollution"*.

Emissions fell by 8 percent in the COVID-19 year of 2020, total atmospheric carbon dioxide increased at the same rate as in the past and there was no change in the weather. As I show later, we are not having *"deadly and catastrophic heatwaves, storms and pollution"* and the opposite is happening.

The Guardian told us that it's now or never to avoid climate catastrophe.[623] Apparently if we waited until after 2020, we would fall over a cliff. We are now past that time. We were told the Arctic Ocean will be ice-free for part of each year starting sometime between 2044 and 2067.[624] Apparently the next speculated reduction in Arctic ice is due to humans whereas previous well-recorded reductions were natural.

Francesca Osowska, a bureaucrat running Scottish Natural Heritage, told us in 2019 that the world will end in 11 years.[625] She stated *"Imagine an apocalypse – polluted waters; drained and eroding peatlands; coastal towns and villages deserted in the wake of rising sea level and coastal erosion; massive areas of forestry afflicted by disease; a dearth of people in rural areas and no birdsong…."*.

Scotland is the home of geology. During the last glaciation, all of Scotland was covered in ice. Now that the ice has melted and Scotland has shed extra weight, it is rising. Coastal towns and villages are undergoing a sea level fall. Maybe Ms Osowska should have spoken to a geologist or read some history before her hyperbolic emotional exaggerated activism.

It appears that if we didn't sign the Paris Accord and stop global temperature rise by 1.5°C, then 86 percent of Emperor penguins would have snuffed it by 2100.[626] Why didn't they snuff in the Holocene Maximum 4,000 years ago, the Roman Warming 2,000 years ago or the Medieval Warming 1,000 years ago when temperature was 5°C warmer than now?

The year 2050 is now the new D-day. Coastal flooding will rise one or two feet due to global warming, the number of fish and wildlife will decrease and hurricanes and storm surges will become more common and severe by 2050.[627] Rising seas from global warming yet again are going to wipe out major coastal cities in the US, China, Vietnam, Iraq, Egypt and India by 2050.[628] There is no record of such rises and extreme events during the last 5 warming events over the last 5,000 years so why should it happen in 2050?

The Global Warming Policy Foundation had a competition in 2018 to more accurately predict the following year's temperature than the UK's Met Office.[629] Well informed amateurs trounced the multimillion-dollar computers of the over-paid experts of the Met Office. If weather forecasters could predict the weather accurately today for the first Tuesday in November each year, then I might believe them about predicting the climate in 50 to 100 years time.

The CEO of Environment Victoria announced in January 2020 on radio 3AW Drive that Melbourne is at risk of running out of water by 2050.[630] We've heard this before from those of similar ilk such as Flannery. Victorian greens are against the building of dams. You can't have it both ways. The CEO of Environment Victoria is a key person deciding if and when new dams are built.

Greta Thunberg[631], misquoting the IPCC, stated *"With today's emissions levels, the remaining budget is gone in less than eight years"* and *"These aren't anyone's views. This is the science"*. She needs to go to school, learn about the carbon cycle, history and the nature of science. One simple question such as *"What is the molecular weight of the gas emitted?"* would show that these are not her words but those of her radical activist minders and parents. More about this poor abused child later.

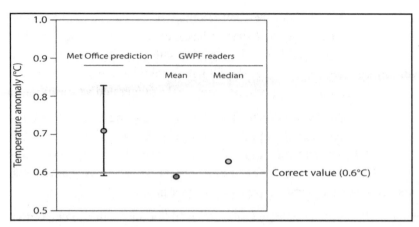

Figure 23: *UK Met Office temperature predictions for 2019 compared to the scores of Global Warming Policy Foundation reader' predictions.*

The Australian Bureau of Meteorology (BoM) published a map on 30th January 2020 showing little chance of heavy rain in Australia in the period February-April 2020. A week later the area with the lowest chance of heavy rain (Pilbara, Western Australia) was hit by a drenching cyclone and eastern Australia was enjoying the normal conditions of drought followed by flooding rains. It is the same BoM that changes historical records[632,633] and then tells us that in 50 years or so we'll all fry-and-die. More of that later.

If you can't rely on public-funded institutions such as the CSIRO, BoM, UK's Met Office and university researchers to provide unsullied data about climate, then you can do it. The information is hidden out there. Are we really in a climate emergency? The UK Met Office's Central England Temperature Record shows that there have been warmings and coolings but no significant warming in the UK since 1659. However, there has been a significant increase in climate hysteria over the last 30 years.

Oh no, on 13th July 2021 we were told that the next lunar wobble will be in the mid 2030s.[634] Every major media network on the planet covered this scare story uncritically. A NASA study told us that sea level will rise, coastal cities will be inundated with flooding 10 to 15 times a month, *"seeping cesspools will become a public health issue"* and coastal flooding will change from *"a regional issue to a national issue a majority of US coastlines being affected"*.

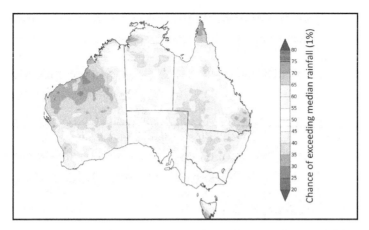

Figure 24: *BoM 30th January 2020 prediction for Australian rainfall for February-March 2020.*

These lunar tidal wobbles occur every 18.6 years and yet they've never produced cesspools or flooded coastal areas over the past decades or centuries. The only lunar wobble I have experienced is during a big night on the red when I'm at high tide. I can't remember whether they were every 18.6 years or with much greater frequency.

We've seen global warming scares before. The *Washington Post* wrote *"The Arctic Ocean is warming up, icebergs are growing scarcer and in some places the seals are finding the water too hot, according to a report to the Commerce Department yesterday from the Consulate at Bergen, Norway.*

Reports from fishermen, seal hunters and explorers all point to a radical change in climate conditions and hitherto unheard-of temperatures in the Arctic zone. Exploration expeditions report that scarcely any ice has been met as far north as 81 degrees 29 minutes. Soundings to a depth of 3,100 meters showed the Gulf Stream still very warm".

The report continues *"Great masses of ice have been replaced by moraines of earth and stones, the report continued, while at many points well known glaciers have entirely disappeared.*

Very few seals and no white fish are found in the eastern Arctic while vast shoals of herring and smelts, which have never before ventured so far north, are being encountered in the old seal fishing grounds.

Within a few years it is predicted that due to the ice melt, the sea will rise and make most coastal cities uninhabitable".

This article was dated 2nd November 1922 during a warm period that may have exceeded any since that time. But the predictions of disaster failed to eventuate. We've heard it all before. It takes a certain strange personality to repeatedly make predictions of doom and gloom that are continually wrong. We see it again and again today with those in the climate industry.

By not eating meat, not driving and not flying we can stop Armageddon. Or so we are told. Human-induced climate change is the biggest scientific and financial fraud in the history of time. There are millions living off the hundreds of billions of dollars floating around each year for alternative energy and financial climate scams, bureaucracy associated with treaties and agreements and climate "research".

Those I have met during my geological field work in the poor parts of Africa, India, China and South America have no interest in climate. Their interest is their next meal and the future of their children.

Climate hysteria will end in tears and we will all pay dearly. It's already happening. You will be poorer, have fewer employment opportunities and will not have enough cheap reliable energy to stay warm during the next inevitable cooling. Besides unreliable expensive energy and energy shortages, you might even have food shortages. Countries could go broke, especially those that have sold off their sovereignty to China. If the world becomes zero carbon, there will be poverty, starvation and a population decrease. This is what the greens want.

Climate cycles and the past cannot be ignored. They are the key to the present. Climate soothsayers need not be ignored if you have a bucket of cold water at hand. Greens have a herd immunity not admitting mistakes, irony and humour. What a glum life. How can one get experience without admitting to the numerous inevitable mistakes that make life rich yet green activists try to tell us that they are perfect?

It's not hard to see where Hollywood sources ideas for apocalyptic films.

Sinbad silliness

We are continually told by alarmists that the Arctic ice is melting. This is wrong. The ice has grown and reduced many times. As Sherlock Holmes said "*I have no data yet. It is a capital mistake to theorise before one has data. Insensibly one begins to twist facts to suit theories, instead of theories to suit facts*". No wonder Sherlock Holmes solved so many crimes.

The Pacific Decadal Oscillation (PDO) is a switch between two circulation patterns that occurs every 30 years. It was originally discovered in 1997 in the context of salmon production. The warm phase tends to warm the land masses of the Northern Hemisphere.[635] The Atlantic Multidecadal Oscillation (AMO)[636] and PDO data sets are not similar and cannot be added or averaged. In the 1930s, the AMO and PDO warm phases were coincidental. This was a period of time of record temperatures and dustbowls in the USA. Both cold phases came together from 1964 to 1979, the time of panic that the planet was entering another ice age.

For a decade from the mid 1990s, the warm phases again coincided and this was the last time the Earth was warming. The PDO has now turned cold, the AMO has peaked in its warm phase. The oscillations are out of phase and temperatures are neither increasing nor decreasing. From the mid 2020s to the mid 2030s, both the AMO and PDO will be in their cold phases. Solar activity is also declining and some solar physicists are predicting a Dalton Minimum or even a Maunder Minimum level.

Sir Francis McClintock wrote that in 1854 he was in Barrow Strait (Canada) and doubted whether he could ever escape. He also noted that in 1860 there was absence of ice. When the sea ice was reduced, Roald Amundsen sailed through the North West Passage in 1906 and discovered that the magnetic north pole was moving. Using green activist ideology, the polar ice had obviously retreated because T-Model Fords were swamping the market and emission of carbon dioxide had increased. Ford didn't need a tax on horses to make T-Model sales.

At other times sea ice prevented polar travel. Fellow Norwegian Henry Larsen also sailed through the North West Passage in 1944. The US submarines *Nautilus* and *Skate* were able to surface at the North Pole through thinned ice in 1956 and 1959 respectively during the previous AMO thinning. In late August 2020, the sea ice charts from the Danish Meteorological Institute showed this voyage would be impossible today.

Both these times of thinner Arctic sea ice were when the atmospheric carbon dioxide content was less than at present. There is no relationship between global warming and ice thickness when ice was thinner at the beginning and middle of the 20th Century. Trillions of dollars are being spent in case there is global warming driven by human emissions of carbon dioxide.

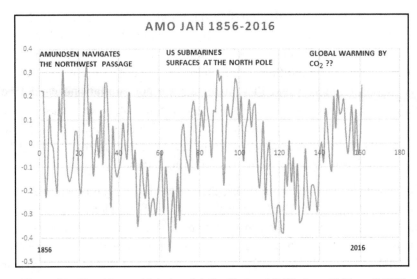

Figure 25: *The Atlantic Multidecadal Oscillation (AMO) from 1856 until 2016 showing sea surface temperatures and successful polar voyages during AMO warm phases.*[637]

Polar ice is the poster child for the climate catastrophists. When the Dark Ages ended (~900 AD) and the planet started to warm naturally in the Medieval Warming (900 to 1280 AD), the Vikings were the first to feel it. They enlarged their fishing grounds and invaded lands to the south and west as illegal boat people.

Although climate catastrophists ignore the Medieval Warming, their narrative is that if the planet was warming due to human emissions of carbon dioxide, then the poles would be the first to feel it. We all know that, because of our sinful emissions of carbon dioxide, polar ice is retreating at an alarming rate and polar bear numbers are declining.

A good green left environmental activist would face heroic hardships and go to the poles to get first-hand evidence. That would be proof. The deluded demonic denialists would be finally silenced. Such trips have attempted to view the reduced polar ice and were to regularly report back horror stories. Such trips were so well planned that ideology was all that was needed. To view the Cryostat satellite data[638] and polar weather station temperature records would not be necessary because this was the sort of data that denialists used.

We are told that the area of Arctic sea ice had been decreasing for some years. We were all doomed. However, the European Space Agency[639] reported that the Arctic sea ice had increased by a staggering 33% in 2013 and 2014. Canadian researchers on an icebreaker in Hudson Bay had to suspend their activities because the ice was thicker than it had been for the last 20 years. Public record Greenland temperature data is freely available from NASA and NOAA.[640]

In 2008, Lewis Pugh and two colleagues were sponsored by a climate risk insurance company and backed by Al Gore, the BBC and the Prince of Wales. They left Svalbard (Norway) for a leisurely 1,200 km paddle across the Arctic Ocean to the North Geographic Pole. The intention of the expedition was to electronically measure ice thickness and show how much Arctic ice had vanished. This is not science. One traverse is a snapshot in time and shows absolutely nothing about a dynamic system. Only measurements taken at the same location with the same equipment over many decades of the complete Arctic Ocean might show a trend.

The equipment froze after a few days and a tape measure was used. After a few weeks, the courageous expeditioners had to be helicoptered back to a rescue ship because the constantly moving ice was too thick. Both the helicopter and ship used fossil fuels. The expedition was abandoned 135 km from Svalbard. Pugh and his fearless fellow fanatical fools couldn't find a gap in the ice despite many tries. If Pugh had used existing data from satellites and Greenland temperature measurements or even been rational, he would never have attempted his ill-fated trip. The media was so busy reporting disasters that it forgot to report this monumental failure.

Clearly Englishman Pugh does not watch quality British television. If he did, he would have seen that in 2007 the *Top Gear* crew took nine days to drive a Toyota Hilux from Resolute Bay (northern Canada) across sea ice to the 1996 Magnetic North Pole (then at 78°35.7'N, 104°11.9W). The ice was so thick that the vehicles did not end up on the bottom of the Arctic Ocean.[641] The North Magnetic Pole is moving northwards and westwards and is currently about 1,700 km from the North Geographic Pole (90°N, 0°E or W).

Pugh has form. In July 2007 when the Arctic sea ice was at its lowest minimum, he swam one kilometre across an open patch of sea at the North Pole without a wet suit in order to bring attention to climate change. He has also publicised short swims in the seven seas and across the Maldives in order to bring attention to climate change and the health of the oceans. It didn't work. In 2010, Pugh swam across a glacial lake near Mt Everest to show that Himalayan glaciers were melting and that the impact of reduced water supply will have an effect on world peace. Himalayan glaciers melt a little in summer and grow a little in winter. His actions provided no interest, except possibly to psychiatrists.

Pugh is such a successful pin-up boy for the green movement that the COP26 web site promoted his August 2021 swim of 7.8 km over 12 days in Greenland's Ilulissat ice fjord at the toe of the Sermeq Kujalleq Glacier, also known as the Jakobshavn Glacier. The site claims that he overcame falling icebergs to show the dramatic impact of the climate crisis.[642] I'm a bit confused. How does a heroic swim in the fjord at the toe of the fastest moving glacier in Greenland have anything to do with climate?

The 2,400 km-long, 1,100 km-wide glacier that covers 80 percent of Greenland has grown and shrunk many times since observations began 250 years ago.[643] Surely if it were melting due to global warming, it would only shrink? Much of the glacial growth and shrinkage could be from heat below and not climate change.[644] COP26 certainly does not "believe the science" and it is misleading and deceptive for COP26 to infer that calving of the glacier has anything whatsoever to do with climate change. The COP26 site also shows that their principal partners for the UN- and UK-sponsored Glasgow talkfest are renewables energy companies, banks and woke retail companies. All, of course, who have absolutely nothing to gain by COP26 participation.

An attempt at niche summer tourism to Antarctica in January 2014 for climate alarmists ended in farce. It could have been a tragedy with multiple fatalities. It was promoted as an *"expedition to answer questions about how climate change in the frozen continent might already be shifting weather patterns in Australia"* by retracing the steps of Sir Douglas Mawson 100 years earlier. The tourists on this largely taxpayer-funded jaunt that cost $1.5 million found no flowers growing in meadows around Mawson's Hut in Antarctica and were not able to return as heroes with the proof that human-induced global warming is warming Antarctica.

Chris Turney, plus wife and children, mustered paying tourists and a sympathetic free-loading media from the BBC, ABC, *Guardian* and Fairfax Press on to the *Akademik Shokalskiy* for this junket. They all watched, with bated breath, the heroic planet-saving scientists battling against the elements to measure the thinning ice because we all know that human emissions of carbon dioxide from burning fossil fuels drive global warming and this results in melting of polar ice.

Never mind the huge amount of fossil fuels burned to get to Antarctica.

No wind- or solar-powered transport was used. The activist "scientists" ignored measurements made far more easily from satellites and history that show that over time the Antarctic ice sheet both expands and contracts.

The Russian gin palace became trapped in ice, a fossil fuel-burning Chinese ice breaker sent to rescue these heroic adventurers also became trapped in ice. The real Antarctic research from bases was interrupted as an Australian fossil fuel-burning ice breaker supply ship was diverted and the climate catastrophe tourists were eventually flown by a fossil fuel-driven helicopter to the warmth of a fossil fuel-heated ship well away from the ice.

The Americans, Australians and Chinese all ran up huge costs and burned a huge amount of fossil fuels to rescue the passengers from the ship of fools. All sorts of excuses were invented to show that the climate "science" activists on the ship were not ill prepared, incompetent, ignorant or unaware of past ground and satellite measurements. When questioned about their failure, they resorted to obfuscation.

The climate activist community was silent, the normal suspects in the media became very creative with excuses and the journal *Nature* showed that it was a magazine of political activism rather than one of scientific independence. The expedition was to show that this area was warmer than when Sir Douglas Mawson was in the exact same place 100 years ago. The farce showed the exact opposite. Mawson was able to get much closer to the land in his coal-fired steam driven yacht the *Aurora* because of the lack of ice.

Dozens of tourist vessels visit the Antarctic without becoming trapped in ice. The navigators look at satellite data and the extent of the sea ice. It appears that the only tourist ship ever to be trapped in ice in summer was one with climate "scientists" trying to show that the ice was disappearing.

If Chris Turney did not live off research grants and was not employed by a university, he would have been sacked for gross incompetence, breaches of safety protocols and misleading and deceptive conduct. The taxpayer still keeps paying him. However, the public was not fooled. Turney showed that whatever comes out of University of New South Wales' Climate Change Research Centre is laughable.

In 2020, the German icebreaker and research vessel *Polarstern* was stuck in two-year Arctic drift ice that was too thick. The German polar research vessel *Alfred Wegener* sailed northwards to bring home proof of a *"dying*

Arctic Ocean" which would be followed by ice-free summers. Much data was collected, the mission failed and no proof was provided.[645]

In September 2019, the *Maritime Bulletin* reported that the Arctic fossil fuel burning tour ship MS *Malmö* became stuck in ice off the Svarlbad Archipelago (Norway).[646] On board were a climate change film crew and tourists concerned about the melting of Arctic sea ice. Passengers were evacuated by a fossil fuel-burning helicopter.

The Arctic is very disobedient. It was ordered to melt by the EU, IPCC, Al Gore, John Kerry and an army of green activists. Instead, it captured and imprisoned a ship full of climate change warriors charging polewards to change the world. There was no comment about this incident from an abused Scandinavian child.

Both poles have a strange effect on climate activists. There must be something in the organic water[647] that climate activists drink whereby they ignore measurements from satellites, ground observatories and balloons, ignore climate cycles and just follow their misguided dangerous ideology. Surely this is the behaviour of religious fundamentalists and not that of rational people?

Climate research vessels and green alarmists have developed an amusing art form with their entrapment in sea ice concurrent with investigating ice-melting global warming.

God has a sense of humour.

Gored by bull

Scaremongering is stock in trade for climate activists. In Gore's 2006 movie *An Inconvenient Truth*, his promotion of fear was the business model that has made him a billionaire even though every one of his apocalyptic scenarios has never materialised. He told us nearly two decades ago that "*Unless we take drastic measures the world would reach a point of no return within 10 years*".

It's easy for teachers if Al Gore's "*Inconvenient Truth*" is shown to the class rather than standing up and teaching. The 2006 movie is promoted as predicting unprecedented problems for our planet within 10 years. "*Humanity is sitting on a time bomb. If the vast majority of the world's scientists are right, we have just ten years to avert a major catastrophe that could send our entire planet's climate system into a tail spin of epic destruction involving extreme weather: floods, droughts, epidemics and killer heat waves beyond anything we have ever experienced, a catastrophe of our own making".*

Those 10 years have been and gone, the planet is fine and we are not at the point of no return.

In Gore's 2006 film, he talks about Antarctic ice cores covering the last few hundred thousand years of climate change. He states *"When there is more carbon dioxide the temperature gets warmer."* (3'40"-3'45"). In college, Gore often struggled academically in science and technology.

In a video by John Harper, former director of the Canadian Geological Survey, he states *"The carbon dioxide concentration is a consequence of the Earth's climate, not a cause of the Earth's climate"* (1'43"-1'52"). One is by a person with a bachelor's degree in government, the other is by a person with a PhD in geology. These two statements cannot both be correct. Gore did not state that there is a well-documented lag in ice cores of 650 to 1,600 years between a rise in temperature and the later rise in carbon dioxide. Why not? This is fraud.

In Britain, the Department of Education and Skills, apparently ignorant that the film was scientifically defective, announced that all secondary schools were to be provided with a climate information pack that contained a copy of Gore's film. One school governor mounted a legal case and in October 2007 and the judge stated that Gore was on a *"crusade"*, the film had been used *"to make a political statement and to support a political program"* and that the film contained *"nine fundamental errors of fact"*. Al Gore has never been held to account and has made no corrections. These fundamental errors were[648]

(i)   Gore incorrectly claimed that low-lying Pacific atolls *"are being inundated because of anthropogenic global warming"* (They are not, they are increasing in size).

(ii)   that the Gulf Stream was shutting down (It is not, it changes for a great diversity of reasons).

(iii)   that there was an exact fit between the rise in temperature over 650,000 years and the rise in carbon dioxide (It is exactly the opposite. Furthermore, some 650 to 1,600 years after each natural warming event atmospheric carbon dioxide increases showing that warming is not driven by carbon dioxide).

(iv)   the disappearance of snow on Mt Kilimanjaro was due to human-induced climate change (Land clearing has reduced precipitation giving less snow and less ice).

(v)   the drying of Lake Chad was an example of climate change (It is an example of the locals taking too much water for crops).

(vi)   hurricane Katrina was due to global warming (The number of hurricanes hitting the US has decreased for 100 years and is unrelated to temperature cycles).

(vii)   polar bears had drowned because of swimming long distances to find ice (Like all animals, polar bears die, they swim hundreds of kilometres out to sea and their population is increasing).

(viii)   coral reefs all over the world were bleaching because of global warming (Coral bleaching events have been happening for millions of years and are unrelated to humans).

(ix)   a sea level rise of up to six metres would be caused by melting of either west Antarctica and Greenland in the near future (The Antarctic and Greenland ice sheets grow and shrink, polar ice is a rare in the history of the planet and over 150 hot spots and volcanoes have been identified under the Antarctic ice which melt ice).

These fundamental errors were known by teachers. Yet in 2010, Al Gore's film was to be included in the new English Curriculum in Australian schools. Why should scientific fraud and political propaganda be taught in an English course? Why not teach children to write, spell and critically analyse some of the great literature of our culture rather than feed them with fraudulent propaganda?

Green Al Gore is another who says *"Do as I say"*, not *"Do as I do"*. He made a fortune from mining royalties on his Tennessee farm[649,650], he flits around the world in a private jet belching out carbon dioxide so he can give speeches for fees north of $300,000 about our evil emissions of carbon dioxide.

He sold his media business to a Qatar government company Al Jazeera and ended up embroiled in litigation.[651] Qatar is wealthy because of fossil fuels and produces nothing else. At his Nashville home, he consumed more than 20 times more electricity than the average American household per year and in his California mansion, his electricity use can be up to 34 times the national average.[652,653]

Despite him cautioning the world that sea level was going to rise six metres, he spent $9 million on a waterfront mansion in California.[654] He founded a carbon trading company that has made hundreds of millions in profits and has made massive profits in his promotion of the global warming fraud.[655] Bill Gates, Barack Obama and Tim Flannery also chose to live at the waterfront despite dire warnings about sea level rise. Do they know something that we don't?

He has managed to persuade George Soros to fund his Alliance for Climate Protection[656] and Gore extracts funds from the Open Society Foundation[657] chaired by George Soros and supported by John Kerry. Gore refused to sign a personal ethics pledge to consume less energy than the average American household.[658]

In a 2018 interview[659], Gore claimed "*the climate crisis is now the biggest existential challenge humanity has ever faced*". He boasted "*I have been fortunate enough to be able to pour every ounce of energy I have into efforts to contribute to the solution to this crisis*". He did not state that he personally uses vast amounts of energy with private jets, limousines and SUVs and lives in mansions that use up to 34 times more electricity than the average American home. Gore claimed that Pope Francis was "*a moral force for solving the climate crisis*" and praised the Pope's socialist environmental encyclical.[660]

Since this 2007 UK judgement, Gore still travels the world promoting *An Inconvenient Truth*. This is fraud. Al Gore came back to Australia in 2017 to promote his movie *An Inconvenient Sequel: Truth to Power.*[661] At Al Gore's 2019 conference in Brisbane, he charged $320,000. There were 37 people who flew from abroad for this conference, as did Al Gore.[662] Only private jets of course. No such thing as sitting up the back in cattle class.

Taxpayers forked out a fortune so Gore could enrich himself and tell lies.[663] How much carbon dioxide was emitted such that all these people could agree with each other? He plans to come again. There was an "*inconvenient pause*" in warming for 18 years and recent cooling, something Gore does not tell his uncritical audiences. As electricity prices go through the roof, people are starting to wake up to those crying wolf.

The normal fawning suspects attended Gore's talks, Labor premiers, Green politicians and the fairies at the bottom of the garden gang. With a beaming Gore, South Australia, Victoria, Australian Capital Territory (ACT) and Queensland pledged that they would embrace renewables to produce zero emissions by 2050. South Australia can't even survive today on gas- and diesel-supported renewables and batteries with minutes of life let alone with zero emissions.

With zero emissions, no steel, concrete and metal could be made. There would be no transport system to bring food from the country into the city. It would not be possible to manufacture and construct wind and solar energy industrial plants. If fuel poverty, increased mortality and increased unemployment is a problem now, the country wouldn't have the taxation base in 2050 for health, education, pensions, welfare and defence.

Humans breathe in 0.04% carbon dioxide and breathe out at least 4% carbon dioxide. In a zero emissions world then all humans would have to be sacrificed. This is the green anti-human agenda.

Al Gore lobbies for climate policies that limit the consumption of meat. One of his companies has invested $200 million in a meat substitute company Beyond Meat. He also sits on the board of Apple. Political folk in the US who support the Green New Deal and other proposals to reduce carbon dioxide emissions are up against some very inconvenient facts that can be immediately accessed through a smart phone app. Users of Apple's iPhone can no longer access the app.

In 2007 Gore told lies, misled the world, did not lead by example, used huge amounts of energy and emitted far more carbon dioxide than the average American in his quest to get filthy rich by frightening people witless. He still does. Gore continues to lie about his relationship with the father of US climate science Roger Revelle in order to big-note himself.[664]

What words come to mind? Huckster, fraud, hypocrite, liar, charlatan, economic terrorist, oxygen thief, just to name a few.

And some lazy teachers still take the easy way out and show the scientifically incorrect film to pupils made by a hypocrite so that the teacher can take a break. Showing a propaganda film is not teaching.

Every one of Gore's previous apocalyptic scenarios failed to materialise yet he tells us that mother nature is screaming and the world would descend into *"political disruption and chaos and diseases, stronger storms and even more destructive floods"* unless we buy his snake oil.[665] With such a woeful record of failed predictions, one wonders why he even bothers to show his face. Unless, there is a squillion to be made. He cries wolf while our electricity costs go through the roof.

If Australia had zero emissions, what effect would this have on global emissions of carbon dioxide? None. A zero emissions Australia would be replaced by a few hours of new emissions from the developing countries such as China and India. Simultaneously, South Australia is pretending that the world's biggest battery built at huge taxpayer expense by another global green huckster, Elon Musk, is going to save the day.

Hospitals for the mentally ill are full of people less delusional than those who advocate zero emissions.

Caveat etian sentina[666]

The reason why scepticism is on the rise is the average taxpayer has realised that the climate "science" is exaggerated, embellished, wrong or fraudulent. Time and time again models and predictions have been wrong and the consequences of failed politicised science have started to become too expensive for the average person. They have become sick of being told that the world's going to end.

Print, television and radio are now no longer the only source of information. The green left environmental alarmists have been crying wolf for four decades, the climate has not changed noticeably, people are struggling even more to pay their energy bills and the community is starting to close its ears.

People have not been influenced by their political leaders. Political leaders have responded to the community scepticism and have acted accordingly. In the UK, Australia and elsewhere some people have changed political allegiances because a particular party has uncritically embraced a policy on climate change.

Independent thinkers do not accept being told what to think, need reproducible validated evidence and do not accept pronouncements made upon high by authorities. They can actually think for themselves. Sceptics tend to be older. Maybe older people are more experienced with life and had a more rigorous education where facts and thinking processes were paramount rather than feelings, beliefs and environmental ideology.

Just because something is published in the peer reviewed literature does not make it a great piece of work or correct. The peer review occurs after many have read the published work and tried to validate the conclusions. For example, Jan-Hendrik Schön managed to publish a paper every eight days in major scientific journals between 2000 to 2002 on nanotechnology and single molecule behaviour. Every paper was a peer reviewed paper, every one was bogus and yet Schön's great works were published by *Nature*, *Science*, *Applied Physics Letters* and *Physical Review*. Schön won a number of prestigious prizes for his published work. This fraud was later detected by a student, not by other scientists or the peer reviewers. Schön showed that if you have influential mainstream mates, fraud can easily be published in the best peer-reviewed scientific journals.[667]

This fraud was discovered and corrected by the post-publication evaluation of peer reviewed papers. Peer review is not the gold standard that assures the quality of science publishing, which is why a number of journals and institutions have a science citation index that attempts to quantify the scientific impact of a paper. Again, this is not the gold standard because in a small field the impact can be different from a big field and the various tribes are normally unwilling to cite the work of competitors.

Sociology journals have form. How can we forget that the physicist Alan Sokal submitted a sham article generously salted with nonsense, pseudo-babble and obvious howlers to the cultural studies journal *Social Text*? It was published in 1996.[668] Sokal exposed his article himself, it was not discovered by readers, reviewers or editors.[669]

The journal editors tried to justify publishing the article because it was "...*the earnest attempt of a professional scientist to seek some sort of affirmation from postmodern philosophy for developments in his field*". It was no such thing. It was a published paper showing the lack of scholarship in a major journal of sociology. One of the editors even claimed that "*Sokal's parody was nothing of the sort and that his admission represented a change in heart or a folding of his intellectual resolve*".

It was no such thing, despite the editors of *Social Text* making valiant weak explanations.[670] It was not a parody, it was an exercise to show the lack of scholarship in sociology. Sokal was not admitting to being dishonest or changing his view. He was a rigorous physicist who showed that some sociology journals accept whatever nonsense is submitted to them as scholarship couched in post-modernists deconstructionism language. The title of the paper was absolute garbage and any educated person would have seen that the text was nonsense.[671]

Two US academics, Peter Boghossian and James Lindsay had a peer-reviewed hoax paper published in 2017 in a social sciences journal. They argued that a penis is not in fact a male reproductive organ but merely a social construct and that, furthermore, penises are responsible for causing climate change.

"*The conceptural penis as a social construct*" had two peer reviewers who were allegedly experts in gender studies, one of whom praised the way it captured "*the issue of hypermasculinity through a multidimensional and nonlinear process*" and the other who marked it as "*outstanding*" in every applicable category.[672]

The conclusion should have set the alarm bells ringing: "*We conclude that penises are not best understood as a male sexual organ, or as a male reproductive organ, but instead as an enacted social construct that is both damaging and problematic for society and future generations. The conceptural penis presents significant problems for gender identity and reproductive identity within social and family dynamics, is exclusionary to disenfranchised communities based on gender or reproductive identity, is an enduring source of abuse for women and other gender-marginalized groups and individuals, is the universal performative source of rape and is the conceptural driver behind much of climate change*".

Most of the cited references are quotations that make no sense in the context of the paper, others were obtained by searching plausible-sounding keywords and the authors stated that they did not read a single reference cited. Fake journals such as *Deconstructions from Elsewhere* and *And/Or Press* are cited as is the fictitious researcher S. Q. Scameron.[673]

They also published other hoax papers such as on gender and dog rape in public parks. Once the sociology journal *Gender, Place & Culture* (a journal of feminist geography) realised it had been conned, they cancelled the paper. They should have realised the paper was not scholarly from statements such as "*Dog parks are petri dishes for canine rape culture*".

Another paper by the same crew of James Lindsay, Helen Pluckrose and Peter Boghossian was published in a feminist journal of social work entitled "*Our struggle is my struggle: Solidarity feminism as an intersectional reply to neoliberal and choice feminism*" comprised scattered passages from Hitler's "*Mein Kampf*" and was published in *Affilia*.

Another paper published in a journal called *Sex Roles* was investigating why heterosexual men prefer to eat at the restaurant chain Hooters in a paper entitled "*An ethnography of breastaurant masculinity: Themes of objectification, sexual conquest, male control, and masculine toughness in a sexually objectifying restaurant*".[674] The word breastaurant in the title was a dead give-away but was not noticed by editors, reviewers or readers.

Other gems by the hoax crew were "*Who are they to judge? Overcoming anthropometry and a framework for fat bodybuilding*" published in *Fat Studies* and "*Going through the back door: Challenging straight male homohysteria and transphobia through receptive penetrative sex toy use*" published in *Sexuality and Culture*. How on Earth do such journals survive? Who reads them?

How about these. *"When the joke is on you: A feminist perspective on how positionality influences satire"* published in *Hypatia* and *"Moon meetings and the meaning of sisterhood: A poetic portrayal of lived feminist spirituality"* in the *Journal of Poetic Therapy*. Who on Earth would subscribe to such journals? Boghossian, Pluckrose and Lindsay must have consumed quite a few bottles of wine and had many laughs writing such tripe to show how peer review operates in the social sciences.

Readers, reviewers and editors of these peer-reviewed journals did not realise the hoaxes until the authors published their exposure of the hoaxes.[675] Why were such hoaxes not spotted immediately? Sokal[676] argues that if ideas are fashionable, then critical faculties required for peer reviewing allow total nonsense to be published as it promotes certain values. Is climate "science" fashionable nonsense?

A recent invited paper was published by a mythical author from a mythical institute based on a whole episode of the television sit-com Seinfeld (*The parking garage*, 1981).[677] References were hoax citations using the names of Seinfeld characters. The mythical medical affliction called uromycitisis poisoning was pulled out of thin air to describe urinating in public. The paper was peer reviewed and quickly published.[678] The conclusions gave it away: *"...a national reciprocity program of public urination passes [...] so that people with uromycitisis can be free to urinate, if medically necessary, wherever and whenever they need to not be burdened legally (or, indeed, psychologically) by existing local or state laws and regulations against public urination"*.

Now just tell me again about the gold standard of the peer review process. There is an enormous amount of garbage that is published as "scholarship" in peer reviewed soft sociology journals. Sokal and those hoaxers who followed showed that the hard sciences have very different standards and, although it would be almost impossible for a hoax paper to pass muster in the hard sciences, many hard science papers are weak, have pal review and not rigorous independent peer review. Many journals are controlled by the climate catastrophist clique who find every method of rejecting a paper that does not follow the party line.

Every active scientist is constantly involved in scientific disputes because science is never settled on anything, new data and ideas are continually being aired and any new idea is generally critical of previous work. That is the nature of science. It evolves over time.

Following concerns that the so-called climate consensus was not reaching the public, a comprehensive opinion poll of 10,000 Europeans in 10 countries was conducted to establish levels of awareness, concern, and trust amongst different demographic groups and nationalities.[679] A majority (54%) of Europeans and two thirds of Britons rejected the claim that climate change is mainly or entirely caused by humans. Europeans remain sceptical and unprejudiced, despite decades of climate alarmism and green indoctrination.[680] Maybe the public is not as stupid as climate "scientists" would like to believe.

The whole issue of human-induced global warming has become a gravy train. Even sociologists have swooped in to get funds to study why people may be sceptical about human-induced climate change. For example[681] *"Climate scepticism persists despite overwhelming scientific evidence that anthropogenic climate change is occurring (IPPC, 2013). The reasons for this are varied and complex. Understanding why climate scepticism endures or is even on the rise, and why levels of scepticism vary across countries, requires accounts that recognise that biased information assimilation is mediated by cross-cultural and intra-national differences in values and worldviews. Fruitful explanations of scepticism must also account for the way in which partisans are influenced by their political leaders. Integrating such accounts may provide a way to both understand and address the social problem of climate scepticism"*.

This is post-modernist gobbledygook. Climate always changes hence there is no such thing as climate scepticism. All real scientists are sceptical. About everything. There is not overwhelming evidence that anthropogenic climate change is occurring, the only reference used to support this statement is by an activist group with self-interest.

There is a huge literature in the integrated interdisciplinary world of science (as I outlined in *Heaven and Earth. Global warming: The missing science*)[682] that has well-founded but different conclusions. One doubts whether those with a contrary scientific view read the breadth of the scientific literature.

It is little wonder that a *Scientific American* poll showed that 81% of those surveyed thought that the IPCC is corrupt with group-think, and with a political agenda; 77% said that they did not want to pay anything to stop catastrophic climate change; 75% thought that climate change is caused by solar variation or natural processes; 65% thought that we should do nothing about climate change as we are powerless to stop it; 65% thought that science should be kept out of the political process and only 21% thought that climate change was due to human emissions of greenhouse gases.[683] This was an October 2010 survey comprising 5,190 respondents. Readers of *Scientific American* could hardly be called scientifically illiterate.

The views of German ex-Chancellor Helmut Schmidt echo the *Scientific American* poll. In a speech in early 2011 at the Max-Plank-Gesellschaft, Schmidt stated *"The documentation delivered so far by the international group of scientists (Intergovernmental Panel on Climate Change) has been met with scepticism, particularly that some of the involved scientists have proven to be frauds. In any case, targets set by some of the governments have turned out to be less based on science, and based more on politics. In my view it is time for one of our leading scientific elite organisations to take a hard, critical and realistic look at the work of the IPCC, and thus explain in a clear manner the yielded conclusions to the public"*.

Time and time again when politicians retire, they tell us that they didn't really believe the political policy they were pushing.

A petition of 11,000 scientists in 2019 claimed that unless there was a massive reduction in emissions, the end of the world is upon us. Famous people like Professor Mickey Mouse signed as well as some 400 Canadians who were undergraduates, people with phoney qualifications, yoga teachers and experts in reincarnation, romance and hypnosis.[684] There were no recognised Canadian climate scientists. Only 1% of the signatories had the word climate in their job title, most of these being "experts" in mitigation and adaption.

Why is it that climate change activists don't do basic checks before they circulate their mud-on-your-face petitions? This is in contrast to the Oregon Petition of a decade ago of 31,487 American scientists, including 9,029 with PhDs, that said there was no convincing evidence of catastrophic man-made warming.[685]

The punter's natural scepticism is reinforced by a major survey in the US. The Rasmussen Poll shows that a large majority of Americans think that human-induced global warming has not been proven and that warmist scientists may falsify information trying to prove their theory is correct.[686] Since the Climategate fraud was revealed in late 2009, belief in human-induced global warming has dropped by 10 percent.

Americans are very much aware that warmists exaggerate and that the doom-and-gloom picture is not the reality they experience. The US punters are awake to the fact that they are being conned. The environmental activists' standard procedure of attacking the man rather than answering simple questions of science has not impressed those surveyed. They want to hear the arguments and wonder what the climate industry is trying to hide. In short, the public has woken up to the hypocritical unctuous sanctimony and self-righteousness of high-profile environmentalists.

Most media, academics, politicians, bureaucrats and elites think the public is stupid. Election results, which are apparently a surprise, show that they are not. Surveys also confirm that the public is aware that the global warming mantra is fake science. There is no substitute for common sense.

The history of predictions shows the average person is being treated as a dope. Don't let it happen to you.

The past is the key to the present
The story of planet Earth is a marvellous incomplete chronicle written in stone.

We have enough empirical evidence from history, archaeology, glaciology and geology to show that past climate changes have never been driven by trace additions of carbon dioxide into the atmosphere. There is no reason to conclude that present human emissions of carbon dioxide will be any different. The only way to understand climate is to read the rocks because the past is the key to the present.

To view modern weather and climate in isolation from the past is nonsense and non-science. And this is exactly what is done by those who claim to be climate "scientists". To make predictions based on computer models can only create wrong answers. There has been testing of enough models to show that not one has predicted what we have experienced or measured.

Past climate changes have been very complicated in a chaotic multivariate non-linear system with sporadic randomness. These behaviour of those systems are poorly understood and it is only by looking at the past and integrating with what we know about the present that we can hope to understand major natural processes. That understanding is a long way away.

For scientists to argue that traces of a trace gas emitted by humans into the atmosphere are the main driving force for climate changes on planet Earth is fraudulent. To argue that every change on a dynamic planet is due to human activity ignores the rich past that the chronicle of planet Earth gives us.

Just because we are alive today does not mean that we are changing the planet's climate. Just because we are alive today does not mean that we understand all natural processes. Nature rules. It always has. Although humans may have a slight effect on the Earth's atmosphere, carbon dioxide in the Earth's atmosphere has never in the past driven global climate and there is no convincing evidence to suggest it does now. Human effects are swamped by the enormous natural changes on Earth. The decision is in your hands. You vote.

My predictions

For thousands of years there have been doomsday predictions. Not one has been correct. If just one of the zillions of end-of-the-world scenarios had occurred, then we wouldn't be here.

Complex spatial mathematical forecasting systems have been shown to be far better than expensive computer climate models used by the IPCC but such a sober look at the world does not see the light of day in the popular sensationalist media.[687]

Scientists, religious leaders, your neighbourhood nutter and even domestic animals all have made predictions about forthcoming disasters and prey on the insecure and ignorant.[688] Environmentalists have now taken over the field that was once the preserve of colourful unbalanced characters.[689]

Global warming is just one of thousands of doomsday scenarios that have littered the past. I predict that:

(a) Predictions about apocalypses will continue and all will be wrong. Time will show that I am correct because history and geology are on my side.

(b) Ocean degassing, which adds most carbon dioxide to the atmosphere, will continue until the next ice age no matter how many COP climate catastrophe conferences are convened for the comrades or how many agreements are signed. The history of climate is on my side.

(c) Human ingenuity will prevail as we continue to adapt to our dynamic Earth and work through real challenges and not confected cons. The history of evolution is on my side.

4

FIGHT FIRES WITH
FACTS AND FIRES

The Greens claim that whatever happens today is unprecedented and can only be due to climate change. This is deliberate deceit.

Chronicles of past more catastrophic events such as bushfires can be found with a 30 second smart phone search. Green activists knowingly ignore information to pursue political rhetoric.

Regular Aboriginal firestick burning of temperate forests prevented catastrophic bushfires for thousands of years in Australia. The same method was used in Africa, Canada and the US.

Almost 100 enquiries after catastrophic fatal bush fires have recommended fuel load reduction by regular winter small cold burns. This has not been done because of green activist regulations, deliberate inaction and green pressure resulting in catastrophic high intensity fires in summer.

Far too many fires start in inaccessible National Parks but don't stay there resulting in the inevitable killing of people, wildlife and forests and property destruction.

Most fires are started by arsonists and lightning. Human stupidity has also started many fires. There is no relationship between climate change and forest fires. Some fires have been deliberately lit to clear land for wind turbines and solar panels.

Green activism has turned forests in the US and Australia into blackened wastelands with billions of trees and at least a billion animals killed with each catastrophic fire. This shows the green activists have no concern for the environment.

Why were no Greens Party politicians hugging trees to save them in bushfires? Why are there no green activists in volunteer firefighting crews trying to save human life and property and forests and their animals?

Why don't green activists donate to help those who have lost everything in bushfires? Why don't they have soup kitchens for firefighters? Why don't they open up their houses to people who have lost everything? Has a green activist ever saved anybody's life?

During the 2019-2020 Australian bushfires, green activists held a demonstration about climate change pulling police away from bushfire duties and, while people were being killed in these fires, the Greens Party was appealing for funds to continue their great works.

Bushfires show that greens activists have no concern for the environment, people or property and are only interested in trying to gain political points and more funding from disasters. The blame for catastrophic intense bushfires rests with green activists. Own the fuel, own the fire.

Green activists have blood on their hands.

Taxpayer-funded media ignore the past and blamed the 2019-2020 bushfires on the Liberal-National Party Prime Minister but gave no such criticism when there were bushfires under Labor Prime Ministers. The words unprecedented and climate change were intertwined with what was claimed to be news.

At the 2005 Montreal UN's COP11 conference attended by 10,000 green activists, Greenpeace's Steven Guilbeault stated[690] *"Global warming can mean colder, it can mean drier, it can mean wetter, that's what we're dealing with"*. If he didn't even know what he was dealing with, why was he there?

Conference delegates flew into Montreal in fossil fuel-burning aeroplanes, were driven around in fossil fuel-powered cars, stayed in centrally heated buildings and had hot meals and drinks. All thanks to fossil fuels. Delegates want to minimise our burning of fossil fuels but not theirs because they are the hypocritical elite.

According to green activists, it looks like whatever happens on planet Earth is due to human-induced global warming, humans are to blame, natural processes do not exist and the past has been deleted.

According to green activists, whatever happens at the present time can only be due to humans whether it be bushfires, climate change, sea level change or ice sheet changes. If whatever happens is spectacular, then the fires, floods, rain, drought, wind and cyclones are unprecedented.

The list of extraordinary claims by green activists is very large. There have been huge droughts in Australia before and during settlement. I stress fires in this book because they were recent and received world-wide attention. The catastrophic 2019-2020 forest fires in south-eastern Australia had all the green activists and media networks foaming at the mouth about climate change yet there is no evidence to show that climate change played any part.

NASA reported in 2019 that the area burned by global wildfires has dropped by 25% since 2003. NASA also showed that, the despite the increase in farm output, the worldwide forest area grew by 2.24 million square kilometres from 1982 to 2016.[691,692]

Fires before humans

After land-based vegetation appeared on Earth 470 million years ago, in the cyclical periods of high atmospheric oxygen there were wildfires; probably initiated by lightning, electrostatic discharge from dust storms and volcanic eruptions. Charcoal in lake[693] and marine sediments[694] suggests that the fires were large and widespread. Pieces of charred wood are common in coal.[695,696]

Charcoal, pollen and spores in sediments have also been used to document plant evolution and past climate changes.[697] Wildfires have been affecting the Earth's surface and atmosphere for at least the last 360 million years.

Fossil pollen, spores, resins, wood and micro-charcoal show that at the time of the dinosaurs in Wyoming (USA) 120 million years ago, the pine-cycad-cypress forests had an understorey of ferns, mosses and small flowering plants. There were scattered fresh-water ponds. Micro-charcoal was the most common particle in the sediments encasing a dinosaur fossil. The climate was warm and humid, widespread bushfires were very common and this resulted in a blanket of micro-charcoal over the land at that time.[698]

Eucalypts appeared in the fossil record in Australia 80 million years ago. With them were charcoal fragments showing the long association of fire with Australian plants. For example, at Lake George (NSW), lake sediments show climate change and the resultant change in vegetation.[699,700]

Whenever there was a shift from a cool temperate climate to a dry arid climate, there was an increase in charcoal in the lake sediments because of the increased abundance of eucalypts. Between 200,000 and 250,000 years ago, there was a great increase in charcoal from bushfires and this was well before humans were in Australia.

Monstrous forest fires were present well before humans were on Earth and still occur today. Charcoal also exists in the geological record well after humans spread across the planet.[701]

Satellite observations over the last 20 years reveal a decreasing trend in wildfires. In 2017, the global area burned by wildfires decreased by 24% since 2000.[702,703]

There is no evidence to show that modern day wildfires are any different from those in the past. In fact, there is evidence to show the opposite because of the cyclical higher oxygen content of past atmospheres.

Starting a fire

Fires need ignition, fuel and oxygen. Forest fires were ignited by lightning (mostly), cigarette butts thrown from cars; power lines; car crashes; use of farm machinery such as angle grinders, welders, metal drills, sharpening wheels; abandoned compressed gases and highly flammable chemicals; grass caught in vehicle exhaust systems; escape of embers from camp fires; back-burning and crop stubble burning gone wrong; wind turbines; re-ignition of previously controlled fires and deliberate lighting of fires by children and adult thrill seekers. Forests don't self-ignite.

In Italy, 57.4% of wildfires are caused by arsonists and the head of the Sicilian Anti-Mafia Commission, Claudio Fava, accused mafia groups of causing wildfires in order to make way for massive solar and wind facilities as determined by the EU Green Deal.[704,705] Why am I not surprised? I wonder if some of the deliberately lit fires in California and Australia were by green activists?

There are cases of people caring for the environment by burning rather than burying soiled toilet paper, a product made from wood, presumably such that they would not pollute the environment. The burning paper started bush fires resulting in property damage and a fatality.[706]

The is no record over the history of time of fires starting because of climate change. Weather conditions such as drought, wind and temperature don't cause fires.

Firestick hunting

Fossil and genetic data shows that *Homo sapiens* was in Africa for at least 315,000 years. Archaeological evidence shows an increasing complexity of behaviour took place over the continent and fire played an important role in exploiting previously uninhabited or extreme environments. In south-central Africa, humans have been using seasonal fires for at least 92,000 years to alter their environment.[707]

What is wilderness in Australia? Almost all of our forests and grasslands have undergone substantial change by Aboriginal fire practices over the last 60,000 years hence it is not really known what represents a pristine wilderness. How can we preserve a wilderness that is dynamic? Wilderness is a renewable resource and is far too important to be left to green bureaucrats and green activists to control. Their track record is one of death and destruction.

Well before settlement, Australia had changed considerably due to fire. The once dominant casuarina is not resistant to hot fires and has been slowly replaced by eucalypts. Australia has ecologically adapted to fire and eucalypts use fire for regeneration. Eucalypts love fire, the oils in leaves are flammable and actively encourage fire to spread which helps them propagate.

Eucalyptus leaves don't decompose easily and with dead branches create a fuel load on the forest floor. Some species of eucalypts hold their seeds inside small insulated capsules. Fire triggers a massive drop of seeds to the ground where competition has been destroyed by the forest fire. A eucalypt forest fire is inevitable sooner or later. If fires are frequent, the less forest and fewer animals are destroyed.

A forest management policy to avoid very regular eucalyptus forest fires makes the inevitable fire only bigger. Such fires crown with the explosive burning of highly flammable vapour above the canopy, a crowning fire can move faster than any animal can run, the crowning often precedes burning of the undergrowth and the downward radiative heat give no hope to wildlife.

Before European settlement, the aboriginal people managed fuel loads for at least the thousands of kilometres of coastal forests in eastern Australia from Tasmania to north Queensland. The fire stick was the most powerful tool the aboriginals took to Australia. Fire destroyed undergrowth, opened up the forests to ash-fertilised grasslands which were used as hunting and trapping grounds. Fire was also used to fight enemies, send smoke signals, fell firewood, keep warm, cook, drive away biting insects and, at times, fire had a religious and ceremonial purpose.

When the early explorers sailed along the coastline of Tasmania and eastern Australia, they were amazed to see small puffs of smoke everywhere. They assumed that these were camp fires and that Australia was far more populated than it was. These were traditional cold fire controlled burns undertaken in cool wet weather and not during hot windy weather.

Small cool fires encouraged animals to move a short distance out of the way from the fire and did not allow the fuel loads to built up to enable a catastrophic explosive forest fire. Grassy land made aboriginal hunting efficient because the food on the menu was grazing on the grassy patches.

Abel Tasman (1642) and James Cook (1770) learned that the aboriginals deliberately lit fires and never tried to put them out. In Captain Cook's 1770 journals, he wrote how the east coast of Australia was covered in smoke from bushfires. When the French ship *Geographe* came to anchor in D'Entrecasteaux Channel (south-eastern Tasmania), the French scientist Francois Peron commented on 5th February 1802 "*In every direction, immense columns of flame and smoke arise; all the opposite sides of the mountains were burning for an extent of several leagues*".

Early white settlers were amazed to see that the country in and around Sydney resembled what they described as an English park. The trees were more widely spaced and the land beneath them was grassy and easily passable for people. There was no tangled undergrowth with plants with needle-like leaves. Many larger trees had blackened trunks.

By the time the first European settlers arrived, fires were being lit throughout the year. In the dry season in northern Australia they burned unchecked until they ran into an area that had been previously burnt. Fires were not intense even on hot windy weather because the fuel load was light. Within 50 years of settlement, parts of the bush were difficult to penetrate because the settlers had stopped the Aboriginal people doing what they'd always done.[708]

Cheney[709] wrote "*Historical accounts record that Aborigines burnt extensively and often. Although they had little capacity for fire suppression there seems little doubt that they had a very extensive knowledge about when and where particular areas would burn and the biological consequences of their burning. They burnt some areas early in the fire season, before fires would spread extensively, to protect them from fires later in the season'."*

City-based green bureaucrats and politicians with no connection to the land apparently now know far more about the Australian bush than indigenous people. However, green bureaucrats need water bombers to try to quell the fires for which they are responsible and don't seem to want to face accountability. Aboriginals did not need water bombers because they kept the fuel load low.

Now green bureaucrats seem to know far more about our eucalypt forests than the Aboriginals. Steffensen describes[710] how National Parks employees are so steeped in green bureaucratic nonsense that they would step in and stop Steffensen and Aboriginal elders doing small cool burns.

They favoured hot "fuel reduction" burns that resulted in the deaths of many animals and trees and, at times, these hot fuel reduction burns got out of control to become major bushfires. This is why in parts of rural Australia, the National Parks and Wildlife Service has earned the affectionate name of the National Sparks and Wildfires Service.

Fires after settlement

At present, Australia has six different vegetation zones. About 91 percent of Australia is covered by native vegetation.[711] During the last glaciation, the deserts expanded and the grasslands and forests retreated. Much of today's inland Australia is sand, the remnant from the expanded deserts during the last glaciation.[712] The desert sand is now stabilised by the current interglacial vegetation.[713] Grasslands have regular fires. The catastrophic fires occur in the south-eastern and south-western parts of Australia. This is where most Australians live.

In the 19th Century, forests were cleared for farmland. Timber was used for fence posts, slab huts, drays, steam boilers at mines and factories, pit props in mines, mine headframes and shaft timbering, buildings in town and for domestic heating and cooking. Almost every forest had a few sawmills with timber cutters mainly harvesting hardwood eucalypts. Some exotic timber such as cedar and silky oak were cut in sub-tropical forests.

White ant-proof timber from drier areas, such as cypress, was prized for posts, building stumps and slabs. Red river gum was prized for buildings and furniture. It still is. Small distilleries appeared in some of the temperate forests to make eucalyptus oil. Land clearing created piles of unwanted timber that was burned. The ash was used as a fertiliser.

In the 19th Century it was known that if they didn't burn the bush, it would burn them. Management of forest lands by settlers commenced. The only way to make bushfires less intense was to take out the undergrowth that burns and converts a small fire into an uncontrollable monstrosity. It still is.

Management of forests in Australia is a State and not Federal responsibility. There is nothing in the Constitution which requires the Federal government to be involved. Fire control and controlled burns is a State and to a lesser degree local government responsibility. The ultimate responsibility rests with State Premiers not the Prime Minister.

The Commonwealth of Australia has no personnel or equipment of its own for fighting bushfires. This is the responsibility of the States under the Constitution. The Commonwealth (i.e. Federal government) is not responsible for resources and land management. This is a jealously guarded responsibility of the States. Victoria and NSW control budgets totalling $150 billion and emergency services, including rural fire services, are built into these budgets.

The States can request Australian Defence Force support from the Commonwealth or other logistical support but initiative for their use comes from the State authorities. There is a long-standing Federal-State cost-sharing agreement for natural disaster recovery. The media is fully aware of how the Federal system of government works and that the Federal Government is not responsible for everything that matters to us but, if anything that goes wrong, they pile on to blame the Federal government.

Native hardwood forests were once managed by foresters from the various State Departments of Forests. Bushfire management comprised preparation, response and recovery. Preparation required thinning of dense vegetation and the best way to do this was to harvest the forests thereby reducing the shrub layer. Forest tracks were cut and trained foresters on site controlled the whole shebang. Foresters know how to manage and protect the forests; not Greens Party politicians and activists in local, State and Federal governments sitting on their backsides miles away and giving us the benefit of their inexperience, lack of knowledge and ideology.

Management included selective logging, controlled burns, destruction of invasive plants and animals, building and maintaining forest roads and bridges, manned fire lookouts, quarrying permits, grazing permits and apiary permits. They maintained their own fire-fighting crews, cleared flammable litter from the forest floor, killed invasive introduced plants, were key personnel in search and rescue operations, and maintained recreation trails. The forests were managed and not locked up.

Native forests were self-regenerating whereas plantation forests, mostly softwoods, were not. Foresters logged slow-growing mature trees which were essentially dead wood. This allowed light penetration and space for younger trees to grow after some clearing of the undergrowth.

Without management, exotic plants and animals move in such as lantana, bitou bush and feral pigs. Revenues generated from timber harvests provided funding for all these activities. District foresters were also the local fire control wardens as well as managing controlled burns in native forests. They would determine when local farmers could burn off their properties using local knowledge and weather conditions.

Local area fire control was local rather than centralised in a city hundreds of kilometres away. Then there was the long march into the institutions by greens activists, bureaucrats and petty green power-hungry politicians. Foresters and timber harvesters were demonised by urban green activists, their tame bureaucrats, academics and the ABC. Managed State forests were converted to National Parks and Wilderness areas. They influenced governments to stop logging and management of forests shifted to government environmental departments. Royalties went to State governments, stayed in the cities and wages were no longer spent in rural areas.

Green-dominated local governments fringing the big cities have insisted that native vegetation be planted around dwellings because native vegetation has been destroyed by the housing site. These plants contain flammable oils and are a fire bomb. This is exacerbating a problem that we have known about for more than 150 years. It would be far better to plant European deciduous vegetation that would operate as a heat radiation shield and wither and die in a bushfire rather than explode.

There were numerous statutes introduced to make native vegetation clearing illegal. Distant governments outlawed actions by landowners to clear bush to protect their family, property and livestock. Until these laws are changed, Australia will continue to lose lives, property, livestock, native plants and animals through the inevitable summer bushfires.

Sawmills were shut down, rural jobs were lost and small country towns struggled. All of this went unnoticed by those in government departments in the cities. Timber imports rose. People in rural Australia were again let down by city-based politicians and bureaucrats. Government environmental departments filled with green activists did not want to harvest, thin out, maintain or control burn forests.

Regulations about exploration, mining, vegetation, fauna, surface water, slopes, soils, rock outcrops, indigenous and heritage sites, firefighting, feed and nesting trees, tracks and roads, health and safety, use of equipment and countless other matters were brought in but the forests were not managed for feral life and fuel load hazard reduction.

To make it worse, Greens Party councils fringing the cities prevented people from protecting themselves against fire. The various government environmental departments had to call on government for revenue rather than generating revenue. Forests were locked up, controlled burns did not take place or were greatly reduced and fuel loads were building up after decades of poor housekeeping

Australia is suffering a shortage of timber.[714] This has increased the costs and building time for home building. In a country that was once an exporter of timber, it is now an importer. The gap keeps widening. Domestic wood production peaked in 2000. Because of the stranglehold on forest production, we now starve in the midst of plenty. Are we sure that the hardwoods we import are legally harvested timber from tropical rainforests? We tap forest resources from areas of the world that have a far greater need for conservation than Australia.

The problems people continue to encounter include the planning rules which are not clear. There is a conflict between bushfire planning intended to preserve life (fire planning officers) and regulation intended to preserve native vegetation (council planning). Planning can take a site-specific approach under regulation and engage bushfire experts to take a risk-based approach to meet the requirements but there is regulatory uncertainty.

Government agencies repeatedly request further information and in some cases refuse to engage with site-specific risk-based assessments prepared by bushfire experts because it appears that any risk is too much risk yet no bureaucrat carries any accountability. People buy a block of land yet are prevented from building because of the maze of contradictory regulations. Their property rights have been confiscated.

There are few risk-free home sites. Those who build in thick bush or neglect cool-season burn-offs will inevitably suffer from bushfires; those who build on slopes will suffer soil creep, soil expansion and contraction and maybe detachment of soil from the substrate; those who build on flood plains will be flooded; those who build near active faults will be shaken by earthquakes; those who build near the sea risk cyclones, giant waves, king tides and tsunamis; and those that build on volcanic soils risk expansion and contraction or burial under lava and ash.[715]

There is no such thing as no risk however, with all endeavours such as a space shuttle, aircraft, mine, factory or everyday life, risk can be managed and minimised. To minimise the risk of flooding, the best solution is not to build on flood plains, keep them for fruit and vegetable growing. If not, protect flood plain buildings with levees.

With bushfires, the best solution is don't build houses in pine and eucalypt forests, build fire breaks, back burn in the off season and incarcerate arsonists. With seaside damage to property, use seawalls and don't build structures right on the shore line. With drought, droughtproof farms and grazing businesses with extra soil retention, turn around rivers and create irrigation canals and where exposed to high and low temperatures, better building with double glazing, closure of air leaks, better insulated walls and ceilings. In almost all cases of a natural disaster, risk mitigation had not taken place.

Every locked up, unmanaged, unburnt forest inevitably breeds a catastrophic fire when there is a combination of fuel, hot winds, arsonists, poor access for fire fighters and dry lightning. It makes no difference whether it is in the US, Greece or Australia.

Far too many bushfires in Australia start in National Parks and few stay there. The urban green activists have had their day in the Sun. They stole management of forests from foresters, have had 40 years to apply their theories on bushfire management and failed. Why do no green activists chain themselves to trees in the bushfire season?

The current policy of green activists, green bureaucrats and cowardly politicians of all parties is to lock and load National Parks. They are locked and fire trails are unmaintained such that fire tenders have no access. They are loaded full of fuel, feral animals and plants. No wonder many bushfires cannot be stopped. In some places on the south coast of NSW in the 2019-2020 fires, the NSW Rural Fire Service was refused permission to enter National Parks to fight small fires started by lightning before they became uncontrollable catastrophic fires.

I'm not the only one who thinks putting public land into national parks that exclude or restrict access to many land users is the answer to improving our forests. This is the view of the world-renowned expert on bushfires

Kevin Tolhurst who is concerned that bureaucrats and politicians are now controlling our forests and bushfire management from a political perspective rather than qualified foresters and fire experts with decades of experience.[716]

If we want to minimise the effects of wildfires, put local experienced and qualified people back in charge of the forests with costs paid by the orderly harvesting of a crop of trees. This has been successfully achieved in Finland for generations with each generation of forester farmers getting one cut of a tree crop in their lifetime. This would also prevent the importing of pillaged and illegally harvested rainforest trees.

The inevitable green murder of wildlife, native plants and people takes place over and over again in Australia. Management needs to be put back under the control of local foresters rather than city-based green ideologues.

Believers in man-made global warming say that warmer drier conditions and longer fire seasons are preventing hazard reduction burns. A warmer world is not a drier world, it is a wetter world and rainfall trends have gone up and not down.

Australia is a dry continent prone to deadly bushfires and management of Australia's 132 million hectares of forests through frequent mild controlled burns is vital for the survival of our forests.[717]

Bushfires are a part of life in Australia. The impact can be reduced with forest management.

Bureaucratic spite

It is illegal to burn the bush frequently and mildly enough to keep it healthy and safe as was done previously by Aborigines, pastoralists, foresters, apiarists, farmers and anyone else who knows how to manage the environment.

A beef farmer was forced to pay $999,780 in fines and costs in 2016 under the 1999 Queensland Vegetation Management Act and 1959 Forestry Act for making fire breaks too wide and cutting in fence lines and access tracks.[718] The penalty is greater than that imposed on arsonists. The Queensland Government spent $723,780 to prosecute the farmer.

In 2011, the farmer's 9,242 hectare farm had over $300,000 damage to property infrastructure as a result of a bushfire and the farmer was determined to have larger firebreaks to protect his family, property, pasture and livestock.

The Act is unclear. The farmer contacted 32 different government agencies seeking advice on the acceptable width of firebreaks and advised the government that he intended to commence creating fire breaks but received no reply. Some two years later, he was advised that the fire breaks were too large and legal action followed.

This was an appalling action taken against a productive farmer who had tried to do the right thing and work within the law. Why didn't at least one of the 32 government agencies reply to a taxpayer requesting clarification? The amount of conflicting and contradictory guidance, red tape and green tape that farmers have to face is overwhelming. This derives from unelected green bureaucrats I call the brown cardigan brigade.

They live in cities, have no idea of how the productive economy works, seem to be envious of those who have worked, saved and own property and envious of those who take risks, invest money, incur debt and make a profit and arrogantly and facelessly exert power over those who pay them. Environmental laws are made by elected failed unionists in parliament, unelected green activists and faceless bureaucrats who drive normal reasonable people to desperation.

The dark green agendas driving state and local government have robbed many unfortunate individuals of the God-given right to protect their own lives and property. Such decisions divide society even more and destroy the public's trust in the legal system, the public service and politicians.

The Kur-ing-gai Council on Sydney's leafy North Shore fined a pensioner $40,000 for chopping down 74 trees on his property in October 2014 because of the risk of fire, following a long-running legal battle with the local council.[719] Concurrent with the NSW Land and Environment Court handing down its decision in December 2019, the Kur-ing-gai Council issued a catastrophic bushfire warning.[720]

Those responsible for not allowing preventative fuel load reduction lost nothing. Either these authorities are totally incompetent, totally unaware of the realities of rural Australia or don't care about those who pay them and who they are meant to serve. I also suspect they are also spiteful, uncaring and totally untouchable.

In December 2012, The ACT Supreme Court found in favour of Brindabilla farmer Wayne West who had been wiped out by fires. It found that the Rural Fire Service and the green-influenced National Parks and Wildlife Service were negligent.[721] The court case showed how green activist pressure on decision makers infiltrates down as a cascade of bureaucratic obstructions which disempower firefighters and disregard their expertise.

The result was that in 2003, a small fire in the Brindabella Ranges was allowed to rage out of control through the National Park until it became a mega-fire that emerged 10 days later to lethally burn through Canberra suburbs.

Government agencies and employees are protected by statute and don't have to pay compensation and have no personal responsibility. Why do we pay such people? If you own the fuel, you own the fire and are responsible and accountable.

The Top End

Fires in northern Australia are very different from the catastrophic huge forest fires in south-eastern Australia that kill people, wildlife and vegetation and destroy property, infrastructure, employment and hope.

There are two seasons in northern Australia: the wet and the dry. In northern Australia's dry season, lightning strikes start many fires and they burn until there is no fuel left. Cane grass, spear grass and spinifex burn easily and fires can last for months.

Pastoralists also used to burn back the dry tropics just before the wet season and, after the rains, there were green meadows for as far as the eye can see. The fresh greenery is fodder for livestock and wildlife and by the time the dry is finished, there is almost no fodder. Then the whole cycle starts again with the next wet.

Some parts of the Top End burn every year. Year after year. There is no build up of fuel load as every year the vegetation is burned off. It's no accident that the awful devastation in south-eastern Australian in 2019-2020 was not in the zone on the map with annual fires but in a part of Australia where less than 5% of the area burns each year.

A rarely burnt forest is the risky zone because there are decades of fuel load. There were places in the 2019-2020 fires where a whole century of fuel went up to make a stratospheric pyro-convective fire and deliver some burnt leaves and ash to New Zealand.

Much of northern Australia has a 100% chance each year of having a fire. The area burned out each year in northern Australia is three times the area of Europe.

This has been happening well before humans came to Australia and is fundamental for the ecosystems.

Catastrophic past fires

Before Australia had industry, large coal mines, coal-fired power stations and green activists, some of Australia's largest bushfires occurred.

Christine Finlay studied bushfires from 1881 to 1981 in Australia. She found that there was a marked increase in the size and frequency of fires after 1919 when forest management moved away from low-intensity cool burning. Controlled cool burning limits the bushfire's spread and makes suppression easier by reducing the amount of flammable material. Despite warning government and the media, she was ignored.[722] Australia suffers from a collective amnesia when it comes to bushfire preparedness.

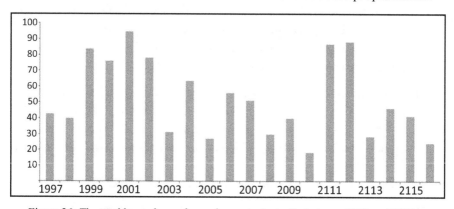

Figure 26: *The total burned area for each year in Australia between 1997 and 2016 in millions of hectares.*

The area burned every year was between 18.2 million hectares (2010) and 94.6 million hectares (2001). On average, the area burned during this time period was 52.9 million hectares. Since there are 769 million hectares of land in Australia, the area burned annually between 1997 and 2016 was 2.4-12.3% of the total land area.[723]

A lot of this land area is in the far north and north-western part of the continent, which is underpopulated and is in the hot, arid or dry tropics. It's not the same as the cool wet corner of south-eastern Australia which has some of the tallest trees in the world. The fuel loads in the north are much lower.

About a third of the average of the affected 52.9 million hectares was due to planned burns, and two thirds was due to wildfires. Since this is so much higher than the rotation of planned burns in the national parks of NSW and Victoria that burnt in 2019-2020, presumably these planned burns were either on private property or in arid zones. The definitions of forest can vary hugely, and in this case also includes vast areas around Kalgoorlie, the Eyre Peninsula, and some sparse trees near Bourke and Cunnamulla.

The Australian bushfire season of 2019-2020 overall was catastrophic. However, it was not exceptional on the country level. It has not been one of the worst seasons using any metric such as the area of burned land or burned forests. The area burned (18.9 million hectares) is about 36% of average area burned annually in Australia and exceeds the minimum burned area year in the 2010 satellite dataset.

The Federal Department of Agriculture provides the *"Australia's State of the Forests Report"* for every five-year period.[724] The latest one has been published in 2018 and covers the years 2011-2016. This report provides details about forest fires in Australia starting with annual forest fires for seasons 2011-2012 to 2015-2016.

Unplanned forest fires burned between 8.9 million hectares (2013-2014) and 21.2 million hectares (2012-2013). The area burned due to the planned burns was between 6.2 million hectares (2013-2014) and 8.2 million hectares (2011-2012). This data correlates well with the satellite burned area dataset.

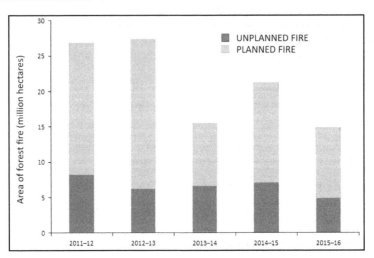

Figure 27: *Annual planned and unplanned area of forest fires in Australia (in millions of hectares).*

Earlier versions of these reports provide similar figures. For example, in 2008 the estimated area of forest burnt in the period from 2001 to 2006 was 24.7 million hectares with an estimated 20.0 million hectares burnt in unplanned fires and 4.7 million hectares burnt in planned fires.

On average, 15.7% of Australian forest land is burned every year and the satellite data from 1998-2020 shows that this area is very variable and has decreased. According to the latest report, the total area of forest in Australia burnt one or more times during the period 2011–2012 to 2015–2016 was 55 million hectares (41% of Australia's total forest area).

Some forests had at least one fire per year during the five different years between 2011 and 2016. Every year a forest was on fire somewhere in Australia.

The Black Thursday fires on 6th February 1851 in Victoria occurred when the temperature was 47.2°C at 11 am in Melbourne. For the preceding two months Victoria had suffered hot winds which dried the fuel load. A quarter of Victoria was immolated (5 million hectares) and at least 12 lives were lost.

Many isolated timber cutters, sawmillers and farmers lived a solitary life in the bush so we'll never really know how many died. Millions of sheep and thousands of cattle were fried and died. The atmospheric carbon dioxide content was 0.0285%.

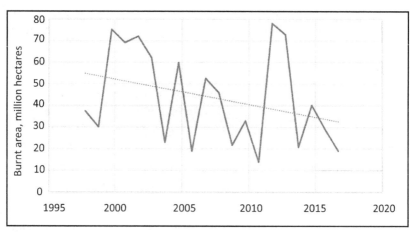

Figure 28: *The burned area of Australia (in millions of hectares).*[725]

Victoria

Geoffrey Blainey wrote *"In our recoded history, there have been no bushfires as spectacular as February 1851, on the very eve of the first goldrushes. They called it Black Thursday. Half of Victoria seemed to be on fire. A wild northerly was blowing, and it drove a column of black smoke right across Bass Strait such that one town near Devonport was so darkened in mid-afternoon that people actually thought the end of the world had come.*[726]*"*

At that time, there were no power grids criss-crossing the country and no human-induced climate change. Burnt out forests made it easier for the prospectors who first discovered gold at Warrandyte (Victoria) in July 1851 and many other places in Victoria later in that year.

In 1898, the Victorian Red Tuesday fires burned 260,000 hectares, 12 lives were lost and more than 2,000 buildings were destroyed. The atmospheric carbon dioxide content was 0.0295%.

The 13th January 1939 bushfires in Victoria (Black Friday) were when there was drought, temperatures were above 45°C and there were huge winds. Some 3,700 buildings were destroyed, 71 lives lost and 5 towns completely destroyed. Some of these towns have disappeared for ever. The fires destroyed 2 million hectares. The atmospheric carbon dioxide content was 0.0311%. The summer of 1938-1939 was hotter than the summer of 2019-2020.

The 1939 Streeton Royal Commission in Victoria concluded *"There had been no fires to equal these in destructiveness or intensity in the history of settlement in this State, except perhaps for the fires of 1851, which, too, came at summer culmination of a long drought."* The Commission concluded that there was too much fuel and the fires were *"...lit by the hand of man"*.

Some comments show that we learn nothing. For example, Streeton wanted *"...to expose and scotch the foolish enmities which mar the management of forests by public departments who, being our servants, have become so much our masters that in some respect they lose sight of our interests in the promotion of their animosities"*. It is now time to drain the swamp and not have forests, wildlife and people killed by green bureaucrats.

For greens and the media to use words like "unprecedented", "climate change", "new normal" and "coal" during the bushfire season shows political cynicism, ignorance and an abrogation of responsibility. Every school child should be taught about Black Friday to give them an appreciation of the extreme danger of the Australian eucalypt forests.

In 1944, the Victorian fires destroyed 1 million hectares, 500 houses and 20 people lost their lives. The atmospheric carbon dioxide content was 0.0311%.

In 1962, fires on the outskirts of Melbourne killed 32 people and destroyed 450 houses. The atmospheric carbon dioxide content was 0.0318%.

In 1965, 320,000 hectares was destroyed in Gippsland (Vic.), 60 buildings were destroyed and 4,000 stock were killed. The atmospheric carbon dioxide content was 0.0320%.

In 1969, there were 12 large fires in January west of Melbourne that burned 250,000 hectares. Twenty-three people died, including 17 on a major highway between Melbourne and Geelong. There were 230 houses, 21 outbuildings and 12,000 stock destroyed. The atmospheric carbon dioxide content was 0.0324%.

In 1977, widespread fires occurred in Western Victoria with 4 people killed and 103,000 hectares, 198,500 stock, 116 houses and 340 outbuildings destroyed. The atmospheric carbon dioxide content was 0.0333%.

Ash Wednesday in Victoria was 16th February 1983. More than 100 fires killed 47 people, and destroyed 210,000 hectares, 27,000 stock and 2,000 houses. The atmospheric carbon dioxide content was 0.0342%.

The Black Saturday fires in Victoria (7th February 2009) were on a 47°C day fanned by strong winds. They killed 173 people and destroyed 450,000 hectares, 3,500 buildings (including 2,039 houses), 11,800 stock, 25,600 tonnes of fodder and grain, 32,000 tonnes of hay and silage, 190 hectares of standing crops, 62,000 hectares of pasture, 735 hectares of fruit and olive trees and vines, 70,000 hectares of plantation timber and over 10,000 km of fences. Four towns were completely wiped off the map. Some fires were deliberately lit.

The Royal Commission in Victoria on the devastating 2009 fires recommended an annual fuel load reduction burn target of 390,000 hectares. The actual burn areas achieved for 2014-2015 was 234,614 hectares (deficit 36,674 hectares), 2015-2016 was 197,940 hectares (deficit 192,060 hectares), 2016-2017 was 125,052 hectares (deficit 264, 948 hectares), 2017-2018 was 74,728 hectares (deficit 315,272 hectares) and 2018-2019 was 30,000 hectares (deficit 360,000 hectares).

Planned burns only achieved 45% of the target over five years and the cumulative deficit over that period totals 1,068,954 hectares. ALP-Greens deals, the Liberal-National Party trying to look green and the infestation of the public service with green activists have resulted in Victoria being a tinder box with an overload of fuel just waiting to burn, kill and destroy.

There is no relationship over time between the atmospheric carbon dioxide content and the frequency and severity of bushfires.

Victoria will be ravaged again by bushfires.

New South Wales

As with Victoria, NSW had catastrophic fires in 1938-1939 when 13 people were killed, an unknown number of houses, outbuildings, vehicles and fences were destroyed and a fire front 72 km wide attacked Canberra. In 1951, Canberra had its own forest fires when 2 people died and some 42 buildings were destroyed including some at the Mt Stromlo Observatory. In NSW, the 1951 fires killed 11 people and more than 4 million hectares was destroyed. In the fire-prone Blue Mountains and South Coast of NSW in 1968-1969, 14 people died, nearly 250 homes and buildings were destroyed and more than 2 million hectares immolated.[727]

Between 1875 and 1984, the Blue Mountains of NSW experienced 25 catastrophic fires. Many were in remote areas and had to be left to burn themselves out. On four occasions between 1951 and 1984, a single fire destroyed more than 50 homes.[728] There are still vacant blocks where houses were destroyed in Leura from the fires of 1968-1969.

In the populated Greater Sydney, Hunter, New England, North Coast and South Coast areas, during the 2002-2003 fire season, the 495 separate fires killed 3 people, destroyed 86 houses, thousands of stock and burnt 1.5 million hectares. Also in the 2002-2003 fire season, 488 houses and more than 100 other buildings were destroyed in fires fringing Canberra, four people died yet the area burned was comparatively small (160,000 hectares). Mt Stromlo Observatory was again destroyed.

In the *1995 Year Book Australia* by the Australian Bureau of Statistics, an article entitled *Bushfires – an integral part of Australia's environment* (by N. P. Cheney, CSIRO Division of Forestry) argued that there is a short-lived memory of bushfires in Australia and that bushfires are a natural Australian phenomenon.

Bushfires are normally within 100 km of the coast in South-Eastern Australia, many fires are in national parks, almost all vegetation types in Australia are fire prone with reduced fires in the tropical rainforests and, in the 1974-1975 fire season, 117 million hectares or 15% of the total land area of Australia was burned. Cheney states "*...the greatest potential for a bushfire disaster is where people have built in close proximity to the tall, wet forests of southern Australia*". Cheney concedes the area burned in NSW (5 million hectares) might be a record for that state but this compares to 4 million hectares in 1951-52 and 4.5 million hectares in 1974-75.

Like Victoria and Tasmania, NSW has had a long history of catastrophic bushfires. Between 1st July 2019 and 31st March 2020, there were more than 11,400 separate bush and grass fires across NSW that burned out 5.5 million hectares of the State. This needs to be put in perspective. The area burnt in just one fire season was more than twice the area of Wales, 60% larger than Belgium and 30% larger than Holland. Can you imagine the kerfuffle in Europe if all of Belgium was wiped out by fire? No more EU!

The 2019-2020 fires occurred after a long period of rains that encouraged rapid growth of forest undergrowth and then dry hot conditions followed. A total of 26 lives were lost including six fire fighters (including two volunteers) and three American crew on a water bomber that crashed. Some 2,448 homes were destroyed, as were at least 5,469 of farmers' outbuildings, countless items of machinery and vehicles, timber bridges and thousands of kilometres of fences.

Some 14,519 homes, 1,486 facilities and 14,016 outbuildings were saved by firefighters. Some volunteer fire fighters were away saving the homes of others while their own houses were burnt to the ground. Firefighters from interstate, the Australian Defence Forces, New Zealand, USA and Canada assisted.[729]

The NSW Rural Fire Service is the world's largest volunteer organisation and they were stretched for months when volunteers gave up family life and their income-earning time to help their fellow man. Some men were away from home for months. Some died trying to help others. Numerous individuals and groups in the community voluntarily fed and watered the fire fighters.

A former conservative Australian Prime Minister is an active member of the NSW Rural Fire Service and was fighting the fires of 2019-2020, as in previous years. No leader from the left political parties, the Greens Party, Greenpeace, the World Wildlife Fund or the Climate Council was to be seen in a NSW Rural Fire Service uniform helping their fellow man, running a soup kitchen or financially and physically helping people rebuild.

They were quite happy to be photographed with a cooked koala or blackened fire-ravaged forest when the fires had been and gone so they could play the political blame game.

Catastrophic disasters bring out the best and worst in people. There were tales of heroic deeds, tragedy and looting. For some families, all they possessed after being burnt out were the clothes on their back. Coastal holiday season businesses were destroyed, there was help from charities, individuals and corporations and assistance from governments was too little and far too late.

Some insurance claims and government help is yet to be "processed" nearly two years later. Green local governments and bureaucrats obstructed rebuilding of homes, added to costs and showed little assistance and sympathy for those who had suffered tragedy. Many who had lost everything were still camping out on their property with no water for a year after the fires, in winter and with no tangible assistance.

They then were hit with the COVID-19 virus, lockdowns and loss of employment. In fact, during the catastrophic fires, green activists, politicians and media networks tried to make political capital by blaming the Prime Minister and climate change.

They still do.

Queensland

Queensland also has a long history of fatal bushfires. In the last century there were fires at Hughenden in 1917 and 1918 with 3 and 2 fatalities respectively. In north-west Queensland in 1918, there were 5 fatalities and there were large fires in Goomeri (1940, 80,000 hectares burned), Julia Creek and Cunnamulla (1941, 120,000 hectares), Charleville (1951, 1 fatality, 2.8 million hectares), Hughenden (1954, 3 fatalities), Muttaburra (1955, 2 fatalities), Thargomindah (1974, 7.4 million hectares), Julia Creek (1979, 420,000 hectares), south-east Queensland (1991, 4 fatalities), south-east Queensland 1995 (333,000 hectares) and Lawn Hill (2001, 1.6 million hectares).[730] In 2002-2003, there were 2,628 fires in Queensland of which the largest burnt out 200,000 hectares. One life and 10 houses were lost. In 2012-2013 a few properties were lost in the Stradbroke Island fires.

Some Queensland fires occurred in winter and in wet rainforest rather than the drier hotter parts of Queensland where there are flammable grasses. *The Courier Mail* of 29th July 1946 reported that bushfires burned in an almost unbroken chain from Brisbane to Townsville. Smoke went as high as 3,000 ft. There were winter bushfires around Brisbane in the winter of 1946 (*Courier Mail*, 6th August 1946).

The *Adelaide Advertiser* of 15th August 1951 reported Queensland bushfires were catastrophic despite occurring in winter. At that time the atmospheric carbon dioxide content was 0.0311%. There was no mention of climate change. Claims by climate activist Joelle Gergis that rainforests in Lamington National Park were burning for the first time[731] were disproved by reports from the spring of 1951 about fire taking out *"2000 acres of thick rainforest country"* in the park.

On 12th May, 1953 it was reported that the Longreach fires burned all year. The *Warwick Daily News* of Tuesday 15th June 1954 reported large winter bushfires.

The ABC was telling us during the summer 2019-2020 fires that the fire seasons were starting earlier, were talking up the fires of 2019-2020 and wanted to blame climate change and the Australian Prime Minister.

However, the fire season in northern NSW and Queensland is in spring and early summer and this is well recorded. A 30 second search on a journalist's smart phone would have shown a history of winter, spring and early summer fires in Queensland. I wonder if someone in the ABC even pondered whether this could be due to latitude?

Why is it that an organisation funded to the tune of more than $1 billion dollars by the Australian taxpayer can't spend 30 seconds and pay no money to look at past newspaper records through Trove which contains old newspaper reports of bushfires? Is it easier for the ABC to promote a Marxist eco-drama?

Tasmania

Like Victoria, Tasmania has a long history of summer bushfires.[732] Fires were recorded around Hobart in 1832. The first large fire was in 1854 in the Huon-Port Cygnet area. There were 14 deaths. In 1897-1898, there were fires around Hobart. It is estimated that six people died and 43 properties were destroyed. Other fires were in 1913-1915 and 1926-1927. In 1933-1934, fires again burned around Hobart and some buildings were destroyed. Again there were fires in 1940-1942 and 1960-1961.

A wet winter and early spring in 1966 led to vegetation growth and over 1966-1967, Tasmania had its driest summer since 1885. On the 7th February 1967, it was 39°C and there were 110 fires in southern Tasmania, either from burn-offs in previous days or accidentally or deliberately lit by humans. The fires burnt 264,270 hectares, they destroyed 1,400 homes and 128 other buildings (including the Cascade Brewery in suburban Hobart), 80 wooden road bridges, 5,400 km of fencing, 1,500 vehicles and at least 62,000 farm animals.

Every summer since 1967, there were fires in Tasmania. In 1981-1982, there were fires in western Tasmania and in central Tasmania with one death, the loss of 46 buildings, farm equipment and fences and at least 3,000 livestock. The 1993 Coal River Valley fires burned 2,400 hectares and took 3 weeks to extinguish.

Reignition of earlier fires south of Hobart in 1998 destroyed 7 homes and 50 people were injured. A deliberately lit fire in late 2006 in suburban Hobart destroyed power systems and 800 hectares of bushland. Later that year, an east coast fire destroyed 54 buildings and one person was killed. Another deliberately lit fire was on King Island in 2007. It destroyed 12,500 hectares.

The west coast had a 18,500 hectare fire in 2008 and a deliberately lit fire in the Upper Derwent destroyed 6,500 hectares and plantations, buildings and farm infrastructure. In the 2009-2010 summer, fires destroyed thousands of hectares in various parts of Tasmania after a dry winter. The wet winter of 2010-2011 led to fewer fires and 2011-2012 was a busy fire season. In 2013, similar conditions existed to 1967 and small earlier fires that had not been completely extinguished resulted in more than 40 large fires that destroyed 90,000 hectares. More than half the buildings in the town of Dunalley were destroyed as well as farm infrastructure. A number of firefighters were killed or injured.

Again, the summer of 2014-2015 had fires from lightning strikes, in 2015-2016 100,000 hectares were burnt. In 2017-2018 there were some 60 fires from lightning strikes causing central Tasmanian fires and the same occurred in the 2019-2020 summer season with fires in the Tasmanian World Heritage Wilderness Area and in populated Tasmania where 36,000 hectares was burned as a result of 406 lightning strikes. The summer of 2020-2021 had many fires, all of which were small. Currently some 16,500 hectares of winter burning to reduce fuel load takes place.[733]

South Australia

South Australia is sparsely populated, has low rainfall and does not have the large forests that characterise south-eastern Australia. It can be very windy, especially in summer. Much of South Australia is unoccupied scrubby desert lands.

There were 44 large bushfires recorded between 1917 and 1945. The biggest fires were 1933-1934, 1938-1939 and 1943-1944. Most fires were in the Adelaide Hills and the Eyre Peninsula. There were catastrophic bushfires in South Australia in 1939 in the Adelaide Hills. There had been heavy winter rain in 1938 and in summer there was hot weather and intense dry winds.

Lightning started the 1951 fires that burnt 450,000 hectares with a loss of stock, feed and fencing. The Black Sunday fires of 1955 destroyed orchards in the Adelaide Hills. Two fire fighters died and 40,000 hectares were damaged. In 1958 in the pine forests of south-eastern South Australia, eight fire fighters lost their lives. In 1959 again in the south-east, one person died in a 28,000 hectare fire and there were fires on the Eyre Peninsula.

In the more remote Flinders Ranges in 1960 and 1961 fires damaged pastoral properties. In 1961 fires were on the Yorke Peninsula and the very remote far north-west where 900,000 hectares was burnt. Between 1966 and 1972, the average number of fires fought annually was 900 and the average area burned each year was 190,000 hectares.

Massive fires in remote pastoral and unoccupied land burnt out 16 million hectares in 1974-1975. Fires in more remote parts of eastern South Australia in 1978 and 1979 destroyed farm infrastructure.

There were four major fires in the 1980s including Ash Wednesday 1 (February 1980) and Ash Wednesday II (February 1983). There were numerous other fires in sparsely populated areas started by lightning that in the 1980s burned over two million hectares.

The worst were the 1983 Ash Wednesday fires which covered the Adelaide Hills, Clare, Yorke Peninsula and Eyre Peninsula resulting in the deaths of 28 people, destruction of a number of famous landmarks, the burning of 200,000 hectares of land and massive property damage. In the 1980s, 830,000 hectares was destroyed by bushfires in South Australia.

In 1990, a fire in the remote Ernabella area consumed 900,000 hectares and in 2001, fires on the Eyre Peninsula destroyed 46 homes in the hamlet of Tulka plus water tanks, caravans, trailers, boats, sheds and vehicles.

Since 2000, there have been the usual summer bushfires. Large fires were on Black Tuesday (2005) on the Eyre Peninsula where 9 people died, 93 houses, 237 sheds, 47,000 livestock and 6,300 km of fencing were destroyed. A total of 890,000 hectares was burned. Farm machinery and vehicles were also destroyed.

For 10 days in December 2007, dry lightning ignited 14 fires on Kangaroo Island and 6 of these grew into uncontrollable major bushfires. In the mid-north a fire raged for a month and threatened a few towns in 2014.

Fires were at Cherryville and Sampson Flat in 2002-2003. The Bangor fire near Port Pirie in 2014 was in inaccessible and difficult terrain. Each time it was considered to be controlled, high winds spread embers up to a month after the fire started, houses, sheds and stock were destroyed. There were also fires that raced through the mid-north wheat belt in 2015 and destroyed property at Pinery.

In October 2019, there were more than 200 fires in South Australia on hot, dry, windy days. Some 211,500 hectares were burnt out on Kangaroo Island. Two lives and 91 homes were lost on Kangaroo Island in 2020.[734] Fires were also in the Adelaide Hills and Yorke Peninsula where one person died and 84 homes, 400 farm buildings and 292 vehicles were destroyed.

The history of bushfires shows that South Australia can expect one large fatal bushfire every six or seven years.[735]

The Climate Council claimed that the South Australian bushfires were a result of climate change. This is contrary to evidence and is callous political point scoring in the wake of disaster.

Western Australia

Perth is often surrounded by summer bushfires. The latest in 2021 at Wooroloo was started by an arsonist. After huge bushfires in Western Australia in 1962, there was a two-decade long campaign of reduction of fuel loads in forests. There were very few bushfires in this period. Once the hazard reduction burning was reduced in 1985, more and more of Western Australia was burned by forest fires.

We were warned

Past catastrophic fires and the recommendations of Royal Commissions and other inquiries were totally ignored.

The quarterly Australian seasonal bushfire outlook published in August 2019 was warning that there were considerable areas of above normal fire potential for the 2019-2020 summer. The Bushfires and Natural Hazards Co-operative Research Centre that prepared the map was correct. Is this map constructed using dud global warming predictions? Or is this map constructed from eucalypt forests with a large fuel load, *El Niño-La Niña* Cycles, the Indian Ocean Dipole[736], rainfall, predicted winds and predicted temperature?

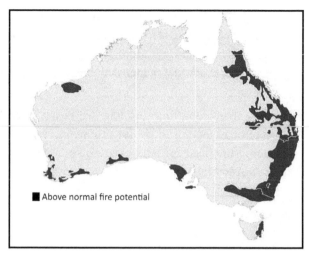

Figure 29: *Seasonal bushfire outlook of August 2019 just before the 2019-2020 catastrophic fires.*[737]

In a 2015 article in *The Age*, former CSIRO bushfire expert David Packham warned that fire levels had climbed to their most dangerous levels in thousands of years and noted that this was the result of misguided green ideology.[738] No one listened. The inevitable huge blaze occurred four years later.

In a 17th March 2015 submission to the Inspector General for Emergency Management by David Packham, the obvious was stated.

> *"1. We have a failed fire management policy and practice in Southern Australia especially Victoria. The failed fire management is resulting in an increasing threat to our forest environment (especially tree deaths), water supplies, life and property loss and a severe decrease in our quality of life.*
>
> *2. The thermal scale of wildfire in Victoria exceeds by a factor of 20 our ability to extinguish.*
>
> *3. Disaster scale wildfires occur about once in every 30 years but will become more frequent as fuel levels continue to rise.*
>
> *4. "Blow up" fire weather causes disaster fires when ignition and fuel are present.*

Packham concludes with reference to the 5% annual hazard reduction burns to reduce fuel loads *"Unless the 5% target is doubled or preferably tripled a massive bushfire disaster will occur, the forest and Alpine environment will decay and be damaged possibly beyond repair and homes and people incinerated"*. Packham was shown to be correct with the 2019-2020 bushfires.

In late August 2019, the Bushfire and Natural Hazards Co-operative Research Centre warned that eastern Australia had an above normal bushfire potential. Australia has had almost 100 inquiries into fires and fire management since 1983 that generated hundreds of non-binding recommendations.

Satellite data shows that global burning has decreased over the last 20 years and forest area has increased by 7% due to increased agricultural productivity reducing the need for land clearing.[739,740] NASA reported in 2019 that the area burned by global wildfires has dropped by 25% since 2003. This matters little if the forests are poorly maintained.

NASA also showed that the despite the increase in farming output, the worldwide forest area grew by 2.24 million square kilometres from 1982 to 2016.

Own the fuel, own the fire

Large forest fires are not constrained to Australia. Speaker after speaker at the International Cardiff Wildfires Conference in 2019 spoke of the need to reduce vegetation in vulnerable area.[741] In the UK, green civil servants have so far ignored this science and are opposed to heather burning, perhaps because it is a practice carried out to improve grouse moors. This is a classic example again of bureaucrats trumping empirical evidence from centuries of experience by shepherds and gamekeepers.

Foresters, farmers and fire fighters know that for a catastrophic fire, the forests need a period of wet weather followed by a drought then hot windy days. Fires need ignition, oxygen and a large fuel load.

Like many parts of Australia, the Sahara is hot, windy and dry. There is lightning for fire ignition and just as much oxygen in the air as in Australia. The Sahara has no bushfires because there is no fuel.

Reduce the fuel load. This reduces the intensity and size of the inevitable forest fires.

South-eastern Australia

Recommendation after recommendation from almost 100 bushfires inquiries argued that fuel load must be reduced and yet this has received opposition from green activists, green bureaucrats and have been ignored by the Greens Party and their green-tinted fellow travellers in the major political parties who know full well that fires kill.

Those who oppose forest maintenance have blood on their hands.

We can't reduce the amount or oxygen in the air from lightning strikes. However, we can reduce ignition by trying to address human stupidity and arson, by power line maintenance and infrastructure. Fuel load can be reduced by cold burns in winter year-in year-out.

More than 200 arsonists have been arrested since the start of the 2019-2020 Australian bushfire season. In just 3 months, 29 fires were deliberately lit in the Shoalhaven Region (NSW) and 27 in the Kempsey area (NSW). Most arsonists were young males 12 to 24 or men in their 60s.

On 8th April 2019, John Hermans of the Gippsland Environment Group sent a letter to the Victorian Minister for Environment, Energy and Climate Change (Lily D'Ambrosio) demanding a pause to fuel load reduction burns in East Gippsland. Later in 2019 and in early 2020, the East Gippsland forests were destroyed by bushfires as were native animals, property and livestock. People died. Hermans and D'Ambrosio have blood on their hands.

In 2019, the Andrews government of Victoria carried out just over one third of the planned fuel load reduction burns that had been recommended by the Black Saturday bushfires Royal Commission. The Department of Environment, Land, Water and Planning (DELWP) approved 246,396 hectares of burning in 251 planned burns but only managed to burn 130,044 hectares. The Royal Commission recommended 385,000 hectares of burning.

If someone in private industry fell so short of such targets, they would be sacked. Why were no bureaucrats sacked? Victoria burned while its bureaucrats fiddled in their city office wilderness and forgot history but never forgot to collect their pay cheques.

The DELWP previously mapped fuel loads which were publicly available but now to view these maps one must make a formal Freedom of Information application and pay thousands of dollars for information funded by the taxpayer. What are they trying to hide?

New South Wales has about 20 million hectares of forests and the current level of prescribed burning is about 200,000 hectares annually. This level of prescribed burning will continue to do little to reduce the risks of catastrophic bushfires.

The hazardous level of fuel loads can be reached within 2 to 4 years from the low intensity prescribed burning in south-eastern Australia. But the prescribed burning practices are not popular among locals. The smoke from the hazard reduction burns is a nuisance and deemed a health issue. A far greater health issue is a major bushfire.

Greens live in concrete city jungles, appear to ignore history and do not have the experience and knowledge to tell farmers and foresters how to protect forests, property, plants and wildlife from bush fires. Equally, farmers do not design office blocks or transport systems in cities.

The green movement has had undue influence on bushfire management over the last few decades. It's called green tape. It is the over-regulation of native forests by green bureaucrats that has created a tinderbox for uncontrollable fires. With governments banning logging that thins forests, provides fire-fighting access roads and forest management, our forests and people living near them are teetering on pyromaniacal oblivion.

The bushfires of south-eastern Australia in 2019-2020 emitted more than 400 Mt carbon dioxide which is only 140 Mt less than Australia normally emits each year. It is green activist policy that adds carbon dioxide to the atmosphere. In a twist of irony, it has been shown that pretty well all of the carbon dioxide released by the 2019-2020 fires has been taken up by algal blooms the size of Australia in the Southern and Pacific Oceans.[742] Why can't algal blooms use the human emissions of carbon dioxide also? Just asking.

Mega-fires occur every 30 years or so when conditions are horrendous in prolonged stinking heat and gale-force winds. Thirty years is enough time for people to forget.[743] And they do.

California

Did climate change cause the recent forest fires in US? US government web sites say no. In pre-industrial times, 140 million acres burned. In 1930, 52 million acres burned (excluding Arizona and Arkansas) and in 2018, 9 million acres burned.[744,745]

In April, 2018 a PBS television[746] series claimed that US wildfires are at an all-time high. This makes for good scary sensationalist television but is false. It can be shown to be false with a 30 second search.

"Scientific evidence reveals there has been no climate effect regards California's wildfires! None! The data below proves it beyond all doubt" stated Jim Steele.[747,748] Some 70% of California's 2020 burnt area was in dry grasslands. The grasses only require a few hours of warm dry conditions

to become highly flammable, so they *"are totally insensitive to any added warmth from climate change"*. The century trends in local temperatures where California's biggest fires have occurred reveal no connection to climate change.

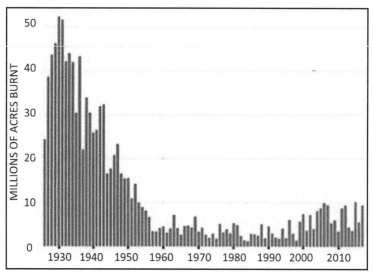

Figure 30: *Area of US forests burned between 1926 and 2017. Since that time, 58,083 fires burned 8,767,492 acres in 2018, 50,477 fires burned 4,664,367 acres in 2019 and in 2020 58,950 fires burned 10,122,336 acres all which were at or less than 20% of the record land area burned.*[749]

In most cases the local maximum temperatures have been cooler now than during the 1930s. Steele presents temperature records from 1900 to 2020 at the sites of the largest fires in California this year as evidence. The state should address fuel management and creating defensible spaces in fire prone California. Fire ecologist Thomas Swetnam wrote *"The paradox of fire management in conifer forests is that, if in the short term we are effective at reducing fire occurrence below a certain level, then sooner or later catastrophically destructive wildfires will occur"*.

The 120-year Australian rainfall record and 125-year Californian rainfall records of NOAA show no systematic change in climate than can be related to forest fires. The super dry year of 2013 in California is not climate, it's weather.[750] California, just like Australia, has frequent catastrophic wildfires.

The worst bushfire for fatalities in the US in recorded history was on a cold windy October 8th, 1871 in Peshtigo, Wisconsin. Some 486,000 hectares was destroyed, twelve communities were wiped out and 1,500-2,000 people died. Was this due to climate change? The area burned is small by

Australian standards.

Californian fires are slowly coming back to their prehistoric state because of the enormous excess fuel load. Putting up solar panels and using biofuels will not make one *iota* of difference. Prescribed burns will. Before 1800, California typically saw between 1.8 and 4.8 million hectares burn each year.[751]

In 2017, the US Geological Survey's Jon Keeley stated[752] *"We've looked at the history of climate and fire throughout the whole state (California), and through much of the state, particularly the western half of the state, we don't see any relationship between past climates and the amount of area burned in any given year."* Keeley modelled 37 regions across the US and found that humans may not influence fire regimes but their presence can actually override, or swamp out, the effects of climate. The only statistically relevant factors for the frequency and severity of fires on an annual basis were population and proximity to development.

Government green decisions made under unrelenting pressure from green groups have converted big forest fires into hellish conflagrations. In California, people like to live around vegetation that every year dried out enough to burn uncontrollably, prescribed burns are now largely prohibited because burning releases the dreaded carbon dioxide, vegetation gets even thicker and every summer disaster is always just around the corner.

The disastrous and tragic 2018 Camp Fire in California thrived on the tremendous amount of fuel lying on the ground and was not helped by drastic reductions in timber harvest in national forests and significant reductions in livestock grazing on Federal lands. More than 45% of California is Federal land. The Californian chaparral and planted Australian eucalypts burn very hot because both plants contain oils. The November 2018 Camp Fire was California's deadliest fire destroying the town of Paradise. It was California's 16th largest fire.[753]

In 2018, Keeley stated that the ignition sources of fire had declined in California except for power lines.[754] He wrote *"Since the year 2000, there've been a half million acres burned due to powerline-ignited fires, which is five times more than we saw in the previous 20 years."* and *"... there's no relationship between climate change and these big fire events"*.

In 2019, wildfires in California consumed 253,214 hectares. Scary photographs of fires in the Californian newspapers in their 2020 summer accompanied claims that these fires were massive, unprecedented and were due to climate change. Again, it only takes a few seconds to access the history of wildfires in California on a smart phone.

This would show that the 2020 fires were not unprecedented and small compared to historical fires. In July 2021, wildfires again destroyed houses, farm buildings and nearly 250,000 hectares in California. Over the past decade, California has seen an annual average burnt area of 314,000 hectares.

People fretting about human-induced global warming clearly have not experienced wildfires sparked by arsonists, lightning strikes and self-immolating wind turbines that are the cause of most of the terrifying wildfires.[755] Wind turbines are not super safe, clean or green and fires in them are ten times more common than the wind industry claims.

Almost all fires are started by humans and are almost exclusively man-made calamities[756]. In Australia between 1977 and 2009, 87% of the 113,000 fires that started in forests and grasslands were man-made. A University of Colorado at Boulder report showed that 97% of home-threatening fires in the US from 1992 to 2015 were started by humans.

Native Americans, like the Australian aboriginals, had small cold burns that cleaned out the undergrowth[757] *"for thousands of years prior to Euro-American settlement, Native American tribes and lightning fires burned as much as 10 to 12 million acres in California every year (five times greater in area than the California fires of 2018)... These fires were typically more benign, burning more often but at lower intensities. The federal government's focus on fire suppression has resulted in denser forests with more continuous fuel to burn in an intensive fire"*. What's new?

The Forest Foundation of Sacramento states[758,759] *"Prior to European settlement in California...[the] generally low intensity fires...helped to clear the understory, keep the forest canopy open, and as a result guarded against mega-fires..."* and *"Starting in the early 1900s land managers were successful in controlling many of the fires, but the result was a tremendous build-up of dense brush and over-stocked forests. We can reduce the risk of wildfires with the help of healthy forest management – which includes forest thinning and the removal of excess fuels that can feed and increase the size of a fire. Historically, our forests have contained 50-70 trees per acre, and today our forests have more than 500-1,000 trees per acre – increasing the risk of catastrophic fire. Without this natural thinning, forests grew more crowded and shade tolerant trees filled the understorey, providing ladder fuels for today's crown fires that jump to the crown of the tree and spread quickly...".* Why do we even bother to listen to the Greens who sprawl on a couch in a city apartment and have no practical knowledge and experience like foresters and who don't read history?

It's all very nice to live in or adjacent to a forest with pines, eucalyptus and other fragrant plants. If the oils in a plant smell nice, it's going to burn hellishly and they want to kill you.

The Californian bushfires stirred Islamic terrorists into action to use massive forest fires as a weapon.[760] Various networks reported terrorists were being trained to create huge forest fires in the US and European forests as a mechanism of jihad.[761] European Mediterranean countries have large areas of flammable native pines and planted eucalypts.[762] Instructions for making ember bombs were issued[763] and arsonists had already targeted a Catholic university as a training run.[764]

Arson is one of the major causes of catastrophic forest fires in California and Australia. One wonders whether fires are deliberately lit for a political or financial purpose as in Italy or done by nutter thrill seekers.

Bush fires are not needed in the US to destroy forests. In the US, some 5% of the global grain crop now goes to clean green ethanol as a fuel rather than food and forest land is cleared for the increased production of ethanol.

Canada

According to the UK Met Office, global warming is leading to record breaking fires in North America. They claimed[765] *"...hot dry weather conditions promoting wildfires are becoming more severe and widespread due to climate change".* However, in Canada wildfires are at the lowest level for decades. Does the Met Office claim that Canada is not part of North America? Maybe the Met Office should consider that the US wildfires have something to do with forest maintenance, power systems and arsonists? Or, perish the thought, maybe the Met Office is just plain wrong. Again.

Canadian environmentalist David Suzuki had a TV series called *The Nature of Things*. There were episodes called *Into the fire* which showed that the indigenous population of Canada used fire stick burning the same as the Australian Aboriginals did. It was done for the same reasons. To make hunting more efficient. This is now no longer done in Canadian forests with the inevitable monstrous wildfires.

The Nation Wildland Fire Situation Report by Natural Resources Canada shows that as of September 9, 2020 the area burned by wild fires was 8% of the 10-year average. By contrast, the September 2019 report shows the area burned was 66% of the 10-year average. The report says *"2020 has been one of Canada's quietist since the 1990s. The national wildfire preparedness level remained at Level 1 most of the summer, indicating fire management agencies had adequate resources to respond to fire events without assistance from other jurisdictions"*. Apparently increasing carbon dioxide doesn't affect forest fires in Canada. Canada's precipitation trend from 1910 to 2019 is an increase of 0.13 mm per day per century. A warmer Canada is a slightly wetter Canada.

Brazil

The usual suspects came out of the woodwork with the Amazon burning season fires of 2019. Many of the fires were on previously cleared land and started by opportunistic landholders rather than big corporations. At the G7 meeting in France, President Macron tried to stop Brazilian land clearing and, to support his claims of fires occurring while he spoke, he used a 20 year-old file photo represented as a photo of the 2019 fires. This is fraud.

Data shows that deforestation and burning in the Amazon has been in decline since 2003. France has burned its forests centuries ago and this increased prosperity. Why does France want to stop Brazilians becoming more prosperous? Some 60% of Brazil is covered by native vegetation and 24% of the country is covered by protected areas. If everyone wants to save the Amazon, who will foot the bill?

There were reports by the international media in 2019 of how the Amazon was burning, allegedly due to man-made climate change. Sao Paulo city was blanketed by smoke. Brazil is around one and a half times the size of Australia. It is wet. Brazil is green right across the country, not just on the eastern edge like Australia. Fires and catastrophic climate change are more newsworthy for the international media than rains and floods.

The New York Times reported regarding the Amazon fires that *"the 2019 fires were not caused by climate change"*. In 2019, fires in four Amazon countries destroyed 2.24 million hectares and in Australia during the same time period, 12.35 million hectares were destroyed

France and other EU countries use biodiesel which derives from palm oil plantations in south-east Asia. About 700,000 hectares of forest is cleared by fire to meet the demand for biodiesel in the EU. Is hypocrisy the word that comes to mind?

Burning forests to save the planet

The EU has deemed that wood-fired electricity generators are carbon-neutral despite being fined for air pollution in the US.[766] Drax (Yorkshire) was the largest coal-fired power station in Europe (3,960 MW) and used 36,000 tonnes of coal per day. It powers 10% of the UK grid. The EU was threatening to close Drax. The solution: burn 70,000 tonnes of wood per day.

This wood, of course would not come from the UK or EU because it would destroy their forests. Not so. It comes from North Carolina and is shipped as pellets over 5,000 km across the Atlantic Ocean from the purpose-built Chesapeake Port in Virginia.

Harvesting 70,000 tonnes per day from the other side of the world is no mean feat. After clear felling North Carolina's forest of maples, sweet gums and oak in the swamp lands, the wood, bark, leaves and sawdust are converted to pellets in giant energy-hungry factories. This pelletising process uses large amounts of energy derived from coal- and nuclear-fired electricity generators in the US.

For the 20-year life of the Drax power station, 511 million tonnes of wood will be harvested using diesel equipment from trees in the US to provide expensive subsidised "renewable" electricity in the UK. The tonnage of wood harvested and transported is more than that of rock most large mines produce in their lifetime.

US environmental groups claim that these forests comprise some of the most biologically important forests in North America and that there are risks to wildlife survival and biodiversity. Otters, pileated woodpeckers and many rare birds are especially threatened. They don't really matter as the EU and UK greens are saving the planet from carbon dioxide derived from burning coal. This is green political policy in action. Destroy the forests and wildlife in another country for the sake of feeling good at home.

Green activists used to demonstrate against the Drax power station burning coal. It was Europe's single largest carbon dioxide emitter. The company now boasts of its "environmental leadership position" and states that it is the biggest "renewable" energy plant in the world. Demonstrations by green activists have ceased.

The passive start to the environmental movement in the 1970s was against harvesting forests for timber. Now, in order to keep the ideological home fires burning, green policies have led to the clear felling of 70,000 tonnes of trees per day in North Carolina for burning in the UK.

No wildfires could do as much damage as Drax in such a short period of time. Clear felled forest destruction for burning wood is now saving the planet. Once it was regarded as destroying the planet.

Europe's renewable energy targets are built on burning trees because, with a stroke of the pen, trees are regarded as a form of renewable energy. Green activist European politicians have deemed that burning compressed 300 million year-old forests to produce steam for electricity generation is sinful and should be punished because of the carbon dioxide emitted.

However, to clear-fell and burn forests in Estonia or the US to produce steam in Europe and the UK for electricity generation is a commendable green energy renewable biofuel project replete with generous subsidies. Almost all European countries have recorded an increase in logging for energy. Nearly a quarter of the trees harvested in the EU in 2019 were for energy, up from 17% in 2000.[767]

Over the last 1,000 years, European forests have been clear felled a number of times for the glass and metals industries and household energy. Burning forest wood was unsustainable then. It still is. Wood burning emits carbon dioxide and toxins that are now deemed good for the environment because language has been changed. All of this was devised and dictated by green activist bureaucrats and paid for by the poor punter.

Pointing the bone

Fires run on fuel. Limited fuel means limited fire. If you own the fuel, you own the fire. For decades it has been shown to greens activists, bureaucrats and politicians that fuel loads must be reduced. Greens, public servants and politicians should be kept well away from forests and National Parks. Forests have now been politicised.

Greens have shown that they have no concern about the environment, people and their property and are only interested in scoring political points and having more control over as many people and policies as possible. Greens kill wildlife, forests and people.

The Australian Labor Party leader Julia Gillard did a dirty deal with the Greens to lock up Australia's forests. This allowed Gillard to have the numbers in parliament to be one of Australia's worst Prime Ministers before her own party threw her overboard. This deal was done a decade before the bushfires of 2019-2020.

The Federal Australian Labor Party leader during the height of the tragic 2019-2020 bushfires said the bushfires were a "national emergency" and stated we should fight fires with a carbon tax. The Australian Labor Party leader did not state how many carbon credits would go to China, how a tax could stop a mega-fire and how a tax would stop arsonists.

If we sent more money to the UN, would bushfires cease? If Australians stopped emitting carbon dioxide, there would be zero effect on bushfires and the history of bushfires shows that there is no relationship between emissions of carbon dioxide and bushfires. The 2019-2020 fire season emitted about 80% of Australians' annual carbon dioxide emissions.

It's true to green activist form that whatever the weather, it's due to climate change. Australia's fires in 2019-2020 we were told were due to climate change as was the extreme rainfall in east Africa. However, at the same time the record cold in India didn't fit that narrative whereby winters were to be warmer so had very little media coverage.

Greens are incensed over suggestions that anything other than fossil fuels and climate change might be turning green California and Australian ecosystems into black wastelands, incinerating wildlife, destroying homes and killing people.

The notion that they and their policies might be a major factor in these fires gets them so hot under the collar that they could ignite another inferno. But the facts are there for all to see. Benjamin Franklin stated *"We are all born ignorant but one must work very hard to remain stupid"*. The Greens Party works very hard.

On 9th January 2020 during the height of the Victorian bushfires when people were dying, forests and their wildlife were being destroyed, and property and livestock were being lost, the four Victorian Greens Party politicians issued a statement wanting an immediate declaration of a Victorian climate emergency and 100% renewable energy and phasing out of coal by the end of the decade.

Greens have no shame, try to politically capitalise on tragedy and have never shown that the 2019-2020 bushfires result from a climate change and that renewable energy and destruction of the coal industry will make the situation any better. Their message is transmitted using coal-fired energy.

Why don't journalists ask Greens politicians simple questions?

Tragic bushfire losses mean nothing to the city-based Greens Party, to green activists and green bureaucrats and politicians. Those with various shades of green blame the Prime Minister, climate change, coal mining and carbon dioxide for the 2019-2020 bushfires in Australia in an attempt to make political capital by exploiting tragedy and spreading misinformation.

Kylie Jenner, a member of the Klan Kardashian, donated $1 million to Australian firefighters and those impacted by fires. When Kylie Jenner is more practical and generous than the whole clan of parliamentary Greens, then we are in real trouble in Australia.

Why were no Greens hugging trees to save them in the 2019-2020 fires? Why was there no green activist firefighting brigade risking their lives trying to save the forests that they claim to care about?

Hypocritical cant from Greens Party politicians shows that they are only interested in trying to score political points in a disaster rather than actually doing something tangible to help people and the environment.

The Greens claim that the most recent bushfires in Australia are due to our use of coal and climate change. Nowhere do they state that almost 200 arsonists have been arrested and that some 85% of all Australian bushfires are from arson or human error. Is the Greens Party claiming that arson is a result of climate change? Do Greens sympathisers become pyromaniacs for the green cause? Nothing would surprise me about green activists.

In the height of the fires on 8th January 2020, the Professor of Terrestrial Ecology at The University of Sydney (Chris Dickman) stated[768] that more than a billion animals have been killed by the Australian bushfires. Images of burned animals were very distressing and the RSPCA, vets and volunteers helping these poor animals are true heroes.

Each year, two million native animals are killed in Australia by feral cats and there is huge opposition by feline fanciers to cull feral felines. There are more than two billion social media video posts by cat lovers. What are green activists doing to stop the major killer of native animals in Australia? Nothing.

In the Australian Senate on 6th February 2020, Sarah Hanson-Young claimed "*A billion animals have been killed in these climate fires....When will the government act to stop the koala being killed by the coal miners and loggers?*" She has shown the end result of the dumbing down of the education system. Why doesn't she complain about the birds and bats being killed by her wind turbines? Why does she object to reduction of fuel loads in forest areas?

Why does Hanson-Young's Greens Party have a lock and leave it policy for the bush which numerous Royal Commissions and Inquiries have shown raise the fuel load for the next inevitable fire that kills wildlife that Greens claim to be concerned about? How can climate create a bushfire when we know that a fire needs ignition, oxygen and fuel? Why are the fires not arsonist fires rather than climate fires? Does she know that loggers open up fire trails and reduce fuel loads?

Is Hanson-Young aware of previous bigger fires such as in 1851 and 1939 that occurred before her mythical human-induced climate change? Is she not concerned that humans have lost their lives, especially those wonderful volunteer firefighters? Is she aware that people lost everything in these fires? Where is her concern for the average person?

Was she aware that while she was rabbiting on about climate fires, there were drenching drought breaking rains elsewhere in Australia? Just run this past me again Sarah, how does a coal miner kill a koala? Is she aware that there is no relationship between atmospheric temperature and coal mining?

As taxpayers, we pay for this Green Party rubbish which is green murder. Murder of people, livestock, native animals and forest plants. Australia once had a Prime Minister who was a volunteer fire fighter. What Greens Party politician is a trained firefighter and is actively engaged fighting fires to save people and their houses, livestock, native animals and our forests?

A Greens-Party-supported climate change protest in Melbourne on 10th January 2020 tied up police and emergency services personnel who were desperately needed in bushfire-affected areas.

A pious email message from the Greens Party leader in Australia about the bushfires in 2019-2020 was almost all narcissistically about his heavy heart. It called for donations to the Greens Party and not the bushfire victims. Senator Richard Di Natale was moralising from his couch, was nowhere to be seen as a volunteer fighting fires, manning a soup kitchen for fire fighters or providing accommodation in his numerous houses for those who had lost everything.

He was claiming that fires that had been occurring for thousands of years are suddenly now the result of Australia burning and exporting coal, human-induced climate change and lack of political leadership by his political opponents. This political leader was nowhere to be seen in the fire-affected areas for a good reason. He would have been lynched.

The Greens believe that they can stop bushfires by importing solar panels and wind turbines from China and putting people out of work in the coal industry. Greens ignore history, take no responsibility for their own murderous policies and have no shame. How can green sympathisers sleep at night in the cities knowing that they are responsible for the loss of lives? A vote for Greens is a vote for forest flora and fauna killing by mega-fires, property loss and killing of humans and their livestock, pets and crops.

After the fatal 2009 Victorian bushfires, Miranda Devine showed that green ideology must take the blame for deaths (*Sydney Morning Herald* 12th February 2009). It was not climate change that killed nearly 300 people in these fires. It was the unstoppable intensity of a bushfire, turbocharged by huge quantities of ground fuel which had been allowed to accumulate over years of drought. It was the power of green ideology in the bureaucracy and government to oppose attempts to reduce fuel hazards before a mega-fire erupts and which prevents landholders from clearing vegetation to protect themselves. She wrote *"It is not the arsonists that should be hanging from lamp-posts but greenies"*.

On January 13th, 2013, Miranda Devine reminded readers of *The Sunday Telegraph* that bushfire catastrophes occur time and time again. She stated that the greens who oppose winter fuel reductions are more concerned about wildlife than humans and property and this wildlife gets totally incinerated during a big bushfire. The big fires of 2019-2020 show that nothing has changed.

For greens to claim that bushfires are due to climate change ignores Aboriginal history and the history of bushfires in Australia. It is a convenient get-out-of-gaol free card for green activists, the Greens Party, governments and the obstructive bureaucracies they create. Green tape, heavy-handed bureaucracies and legalistic and linguistic ploys to rewrite history are used after every great fire.

There is green hostility to proper bushfire management by greens who have infiltrated government decision-making bodies and agencies, green NGOs and Greens Party politicians. It is time for the silent majority to win back control of their own lives. Catastrophic fires will occur again. There are tried and proven ways to make them less catastrophic.

The Greens Party believe that they can stop bushfires by importing solar panels from China and putting people out of work in the coal industry. Bushfire victims don't sit around the remains of their charred home discussing climate change.

Greens ignore history, take no responsibility for their own murderous policies and have no shame. How can green sympathisers sleep at night in the cities knowing that they are responsible for the loss of lives?

The claim by Greens Party politicians that fires are due to climate change, coal mining or carbon dioxide is an abrogation of their responsibility and accountability. The overwhelming evidence is contrary to these claims.

Australia is a continent of fire. The 2019-2020 bush fires in south-eastern Australia were not abnormal. The hazardous massive volume of fuel loads together with arson, an abnormally positive Indian Ocean dipole and the associated drought are the prime reasons for extreme bushfire season in south-eastern Australia. The only way to reduce catastrophic fires is to reduce the fuel load. This is prevented by green activists.

A vote for the Greens Party is a vote for forest flora and fauna killing by mega-fires, property loss and killing of humans and their livestock, pets, crops and livelihood. This is apparently OK because the planet needs to be saved. It is no wonder that those in rural Australia think that the Greens Party are beneath contempt.

Greens murder forests, murder wildlife and murder humans. They are culpable.

What are you going to do about it?

Media monstrosities

The reliably bizarre Moonbat penning for *The Guardian* (8th January 2013) in an article called *"Heatwave: Australia's new weather demands new politics"* shows that Moonbat need to get his prescriptions changed. Apparently through the distant eyes of Moonbat, Australia's 2012-2013 summer was unprecedented, that volunteer firefighter and Prime Minster Tony Abbott was an eco-arsonist and that these fires were due to climate change.

Maybe a bit of reading on previous catastrophic bush fires such as those of 1851 and 1939 and heatwaves such as happened in 1851 and 1896 was just too much of an effort for Moonbat. After all, a 30-second search on a smart phone would have given Moonbat a picture of past fires.

Sir David Attenborough gave us the benefit of his experience from his high horse in the distant UK *"As I speak, south-east Australia is on fire. Why? Because the temperature of the Earth is increasing"*. Attenborough has an army of researchers who in less than 30 seconds would have realised that mega-fires in south-eastern Australia are normal and also took place when the planet was cooler.

In September 2019, ABC Gippsland posted a supportive Facebook story about how locals were protesting against the spring prescribed fuel load reduction burning of 370 hectares at Nowa Nowa (Vic). Protestors held signs *"Spring burns kill baby birds alive"* and *"Stop burning nesting birds"*. It was an ABC free-for-all propaganda parade, they asked no hard questions and made no mention of fuel load. One protestor said *"I'm more worried about climate change than the burns we are having"*.

The original plan to burn 370 hectares was reduced to 9 hectares as a result of the protests. Much of the whole district is now reduced to a post-bushfire wasteland and pretty well all wildlife was killed. The chickens came home to roost. The ABC post quietly disappeared when the 2019-2020 bushfires razed the Nowa Nowa area.

There is something toxic in the air in the hallowed halls of the ABC. Laura Tingle suggested Prime Minister Morrison is personally responsible for the ferocity of the current bushfires as his government refuses to confront *"the realities of climate change."* Using that logic, every previous Prime Minister was also personally responsible for every bushfire since Federation. The same response and a question should be asked of Ms Tingle *"Show me the evidence"* and *"Why didn't you do a 30 second search or do you have another agenda?"*

The 2019-2020 bushfires were only "unprecedented" because commentators did not look at the history of fires, fuel loads and arson. As soon as language is corrupted for a political agenda, we no longer hear facts. The 2019-2020 bushfires were the first fully politicised fires and burnt out an area equal to 80% of the area of the UK. In the past, there were the 1926 Black Sunday fires (60 dead; 1,000 buildings destroyed), 1939 Black Friday fires (71 dead; 5,000 buildings), 1967 Black Tuesday fires (62 dead; 1,300 buildings), 1983 Ash Wednesday fires (75 dead; 3,000 buildings) and 2009 Black Saturday fires (80 dead; 3,500 buildings).

Which one of these catastrophic fires was due to global warming? Why are massive bushfires in areas that had not recently been burnt?

The global media depicted the 2019-2020 fire season in Australia as "unprecedented". It sells newspapers, radio shows and TV broadcast but it is not true. The 2019-2020 fire season ranked fifth in terms of area burned, with about half of the burned acreage of 2002, the fourth-placed year, and about a sixth of the burned acreage of the worst season in 1974-1975. The 2019-2020 fires ranked sixth in fatalities, about half as many as the fifth-placed year, 1926, and a fifth as many fatalities as the worst fire on record in 2009.

With almost every disaster the media get it wrong because they don't look at the past. Politicians and bureaucrats try to blame their failures on climate change. The only failure is that of common sense.

5

THE WET AND THE DRY

Drought is normal, the biggest droughts were before industry emitted carbon dioxide and mega-droughts up to 300-years long changed human history.

Mega-droughts are most common during periods of glaciation, not warm times.

The dustiest, wettest and hottest days took place well before you were born and are unrelated to human emissions of carbon dioxide.

Floods inundate flood plains which are long-term sediment deposits from large floods, all of which exceeded anything experienced in your life. Modern catastrophic floods are inevitable and exacerbated by human mismanagement and are unrelated to human-induced climate change.

A claim that human emissions of carbon dioxide led to ocean acidification is misleading, deceptive and deceitful.

Due to a diversity of causes, sea level rises and falls, as does the land level. Pacific island nations on atolls have increased in land area and have not been inundated. Sea level has dropped over the last 4,000 years.

Rates of sea level changes at present are no different from the past.

Reefs have been on Earth 3,500 million years, coral reefs have been around for 650 million years and past reefs have died due to sea level fall and sediment and volcanic ash inundation.

The 2,000-km long Great Barrier Reef comprises 3,000 reefs built on dead coral 50-100 metres thick. It is dynamic, has migrated, has died and recovered five times in the last 30,000 years and is not threatened by a higher carbon dioxide content in the atmosphere, by sea level rise or ocean temperature rise.

Farming and coal mining inland from the Great Barrier Reef have no effect on the health of the Reef.

Many scientific studies on the Great Barrier Reef cannot be replicated, are of poor quality and vital data is withheld.

Extreme weather events and fatalities have decreased over the last 100 years.

Extinction is normal, there is a constant turnover of species and the number of species, genera and families of life continues to increase over time.

Our planet is in the "Goldilocks Zone" just at the right distance from the Sun. This has enabled liquid water to be present on our planet except for the first few hundred million years when the planet was hot and constantly bombarded by asteroids. For four thousand million years ago, we have not had permanent ice with ice from the top to the bottom of the oceans or the loss of all water by evaporation or blasting away by solar wind. There have been substantial falls and rises in air temperature, far greater than anything climate activists are getting into a lather about today.

Mars has a daily temperature variation of 100°C. The daily temperature variation on Earth is about 10°C and the Western world now has heated and air conditioned houses, vehicles, shopping centres, office blocks and even underground mines. *The Moscow Times* reported that Verkhoyansk, a town of 1,300 residents, is in the Guinness Book of Records for the greatest seasonal variation from -68°C to +38°C.[769] The atmosphere is 0.0001% of the planet's mass and the oceans are 0.022%. The atmosphere and oceans make the planet habitable and some 80% of the planet's surface heat is held in the oceans.

We live on a very boring planet in the "Goldilocks Zone" where surface temperature changes are not unprecedented or catastrophic.

Drought

Get over it. Drought is normal and has been occurring since the year dot. Mega-droughts have changed empires. There is no relationship between carbon dioxide in the atmosphere and drought, let alone human emissions of carbon dioxide.

Over the last 60 years there has not been an increase in drought[770] despite a huge increase in human emissions of carbon dioxide.

In the geological record, drought is more common in colder times than in warmer times. There are massive units of red beds derived from dune sands and of salt layers formed in drought-stricken areas which were not covered by ice in the six major ice ages on Earth.

Drought signature in ice and the oceans

During periods of glaciation, there is less water vapour in air hence there is less rainfall and less vegetation. This results in more desertification and more dust. In the past, global warming has not produced more desertification. It is global cooling that creates desertification. Cold windy times and periods of aridity add dust and sea spray to snow that is later compressed to ice. Dust is also preserved in lake muds and ocean floor muds.

The history of ice on Mt Kilimanjaro gives a story showing that the ice comes and goes. Some 4,000 years ago, conditions became cooler, drier and windier. This is represented by a major dust layer in the Kilimanjaro ice. A period of intense drought started some 4,000 years ago. This 300-year drought was experienced in northern and tropical Africa, the Middle East and western Asia.[771,772] Many civilisations collapsed[773] and the drought was probably global[774] as dust is present in ice cores from the Huascarán glacier in the Andes of northern Peru.

The Kilimanjaro ice sheet grew and contracted in concert with large-scale coolings and warmings in Africa over the last 4,000 years. If there were a 300-year drought from now, the same would happen. There would be a reduction in food, desertification, increased dust, depopulation and a change of the global power structure.

Even the Little Ice Age (1280-1850 AD) dust can be detected in both Greenland and Antarctic ice cores. Ice cores show that the more dust in the atmosphere, the colder the climate and that glacial conditions have over 50 times as much dust in the atmosphere as interglacial times.[775]

Most of the world's oceans are extremely depleted in iron. This is because the oceans are alkaline, and contain dissolved oxygen and so dissolved iron precipitates to the ocean floor. Iron is a micronutrient for photosynthetic micro-organisms. During ice ages, forests and grassland turn to desert. The increased winds blow red iron-bearing desert dust into the oceans.

Much of North Africa, especially the Sahara, contributes iron-rich dust to the oceans[776]. This results in a blooming of micro-organisms in the oceans. The expansion of these photosynthetic organisms withdraws even more carbon dioxide from the air and adds oxygen to the air.[777]

If atmospheric carbon dioxide drives climate change, then during an ice age red dust blown into the oceans would accelerate the removal of carbon dioxide into the atmosphere and the Earth could then not escape from a runaway ice age. This has not happened. Clearly carbon dioxide is not the main driver of climate.

Even though 20th Century dust deposition in Antarctic ice was higher than in the 19th Century, the dust content is still more than 30 times that measured in ice deposited during previous glacial times.[778] The 20th Century dust in Antarctica mainly derives from increasing desertification of Patagonia and northern Argentina as a result of over grazing.

Increased human activity, principally from farming and livestock grazing, has produced up to 500% more dust from wind erosion of surface sediments in the US.[779] There has been a fivefold increase in dust in alpine ecosystems.

Strange as it seems, overgrazing results in the extraction of more carbon dioxide from the air.

Historical droughts

We have good records of mega-droughts over time. When the Western Desert of North Africa changed from grasslands to desert in 4300 BC, it was a result in changes of monsoonal strength[780]. Much of the Middle East had a long drought and the area underwent desertification and people migrated as settlements collapsed.[781,782]

When a prolonged drought began in northern Mesopotamia in 4200 BC, the region was largely abandoned and the population migrated south where there was irrigation-based agriculture. There was a similar collapse of Neolithic cultures around Central China at the same time, with drought in the north and flooding in the south.[783] Climate changed at about 3800 BC from the previous 400 years of cold dry times that led to the collapse of Saharan and Mesopotamian civilisations. Warmer wetter times returned.[784]

Greek mythology refers to deforestation, flooding, siltation of irrigation channels, salination and collapse of the Sumerian city states.[785] Written records dating back 5,000 years describe declining crop yields and decreasing grain production with wheat being replaced by the more salt-tolerant barley. Patches of soil turned white on agricultural lands because of the accumulation of salt by evaporation and there was not enough water for flushing out the surface salt.[786] A mega-drought began again around 2300 BC in both the Northern and Southern Hemisphere. This was a solar-driven drought that led to widespread famine.[787]

Wind-blown sediments deposited in the Gulf of Oman show that starting at about 2562 BC, there was a 300-year period of intense wind. This wind was a characteristic of aridity and the wind-blown dust derived from northern Mesopotamia.[788] The coldness and aridification at this time may have been caused by the weakening of Northern Hemisphere ocean currents.[789]

This was also recorded in the Southern Hemisphere.[790] There was a solar-driven global mega-drought which led to widespread famine.[791] Empires collapsed. The period from 2200 to 1900 BC was a dark age with Dynasty VI Egypt ending in anarchy, the disintegration of the Akkadian Empire, destruction of Byblos and other sites in Syria, and, further afield, the destruction of Troy and the decline of Indus Valley civilisations (e.g. Harappan).[792] The Indo-European peoples migrated to northern Europe, Greece, southern Russia, Turkey, Iran, India and Xinjiang (north-west China).[793]

Increased aridity and wind on the Habur Plains of Syria heralded the onset of this 300-year savage drought.[794] The Akkadian Empire ruled a region from the headwaters of the Tigris and Euphrates Rivers to the Persian Gulf. The northern Mesopotamian civilisation depended upon regular rains and, after four centuries of urban life at Tell Leilan, the city was abandoned.

This led to the collapse of the Akkadian Empire. The synchronous collapse of civilisations in adjacent regions shows that the abrupt climate change was extensive.[795] Although civilisations collapsed during cooling and desertification, humans also adapted to climate change, especially warming.[796]

Ice core records in the Mt Kilimanjaro glacier show abrupt climate changes including a 300-year drought at about 2000 BC.[797] Stalagmites from a cave in West Virginia (USA) provide a detailed record of climate in North America over the last 7,000 years. Cave deposits are undisturbed. They give a better climate record because burrowing animals can disturb the ocean and lake sediments which are commonly used to track past climates.

The cave deposits of West Virginia show that when the Earth received less solar radiation every 1,500 years, the Atlantic Ocean cooled, icebergs increased and rainfall decreased. This led to long droughts, especially between 4,300 and 2,200 years ago.[798] Icebergs would have increased in number.

Clearly the great drought of Mesopotamia was more widespread than just in the Middle East. Climate change is a powerful causal agent for the evolution of civilisation. Global cooling is generally associated with a collapse of civilisations whereas global warming is associated with great advances in civilisations.[799]

In the Indus Valley, in what is now Pakistan and northwest India, between 4,500 and 3,500 years ago, cities such as Harappa and Mohenjo-dar survived by cultivating grain, cotton, melons and dates. Then Harappa just disappeared off the map. No record of war exists. The most logical conclusion is that desertification drove abandonment of settlements. Although the area occupied by the Harappans is now arid, fossils, pollen and lake sediments tell a different story.[800] The area was lush with sedge, grass, mimosa and jamun trees. Jamun needs a rainfall of at least 50 cm a year, twice the modern rainfall in this area.[801] Lake sediments show the area was very wet from 3000-1800 BC and then it was dry until 500 BC.

Sediment layers in near shore marine sediments near Karachi (Pakistan) provide clues about rainfall. Rainfall decreased from 2000 to 1500 BC, coincidental with increasing aridity in the Near East and Middle East, as documented by decreasing Nile River runoff and reduced influx to lakes throughout Turkey to north-western India. These lake sediment records show cyclical drought periods from 200 BC to 100 AD, around 1000 AD and from 1300 to 1600 AD (Late Middle Ages).[802]

Rapid climate change was not only recorded in the Northern Hemisphere. Increased runoff in wet times and drying of lakes in dry times is a feature of many South American lakes.[803,804,805] Similar climate changes have also been recorded in South Africa[806] showing that climate changes were global.

An intense period of aridification starting at about 2000 BC was also recorded in China. This period of aridification was global because it has been recorded in North Europe, Mediterranean Europe, northern Middle East Asia, northern East Asia, East Africa, Middle the East, the Indian Peninsula, the Americas and the Yellow River Valley.[807]

After this intense 300-year drought that caused so much havoc in the old world, there was again a warm period from 1470 to 1300 BC.[808] In this warm period, the Bronze Age, people migrated northward into Scandinavia and reclaimed farmland with growing seasons that were at the time probably the longest for two millennia. The Assyrian Empire, the Hittite Kingdom, the Shang Dynasty in China and the Middle Egyptian Empire flourished.[809]

This warming, in Minoan times, led to a thriving of culture and a growth of empires. The Minoan empire was greatly weakened by the eruption of Santorini and was displaced by the Mycenaean empire. This was a favourable warming period of the Holocene Maximum.

The Bronze Age came to an end with the Centuries of Darkness. Between 1300-500 BC, there was another period of global cooling and associated drought.[810] Glaciers in Alaska, Utah, Scandinavia and Patagonia advanced again and armadas of icebergs again appeared in the oceans. This global cooling may have been a factor in the human upheavals at that time. There were mass migrations, invasions and wars.

The Hittite empire in Anatolia started to decline in 1200 BC and disappeared soon after. The Mycenaean civilisation fell at the expense of the rise of the Assyrian, Phoenician and Greek civilisations. Records from Troy show that it was cold and dry and that there was famine around 1259 to 1241 BC and it was not until 800 BC that Troy recovered. About that time Egypt went into a prolonged decline while Babylonia and Assyria were also weak for most of the period from 1100-1000 BC.[811] At this time, the Jews made their exodus from Egypt when the Nile was consistently running low.

The Sweetwater Canal was built in ancient Egypt for irrigation. However, the canal filled with silt.[812] North Africa had a cool dry windy period from 600 to 200 BC. Grasslands of the Sahara and Arabian Deserts retreated, humans migrated as the deserts expanded due to drought and wind. Tropical rains in Africa caused huge flooding of the Nile and many of the great buildings were inundated. These changes in rainfall, river flow and lake levels were widespread.[813,814] The Dark Ages were a global cold dry period. For example, coastal sediments in Venezuela show that there was very little runoff water suggesting a prolonged drought. This is the same drought that caused the collapse of Mayan cities in Central America.[815]

The Anasazi Indians' agriculture and culture spread in the early part of the Medieval Warming as rains were more consistent. Tree rings from Sand Canyon show low rainfall in 1125-1180 AD and 1270-1274 AD and a 24-year drought late in the 13th Century. These led to food scarcities, internal conflict, the building of fortress-like cliff dwellings and the eventual sacking of the fortresses in the Little Ice Age.[816] By 1400 AD, the maize crop failures had driven the Anasazi from their cliff dwellings and to extinction.

Several centuries later, a similar fate destroyed the Mayan civilisation of Central America which also depended upon seasonal rainfall for irrigation. This culture collapsed during a very severe drought between 800-950 AD.[817,818] The Mayans, like the Akkadians, were not able to sustain their prosperous civilisations in a period of time of prolonged drought.

History shows us that global warming gives us nothing to fear. If we really want to fear something, then the best candidate is a solar-driven global mega-drought concurrent with global cooling. It's happened before, it will happen again.

Pacific Island populations were greatly reduced at the beginning of the Little Ice Age.[819] Other parts of the world were cold and dry, especially during the Spörer and Maunder Minima.[820] Not only was it cold during the Little Ice Age, but there were rapid fluctuations in temperature and precipitation. During the Maunder Minimum, a year of record cold temperatures (1683-1684 AD) was followed by a year of record heat (1685-1686 AD).

An ice age climate change is characterised by drastic changes in temperature, storminess and precipitation without warning. These changes were local, global and rapid. They had a profound effect on human society.[821] A 2015 study found that mega-droughts in the last 2,000 years were worse and lasted longer than current droughts; mainly because of the cold Dark Ages and Little Ice Age.[822]

India is the land of drought, famine and flooding rains. The deadliest droughts in India were before 1924 and over the last 150 years they were in 1899, 2000, 1876, 1965 and 1918. In the 1896 drought, five million people died from starvation in India.[823] That was the same year a brutally hot summer in Australia caused 400 deaths and people fled the inland heat on emergency trains.

Somewhere between one and five million people died a few years later in the next drought in India which was at the same time as Australia's Federation drought. Thanks to modern fossil fuel-driven transportation, the 2015-2018 drought in India did not lead to mass starvation, as they often did in the past. All deadly droughts were linked to the positive phase of the *El Niño* Southern Oscillation and were unrelated to carbon dioxide.

Famine deaths have largely been eliminated in India; mostly thanks to better transport and organisation, higher yields (thanks to manufactured fertiliser, genetically-modified crops and carbon dioxide) and irrigation. Droughts still happen but in a population that has grown from 250 million in 1880 to over a billion. It is extraordinary that more Indians starved when the population was only a quarter of the size and carbon dioxide levels were at the level deemed by green activists as perfect. Weakened people died of cholera, malaria and bubonic plague. Death rates from these diseases often doubled or tripled.

In the US, Prof Roger Pielke noted *"Droughts have, for the most part, become shorter, less frequent, and cover a smaller portion of the U.S. over the last century."*[824]

In their 2013 report, the IPCC predicted that California would have droughts nearly every year from 2030 onwards and with increasing severity until 2100. This scary prediction well into the future that can't be validated was sensationalised by the media. It appears that because 22 climate models were used to make such a prediction[825] out to 2100, then the predictions must be true. It appears that the reduction of carbon uptake during the 2000-2004 drought in California is the villain. Looks like what happens in California stays in California.

In 2017, drought conditions in the US dropped even more as they were limited to only 1.6% of the continental US and what had been touted by climate activists as California's permanent drought came to an end.[826] In 2019, the governor of California claimed the state was in a permanent drought. A short time later all the dams were full.

The normal suspects had screaming headlines such as *"California's terrifying climate forecast: It could face droughts nearly every year."*[827] In a case of nominative determinism, the journalist who wrote the article is called Fears! However, new studies showed that rainfall is expected to increase in California. There were no screaming headlines for this news.

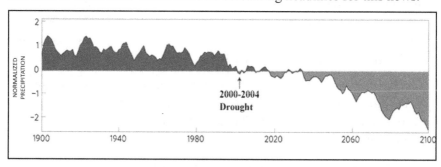

Figure 31: *A rather extraordinary diagram used by the IPCC in 2013 that shows a massive decrease in precipitation in Western North America from 1900-2100 starting just after the 2012 paper used by the IPCC in 2013 was published.*[828] *As with all predictions, it has already failed.*

It now appears that there will be no drought in California due to human-induced global warming and that it's now all due to warmer surface temperatures in the tropical eastern Pacific Ocean 2,500 miles west that will encourage larger numbers of storms in California.

Over the last 120 years, droughts in the US were driven by the Pacific Decadal Oscillation and or by solar activity. Droughts from the 1950s to 1970s were driven by solar activity whereas those in the 1900s and the 21st Century were driven by both solar activity and the Pacific Decadal Oscillation.[829]

Instead of drought increasing over the past century, the data shows that the magnitude and duration of East African droughts have actually declined.[830]

Australia has had eight mega-droughts over the last 1,000 years. Mega-droughts are defined as lasting more than five years. The biggest was a 39-year drought between 1174 and 1212 AD during a century of aridity (1102-1212 AD) during the global Medieval Warming. There was a 23-year mega-drought from 1500-1522 AD.[831] These mega-droughts were continent-wide. Tree ring studies in Western Australia covering the period from 1350 AD show many 30-year droughts during the Little Ice Age.[832] The authors concluded that Western Australian mega-droughts were more common than generally thought.

If drought is solely due to human activities, then we can only blame the Aboriginals for these mega-droughts during the Little Ice Age. But maybe drought is a natural process? Perish the thought! Imagine what a 39-year drought would do to Australia today?

In Australia, global warming alarmists panicked Labor governments into creating subsidised wind and solar power and building massive desalination plants during yet another drought. Four desalination plants built by former state Labor governments in Australia that have since been mothballed costing taxpayers nearly $1 billion each year since 2015.[833]

Australia's climate is driven by the surrounding oceans. This is no surprise as far more heat is in oceans than the atmosphere and rain originates by evaporation from the oceans. The original scientific view was that the *El Niño*-South Oscillation drove Australia's droughts.

It is now accepted that the Indian Ocean Dipole (IOD) is an even bigger factor than ENSO in driving long-term Australian droughts and when the IOD and ENSO combine, Australia suffers the most intense droughts.[834]

Australia has droughts. It is a brown country. It also has flooding rains. In

1911, Dorothea Mackellar was correct in her poem *My Country*. Drought has been and will continue to be a major feature of Australia. Green left environmental activists, the media and city-based politicians are unaware of drought because it does not affect them directly and they've probably never lived and worked in a drought-prone area.

Maybe they should travel in outback South Australia and see the abandoned homesteads and villages where dreams were broken in the 1880-1886, 1888 and 1895-1903 droughts. The South Australian outback ghost town of Farina[835] was founded in 1878, wheat was grown but the town is now gibber and dunes with abandoned ploughs. Settlers lost everything.

There have been 10 major droughts in Australia over the last 150 years[836]. Some droughts lasted a decade, some are in clusters and all had a severe effect on rural Australia. Primary production decreased, banks foreclosed businesses, unemployment rose, rural towns died, farmers committed suicide and government assistance was too little and too late. City life blissfully continued unaware of the tragedy on their doorstep and the only impact was that food prices of selected products increased slightly. Many droughts are broken by "flooding rains."

There was a two-year settlement drought that started in 1791 in eastern Australia.[837] The Australian Bureau of Statistics[838] and Bureau of Meteorology[839] document the major Australian droughts (1864-1866, 1868, 1880-1886, 1888, 1895-1903, 1911-1916, 1918-1920, 1939-1945, 1958-1968, 1982-1983, 1997-2009, 2012-2015), less severe droughts (1922-1923, 1926-1929, 1933-1938, 1946-1949, 1951-1952, 1970-1973, 1976) and droughts where most Australian live (i.e. south-eastern Australia; 1888, 1902, 1914-1915, 1940-1941, 1944-1945, 1967-1968, 1972-1973, 1982-1983, 1991-1995, 2002-2003, 2006-2007, 2017-2019). There is always drought at some time somewhere in Australia.

The three biggest droughts were 1895-1903, 1939-1945 and 2002-2007.[840] However, rather than take advice from government departments that deal with historical information, the last drought led to political panic aided and abetted by green left city-based bureaucrats and environmental activists and cheered along by the hysterical taxpayer-funded ABC and by Fairfax media networks. Australia's most unsuccessful climate forecaster (Tim Flannery) was able to give all sorts of opinions about the 2002-2007 drought without critical questioning from the media.

Were droughts in past geological times due to human emissions? No. Were droughts in ancient times due to human emissions of carbon dioxide? No.

Surely the regular droughts in Australia are due to increases in temperature? No. Dust storms have been taking place since recorded time and have no relationship to climate change or global warming.

Drought is due to a decrease in rainfall. These can be local, regional, continental or global and are driven by solar output and the oceans. Australian rainfall is highly variable but has not decreased over the last 100 years, the longest measured drought in Australia lasted 69 years and, because there is no water to evaporate (and operate as an air cooler), temperatures rose. Warmer temperatures do not create a drought but droughts create warmer temperatures.

The Murray-Darling catchment in south-eastern Australia now contains three times as much water as before the Snowy Mountains Scheme because of water management by dams and irrigation. This has drought-proofed the food bowl of Australia. Recent changes in management of the Murray-Darling catchment resulting from green pressure have destroyed much agriculture and sent many farmers broke. In their eyes, the green activists have done their bit for the world.

History is available and can be read by all. Green ideology cannot change what has happened in the past.

When green activists claim that drought is a result of human emissions of carbon dioxide, they are lying.

Rain

Flood plains comprise minor gravels deposited by massive amounts of fast flowing water with larger amounts of sands and silts deposited by flood waters. Fast flowing water has cut a bed for a river, the flood plain fills with sediment over time and the river changes course as the flood plain broadens. Floods add new fertile soil to flood plains and much of the river water flows underground and can be tapped by wells.

The great civilisations of the world grew on flood plains (e.g. Nile River) because of the availability of foods produced from the flood plain and the availability of water. Great civilisations did not grow on the seaside because of sporadic tsunamis.

The Nile had monstrous floods in the periods 14,700-13,100, 9,700-9,000, 7,900-7,600, 6,300 and 3,200-2,800 thousand years ago. The Egyptian Old Kingdom (2350-2200 BC) thrived on the banks of the Nile during warm wet times, was inundated at times and struggled in cool dry times.

Old records can be very useful. The compilation of 178 years of rainfall data from Sydney, Melbourne and Adelaide shows that it is misleading to use graphs which start in 1970 or 1910 and declare that human emissions of carbon dioxide has an effect on rainfall.[841] Rainfall rises and falls in long cycles. There is no relationship with carbon dioxide. There have been great variability since the 1840s.

There is no relationship in Sydney between rainfall, drought and human emissions of carbon dioxide, the last of which started to increase rapidly after World War II and there is nothing unusual about the last 30 years despite humans emitting 50% of all their carbon dioxide emissions since 1989.

Sydney's wettest year was 1950 and its driest year 1849. In the 1840s there were two very wet events. In April 1841 there was torrential rain with building and roads damage[842] and, in October 1844, there was 518 mm of rain in 24 hours.[843]

In Melbourne, the wettest day was 27th November 1849 when 177 mm of rain fell associated with gales. Flooding of the Yarra River killed several residents and some sheep.[844] The decreased rain days characterise both the continent-wide Federation and Millennial droughts.[845] The driest year was 1967 was at the start of a three-year drought and the 1950s were wet times.

Again, there was no relationship between Melbourne rainfall, drought and increased carbon dioxide emissions since World War II.

Adelaide, Melbourne and Sydney have different climates. Adelaide has more of a Mediterranean climate with cool moist winters and hot dry summers with the occasional storm. The 1960s were dry with 1967 the driest year at the start of a three-year drought. The Millennial Drought was brutal. The wettest Adelaide day was on 6th February 1925 when 141.5 mm fell in 24 hours. The wettest year was 1992. There were prolonged wet periods in the 1850s which stimulated inland settlement. It was also wet in the 1920s.

In Adelaide, there is no relationship between rainfall and atmospheric carbon dioxide. There has been great variability in rainfall since the 1840s, and there is nothing unusual about the last 30 years despite humans emitting 50% of all carbon dioxide emissions since 1989.

The data is clear. There is no relationship between rainfall, drought and increased human emissions of carbon dioxide.

The same applies for California and the UK. In California, there was lower precipitation in the 1930s dustbowl years and there is no relationship between rainfall and increased carbon dioxide emissions since World War II.

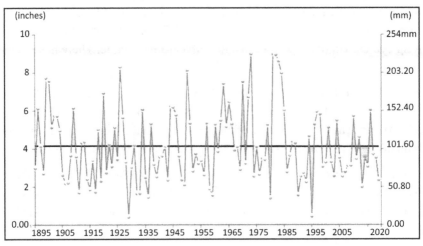

Figure 32: *Rainfall in California between 1895 and 2020.*[846]

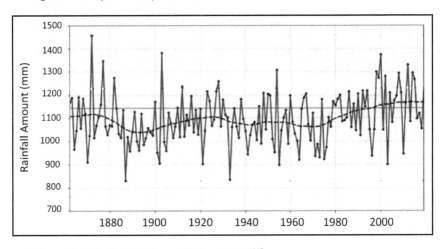

Figure 33: *UK rainfall between 1860 and 1920.*[847]

The IPCC, in its 2012 and 2013 reports shows that there is no evidence of increased extreme rainfall events due to human activity. The 250 years of rainfall records in Britain shows little change nor does the 100-year record in the US. Furthermore, deaths from extreme weather are currently at an all-time low despite increases in the atmospheric carbon dioxide and population.[848]

Why do climate activists and the sympathetic and lazy green media refuse to look back in time and brazenly lie about rainfall and concoct extremes of climate?[849,850]

Floods

The green activist narrative repeated by the media is that flooding is due to human-induced global warming. Is the media uncritical or is there an agenda? The media reported that December 2010 was the wettest on record. Central (e.g. Rockhampton, Bundaberg) and central-western (e.g. Emerald) and south-eastern Queensland (e.g. Toowoomba, Ipswich, Brisbane) were flooded.

The Brisbane floods of 2010-2011 killed 33 people, damaged 28,000 homes and created $100 billion of damage with 2.5 million people impacted[851]. Flood levels recorded at the Brisbane Post Office show that there have been six major floods since 1832 greater than the 2010-2011 floods. The tide gauge at the Port Office in Brisbane shows that the 1840 flood level has never been exceeded.

Were only the 2010-2011 Brisbane floods due to human emissions of carbon dioxide causing global warming? To use the word unprecedented is deliberately misleading and deceptive when there is so much easily accessible information about past flooding.

Previous floods were apparently not due to climate change whereas the January 2011 floods in Lockyer, Bremer and Brisbane Valleys were, according to the green activist narrative. When there is a choice between climate change or incompetence, go for human incompetence every time. Why do green activists lust over disaster and death in order to make political capital?

In 1974, the Wivenhoe Dam was constructed to reduce the severity of flooding in Brisbane. The release of water from the Wivenhoe Dam by authorities followed by downstream flooding led to a successful class action against the Queensland Government, Sunwater and the State-owned dam operator Seqwater (which is appealing the settlement). The partial settlement to 6,500 affected residents was $440 million of the $880 million settlement.[852]

Every Queensland taxpayer is paying for this settlement, which results from mismanagement by State-owned Queensland water authorities as well as paying for rebuilding of infrastructure. This money could have been better spent on Queensland roads, schools, hospitals or police. People lost their lives, homes, stock, employment and businesses. How many bureaucrats and politicians lost their jobs? Where is the culpability if bureaucrats and politicians can call on the public purse to pay for their mistakes?

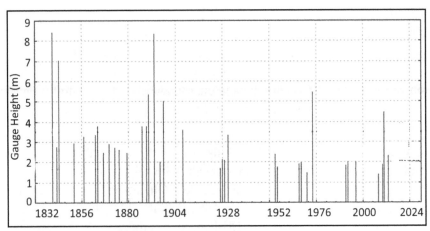

Figure 34: *Flood peaks of the Brisbane River recorded at the Brisbane Post Office over the last 190 years (from BoM).*

In October 2019, the *Sydney Morning Herald* (SMH) uncritically reported that Sydney's dams were 48% full, that dams were emptier than during the millennial drought, that the Bureau of Meteorology's (BoM) climate forecast indicated no foreseeable signs of a reprieve from the current drought with warm and dry weather predicted and, based on these forecasts, the drought was therefore unlikely to break within the next year.

The drought broke with flooding rains soon after. Both the *SMH* and BoM were wrong with their fear-mongering. The *SMH* was once one of the world's greatest newspapers and could be trusted for accurate stories. It is now a rag spreading alarmist panic. In mid-August 2020, Warragamba Dam was 99.3% full.

Contrary to popular hysteria, Sydney's main water storage dam (Warragamba) did not cause the March 2021 flooding downstream along the Nepean River flood plain. It was an east coast low-pressure system that dumped six months worth of rain in six days. Warragamba Dam filled, could hold no more water and water over the spillway filled the already swollen Nepean River. The floods were very damaging as much of suburbia has expanded onto the Nepean River flood plain. A flood plain is called a flood plain for a very good reason.

The largest recorded flood downstream from Warragamba Dam was at Windsor in 1867, almost 100 years before the Warragamba Dam was completed and records show that the river flooded just as regularly before the dam was opened in 1960.[853] The dam was built for Sydney's water storage and not for flood mitigation.

After the 2021 floods, the suggestion to raise the Warragamba Dam wall to minimise flooding[854] met with opposition from green activists.[855] The green activists are not concerned whether people lose everything in a flood, as long as they can keep the narrative that everything is due to human-induced climate change. Just because a big flood occurs in your lifetime does not mean that it is unprecedented or due to climate change.

Following the deadly European floods of 2002, Germany built an extensive flood warning system. During a test in September 2020, most warning measures including sirens and contact by SMS, didn't work.

The European Flood Awareness System predicted the July 2021 floods nine days in advance and formally warned the German government four days in advance[856] yet most people in river towns were unaware that floods were coming. Flood warnings in July 2021 for Europe were ignored. Other flood warnings came from England and they were also ignored. Why didn't the government protect its citizens? This is why taxpayers fund governments.

Before these July 2021 floods, there were a few days of heavy rain. On 15th July, as much as 15 cm of rain fell in a 24-hour period. Flooded streams and rivers washed away houses and cars and triggered massive landslides. At least 196 people died on that day, 165 in Germany and 31 in Belgium. The Netherlands had been expecting floods for decades and built levees around the Meuse River which runs through eastern Belgium and into the Netherlands. Only minor flooding occurred and there were no deaths.[857]

Most of the German flood deaths occurred around the Ayr River on 14th July but river flows were lower than the floods of 1804 and 1910.[858] The normal suspects were looking for a link between extreme weather, flooding and global warming.[859] They failed.

This monumental failure of German early warning systems was due to bureaucrats. Chancellor Merkel blamed the floods on climate change rather than the failure of her bureaucrats and her policies. The Chancellor spent her formative years in communist East Germany. It shows. She uses cynical politics in order to extract more revenue for climate mitigation from hard-working and unprotected Germans.

The floods in July 2021 were below the high-water marks of previous historical floods and it was inevitable that there would be flooding on a flood plain. People have built on flood plains in Europe for centuries leaving excess water nowhere to go. Germany spends €100 million each year on renewables to save the planet but did not build far cheaper low-cost flood protection to adapt to or minimise the inevitable.[860]

Government officials or politicians who give the lame excuse that climate change is to blame abrogate responsibility. Maybe saving the climate from mythical problems is more important than saving their own people.

There is long history of flooding in Europe[861] which should have been known by bureaucrats and politicians. Some large catastrophic floods have entered folklore such as the Cymbrian flood (120 BC), Cucca Flood (589 AD), All Saints' Flood (1170 AD), Magdalen Flood (1342 AD), St Peter's Flood (1651 AD) and the Great Storm (1703 AD).[862] Thousands were killed in these floods. Cities on the big rivers (e.g. Elbe, Main, Rhine and Donau) are regularly flooded because runoff is higher due to roads and buildings.

For example, in 1962, the Elbe River flood at Hamburg was 5.7 metres above the city's dykes.[863] There is high precipitation in the Alps and rivers draining from the Alps flow across Germany and adjacent European countries. In the German alpine foreland, a study of floods since the 13th Century has shown more than 85 floods in the Isar and Lech Rivers at a 31-year average frequency.[864]

There has been no increase in the frequency of floods in Europe[865], there is a seasonality of flooding[866] and there are good historical records of European flooding as many European cities are built on river flood plains and deltas.

The Mississippi River in the US was in flood. Again we are told that this is extreme weather, that this is due to our emissions of carbon dioxide and that this is producing more rain and subsequent flooding. Is this true? This can be easily checked.

Flooding of Old Man River occurs on average once every three years. The Mississippi River has an extensive flood plain, built up over millions of years of frequent flooding. Flood plains are flat, have good soil that is frequently replenished and have access to irrigation.

For millennia, agriculture has been centred on flood plains where canals and protective levees have been built to make agriculture more efficient. Once cities are built on flood plains, the proportion of water soakage to runoff changes and flooding becomes more common. Cities on flood plains sink because of their weight, traffic vibration and extraction of ground water. For example, New Orleans is sinking nine millimetres per year and flooding by the Mississippi River is therefore becoming more intense.

The first recorded catastrophic flood of the Mississippi River was in 1543. More complete records were kept in the 19th Century and floods occurred in 1734, 1788, 1809, 1825, 1844, 1851, 1858, 1859, 1874, 1882, 1883, 1884, 1890, 1893, 1897, 1903, 1908, 1912, 1913, 1916, 1922, 1927, 1928, 1929, 1931, 1932, 1935, 1937, 1944, 1945, 1950, 1973, 1979, 1983, 1997, 2002, 2008, 2011, 2016, 2017 and 2019. By far the biggest measured flood in modern times was in 1927.[867]

A few simple questions. Which one of these Mississippi River floods was due to human-induced global warming? None. Is there a relationship between the emission of carbon dioxide by US industry and flooding? No. Frequent flooding occurred well before the US started to increase carbon dioxide emissions in the late 19th Century. Why have there been fewer floods in the late 20th Century and early 21st Century? Construction of levees to protect subsiding and low-lying areas.

A study of more than 10,000 rivers around the world showed that most rivers flood less now. What used to be a 50-year flood in the 1970s happens every 152 years today because of better flood mitigation measures.[868]

Just because there were floods west of Sydney and in Europe in 2021 does not mean that there is a global process taking place. The floods of Germany and Belgium in July 2021 occurred in your lifetime. They were not as large as previous floods.

Does that make these floods unprecedented or due to human emissions of carbon dioxide?

No.

Ocean acidity

Flood waters, rivers and streams carry detritus and dissolved material into the oceans. The 10 major rivers in the world account for 90% of the plastics in the oceans.[869] Eight are in Asia or the Indian Subcontinent (Yangtze, Indus, Yellow, Hai He, Ganges, Pearl, Amur and Mekong) and two in Africa (Nile and Niger).[870] This is the world's major environmental problem, not traces of human emissions of plant food into the atmosphere. Where are all the Western green activists cleaning up these rivers and helping people in Asia, the subcontinent and Africa?

Most river detritus settles on the continental shelf as sands, silts and muds. Some clay material is carried further and settles to form deep ocean floor sediments. The weathering of rocks in the hinterland release sodium, calcium, potassium, sulphate and chloride which are added to the oceans. These are later removed from the oceans by closure of seas (e.g. in the Mediterranean and Black Seas), then by evaporation and by formation of salt layers that are later sealed by covering with sediment.

Carbon dioxide in the air dissolves in seawater as carbon dioxide, carbonate and bicarbonate. Calcium added to the oceans from rivers combines with carbonate and bicarbonate to precipitate calcium carbonate either in sediments (limestone) or in life such as shells and corals. Carbon dioxide that was once in the air is sequestered in the oceans and has a residence time in air of five to seven years. If this were not the case, then the Earth's atmosphere would be the same as our planet's second atmosphere, as described in Chapter 2.

This process of removing calcium carbonate from the oceans stops ocean water becoming acid. Another process is the circulation of ocean water through the top 5 km of the ocean floor. Chemical reactions take place to prevent ocean water becoming acid. Around some hot springs which are emitting sulphuric and other acids there are small volumes of slightly acid water.

Chemical fingerprints in marine sediments show that there were short times when the ocean surface waters were slightly acid but reverted quickly to the normal state of being alkaline. The measure of ocean alkalinity (pH) is a logarithmic measurement and if all the planet's atmospheric carbon dioxide were dissolved in the oceans, they still would be alkaline.

If we use green activist ideology, increased carbon dioxide in the air makes the oceans more acid. How do the green activists explain the fact that we have fossil shells? Fossil shells are made of calcium carbonate which dissolves in acid and, at the time these fossils formed, the atmospheric carbon dioxide was far higher than now. According to green ideology, the oceans should have been acid, shells would have dissolved and would not be present for fossilisation.

This term ocean acidification is scientific nonsense. It was invented to scare citizens to oppose fossil fuels (which provide more than 80% of the energy for advanced economies). The term ignores billions of years of ocean history, ocean buffering and simple precipitation of carbonates in oceans. Carbon dioxide is a vital part of ocean health and the ocean food chain.[871] Additional carbon dioxide in the atmosphere dissolves in ocean water and allows marine life to thrive.

The ocean food chain is underpinned by microscopic plants and bacteria (phytoplankton) which require carbon dioxide for photosynthesis. These microscopic organisms release oxygen and 80% of the oxygen we breathe derives from ocean phytoplankton.

Dissolved carbon dioxide in surface waters is taken to great depths and upwells far later to enrich surface life again. Shells and marine species live and thrive in widely varying pH conditions from slightly alkaline (pH>7) to slightly acid (pH<7) and the so-called ocean acidification crisis is propaganda masquerading as science.

The capturing of the language by green activists to claim ocean acidification is deceitful. Language capture is political and has nothing to do with facts or reality.

Sea level

Just run this past me again. Human emissions of carbon dioxide are making the sea level rise and this will inundate coastal areas. This is a simple scare story and surely, if true, must be supported by science. However, there are just a few little problems.

This ideology assumes that human emissions of carbon dioxide result in heating the atmosphere. This has never been shown. It assumes that if the atmosphere is warmer, then this will heat the oceans. The problem is that this does not work because air has a very low heat capacity and water has the highest heat capacity of any solid or liquid. The oceans are heated from above by the Sun and from below by a yet unmeasurable amount of volcanic heat.

The story then goes that if the oceans are heated, they will expand and sea level will rise. A parallel story is that a warmer atmosphere will melt ice on land and add more water to the oceans. This story has far too many assumptions and is demonstrably wrong.

The average person might think that to determine sea level rise or fall is simple. Surprisingly, it is complex. When we measure sea level, what are we really measuring? Throw-away lines to scare people by green activists show that again the past and geological processes are ignored. Deliberately. Those scaring people about sea level rise such as green activists Al Gore, John Kerry and Tim Flannery purchased coastal and waterfront properties.

If you live in Sydney, go to Collaroy Beach on the northern beaches. An

1875 fisherman's hut remains exactly two metres above the high tide level where it was 150 years ago. In August 2020, Collaroy Beach suffered massive erosion during a large winter storm. This has happened many times in the past.

Fort Denison in Sydney Harbour has more than 100 years of measurement of sea level. Sea level has changed up and down by as much as 15 cm and those sea level changes are cyclical. It is easy to plot a line using the sea level data to show that sea level has decreased over time or has increased over time. It just depends on where the graph starts and finishes.

Tide gauges at Fort Denison in Sydney Harbour show that sea level has risen and fallen a few millimetres. If you can't measure sea level rise then you have a problem. Building of Fort Denison started in 1841 on Pinchgut Island to prevent a naval attack by the Russian Navy in the Crimean War. It was completed in 1857, a year after the Crimean War finished.

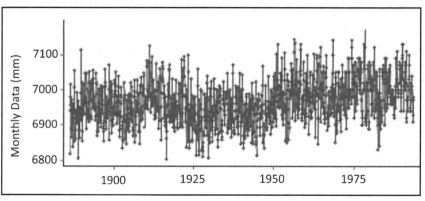

Figure 35: *Monthly mean sea level data for Fort Denison showing cycles of sea level rise and fall.*[872]

However, the IPCC claims that global sea level rose by 0.19 metres between 1901 and 2010 and predicts sea levels will rise by as much as 0.66 to 0.83 metres by 2100. These measurements and assumptions do not consider changes in land level and the fact that many parts of the world such as Collaroy and Fort Denison show no sea level change whatsoever.

Whatever the scary scenario presented, if sea level rises at such a slow rate then it allows societies ample time for adaptation. The Dutch have been adapting to a relative sea level rise for 1,000 years.

From the Gulf of Carpentaria to Tasmania, eastern Australia shows that sea level has both risen and fallen over the last few thousand years. Sea level rose between 11 to 15 metres 9,400 to 9,000 years ago, it then rose another five metres by 8,500 years ago and 3.5 metres between 8,300 and 8,000 years ago. About 6,900 years ago, sea level was a few metres higher than at present.[873,874] Back dune lagoons were left after sea level fell over the last 4,000 years. To suggest that sea level rise is due to human emissions of carbon dioxide requires ignoring a huge body of validated science.

Rock platforms along coastal Australia show that sea level has fallen in recent times. If the rock platform is beneath a cliff, there is a nick at the base of the cliff from waves smashing against the base of it. Piles of collapsed cliff rocks show that waves were undermining the cliffs when sea level was about two metres higher.

From measuring the age of exposed coral microatolls and oysters on rocks above the high water mark at Balding Bay (Great Barrier Reef, Magnetic Island, Queensland), it has been shown that sea level was two metres higher and remained at this level for 1,600 years in the warmest part of the current interglacial some 4,000 years ago.[875]

During our current ice age, sea level goes up and down by 130 metres every 100,000 years. Some past sea level changes have been very fast, others have been slow. During the previous interglacial, sea level was up to seven metres higher than now. Beach and dune sands formed during this interglacial were mined for mineral sands along Australia's east and west coasts.

There are nearly 200 old ocean beaches in the inland Murray-Darling Basin of NSW and Victoria that formed during previous interglacials, some are now 500 km inland from the current shoreline and some of these have been mined for mineral sands (e.g. Ginkgo, Pooncarie, NSW[876]; Kulwin, Victoria[877]).

What's going on? We have evidence that sea level has fallen over the last 4,000 years and we have old beaches hundreds of kilometres inland yet we are told that sea level is rising.

For most of time, sea level has been higher than now. This is why most landmasses are draped with sedimentary rocks that were originally sediments lain down in shallow water.

Sea level is always changing.[878] The land level is also always changing. In the past, sea level has risen or fallen by 600 metres and the land level has risen or fallen by 10,000 metres.

Some 6,000 years ago in the peak of the current interglacial sea level was two metres higher than now. Hobart, Adelaide, Newcastle, Brisbane, Townsville and Cairns airports are built on tidal flats that were covered by water 6,000 years ago. Except for the short north-south 16L/34R runway in Botany Bay and an extension to the parallel long north-south 16R/34L runway into Botany Bay, most of Sydney airport is also built on what were tidal flats 6,000 years ago.

Just because the planet is changing does not necessarily mean that humans are driving the changes. This shows that very large sea level rises and falls have occurred in the recent past. Raised coral terraces (e.g. Huon Peninsula, Papua New Guinea[879]) confirm that there have been great sea and land level changes in the recent past. These changes are unrelated to any human activity.

The geological record shows that there is an increasing biodiversity when there is a sea level rise[880] because of new shallow water environments. If sea level falls, there can be increased extinction of life. In the Sturtian and Marinoan glaciations some 650 million years ago, sea level fell so much that there was no continental shelf[881,882] and life on the continental shelf died (e.g. Arkaroola Reef).

When sea level change is measured, what are we actually measuring? Are we measuring the global sea level change or a local land level change which gives us the relative sea level change?

Between 1764 and 1767, the Harbour Master at Liverpool (UK) recorded the times and heights of tides at what was later to become the Liverpool Observatory. Even later it became a world centre for sea level research. Today the Permanent Service for Mean Sea Level (PSMSL) maintains a worldwide database of tide gauges.

Computer models predicted an acceleration of sea level rise and IPCC technical committees examined the PSMSL records in 1990, 1995 and 2001. There was no acceleration in sea level rise and the IPCC Report (1990) stated *"There is no firm evidence of acceleration in sea level rise during this century"*. Steady on chaps, this is not why we have jaunts all over the world for IPCC and COP committee meetings.

In 2007, the IPCC sea level committee took action. There was a discrepancy between tide gauge data and satellite measurements. It was assumed that tide gauge measurements were not accurate and that satellite data was more accurate. This fitted the narrative better. Suddenly we were told that sea level was rising 30 cm every 100 years.

Older data sets were adjusted downwards to suddenly make a modern sea level rise and newer data sets were adjusted to make a sea level rise appear.[883] This is fraud. Parker and Ollier[114] showed that at all key data collection points sea level change was 0.0 millimetres for the last 50 years. The excitement in the press about sea level rise may be fantasy rather than fact.

The Jason satellites of NASA reported that there had been a deceleration in sea level rise from 2008-2018. The National Oceanic and Atmospheric Association 2016 report based on 200 US tide gauges and some gauges in the Pacific and Atlantic Oceans showed that there was no acceleration in sea level rise.

A satellite altimetry problem was detected and it is the sea level estimates from the flawed Jason satellite that were used by the COP25 in Madrid (December, 2019) and by the various coastal local governments rather than the long-term tide gauge measurements that show no accelerated sea level rise over the last 120 years.

Precise satellite measurements give sea level rise about half that measured from tidal stations.[884] Corrected data for a large part of the globe shows a rise of 1.8 mm a year from 1900 to 1980[885] and this is in accord with measurements from corals and other proxies for the past 3,000 years. Historical records show no acceleration of sea level rise in the 20th Century.[886] During the warming from 1920-1940, sea level rise actually stopped.[887]

Accurate determination of sea level changes from tide gauges is fraught with difficulty. For example, many piers slowly sink[888,889] and the geographic location of tide gauges makes constant measurement unreliable. Tide gauges need a consistency of measurement over a long period of time by having regular high precision surveys of the gauge position and frequent calibration of the gauge. This just does not happen.

Tide gauges only make local and not global measurements and there are many parts of the world very close to each other where the land has risen (e.g. Ephesus, Turkey)[890] or fallen (e.g. Lydia, Turkey).[891] In biblical times.[892] Ephesus was a port city. Strabo recorded that King Attalus Philadelphus of Bergama built a breakwater to protect Ephesus and siltation combined with a land rise destroyed the port.[893] The Romans tried to rebuild the port and failed. Ephesus is now 24 km inland. Lydia, the birthplace of coinage, is now a few metres below sea level.

Plate tectonics, vertical movement of land after loading and unloading with ice, soil and sediment, volcanism, sediment compaction and extraction of fluids all affect tide gauge measurements. Vertical movement of land is also variable. For example, in the Vanua Baluvu island group on the Lau Archipelago of north-east Fiji, some islands have subsided while the whole island group as a whole has risen.[894]

Sea level changes are a result of competing forces. Most of the Northern Hemisphere north of 50°N was covered in ice during the last glaciation. Rocks are slightly plastic and, if a force of loading or unloading of ice is applied over time, they bend. When ice covered these areas during the last glaciation from 116,000 to 14,700 years ago, landmasses sank under the weight of ice. Now that the ice has melted, the landmasses are rising.

At present, land is both rising (e.g. Scandinavia, Scotland and Canada) and sinking (e.g. Holland, north- western Denmark, south-eastern England). The floor of the North Sea is also sinking. This counterbalancing of land level rise and post-glacial sea means that relative sea level changes may be small.

There are well known examples of post-glacial land rise such as Scandinavia[895,896] where there has been a land level rise of 340 metres. The 12th Century castle of Turku (Finland)[897] was built on an island. It is now connected to the mainland as a result of postglacial land level rise. Tide gauges in the Port of Turku are a guide to local uplift, not to sea level changes. Stockholm is no longer an island and is rising at the rate of one millimetre per year.

There is evidence presented in the scientific literature that, as a result of sea level rise, the additional weight of seawater causes uplift in the neighbouring land[898,899,900,901,902] a bit like how a see-saw operates. The loading of the continental shelf with water from melting of the ice sheets pushes down the continental shelf and causes the land to rise. This is dependent upon the thickness of brittle rocks and the plasticity of the Earth's mantle.[903] If the land rises, sea level may actually fall.

In north-eastern Ireland between 21,000 and 11,000 years ago, there were many increases and decreases in sea level in response to ice unloading.[904] The sea level curve is not actually a curve, it is a sawtooth. It shows very strong initial uplift (21,000-19,000 years ago), ice loading, land sinking and a relative sea level rises (19,000-17,500 years ago and 17,000-14,000 years ago) followed by a catastrophic ice loss, rapid uplift of the land

and a relative sea level fall (14,500-13,000 years ago). These very rapid fluctuations in ice, sea level and land level are not factored into geophysical models.[905,906]

The Netherlands is sinking.[907] For more than 1,000 years, the Dutch have been building dykes, pumping out water with wind mills and suffering inundation when storms coincide with high tides. At present, 40% of The Netherlands is below sea level and there are more than 2,000 km of dykes protecting the country from inundation by the North Sea. Since 1200 AD, 580,000 hectares of agricultural land have been lost to the North Sea. The Netherlands is wealthy showing that adaptation to sea level changes can be managed.

Many areas elsewhere are currently sinking due to the extraction of groundwater (e.g. Bangkok, Mexico City and Denver) and petroleum (e.g. Texas Gulf coast). Submergence rates of ~11 mm per year occur at Galveston Bay result from water extraction, petroleum extraction and subsidence which gives a relative sea level rise.[908] Subsidence increases risk of inundation from hurricane activity, as was shown by Hurricane Katrina. Many coastal river plains such as in Italy and China are sinking.[909,910]

A new study measured the vertical coastal land movements of California's coast using the satellite-based interferometric synthetic aperture radar (InSAR). This instrument can detect the land surface vertical movement to millimetre accuracy. The study found that the cities of San Diego, Los Angeles, Santa Cruz and San Francisco are subject to substantial subsidence which increases future risk of flooding. Land sinking has the same effect as a sea level rise.

The San Francisco Bay area is subsiding at rates up to 5.9 mm per year. Santa Cruz is rapidly sinking at up to 8.7 mm per year.[911] The Los Angeles area shows subsidence along small coastal zones, but most to the subsidence is occurring inland. The coastal land north of San Francisco is experiencing uplift of 3-5 mm per year which is the same effect as a sea level fall.

Tide gauges at London show a sea level rise. Not so. South-eastern England is sinking. At the end of the last glaciation 14,700 years ago, sea level rose. A river valley was filled with water. This valley is now the English Channel. What was a small fresh-water river (Thames River) was inundated and became tidal with marshes in a wide open valley.[912]

The post glacial rise of Scotland is pushing south-eastern England down and parts of the east coast have sunk six metres in the last 6,500 years.[913,914]

Western England is rising and eastern England is sinking. These events of inundation are also well recorded after Roman times.[915,916,917,918]

The extraction of groundwater from gravels and sands on the banks of the Thames River, traffic and large buildings have resulted in compaction which increased the sinking of London.[919] After two thousand years of river flooding and inundation from the North Sea, the Thames Barrier was built and opened in 1984.[920] The Thames Barrier is designed to protect London until 2030. London is still sinking. That does not stop some media networks ignoring history and claiming that London will be inundated due to climate change.[921]

We humans can adapt to climate change or land subsidence, equivalent to a sea level rise. In the past we have moved or used our ingenuity such as dykes in Holland and flood barriers (e.g. Thames River, UK; Venice's Modulo Sperimentale Elettromeccanico with floodgates at the Lido, Malamocco and Chiaggia).[922]

These dykes and flood barriers protect human settlements from a problem that is centuries old and show that human ingenuity and adaptation solve problems. Studies by Lomberg and others on the effects of a speculated climate change show that adaptation is far cheaper than futile naïve attempts to change a global planetary system.

Sea level and land level changes force shorelines to advance and retreat. Towns have been lost to the sea by shoreline changes.[923] Around The Wash (Yarmouth and Lowestoft), the coast is retreating.[924] The Romans tried and failed to stop coastal erosion here and the shoreline has retreated 1.8 metres per year since Roman times.[925]

Other areas have undercutting of cliffs (e.g. Lyme Regis, Dorset) and elsewhere the coast is advancing with headland growth of 12 metres per year on the Yorkshire coast.[926] There is a perception that geological processes are very slow and that they play little or no role in modern processes. Not so. History shows the opposite and that some geological processes are fast.

Believe it or not. We are going to be inundated as sea level rises. However, full tidal records from geodetically stable tide gauges, such as those in Newton, Cornwall (UK) show that satellite measurements indicate a greater rate of rise than tide gauges.[927] Maybe this is why Barack Obama, Al Gore, Bill Gates, Tim Flannery and others bought waterside properties because they knew the satellite data was flawed.

Sandy beaches are much less vulnerable to rising seas than was claimed in a European Commission study which caused "unnecessary alarm" recent subsequent research found.[928] Scientists from 12 universities around the world re-examined the data and methodology that underpinned the study and found it was based on flawed computer models[929] and was "arbitrary and unjustified". Analysis of satellite derived shoreline data indicates that 24% of the world's sandy beaches are eroding at rates exceeding 0.5 metre per year, 28% are getting bigger and 48% are stable. Never mind that the sequence stratigraphers in industry and academia who find new coal and oil deposits and who showed 100 years ago that marine regressions and marine transgressions are cyclical.

One has to be extremely optimistic to use tide gauge measurements to determine long-term sea level changes. The chance of long-term consistent accurate measurements is remote. Most of the reliable long-term tide gauge records come from Europe and USA. For example, the annual mean relative sea level rise at New York City from 1893 to 1995 was 2.9 mm for the entire 103-year span. This is a relative sea level rise and it means that sea level could have risen, land level could have fallen or both.

Increasing urbanisation leads to land subsidence. We see that in Mumbai, Singapore, Shanghai and Mexico City. New York is no exception. It also depends on the way the data is viewed. Blocks of 20-year intervals vary widely with sea level rise from zero to six mm per year and blocks of 40-year intervals vary from 0.9 to 3.5 mm per year.[930]

There are inter-annual and decadal cycles of sea level change hence only very long records can accurately give a trend. Sea level change is related to the presence or absence of ice caps, the shape of continents, the shape of the sea floor and the temperature of the oceans. The gravitational pull of mountain chain near the sea (e.g. the Andes) and large ice sheets gives a local sea level rise.

Some past changes were dramatic, such as the sea level drop associated with the initial expansion of the Antarctic Ice Sheet some 37 million years ago in our current ice age[931] or in the Sturtian and Marinoan glaciations. Other sea level changes occur during the 18.6-year lunar cycle and because of *El Niño* events, earthquakes, volcanoes, ocean floor subsidence, during high winds and during foul weather associated with large low-pressure cells.

The whole of the Texas Gulf area is subsiding. In the three years before Hurricane Katrina devastated New Orleans in August 2005, the city and surrounding area had undergone rapid subsidence of about one metre.

If we load up the oceans with water, the ocean floor sinks. We can measure the same process every time a large water storage dam is filled. The substrate beneath the dam water sinks. If the dam is dry, the substrate rises.

If we unload the oceans, molten rock at depth has less weight on top of it and can rise. In the Red Sea, as soon as sea level was lowered during glaciation, submarine hot springs became active as the weight of overlying water was reduced.[932]

A lowering of sea level may initiate volcanic eruptions.[933] The eruption of Santorini in the Aegean Sea is controlled by sea level with huge eruptions occurring when sea level was lower.[934] Sediment cores show that 208 of the last 211 eruptions occurred when the Aegean Sea was more than 40 metres lower than now. The cocktail of a low sea level during glaciation and increased aerosol dust from explosive volcanoes increases cooling.

During the last 2.67 million years of glaciations and interglacials with rapidly rising and falling sea level, there has been an increase in explosive volcanicity. There may be a correlation between explosive volcanoes and changes in sea level[935] and a correlation between explosive volcanicity and changed climate.[936] There have been and still are numerous submarine plumes of molten rock beneath the ocean floor.

These generate broad domal uplift and the ocean floor for an area of over 1,000 km in width can be lifted 500 to 1,000 metres.[937] Not only does this have a profound effect on ocean currents and the addition of monstrous amounts of heat and carbon dioxide to the oceans, but it creates a sea level rise because seawater is pushed aside.[938]

The last scale used was the historic record of the last few hundred years based on tide gauge measurements. The conclusions were that the historic record was *"a continuation of the past rather than a perturbation"*, that there was no indication of a pronounced temperature rise in the 20th Century and there is no relationship between atmospheric carbon dioxide and temperature.

El Niño events can cause sea level to rise by half a metre.[939] The 1982-1983 *El Niño* raised sea level 0.35 metres along the west coast of USA because there were no easterly winds to force the water back into the Western Pacific.[940,941,942] Higher sea levels were measured during the 1997 and 2015-2016 strong *El Niño* events.[943]

If plate tectonics deepens ocean trenches and the medial rifts of mid ocean ridges, then sea level falls. As continents undergo weathering to form soils, erosion removes them, there is a rise in the land (especially in alpine areas) and a fall in the continental shelf under water as it is loaded with sediment which later becomes compacted to rock. This produces a relative sea level fall.

Measurements of sea level change by satellite may be more accurate. However, the cause of a measured sea level change is no closer to being resolved. Altimetric satellites can measure the absolute sea level change for the whole planet. Satellite measurements need correction using a geophysical model that utilises the fact that the crust of the Earth is both elastic and brittle. These corrections are very rubbery because we don't know the geology of the crust beneath our feet well enough.

The Pacific Ocean floor around Tuvalu is sinking giving the appearance of a sea level rise. For the last 20 years, Tuvalu has been the symbol of sea level rise and inundation of Pacific island nations. Tuvalu is still there. It has not been inundated.

Pacific island nations claim that a rise in sea level and destruction of island nations is due to Australia's coal industry. However, the opposite is happening. Despite the sea floor sinking, hundreds of Tuvalu islands are growing with the total land area of Tuvalu increasing by 2.9% over the last four decades[944] and a couple are sinking[945], which is normal for volcanic islands.

At the 2019 South Pacific Forum in Tuvalu, Australia was berated for refusing to commit economic suicide and close its coal industry despite handing out $500 million for climate change and sea level rise as compensation to the region.[946] This shows the essence of the politics of climate change. It is a cash grab underpinned by a lack of logic and ignorance of science. It is a very long bow to draw to suggest that the burning of Australian coal (and not Chinese coal) will cause a rise in sea level that will affect Tuvalu. I'm at a loss to understand how giving away money will stop a sea level rise.

Time magazine on June 2019 has a cover article entitled *"Our sinking planet"* featuring UN chief António Guterres waist deep in water at the Pacific island nation of Tuvalu. He was the secretary-general of the Socialist Party of Portugal (1999-2005). He claimed that Tuvalu was one of the world's most vulnerable countries to global warming. His narrative is that human emissions of carbon dioxide drive global warming, that the global warming is melting the ice caps and causing sea level rise and that the Pacific island nations are being inundated.

Every single aspect of this narrative is scientifically wrong. It has never been shown that human emissions of carbon dioxide drive global warming. Ice melting and sea level rise are very complex and are not only related to warming. Charles Darwin in 1842 showed that Pacific island atolls grow with a sea level rise. Recent studies have shown that the Pacific island nations have grown in size and some 90% of coral atolls are either stable or have increased in size. This is the validation process of science at work.

Guterres flew to Tuvulu in a carbon dioxide emitting jet made out of aluminium, a metal that requires huge emissions of carbon dioxide during mining, processing, smelting and refining. I guess it's OK to be a hypocrite and fraud for a cover photograph on *Time*, a magazine that had a cover article of 1977 entitled *"How to survive the coming ice age"* and a 2007 article *"The global warming survival guide"*. No point in changing magazines, *Newsweek* warned us of the forthcoming ice age in 1975 and later warned us that we would fry-and-die. The media does not inform, it sells sensation to make money.

Studies of modern sea level changes provide results contrary to popular concern about a sea level rise. Ice shed from the giant ice sheets covering Antarctica and Greenland is responsible for only 12% of the current sea level rise.[947] There is a major obstacle in predicting the future of ice sheets: we do not know what is going on beneath them[948], especially as there are more than 150 volcanoes and hot spots beneath the Antarctic ice sheet.[949] Although there have been claims that the rate of sea level rise has been accelerating in tandem with the rate of rise of the air's carbon dioxide concentration and temperature[950], this claim now has been tested.[951] It's not true.

The 130-metre sea level rise over the last 14,700 years means that much of the West Antarctic Ice Sheet is now not underpinned by land. Two thirds of the West Antarctic Ice Sheet has collapsed into the ocean, ice bergs floated and slowly. There is another third yet to melt and this would create a seven-metre sea level rise.[952] Sea level was seven metres higher during the previous interglacial and even if sea level did increase, it would not be unprecedented. This is in fact a very slow rate of sea level change.

Sea level changes and the rate of heating are not cause and effect. If sea level rise is only from thermal expansion of the oceans and from melting glaciers, then it takes a long time to heat the oceans. If indeed we are experiencing a sea level rise at all, it could be due to a lag from the Medieval or Roman Warmings and certainly not due to contemporary processes. Sea level during

Roman times was about 1.4 metres lower than now.[953] Melting sea ice does not change sea level and it is only terrestrial glaciers that can raise sea level. Again, this is a process that takes a long time.

There is a suggestion that sea level will fall and not rise. Oceans are getting deeper and sea level has fallen some 170 metres over the last 80 million years and this may be followed by a further 120-metre sea level fall over the next 80 million years.[954] Using chemical fingerprints, it was also shown that over a 550 million-year period, sea level has actually been falling as a result of the deepening of the ocean floors.[955]

Between 542 and 251 million years ago[956], at least 172 significant sea level changes have been recorded varying in magnitude from 10 to about 125 metres.[957] The long term rises and falls of sea level are vital for understanding marine basins prospective for oil exploration[958,959] and sea level change curves[960] over the last 500 million years have been constructed by the petroleum industry to aid exploration.

The International Union for Quaternary Research deals with the last two million years of environmental and climate change. The former president of their Sea Level Commission stated that there is no trend in sea level over the last 300 years and satellite telemetry shows virtually no change over the last decade.[961] The IPCC suggests that sea level rise will be in the range of 0.09 to 0.88 metres between 1990 and 2100 whereas the Sea Level Commission states that sea level rise will be 10 ± 10 cm over the same period thereby suggesting that a sea level rise cannot reliably be predicted.

During the last interglacial period (130,000-116,000 years ago), global mean surface sea temperature was at least 2°C warmer than at present[962] and mean sea level was four to six metres higher than at present.[963,964,965] These figures have been used to analyse the effects of a modern interglacial sea level rise.[966] At 123,500 years ago, the rate of sea level rise was 1.5 to 2.5 metres per century whereas at 119,000 years ago, the rate of sea level fall was 1.3 to 1.6 metres per century.

Over the duration of the current interglacial, the average sea level rise over the last 14,700 years has been 1.0 metre per century. This is slightly more than the worst-case scary scenario sea level rise predicted by the IPCC. If sea level did rise of one metre by 2100 AD, then this is exactly what would be expected at the current post-glacial sea level rise rate. The IPCC's own computer models show that their predicted sea level rise is probably not due to increased additions of carbon dioxide by humans into the atmosphere and is due to the continuing post-glacial sea level rise.

The sea level rise of one metre per century over the last 14,700 years must be placed in context. Most of this sea level rise was from 14,700 to 8,000 years ago. By 8,000 years ago, sea level was three metres lower than at present and sea level attained its current position by 7,700 years ago.

This means that sea level rose by two metres a century during that earlier period. Sea level continued to rise and, by 7,400 years ago, sea level was at least 1.5 metres higher than at present and this was followed by a high static sea level that lasted until about 3,000 or 2,000 years ago.[967] Sea level has dropped 1.4 metres since Roman times.

In the geological literature, it has been known for more than two centuries that sea level on a local and global scale rapidly rises and falls for a great diversity of reasons. Over the last 6,000 years, sea level rises and falls of two to four metres over a period of two decades are common.[968,969,970] Sea level changes were both global and regional[971] and, in some places, there is widespread synchronous rapid sea level change.[972] Climate change modelling requires a stable system upon which the variable of man-made carbon dioxide can be placed. The history of sea level changes shows that there is no stability in surface processes and that atmospheric carbon dioxide is not a destabilising force.[973]

Measurement of sea level is incredibly difficult. Satellite measurements need corrections based on models. Tidal gauges are fraught with difficulties. Both the land and sea level go up and down like a yo-yo and unless we marry the modern slight sea level change with past sea level rises and falls, then few conclusions can be made.

There is still a lot we don't know. For example, there is a relationship between sea level and solar activity and a lag between solar activity and sea level.[974]

It is not possible to discuss sea level unless it incorporates a discussion of land level. Land levels rise and fall as quickly as sea levels. If ever you are asked whether sea level is rising, the only answer a thoughtful well-read person can give is "ask me again in a few thousand years".

Meetings, treaties, agreements, laws and gifting billions to the UN cannot change sea level. Even the worst-case sea level rise predictions by the IPCC and others are little different from normal post-glacial sea level change.

Reefs

The Great Barrier Reef is as big as Germany.

It is healthy and is doing what reefs have always done. Green activists have picked the wrong symbol if they want to claim that the planet is dying because of human actions.

If green activists really want to worry about what's happening in the oceans, they should focus their attention on plastics.

Planet Earth has had reefs for 3,500 million years. They come and go. Coral reefs have been around for 650 million years. They also come and go. Reefs need sunlight, nutrients and warm water. They thrive when sea level is high and the sea temperature is warm. Reefs are composed of calcium carbonate. The calcium derives from runoff waters from the land and the carbonate derives from atmospheric carbon dioxide dissolved in sea water.

In Australia, there are some huge great barrier reefs in the fossil record such as the Gogo Formation (Late Devonian, 380 million years old) in north-western Australia. It was a tropical reef up to 700 metres thick, 1,400 km long and tens of kilometres wide.

New scientific work assessed the number of coral colonies in the Pacific Ocean and evaluated their risk of extinction. There are roughly half a trillion coral colonies. This is about the same as the number of trees in the Amazon or birds in the world. The eight most common coral species in the Pacific have a population size greater than the 7.7 billion people on Earth.[975] The global extinction risk of corals is far lower than previously estimated.

Each time sea level fell over the last 30,000 years, the Great Barrier Reef migrated seawards. It was at its most eastern position at the zenith of the last glaciation 20,500 to 20,700 years ago. The reef then migrated landwards as sea level rose and ocean temperature increased during deglaciation (20,000 to 10,000 years ago).[976]

Regrowth of the Great Barrier Reef was interrupted by five reef-death events caused by exposure of the reef or very rapid sea level rise. Each time the Reef died, it later recovered. The reef does not grow on rock on the sea floor. Each of the 3,000 individual reefs over the entire 2,000 km length is built on a 50-100 metre high pile of dead coral rubble that has built up over millennia. The reef is fine. It needs periodic periods of death to build up the thick substrate.[977]

Coral polyps are in the water. If there is a substrate such as a ship wreck, under water cameras, monitoring equipment or a dumped pile of rocks, corals quickly grow and, within a few years, the underwater feature is covered by coral. This would not happen if the Reef was on its last legs.

The Great Barrier Reef has been far more resilient to sea level and temperature changes than previously thought. It is tens to over a hundred kilometres in width today. During glaciation, it was reduced to a collection of narrow fringing reefs on the edge of a vast coastal plain.[978] Back then, it was neither "Great" nor a "Barrier". The current Reef is built on an old dead coral reef substrate 120,000 years old that died during the last glaciation when it was exposed to the elements.

During interglacials, the Great Barrier Reef is one of the largest reefs that has existed in the history of time. Reefs are colonies of animals that compete and there is constant biological warfare. At times, the Reef gives the impression of dying with coral the area of Belgium killed by cyclones, native starfish plagues or bleaching.

There is data from drill cores for Great Barrier Reef coral growth covering the last 300 years that shows coral growth is stable. It was claimed that there was an unprecedented dramatic collapse in growth in 1990. The Australian Institute of Marine Science (AIMS) has admitted this was a mistake because they changed their measuring methodology. AIMS claimed that growth should now be 30% lower than in 1990.

There has been no data released for coral growth since 2005. Why not? Taxpayers fund AIMS yet there is no way to find out if such funds are being spent wisely. And then, in 2021 under pressure from the Senate, data appeared that showed that the Reef had recovered from an *El Niño* bleaching event and the catastrophic narrative was destroyed. We have been drip fed with BS.

The world has been convinced that the Great Barrier Reef is on its last legs. Some tourists have visited the Reef to see it before it dies and the scare campaign has had a severe economic effect on local tourist operators. The science behind this scare campaign is wrong and no one wants to remedy the problem. The International Union for Conservation of Nature (IUCN) has raised the stakes by claiming that the Great Barrier Reef has gone from "*significant concern*" to "*critical*".

It blames everything dear to green activists such as climate change, agriculture, coastal development, industry, mining, shipping, fishing, disease, coal dust – you name it and it cops the blame. However, the IUCN uses old, mostly wrong or misleading information produced by untrustworthy scientific institutions with an activist agenda and no commitment to quality assurance.[979]

There has been huge activist pressure about potential new coal mines hundreds of kilometres inland from the shoreline. According to green activists, these mines will destroy the Reef and the coal exported to India to give poor people electricity will produce carbon dioxide which will kill the Reef. Politicians struggled with this tortured thinking, were frozen and could make no decisions using common sense.

Politicians then, and as with all environmental decisions now, have one eye on the green vote; and to try to make a compromise between uncompromising green activists and those who produce wealth by employing people. The end result is political paralysis, indecision and economic losses.

The main reef is 100 km from the shore and is totally unaffected by farming, mining or any other activity on the land. Some scientists focus on the inner reef which comprises numerous small fringing reefs that contain less than one percent of the coral in the Great Barrier Reef. They are occasionally affected by flood waters from the big rivers; mud from farms is about one hundredth that from natural processes.

We have been led to believe by green activists that mud kills off reefs but research on the Great Barrier Reef shows that coral has been growing in muddy conditions for thousands of years.[980] The main damaging process is resuspension of mud by waves stirring the seabed, especially during strong south-easterly winds and cyclones[981], a process that's been taking place for thousands of years.[982] In fossil reefs, bands of mud between coral layers is common showing that the same process has been taking place for hundreds of millions of years. Water quality studies of fertilisers and pesticides show that outstanding quality water flows into and out of the Reef.

The Australian Senate inquiry into the regulation of farm practices impacting water quality on the Great Barrier Reef was the first time many of the scientists have been asked difficult questions and publicly challenged by hard evidence. Paul Hardisty of AIMS stated that only three percent of the Reef, the inshore reefs, is affected by farm pesticides and sediment. He also stated that pesticides are a low to negligible risk, even for that three percent. This is contrary to the media hysteria never corrected by AIMS.

The other 97 percent, the true offshore Great Barrier Reef, mostly 50 to 100 km from the coast, is effectively totally unharmed by pesticides and sediment. Why was this fact not mentioned in the Great Barrier Reef Outlook Report produced by the Great Barrier Reef Marine Park Authority? Omission is deceit.

At the Australian Senate inquiry, AIMS stated that coral growth rates show no impact from agriculture. Large corals live centuries, have annual growth rings like trees and record their own rate of growth. If farming, which started about 150 years ago on the reef coast, was damaging it, there should be a slowing of the growth rate. The records show no slowing when agriculture started more than 150 years ago or when large-scale use of fertiliser and pesticides began to be used in the 1950s. Why didn't AIMS correct the media hysteria?

Professor Peter Ridd has written that AIMS has been negligent in not updating the Great Barrier Reef average coral growth data for the past 15 years. There is data going back centuries but nothing since 2005. AIMS claimed coral growth rates collapsed between 1990 and 2005, due to climate change; however, there is considerable doubt about this result because AIMS changed the methodology for the data between 1990 and 2005. At the Senate inquiry, under some duress, AIMS agreed it would be a good idea to update this data if the government would fund the project.

They did update and their most recent data showed that the coral had recovered and that the collapse was only for a few years. Why was this important data not released earlier? Was keeping the community perpetually scared about the demise of the Reef a method of keeping research funds flowing?

The IPCC became fully apocalyptic and suggested *"Coral reefs would decline by 70% to 90% with warming of 1.5°C..."*. Armageddon day is around 2040 despite no data to show complete reef destruction. The Great Barrier Reef occupies 348,700 square kilometres and the IPCC is speculating that 243,000 square kilometres will die. The IPCC are fed data by AIMS. This data is, at best, dodgy.

They don't tell us that in the past coral reefs have thrived when water was far warmer and that coral have survived catastrophes that nature has thrown at them for the last 500 million years. Nor that over the last 5,000 years, there is no sign that storms have been getting worse; and that, today, over the 2,000 km extent of the Great Barrier Reef there is no destruction due to warmer water. Corals live in water from 27-32°C, the water is naturally warmer over the northern section of the Reef, and that corals further north around Papua New Guinea live in even warmer waters.

As a result of dodgy data, the IPCC and political games, probably by China,[983] UNESCO announced in 2021 that it would place the Great Barrier Reef on the danger list for 2021. This was punishment to Australia for its alleged climate change breaches and it would have profound economic consequences. It was only intense lobbying and political pressure by the Australian Federal government that stopped this game which is only at half time. The problem will reappear, Australia has been warned and the Great Barrier Reef is the focus for political games of power that have nothing to do with the environment and health of the Reef.

The land along the east coast of Australia is rising and microatolls at the low tide mark have been slightly raised and have died from exposure. I guess if green activists knew this, they could circulate photographs showing dead shoreline microatolls and claim that this was due to climate change,[984] agriculture, fishing or wearing size nine shoes.

The latest scare campaign is that coral bleaching is the signal that reefs are dying as a result of global warming. This global warming, we are told, is due to human emissions of carbon dioxide. What we are not told is that reefs would not exist without carbon dioxide. Solid coral reefs contain 44% carbon dioxide.

In the sub-tropical south-western Atlantic, reefs were subject to a heatwave in 2019. About 80% of coastal coral experienced bleaching. However, the mortality rate was less than 2% and far less than for bleaching events of similar magnitude in other areas.[985] Reefs show a remarkable tolerance and low mortality to intense bleaching events and easily adapt to rising sea level (or land level fall).[986]

On 8th January 2020, *Nature* published a paper *"Ocean acidification does not impair the behaviour of coral reef fishes"*. The authors studied more than 900 fish from six species over a period of three years in an attempt to validate findings published in eight papers by a research team at James Cook University. Validation attempts failed. Not one of the findings in the eight papers was found to be correct.

One author was Oona Lönnstedt who had published another paper claiming that baby fish in our oceans preferred eating plastic microbeads to their natural diet. She got hysterical and fawning international media coverage. The Lönnstedt paper was fraudulent and so riddled with errors that it was withdrawn. There was no great media fanfare about how the microbead fish diet was fraud.

Professor Peter Ridd was fired by James Cook University for academic misconduct because he publicly criticised the university's Australian Research Council's (ARC) Centre for Excellence for Coral Reef studies for producing science that was not *"properly checked, tested or replicated"* and he stated that many studies were just plain wrong or greatly exaggerated. James Cook University has now appointed a committee to look for evidence in the PhD work done by Lönnstedt between 2010 and 2014. Three of her PhD supervisors were authors of five of the eight discredited papers.

Can we expect the ARC to stop giving taxpayers' money for coral reef research at James Cook University? Can we expect James Cook University to treat the discredited researchers the same as it treated Peter Ridd? A series of court cases took place and James Cook University of North Queensland is spending a huge amount of taxpayer's money defending the indefensible. Benefactors beware. Freedom of speech no longer exists in universities and some of the community has lost faith in institutions such as schools, universities, courts, parliaments, police and the public service.

The one author in common in all eight papers was Philip Munday who has contributed to IPCC reports on ocean acidification, hence the IPCC pronouncements about ocean acidification and tropical fish are not worth a pinch of anything. The Great Barrier Reef is in good health but the adjacent James Cook University is in poor intellectual health.

Peter Ridd has written an outstanding book[987] on the Great Barrier Reef showing that all highly publicised hysterical crises are underpinned by poor science, much of which could not be replicated. He showed that all the sensationalist crises (coral-eating crown of thorns starfish, dredging of shipping ports, agricultural soil erosion smothering the Reef with mud, nutrient pollution from agriculture and climate change) are all underpinned by hype, unreliable science, vested interests and are just plain wrong.

Ridd shows that the only real crisis regarding the Great Barrier Reef is the quality assurance and quality control of Reef science and the peer review process. This one book shows the taxpayer-funded scientific community needs checks and balances, that almost everything that one hears about the declining health of the Great Barrier Reef is wrong or highly exaggerated and that this is done to keep the taxpayer funds rolling in and to provide a diet of catastrophism for green activists.

The only hype that should be listened to about the Great Barrier Reef is that it is in good shape, should be visited many times and should be viewed as proof that our dynamic planet is healthy.

Extreme weather

Why is the occasional bad weather due to global warming and not the long periods of good weather?

At some time in history, a day, week, month or year will be the wettest, hottest, coldest, windiest or driest time recorded. Maybe if you ignore history then you could claim that a weather event is unprecedented because it happened now. If it happens in your life, does this really mean anything?

These speculations about the hottest year don't tell us about the natural thermostats such as precipitation and evaporation and ocean heat. They don't tell us about the past when there was far more carbon dioxide in the atmosphere than now yet there was no global warming as a result. They don't tell us about the changes in the main greenhouse gas in the atmosphere (water vapour).

They don't tell us that we had a solar maximum event coincidental with warming. They don't tell us that in recent years there were fewer typhoons, more Arctic ice, more Antarctic ice, a decrease in sea surface temperature and that sea surface temperature drives atmospheric temperature, not the inverse.

It is claimed that countless lives are being lost by climate-related disasters worldwide. The International Disaster Database shows that in the 2010s, 18,357 people died each year from climate-related impacts such as floods, droughts, storms, wildfire and extreme temperatures. That is the lowest death count in the past century, a 96% decline since the 1920s, despite a larger global population. In 2020, the death rate count was 8,086. The data shows that global climate-related disasters are trivial compared to malaria, cooking smoke inhalation, TB and numerous curable tropical diseases.[988]

The Global Warming Policy Foundation published a June 2020 report by the physicist Ralph Alexander that looked at trends in hot and cold weather extremes, floods and droughts, hurricanes and wildfires and finds only a minor increase in cold weather extremes.

The most extreme weather, such as heat waves, droughts, floods, hurricanes, occurred many years ago. The recent heat waves in western Europe are nothing compared to the soaring temperatures of the 1930s, a period when three of the seven continents and 32 of the 50 US states set all-time high temperature records, most of which still stand. There is no evidence that droughts or floods are becoming worse or more common.

Hurricanes currently show a decreasing trend around the globe. The average number of strong tornadoes annually from 1986 to 2017 was 40% less than from 1954 to 1985. Global hurricane activity was below average for 2020. Much of the data for hurricane activity derived from eye-witness accounts and there was breathless media reporting that hurricanes in our time are an impact of global warming. Satellite and ground measurements tell a different story. In fact, an analysis of records since 1851 show that the intensity and frequency of modern hurricanes is no different from the past.[989] Everything we have heard about Atlantic hurricanes is wrong.

Countless claims of catastrophes due to human emissions of carbon dioxide have been debunked. Wildfires produced by climate change, for instance, were supposed to be raging around the world. As *Forbes* magazine pointed out recently, the number of wildfires has plummeted 15 percent since 1950, and according to the US National Academy of Sciences, that trend is likely to continue for decades. On hurricanes and tornadoes, which green activist alarmists assured us were going to get more extreme and more frequent, it probably would have been hard for "experts" to be more wrong.

"When the 2014 hurricane season starts it will have been 3,142 days since the last Category 3+ storm made landfall in the U.S., shattering the record for the longest stretch between U.S. intense hurricanes since 1900" noted professor of environmental studies, Roger Pielke Jr, at the University of Colorado. On January 8, 2015, meanwhile, the Weather Channel reported *"In the last three years, there have never been fewer tornadoes in the United States since record-keeping began in 1950"*.

Lomborg[990] shows that deaths due to cold weather are far greater than extreme heat which is in accord with other studies[991]. Worldwide, both human mortality from natural disasters have dropped by more than five times since 1980 and economic consequences of natural disaster has also decreased.[992] What we hear or read in the media is not in accord with the measured facts.

Deaths due to natural disasters fell to an all-time low in 2019 and property damage due to natural disasters was about the 30-year average in 2019. Munich Re also showed that the 820 natural disasters in 2019 caused losses of $150 billion which are broadly in line with the inflation-adjusted average of the past 30 years.

Munich Re does not mention another factor. There is a general shift in population towards the coasts as people become wealthier and this is where they can be affected more seriously by major storms. Tropical storms making landfall on highly developed coastlines cause the most damage.

The Munich Re report is validated by a 2020 paper by Pielke that reviews 54 normalisation studies published between 1998 and 2020 and finds little evidence to support claims that any part of the overall increase in global economic losses documented on climate time scales can be attributed to human-caused changes in climate.

Normalisation seeks to adjust historical economic damages from extreme weather to remove the influences of societal change from economic loss to estimate what losses past extreme events would cause under present-day societal conditions. Adjustment factors include inflation, population, gross domestic product, construction standards and others. An unbiased economic normalisation will exhibit trends consistent with corresponding extreme event trends.

There are no upward trends in the frequency or intensity of US hurricane landfalls since 1900, so we should expect that an unbiased normalisation would show no upward trends. Seven normalisation studies of US hurricanes find no trends but three do find increasing trends. The lack of trend of US hurricane data indicates biased results from the normalisation methods used by the three studies that reported those increasing normalised damages.

Temperature

Shock-horror revelations that a particular year is the warmest on record implies that it is due to humans emitting carbon dioxide and, as a result, the planet is getting warmer and warmer. But what does this really mean?

We measure temperature by thermometer but sediments, fossils, ice, caves and soils also provide us with a record of past temperatures. It was far warmer for most of the history of planet Earth than now. In the last interglacial some 120,000 years ago, it was warmer than now. Sea level was seven metres higher than now. Most of the past 10,500 years have been warmer than the present. With the exception of a few short sharp cold periods, Greenland temperatures were considerably warmer than now for at least the last 9,100 years.

In the last 10,500 years of the current interglacial, 9,099 were warmer than now. Some 6,000 to 4,500 years ago in the Holocene Maximum, it was warmer than at present and sea level was about two metres higher than at present. This was the peak of the interglacial that we now enjoy. It was only 8,000 years ago that there was no summer ice in the Arctic.

Many solar scientists are now predicting that we are on the downhill run towards the next cold cycle, glaciation or ice age. I hope not. Some of us like it hot and colder times have in the past led to starvation, conflict and depopulation although we are now better buffered with more technology, wealth and knowledge.

Are today's heat waves really heat waves? The media in Australia is now calling any day above 30°C a heatwave. In outback times decades ago, there were weeks on end when it was above 40°C during the day and night-time temperature plummeted to a freezing 35°C. We were careful not to walk on bitumen roads in bare feet as molten bitumen stuck to the soles of our feet and we were burned. We called it summer. It happened every year. The warm weather in California in August 2020 was touted as unprecedented yet the highest measured temperatures were in 1931 and 1933.[993,994]

A warm July in 2019 in Britain had people in a lather about global warming. The UK also had warm weather in July 2020 and July 2021. It's called summer. The mean Central England temperature measurements since 1659 AD show a different story. July 2019 was the 45th warmest since 1660 and the same as 1847, 1870 and 1923. It was also 1°C cooler than July.[1763] The average mean July 2019 temperature was 17.5°C was completely unremarkable.

The headlines scream about the hottest summer on record and use words like unprecedented despite the fact that in July summer had not finished. Unprecedented means that something never happened before. There have been many summers hotter than 2019. You don't have to look far to find something unprecedented. Cricket commentators are obsessed with statistics and can always find something unprecedented.[995]

Australia has had some hot weather. When temperatures reach 45 or 50°C in Australia, history shows us that it is not the first time. Explorers Charles Sturt[996] recorded temperatures of 53°C, 54°C and 56°C, Mitchell[997] recorded 55°C as did Stuart.[998] These were at times when atmospheric carbon dioxide was lower than at present.

Historical high temperatures have been adjusted downwards[999] but that didn't help people who died like flies in the heat of the Federation Drought in the late 19th Century. In the warm times of 1910 to 1930, the raw data of the ACORN-SAT measuring sites had "homogenised" the minimum temperatures downwards[1000] resulting in a temperature trend increase of 40%.

According to the "official" records (ACORN-SAT), the hottest day ever in Australia was 50.7°C on 2nd January 1960 at Oodnadatta (SA). The hottest ever temperature in Australia did not occur in the inland deserts at Marble Bar (WA), Marree (SA), Oodnadatta (SA) or Birdsville (Qld). It was on the coast at Albany in south-western Western Australia. On 8th February 1933, the good folk of Albany enjoyed a summer day of 44°C. They were not to know that the BoM some 80 years later decided that the temperature was really 51°C. The ACORN revision clearly introduced errors that have not been corrected. There is clearly no quality control at the BoM and that institution needs a quality assurance/quality control audit.

Explorer Sturt commented *"The thermometers sent from England, graduated to 127 degrees only, were too low for the temperature into which I went, and consequently **useless** at times, when the temperature in the shade exceeded that number of degrees"*. Some thermometers burst. Greens do not read history because it disproves their sensationist false ideology.

Did you know that the hottest day on record in NSW was not in 2020, 2019 or any day in your lifetime. In 1828, the explorer Charles Sturt recorded a blistering 53.9°C at Buddah Lake (Trangie, NSW). Cloncurry (Qld.) recorded a temperature of 53.1°C in January 1889 which a sign in the town claims is the hottest temperature recorded in Australia. Whatever it was it was not in recent times.

From the scientific perspective, if there is a claim that a certain year was the hottest year, then questions must be asked. Hottest day since when? What is the order of accuracy? Normally the order of accuracy is greater than the suggested temperature rise hence the claim is scientifically invalid.

The suggested annual temperature rise for Australia of 0.17°C is less than the order of accuracy of many measuring stations used to deduce this trend because older measurements are ± 0.5°C. This is meant to be an average temperature rise but how does one construct an average with few measuring stations and even fewer in remote areas? This is mathematical nonsense.

These speculations don't tell us about the past when there was far more carbon dioxide in the atmosphere than now yet there was no global warming as a result. They don't tell us about the changes in the main greenhouse gas in the atmosphere (water vapour). They don't tell us that we had a solar maximum coincidental with warming.

They don't tell us that there have been fewer typhoons, more Arctic ice, more Antarctic ice and a decrease in sea surface temperature. Sea surface temperature drives atmospheric temperature, not the inverse.

The BBC breathlessly claimed *"Last month was the hottest May on record globally as well as in the UK"*. It would only have taken the BBC a minute to check the on-line Met Office records that show that May 2020 was the 15th hottest since 1884 and was nowhere near as warm as 1893 and 1911. Why the lies?

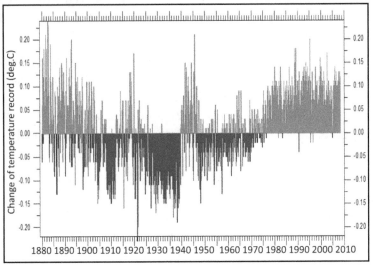

Figure 36: *The net change since 17th May 2008 for the global monthly surface temperature using data extending back to January 1880 (from GISS).*

Hurricanes

Scary footage of hurricanes generally accompanies alarmist stories that a perfectly natural phenomenon is due to human-induced climate change.

The intensity of hurricanes is not increasing. The strongest was in 1935 (Labor Day, 892 hPa) with weaker hurricanes in 1969 (Camille, 909 hPa), 2018 (Michael, 919 hPa), 2005 (Katrina, 920 hPa), 1992 (Andrew, 922 hPa), 1886 (Indianola, 1886, 925 hPa), 1919 (Florida Keys, 927 hPa), 1928 (Okechobee, 928 hPa), 2017 (Irma, 929 hPa), 1960 (Donna, 930 hPa), 2021 (Ida, 930 hPa) and 2017 (Harvey, 938 hPa). Florida has had 119 hurricanes since 1850. The last were due to climate change whereas numerous more intense hurricanes were not. Or so we are told.

NOAA data shows that 2016 was the eleventh consecutive year without a major (Category 3 or above) hurricane strike. This is the longest period without a major hurricane since record-keeping started in 1851.[1001] On a global scale, there has been no trend in the global accumulated cyclone energy over the past 30 years, a time when there was alleged global warming and increased cyclonic activity.[1002]

In 2017, the nail was driven into the coffin by NOAA which reported "*It is premature to conclude (that man made global warming has) already had a detectable effect on hurricanes*".[1003] The 2016-2017 cyclone season in the Southern Hemisphere was the quietest on record.[1004]

Global hurricane numbers in 2020 are low as are the number of major hurricanes. There is no evidence for an increase in hurricane intensity. In 2020, there were just eight Atlantic hurricanes, including two major ones, Laura and Teddy. There have been a large number of tropical storms at sea that would never have been spotted in the pre-satellite era. The total number of Atlantic hurricanes in 2020 was unusually high but the number of major hurricanes looks like being relatively low.

We know the President of the US is a powerful person and packs a mean punch but the *Washington Post* in September 2018 on Hurricane Florence stated that "*When it comes to extreme weather, Mr. Trump is complicit. He plays down human's role in increasing the risks, and he continues to dismantle efforts to address those risks*".

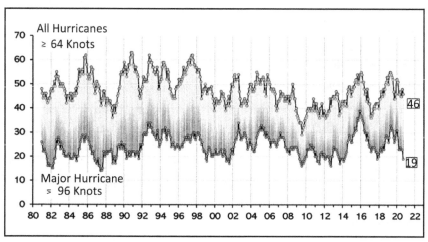

Figure 37: *Frequency of major hurricanes and frequency of global tropical accumulated cyclone energy.*[1005,1006]

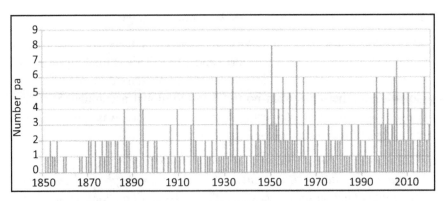

Figure 38: *Major Atlantic hurricanes 1851-2019 showing that we do not live in a time of extreme climate.*

This is no different from blaming witches for a crop failure. Florence was intense but not nearly as intense as other hurricanes that hit the US when it had Democrat Presidents.

The impact of hurricanes might seem more severe because of the blanket instantaneous news coverage and because more people now reside in hurricane-prone areas hence more property damage and loss of life.

Extinction

Species conservation is a romantic non-scientific view of planet Earth.[1007]

Over the geological time period since there was multicellular life on Earth, there have been five major mass extinctions, scores of minor mass extinction and a constant species turnover by extinction. All of this occurred before sinful humans emitted carbon dioxide. More than 99.9% of all species that ever existed on planet Earth are now extinct. New ecologies produced by extinction allow other species to fill the vacated space and thrive. Extinction is normal, conservation of species is not.

Animal life is classified according to the kingdom, phylum, class, order, family, genus and species. We humans are in Animalia (kingdom), Chordata (phylum), Mammalia (class), Primate (order), Hominidae (family), Homo (genus) and Sapiens (species). There is, of course, much argument in science about classification systems. In studies of extinction, normally the families and genera are considered rather than the individual species.

There have been mass extinctions of multicellular life 450-440 million years ago when 25% of families and 57% of genera disappeared. That mass extinction was actually two events. In the end of the Devonian mass extinction that may have lasted for up to 15 million years (375-360 million years ago)[1008], 19% of families and 50% of genera disappeared. In the biggest extinction of all time 251 million years ago, 57% of families, 83% of genera and 96% of species disappeared[1009] and, again, there may have been two events.[1010]

A less devasting mass extinction took place 205 million years ago with the loss of 23% of families and 48% of general.[1011] The extinction we all know about which led to the demise of the dinosaurs was 66 million years ago and 17% of families and 50% of genera disappeared. The emotive argument used by green activists is that we are in the planet's sixth mass extinction. Nothing could be further from the truth. This is fraud.

There is a great debate in the scientific literature about extinctions and the science, as in all fields, is certainly not settled. Most likely causes were toxic gas release associated with continental basalt volcanoes, asteroid and comet impact, continental drift, sea level fall associated with glaciation, habitat destruction, competition, disease pandemics and a loss of oxygen from ocean waters.

In the past, neither major mass extinctions nor minor mass extinctions were influenced by global warming or a high carbon dioxide content in the air. For 73% of time over the last 450 million years, the carbon dioxide content of the air has been higher than now. Some mass extinctions occurred instantaneously whereas others were multiple events over millions of years during which time there were repeated extinctions.

The history of time shows us that animal and plant populations boom and crash. That's the normal natural disorder. Over time, we see that the number of marine genera and families have increased despite mass extinctions.

Counting the number of species on our planet is not as easy as it seems. We don't know how many plant and animal species exist. Each new plant and animal species is described and internationally accepted. Each year, some new species that were previously described and accepted are discredited and expunged from the list of known animals and plants, mainly on molecular biology grounds.

Some 100 new phyla of animals appeared in the Cambrian explosion of life. No new phyla have since appeared. Why not? Phyla have disappeared during mass extinctions. After the Cambrian explosion of life, the seas were teeming with nutrients, bacterial and multicellular life.

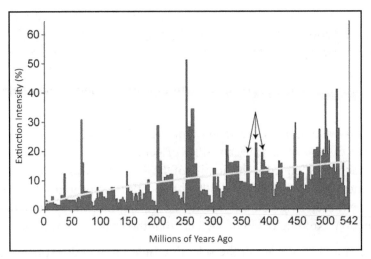

Figure 39: *History of extinction of marine genera over the last 542 million years showing five major and about 20 minor mass extinctions. The total number of marine genera has decreased over time and this is unrelated to the presence of humans.*[1012]

The fossil record is an incomplete record and the chances of an organism becoming fossilised are slim. For example, if you drop dead in the desert, what is your chance of becoming fossilised? You undergo bacterial decomposition, carnivores eat your flesh, your bones are eaten, disarticulated, spread and bleached and it is a case of atoms-to-atoms. The best chance of being fossilised is if you die in an oxygen-free swamp that quickly gets covered by sediment.

Some individual cloud mountain Central American rainforest trees have hundreds of endemic insects. If this one tree dies, what are the chances of these insects becoming fossilised and escaping bacterial decomposition in the rainforest leaf litter? Zero.

It's difficult to determine whether the number of species in modern times is increasing or decreasing.[1013] There is a good chance that most species that exist today or existed in the past leave no trace in the fossil record.

There is a huge amount of emotion associated with extinction.[1014,1015,1016] Some speculations suggest that a mere 0.8°C temperature rise over 50 years will result in extinction of 20% of the world's species. This is far less than daily and seasonal temperature changes.

If this were the case, we should have seen a mass extinction of life in the Minoan Warming, the Roman Warming and the Medieval Warming. We did not. We may actually be living in a period of low extinctions, with relatively few species becoming extinct over the last 2.5 million years.[1017] Current projections of extinctions may be an overestimation as we focus on terrestrial vertebrates and not the spectrum of life on Earth.

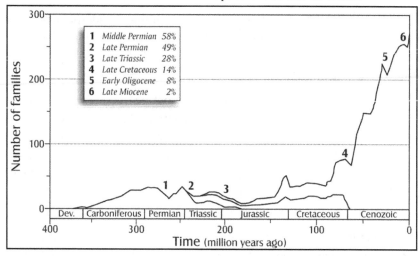

Figure 40: *Increase in diversity of life and mass extinctions over time as shown by preserved families of tetrapods. Minor and major mass extinctions are numbered 1 to 6 with percentages of tetrapods rendered extinct. The three lines represent the three dominant tetrapod faunas.*

The IPCC has promoted the view that global warming creates extinction. Suggestions that human-induced global warming results in extinction is, at best, scientifically flawed. It was suggested that rising sea surface temperature in the equatorial Pacific Ocean led to the disappearance of 22 of the 50 known species of frogs and toads in the Montverde cloud forest of Costa Rica.[1018]

However, the authors also suggested that lowland deforestation may have a major influence on preservation of the cloud forests. This reservation was ignored as human-induced global warming was given as the reason for extinction. Although 21 of these species are known from elsewhere, one species (the golden toad) lost its only habitat and became extinct.

It was the loss of this species that led to the conclusion that human-induced global warming could create an extinction of 20% of species over 50 years with a 0.8°C temperature rise. However, the trade winds that bring moist air from the Caribbean spend 5 to 10 hours over lowlands before they reached the golden toad's habitat. By 1992, only 18% of the lowland vegetation remained after land clearing, resulting in an increase in the altitude of the cloud base thereby depriving the cloud forests of their moisture.[1019] Land clearing created the extinction of the golden toad, not global warming. Habitat change and competition have long been a cause for species extinction.

The emotion about human-induced global warming is underpinned by the assumption that a future climate change will be so rapid that plants and animals would not be able to adapt to keep up with the rate of temperature change. This view ignores the past.

For example, the Fremont Glacier in Wyoming records a substantial warming from 1840 to 1850.[1020] A substantial warming in less than a decade is far faster than the most speculative catastrophist models for human-induced global warming, yet there is no evidence that there was an extinction in North America at that time. Multicellular plants and animals have been on Earth at least 500 million years, so they have enjoyed and survived at least 20 major climate changes.

If we took the emotive argument to its logical conclusion, then there would be no multicellular life on Earth, as previous global warmings would have rendered life extinct and planet Earth would be a moonscape. Why is it that only the late 20th Century-early 21st Century Warming will produce extinction whereas in previous times when it was far warmer there was not extinction?

Furthermore, when extinctions are measured there is a different story. Computer predictions are the opposite of measurements. The measured red list of all species has decreased between 1870 and 2009.[1021,1022] This estimate is based on measurement which has limitations. Were these macro species? Surely an analysis of only 529 species is incomplete? I take such surveys with a large grain of salt.

The diagram on the next page shows that if you were really concerned about extinction of native animals in Australia, you would stock up on guns, knives, poisons, traps and genetically engineered weapons, go bush and destroy every feral introduced animal. Why do green activists moan and not take action?

A key argument is that plants are immobile, hence a rapid global warming will push them into extinction. The scenario is that when plants become extinct, then animals that feed off plants will also become extinct. Because this was not seen in previous warmings, the alarm bells should have been ringing for those speculating about extinction due to warming.

What is observed is that plants in the Arctic have adapted to the frigid conditions but their distribution is rarely limited by warm conditions.[1023,1024,1025,1026]

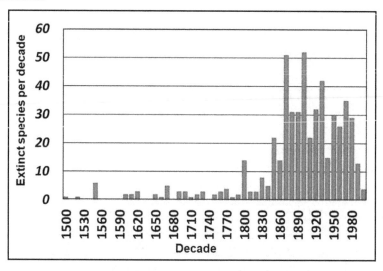

Figure 41: *The "red list" of all extinct species by decade from 1500 to 2009. The extinction peak coincides with the introduction of non-native species, primarily in islands (including Australia).*[1091]

Geology shows that in times of global warming, there is an explosion of life, diversity increases and speciation is rapid. The Cambrian explosion of life took place in the post-glacial warm times when atmospheric carbon dioxide was at least 25 times greater than today. Other great diversifications have taken place in the past. Species-rich forests existed during the warm Tertiary times in the western USA where many mountain species grew amongst mixed conifers and broad leaf sclerophylls.[1027,1028] It is only by completely ignoring the history of the planet can it be claimed that global warming can produce extinction.

In modern settings, it is also suggested that increased temperature will bring more species diversity[1029] by extending the ranges of plants and animals. Replacement of high-altitude forests by mixing with low altitude forests to create greater species diversity has happened in previous times of warming and would be expected in another warming event.

Furthermore, if a future warmer climate had a higher atmospheric carbon dioxide content, plant life would be far more vigorous because increased carbon dioxide enables plants to grow better in nearly all latitudes, especially at higher ones.[108]

Both animals and plants are limited by the latitude and altitude cold-boundaries of their range and are not limited by the heat-limited boundaries of their range.[1030,1031,1032,1033,1034] If atmospheric carbon dioxide is doubled and temperature is at 10°C, plant growth doubled and woody plant growth increased 50 to 80%.

In the European Alps, there are plant species counts from 1895 to the present. Mountaintop temperatures have increased by 2°C since 1920 with 1.2°C of that rise over the last 30 years. Of the 30 mountaintops, nine showed no change in the species count, 11 gained 59% more species and one had a 143% increase in species. The 30 mountaintops showed a mean species loss of 0.7 out of an average of 16 species.[1035]

The loss of a species from a particular mountain does not mean extinction but shows local mobility of plants. There are numerous other studies of plants[1036,1037,1038], lichen[1039], butterflies[1040], birds[1041], plankton[1042], marine systems[1043,1044] and fish[1045] that show that a slight temperature rise induces species diversity, species migration and adaptation.

Detailed studies in a specific area may record a local extinction, as an extinction is the total loss of a species whereas local extinction could mean that a species has migrated to another area. Some species that were thought to be extinct in one area have been found decades later in another area.

Polar bears are cute, people care about them and they sell. Just ask the marketing executives of Coca-Cola and Bundaberg Rum. Well, they are not really that cute. Try to hug a polar bear and suffer the consequences. The polar bear has become an icon of the global warmers.

Claims that the polar bears are facing extinction due to human activity packs an emotional punch, as we can imagine a cuddly furry polar bear painfully drifting into extinction, whereas we don't suffer too much emotional damage if bacteria, scorpions, snakes and spiders become extinct.

However, the polar bear has survived the Medieval Warming, the Roman Warming, the Minoan Warming and numerous previous warmings when the temperature was higher than today. If the ice distribution changes, polar bears move. They are not glued to the one place on a static planet.

The US Fisheries and Wildlife Service[1046] showed that there are some 22,000 polar bears in about 20 distinct populations worldwide. Only two bear populations, accounting for 16.4% of the total, are decreasing, and they are in areas where air temperatures have actually fallen, such as the

Baffin Bay region. By contrast, another two populations, about 13.6% of the total number, are growing and they live in areas where air temperatures have risen, near the Bering Strait and the Chukchi Sea.

As for the remaining 10 populations comprising 45.4% of the total, four are stable and the status of the remaining six is unknown. The US National Biological Service found that polar bear populations in western Canada and Alaska were thriving to the point that some were at optimum sustainable levels.[1047]

In yet another irrational scare story, we get yet another version of the *"polar bears will be extinct unless we stop driving cars and using coal-fired electricity"* which has saturated the media for decades. Models using discredited climate scenarios use a 600% increase in coal burning and a 6°C rise in polar temperatures. These climate activists fail to inform the readership that polar bears in the Arctic have survived numerous periods of warming when temperature rises of 6°C were not uncommon.

Polar bear observations on a small population in one part of Canada are not representative of all the Arctic. The world expert on polar bears shows the climate activist narrative that polar bear populations are declining is not supported by data.[1048] It's not a myth: 2020 was another good year for polar bears.

Dr. Susan Crockford, former adjunct professor and zoologist at the University of Victoria released a 13 minute video that shows "...*the strong polar bear component to the terrorization of the world's children about climate change, which began for many youngsters in 2006 with the BBC and Sir David Attenborough's commentaries about the dire future of polar bears – and continues to this day"*.

In the BBC Frozen Planet series called '*On Thin Ice*', Sir David Attenborough falsely claimed that polar bear numbers had been falling *"in many regions"*. The 2019 BBC's '*Seven Worlds, One Planet: Asia*' shows several allegedly hungry polar bears driving walruses over the cliff, all falsely blamed on climate change. The walruses were driven off the cliff by the camera crew and sensationalist emotive TV footage captured the walrus hurtling to their death.

Based on the evidence, there is little reason to suggest that polar bears are on the path to extinction. Polar bears have survived for thousands of years, during both colder and warmer periods, and their populations are by and large in good shape. Polar bears suffer many threats, but global warming is not one of them. The main threats are ecotourists, bureaucrats and hunters.

Most modern extinctions have occurred by the introduction of predatory species on island populations that had no natural predators. Past human-induced extinctions were due to over hunting or maybe excessive hunting after a natural environmental catastrophe such as an impact.[1049,1050,1051,1052] Some modern extinctions, especially of large animals, occur due to clearing for agriculture, especially large animals and to logging of timber trees in poor countries.

The International Union for the Conservation of Nature (IUCN) has shown the number of extinctions of threatened species has decreased for more than 100 years. They also publish a red list of threatened species.[1053] The IUCN is independent from the UN who claim that there is an "unprecedented" decline in the number of species.[1054] Once the word "unprecedented" is used, you know you are being conned.

There is great concern about the decreasing forest habitat and possible future extinctions. Why? Because the demand for cheap crop-based biodiesel, such as palm and soy oil, has increased. Forests in Asia and South America have been cleared for crops. As countries race against time to meet their Paris Accord targets for reduction of carbon dioxide emissions by 2030, biofuels have emerged as a preferred alternative to fossil fuels. The EU's Renewable Energy Directive (RED) in 2010 set a 10% renewable energy target for transport by 2020.[1055]

About 10% of the world's remaining orangutan habitats have vanished because forest was clear felled for biofuel crop production. Rainforests are being clear felled to produce the 1.9 million tonnes of palm oil that end up in European diesel fuel tanks each year. The EU claims that palm oil is sustainably produced yet the EU's biofuels policy has wiped out forests the area of The Netherlands.[1056] It is policies pushed by green activists that destroy forests and push threatened species closer to extinction.

Plantations covering an area of four million hectares of land are needed for Europe's biofuels. This land was rainforest habitat for 5,000 endangered orangutans. Biodiesel from palm and soybean oil and also from European-grown canola has a far larger carbon footprint than diesel from fossil fuel sources. In the EU, biofuels are subsidised to the tune of €10 billion. A policy to save the planet ended up trashing it and putting orangutans under extinction pressure.

Biofuels for automobile alcohol make even a larger dent on the environment.

Increased land clearing, food shortages and food costs and carbon dioxide emissions are the norm for alcohol production.

In the full cycle of biofuels (ploughing, planting, weed suppression, growing, harvesting, transport, conversion to biofuel and finally burning in a combustion engine), biofuels emit as much carbon dioxide as petroleum products as used in internal combustion engines. Biofuels require enormous amounts of land, water, fertiliser, insecticides and energy.[1057,1058] None of this is renewable or sustainable and uses land and water that could be used to grow food.

We have been bombarded with catastrophic stories about how global warming will kill off the Antarctic's emperor penguins. However, data gives a different story and shows that the penguin population has grown by up to 10% between 2009 and 2019.[1059] It's climate models that attempted to send the emperor penguins to extinction, not the climate.

In 2015-2016, maybe a million common murres (a North Pacific Ocean seabird) died. It was suggested that it was due to a blob of warm water in the Pacific Ocean in 2013-2014 affecting bird feeding and breeding and implied that this was due to alleged global warming. No climate catastrophist or media organisation ever thought to consider that submarine volcanicity may have warmed an ephemeral blob of ocean water.

Extinction, like sea level change, drought and climate change, is a fascinating and complicated field of learning for those with an intellectual curiosity and cannot be understood by simplistic chanting, slogans and signs.

Extinction Rebellion is a group of extremist climate change anarchist thugs with no concern about extinction, climate change, fossil fuels or the environment. Because they have not been able to win arguments using logic and science in the public arena or be elected by democratic means, they resort to violence. There have been some horror stories about the internal violence and rape within this group.

They are totally intolerant of competing opinions yet exploit the tolerance of a democratic system. Maybe the next time they attach themselves to a road made from fossil fuels and using glues made from fossil fuels, we should respect their decisions and not use fossil-fuel derived solvents to dissolve glue for their release.

One of the founders of the Extinction Rebellion is promoting murder of those who do not agree with him. Roger Hallam, founder and still an active member stated "*Maybe we should put a bullet into the head as punishment*" for all those he deems responsible for an alleged planetary extinction.[1060] Why were green, environmental and conservation groups not up in arms in public about such fellow travellers?

Extinction Rebellion wants to silence voices it disapproves of and even targeted the BBC, an organisation that itself does not allow questioning of the mantra of human-driven climate change. Extinction Rebellion can easily exist in a society where politicians and the media do nothing to engage in healthy debate about climate change. The huge hole they have left is filled by Extinction Rebellion.

In 2020, Extinction Rebellion in the UK started to split amid allegations of bullying, racism, financial mismanagement and power trips. One of their leading lights, Zion Lights, expressed misgivings about numbers pulled out of thin air by the founder, Roger Hallam. She left Extinction Rebellion, received the normal toxic abuse for thinking for herself and has realised that there can be no reduction in carbon dioxide emissions by humans unless there is a massive shift to electricity generated by nuclear reactors.[1061]

As a reward for defacing buildings, splashing paint around and being a public nuisance, an ACT magistrate in Canberra rewarded various arrested Extinction Rebellion thugs with a $20 fine for their calculated vandalism. One thug had a pat on the back and was let off with no conviction whatsoever. This shows that some magistrates and other elites are totally out of contact with the community who pay them.

As a result of the violence, destruction and anarchy by Extinction Rebellion trying to close printworks giving the public news that Extinction Rebellion deems wrong, some have called for the group to be treated like an organised crime gang.[1062,1063] There was no action against Extinction Rebellion and they continue to exploit the tolerance of a democratic society by blocking roads to industries that they deem wrong.[1064] Extinction Rebellion thugs do not use the ballot box to try to make changes because they are aware that they would harvest almost no votes. Instead, they use violence to try to achieve their aims.

The co-founder of Extinction Rebellion, Gail Bradbrook, uses violence to advocate for the total end of fossil fuel use by 2030. She claimed in a now deleted tweet[1065] that "*This movement is the best chance we have of bringing down capitalism*".

She drives a diesel-powered car because she claims that she can't afford to buy an electric car. She also used fossil fuels to fly nearly 20,000 km for her capitalist's holiday in Costa Rica. She claimed that she needed to fly for a health issue that couldn't be treated in Britain.[1066]

Of course. Silly me. It was for a Pinocchio nose reduction operation.

6

HOW TO RUIN A COUNTRY

Renewables use concrete, metals, plastics and rare earth elements. Production of these commodities results in massive pollution and more carbon dioxide emitted into the atmosphere than renewables would save.

The energy used to make wind and solar generators is far more than they will produce in their lifetime.

Coal-fired generators emitting carbon dioxide must operate 24/7 as backup for renewables for when the weather changes and at night resulting in no reduction of carbon dioxide emissions.

Turbines change local weather, change rainfall, destroy bird and bat populations, start bushfires, damage human health and don't produce electricity when needed.

Turbine blades and solar panels are dumped after their short life resulting in heavy metals and toxins leaking into the soil, water and air. Renewables companies don't have environmental bonds for cleaning up their mess at the end of the project life.

If any industry killed wildlife, destroyed forests and polluted the air, water and soil as much as the renewables industry, they would be closed down. Opaque sweetheart deals with bureaucrats and governments allow the renewables industry to continue.

Forests have been clear felled and food-producing land has been sterilised for wind and solar generation.

Renewables

How green is my energy?

Let me get this right. Carbon dioxide emissions from burning coal to make electricity are creating global warming. This must be bad for the planet. If we stop coal burning to produce electricity and produce electricity from the Sun and wind, then we are doing our bit to save the planet. After all, the wind and sunshine are free and we can make clean green solar and wind power to replace coal-fired electricity. Wrong, seriously wrong and hopelessly wrong.

Too often when green activists are shown to be wrong, they invoke the "precautionary principle". There is no such principle in science.

This is translated simply: When green activists have lost the economic, scientific, moral and logical arguments, they still want to control all policy because they claim that there is a threat from an unknown unknown. Logic, science and facts are beyond the ken of green activists.

It has never been shown that human emissions of carbon dioxide drive global warming. Even if a small country like Australia completely stopped emissions of carbon dioxide, it would have no effect. This shows that the whole issue of climate change is unrelated to science and, as I showed in earlier chapters, measures to supposedly address it destroy the environment.

China emits in 16 days what Australia emits in a year. Unless China, India, USA, Brazil, Indonesia, UK, EU and other large emitters completely stop emitting carbon dioxide from their industries, any actions Australia takes are pointless.

Higher carbon dioxide in the air is good for the planet. Plants grow better and crop yields increase. When the planet is in one of its warm cycles, then growth and yields are even greater.

Every year, the International Energy Agency produces a World Energy Output that shows that the green energy transition just isn't happening[1067]. The facts are simple. The world is fuelled by fossil fuels and this will not change in the near future. Politicians may sign feel good popular agreements but when the rubber hits the road, they are reticent to damage their chances of re-election.

The landmass and territorial waters of Australia adsorb far more carbon dioxide than Australia's industry and domestic activities emit. Supporters of solar and wind electricity try to claim the high moral ground by claiming that renewable electricity generation methods are cheap and environmentally friendly.

Barely 25 years ago, Australia was regarded as an energy superpower with abundant uranium, coal and gas reserves that ensured reliable, secure and affordable electricity. Our economy was then, as it is now, underpinned by burning coal to make electricity. Business then was about taking that electricity and using it to create relatively low-cost goods and services. At that time, Australia had a vibrant manufacturing industry which depended upon cheap reliable energy. In that environment, business flourished.

In September 2021, we had the UN interfering with Australia's internal affairs and giving Australia a 10-year deadline to kill off its coal-mining industry.[1068] Their official flew to Canberra from abroad to give us the message and in the process burned huge amounts of fossil fuels. Why should unelected bureaucrats from elsewhere in the world dictate policy to an elected government? Has the UN told China to stop using coal? No. Why not?

Coal mining added $50.6 billion to the Australian economy in 2019-2020 and directly employed 40,000 people and indirectly employed at least 200,000 people.[1069] About 80% of the coal mined in Australia is exported to other countries to underpin their escape from poverty. The UN should be rewarding Australia.

To close coal mining in Australia would economically cripple Australia and allow China, Russia, Indonesia and India to increase production. Closing coal mining in Australia would not make one *iota* of difference to global human emissions of carbon dioxide. The taxation base would decrease and there would have to be savage cost cutting of all public spending and social services. To have grid electricity would be a rarity and unemployment would skyrocket. A good start to cost cutting would be to cancel membership to the UN, a discredited bloated bureaucracy that has been captured by unrepresentative elites that are no friends to the West.

Australia used to have safe, cheap, reliable power, mainly coal, until the arrival of the unproven global warming hypothesis decades ago. Our present power system is neither cheap nor reliable and lobbying by the likes of Musk will only make it worse. Given our economic needs and the trifling impact our emissions have on the planet, security and affordability must come first. Businesses and households are being directly attacked by green activists, governments and bureaucrats. The attack is most savage on the poor. Electricity prices have risen alarmingly and jobs have been exported.

Australia is heavily dependent upon coal for electricity, more so than any other developed country. About 60% of Australia's electricity generation is from coal, natural gas is increasingly used for it, especially in South Australia and Western Australia and, after many years of low investment, there is a major challenge to build more dispatchable generating capacity.[1070] Base-load power needs grid reliability which is destroyed by wind and solar.[1071]

In 2018, 46% of Australia's electricity generation was sourced from black coal, 19% gas, 14% brown coal, 6.7% hydro, 6.2% wind, 4.6% grid and rooftop solar and 2.06% diesel and 1.39% biomass. Wind and solar contribute only 10.8%, hydro requires a mountainous topography and high rainfall and the leavy lifting is done by fossil fuels (70%). Base load power is about 64% of the total demand.

Australia could not survive on sea breezes and sunbeams, we could not even keep refrigerators on overnight in supermarkets if we relied on wind and solar for electricity. Anyone claiming that wind and solar have a serious contribution to an industrial nation's energy mix has never seen a sunset, experienced a period of dead calm weather or suffered numerous blackouts.

Did you know that not only do the Australian Greens Party want coal-fired electricity banned (60% of our electricity) but they also want gas (19%), hydro (7%) and diesel (2%) banned. Renewables currently produce 12% of the total energy at present but not when and where we want it and the Greens Party want this to be the only source of energy for an urban industrial modern society. Nuclear, of course, is not on the Greens Party menu.

Australia has one small 20MW nuclear reactor at Lucas Heights (NSW) which provides medical isotopes for Australia and our near neighbours. If a Greens Party politician has cancer, will they refuse treatment because they have an objection to nuclear reactors?

When hypocritical Greens Party politicians lead by example and have wind turbines in every park and back yard in their electorates, reject modern nuclear medicine and live naked in caves, then we might listen to them. The COVID-19 crisis has given people a taste of green nirvana and it isn't pretty, especially if you have been made jobless and end up broke, depressed, suicidal, locked down and isolated.

In 2007, Australia had one of the lowest electricity costs in the world.

Australia had a vibrant manufacturing industry. From 2007, it only took a few years under a Federal Labor/Greens Party coalition for Australian energy prices to increase such that they ranked just behind Denmark and Germany which have the highest-priced electricity in the world. The Labor/Greens Party coalition was a huge supporter of renewable energy schemes using taxpayer's money with costs passed onto the consumer.

Australia's energy regulator figures reveal almost 60,000 households are on electricity hardship payments and another 151,862 customers are on electricity payment plans.

In the 12 months before COVID-19 struck, 75,000 Australians were disconnected from the energy grid. In the early months of 2021, disconnections were on the rise again.

The energy companies do not answer phone calls from distressed customers and in November 2020, the Australian Energy Regulator was ordered by the Federal Court to pay $1.5 million in penalties after it had breached energy laws and wrongfully disconnected eight customers in financial hardship.[1072]

The Labor/Green Party alliance increased the Federal Renewable Energy Target (RET) to 45,000 GWh driven by ex-unionist Labor environment minister Greg Combet who, along with his old union mate Gary Weaven, used the union superannuation money they controlled to invest in wind power outfits like Pacific Hydro.

Once upon a time, the unions and their Labor Party would have come to the rescue of the poorest and most vulnerable in the community. Now they are a business to scam the poor and vulnerable. With three trillion sloshing around in superannuation funds, there is a lot of money available to invest in woke projects rather than investing to make money for superannuants to have a comfortable post-retirement life.

Labor is pushing for a 50% Renewable Energy Target (RET), maybe to protect city electorates from the socialist Greens Party which is competing against it for seats or maybe to increase income for their businesses. Whatever the outcome, more of the poor and vulnerable will continue to have their electricity disconnected. Labor wants businesses that can't meet their net zero carbon balance to buy carbon credits of bushland or create bushland. This is called taxation and is yet another scheme to attack farmers and reduce the area of land for food production.

A report by the government's chief scientist (Finkel Report[1073]) suggested a renewable energy component of all electricity at 42.5% yet the system cannot cope with half that. The Report did not state that the poor would bear an increasing burden for a basic commodity although it recommended an increase in subsidies. Wholesale electricity prices were around $40 per MWh until 2015. They now exceed $80 per MWh because of subsidies. In the past, the coal-fired electricity wholesale price was less than $30 per MWh.

The Finkel Report told us what we already knew. *"The past few years has seen the retirement of significant coal-fired capacity from the National Electricity Market while there has been no corresponding reinvestment in new dispatchable energy"* and *"If new dispatchable capacity is not bought forward soon, the reliability of the National Electricity Market will be compromised"*.

This is bureaucratic speak for: Wind- and solar-generated electricity has been a total failure. We have a huge problem unless we build more coal-fired power stations. Now. Neither coal-fired nor nuclear power stations are being built in Australia.

Matters can only get a lot worse before it gets better for Australian electricity consumers. A way out depends upon whether both State and Federal governments make decisions on behalf of consumers and not for green activists and subsidised carpetbaggers.

Finkel's suggestion that we keep coal-fired power stations operating, despite them rendered unprofitable and bleeding cash by heavily subsidised wind and solar power, is as ridiculous as his suggestion that adding more wind and solar power will lead to lower price rises. Irrespective of their age, Australia's coal-fired plants with a life of about 60 years deliver reliable and affordable electricity, whether it's night or day and whatever the weather.

The Australian electricity market has become a stinking swamp covered with a net of treaties, laws, rules, obligations, prohibitions, targets, taxes and subsidies. It has been made deliberately complex with all sorts of acronyms. We have an alphabet soup of regulatory agencies analysing, advising, fiddling, gouging and enriching themselves.

At the Commonwealth level, we have the ESB, AEMO, AEMC, AER and ACCC seeking a place in the Sun. On top of this are State regulatory agencies; taxpayer-funded research by the BoM, CSIRO, CRCs and universities and bucketloads of money used as aid that is gobbled up by the UN and spent on jaunts to climate conferences and IPCC meetings that handcuff Australia to high costs and unreliable electricity.

Energy in Australia is now so complicated that it is not transparent. This allows corruption to be paid for by the consumer and taxpayer. We need another Alexander the Great to cut the Gordian knot and to deliver the taxpayer from spivs.

Any policy that lowers carbon dioxide emissions increases costs. This has been shown time and time again in many countries. The environmental damage from renewable energy is eye watering.

Renewables just cannot power Australia because no legislation can stop the Sun setting every evening. No legislation can move the frequent windless high pressure weather systems that hover over south-eastern Australia. Grid scale battery storage and grid scale pumped hydro is as believable as unicorns; there is no grid nuclear power in Australia and, unlike European countries, there are no neighbours to fill the shortfall when the wind and sunshine go walkabout.

Subsidies are the only thing renewable about renewable energy. Solar and wind power generation does not just harvest the Sun and wind. They harvest bucketloads of other people's money. Forget the voluble pomposity, saving the planet and faux moral piety.

In what is the greatest state-sanctioned wealth-transfer in human history, hundreds of billions of dollars are being shovelled from the pockets of taxpayers and consumers into the coffers of cynical and conniving rent-seeking global wind and solar corporations.

Australia's energy future does not look good. Everyone apparently loves renewables if the costs are modest. If they can't deliver the electricity as and when required and when the costs skyrocket, opinion changes. Over the next 20 years, some 8,500 MW of cheap reliable coal-fired electricity will be retired in NSW.

The obvious solution is to place modular nuclear reactors on the sites of retired power stations to utilise the existing grid, workforce and engineering infrastructure. To replace this with the equivalent solar generation would require around 29,200 MW of nameplate solar power capacity employing 387 million solar panels spread over 71,450 hectares. This is 30% of the area of the ACT.

Energy will not be produced at night for any industry needing it 24/7 such as for refrigeration of food, hospital, fuel stations, mines sites, computers, traffic lights, airports etc. To this must be added the extra capital costs for transmission of electricity from its inland area of generation to the regional and coastal areas of high consumption. There would be no excess electricity in NSW to feed other states or the ACT.

The hidden costs of climate policies and renewables in Australia attack the poor. Costs are at least $13 billion ($1,300 per household) and these hidden costs account for 39% of household electricity bills. This causes a net loss of jobs to the economy and with every green subsidised job created, 2.2 real jobs are lost.[1074] Forever. Meanwhile, green activists are employed by taxpayers money or by donations from mugs.

If green bureaucrats and politicians thought they could generate energy using the weather, they should create legislation to stop high pressure systems stalling over south-eastern Australia at night when no electricity can be generated from renewables. I wouldn't fly in an aeroplane piloted by a know-all idealogue and, by the same token, we should not have complex electricity generation and grid systems run by green activist know-alls.

Complex power and transmission systems are now driven by green activists. Do they have the skills to do a due diligence on the wind supply with their plans to pursue net zero emissions? Previously one needed a degree in electrical engineering to provide the community with electricity. Now one with degrees in law, environmental science, political science or perhaps even gender studies is guaranteed employment in government energy agencies.

We are continually told that we are just a heartbeat away from a future powered by cheap wind and solar. The experiment has failed and been subsidised for 30 years. Time to quit. Not one modern industrial economy powers itself on renewables but trust me, say green activists and their carpetbaggers, it's just around the corner because we will have grid-scale battery storage of electricity.

Wind and solar can't save the planet so why let the wind and solar industries keep destroying it? If wind and solar are so cheap, why are they subsidised by the consumer? Renewable energy subsidies have poisoned electricity supplies and the environment[1075] and green activists are now claiming that new energy can be created by using renewable energy.[1076]

A week of cloudy weather is not uncommon hence solar-charged batteries need twenty-one times the solar panel generating name-plate capacity plus a little bit more for charging. To achieve this a very large amount of productive farmland is sterilised. If the solar array is remote, then the voltage losses require thirty-times the nameplate capacity.

Western democracies are evolving into societies so heavily regulated by governments that renewable energy has raised costs and increased regulation to solve failed policies will continue to increase costs. No country can become prosperous by more taxation, more regulation and employing ever more bureaucrats.

Future generations will wonder why we have destroyed the planet while claiming to save it by installing completely useless industrial wind turbines and solar panels. Whether it is dumping hundreds of thousands of toxic solar panels and windmill blades into landfill, or the creating of toxic lakes in China where rare earths are processed to make them, the so-called green energy revolution is anything but green.

Why don't green activists, green politicians and green corporations disconnect from the grid that has predominantly coal-fired electricity? They could lead by example and run on renewables. They don't. Because renewables can't do the heavy lifting.

Pity if we had to rely on renewables and not coal. The renewable contribution at evening meal time in south-eastern Australia can't do the job. For example, in August 2020 it was 5.8% on Wednesday August 5th, 6.2% on Thursday 6th, 14% on Friday 7th, 6.6% on Saturday 8th, 9.3% on Sunday 9th, 6.6% on Monday 10th, 10.5% on Tuesday 11th, 9.3% on Wednesday 12th, 6.2% on Thursday 13th, 4.4% on Friday 14th and 7.4% on Saturday 15th. Most of the time reliance on renewables would only result in you having cold meals, no TV and no heating or cooling.

When renewable-loving green activists eat bacterially-infected food kept at room temperature for days because they don't use a coal-fired fridge or fossil-fuel driven stove, then they no longer will be carping hypocrites. If they become ill from poor food quality, then no fossil-fuel driven vehicle could take them to hospital, no hospital stay could occur because medical equipment is driven by fossil fuels and no drugs could be taken because they are manufactured using fossil fuels.

If green activists want to go back to the times of 500 years ago when big bacterial blasts controlled population growth, then we can only encourage them to live by their principles.

If you are looking for a reason why wind and solar power have never worked and never will, it is simple. It is the weather.

The great socialist reset to transition to the 'inevitable' wind and solar mean no future. And that means for you. Net Zero is the start

Green activists tell us that wind and solar electricity is free. That's until we need to purchase electricity which does not get produced by weather-dependent generators 24/7. Solar energy is very dilute. Solar collections usually cover huge areas of flat arable land, stealing farmland, starving wild herbs and grasses of life-giving sunlight thus creating "solar deserts" where wildlife cannot survive.

Wind turbines steal energy from winds which often bring moisture from the ocean. These whirling scythes slice and dice birds and bats. Walls of turbines create rain shadows, producing more rain near the turbines and more droughts down-wind. The wall of wind towers offshore means that less wind and rain reach the shore.[1077] Is this green?

German offshore wind turbines cannibalise wind from each other resulting in a reduction of load hours achieved[1078]. Offshore wind in Germany is inefficient unless half the continental shelf of the North Sea is covered with wind turbines separated from each other by miles of sea.[1079]

In the UK in winter, the prevailing weather patterns are such that the warmer periods are windier while the colder times are calmer. As temperatures fall and electricity demand increases, the Met Office states that average wind energy supply reduces. *"Finally the study highlights the risk of concurrent wide-scale high electricity demand and low wind power supply over many parts of Europe. Neighbouring countries may therefore struggle to provide additional capacity to the UK, when the UK's own demand is high and wind power low."*[1080] What the UK Met Office study is saying is that wind and solar are totally useless most of the time.

Do those in the UK know that 40% of their solar panels in the UK were built by slave labour camps in Xinjiang Province (China)?[1081]

A large proportion of solar panels in Western countries are made in China. Solar means slavery sourced, green endorsed and tax subsidised. Greenpeace says[1082] *"Climate change is a human rights issue"*. Does Greenpeace mean that human rights are important for an ideology but not for the million slaves in China producing solar panels to support this ideology?

Does this mean that green activists support slavery? I am not aware of any green activist demonstration against slave-built solar panels. Why not?

Solar is an economic issue resulting in more money for greens and their friends and has nothing to do with the environment, climate or human rights.

Markets drive electricity generation, distribution and pricing and power engineers build systems to deliver electricity when and where it is needed. Once a system is propped up with subsidies based on an ideology and driven by bureaucrats, it is doomed to failure.[1083]

It is only by mandating and subsidising increasing proportions of electricity consumption from the highest cost "renewable" sources, thereby driving up electricity prices, that governments have made cheap coal-fired energy more expensive and comparable in price to wind energy.

According to the International Renewable Energy Agency, the world doles out billions each year in direct energy subsidies. This amount would assist Africa to becoming wealthy and solving their problems of pollution, low clean energy availability and of poverty. The actual figure is probably twice or three times as much if price gouging from the poor consumer is considered.

Since 1979, more than $103 billion *per annum*[1084] has been spent on subsidies for wind and solar power generation in the US. No worries, it's someone else's money. The combined 2020 budgets of the US Department of Homeland Security, Department of Energy, Department of the Interior and the Environment Protection Agency did not exceed $100 billion.

Costs are passed on to the consumer and distort the grid. I wonder if the poor in the US might feel that this money could have been better spent. Fiscal discipline, whether for a household or for a nation, should be a culture based on *"look after the pennies and the pounds will look after themselves"*.

Wind and solar in the US provided less than 3% of the total electricity and yet received 50 times more subsidy than coal and gas combined.[1085] Not only are wind and solar costly, they are useless. These additional costs are borne by the taxpayer, thanks to renewable energy mandates in 29 states and the District of Columbia that are designed to guarantee a market share no matter what the production costs for wind and solar might be.

In an attempt to display his green credentials, President Biden wants to increase offshore wind electricity generation and squander untold billions to put Atlantic fishermen out of work, despoil the environment and create excessively expensive and unreliable expensive electricity.[1086]

A 2009 testimony about Spanish renewable energy to the US House Select Committee on Energy Independence and Global Warming[1087], showed that for every green energy job financed by Spanish taxpayers, 2.2 jobs were lost. Only one out of 10 green jobs were in essential maintenance and operation of already installed "alternative" energy plants and the rest of the jobs were only possible because of high subsidies.

Each green job in Spain cost the taxpayer $750,000 and green programs led to the destruction of 110,500 jobs. Each green energy mega-watt installed destroyed 5.4 jobs elsewhere in the economy. I'm sure those pushed into unemployment by green activism feel that they have made a sacrifice for a higher cause.

On 16th January 2009, during a visit to an Ohio wind turbine component manufacturing business President Obama stated *"And think of what's happening in countries like Spain, Germany and Japan, they're making real investments in renewable energy. They're surging ahead of us, poised to take the lead in these new industries"*.

President Obama did a pretty good job of increasing unemployment. Spain has extraordinarily high cost electricity and subsidies. Thanks to the green activists.

The US Senate Finance Committee[1088] had a $23 per MWh production tax credit for generating electricity from the wind. It is now $18 per MWh. This is the eighth time since 1992 that a subsidy has been granted to help *"the industry competes in the marketplace"* and there have been other "temporary" federal subsidies since 1978. In 2012 the wind capacity additions were 13.3 GW and in 2019 12.7 GW.

One would have thought that after more than 40 years, wind power would have been able to compete and would not need to be propped up by subsidies which are paid by taxpayers and consumers. Unless, of course the industry is so hopeless that it cannot survive without massive subsidies.

Wind turbine manufacturers are facing a quadrupling of transport costs and increases in steel, copper, aluminium and carbon fibre prices making wind turbines more expensive[1089] with some wind turbine makers struggling to make a profit.[1090] This is in sharp contrast to the wind lobby who claim that costs are falling.

Why should the US provide a subsidy to uneconomic wind and solar electricity generation that increases costs to US manufacturers and households, especially as the US has abundant oil, uranium coal and gas? China, a competitor for so many markets, does not subsidise wind and solar power and has captured markets from the US because of cheap reliable coal-, nuclear- and gas-generated electricity.

In the UK, thanks to smart meters, the government is planning to cut off power without warning when wind and solar output collapses and without compensation.[1091] The hidden victims of green renewable energy virtue-signalling are not climate activists but the working poor who are becoming increasingly crushed by costs. We can all become North Koreans by being green.

UN-promoted stupidity means forests are logged in the US and shipped across the Atlantic in fossil fuel-burning ships for burning in a British power station. This is green energy at its best. Native forests are cleared in SE Asia, Indonesia and Brazil to grow palm oil for bio-diesel and food grains are distilled to make ethanol fuel for motor vehicles. Nothing green about this.

Now the green activists want to electrocute nine tonnes of water to make one tonne of hydrogen that needs huge amounts of energy for compressors and heat exchangers to store at 700 times atmospheric pressure and a temperature of below minus 252.87°C.[1092]

Prime Ministers who go green and woke end up with a shortened political career.

Just a flick of the switch

It's not too complicated. All we need is what we had before renewables destroyed a system that worked well for decades. Not too long ago, we were able to flick a switch at any time of day and in any weather and have cheap reliable electricity.

The power system was designed by practical people who knew what they were doing and was not based on ideology. The model was to have large, centralised and reliable economic power stations as close to the load centres as possible. These load centres are cities, smelters and heavy industry.

Electricity was reticulated in a one-way system using various voltages which were progressively stepped down for ultimate consumption. Our houses use 240 volt systems. Transmission lines run at 132,000, 256,000 or 512,000 volts to reduce power losses due to line resistance.[1093]

When electricity gets to our homes we want 240 volts. The local system draws from a high voltage feeder, usually 11,000 volts, through a nearby transformer to get 240 volts. Power poles carry three wires at the top (usually 11,000 volts) and four underneath comprising the 415 volt three-phase distribution system. The 240 volt supplies are taken off the lower wires and every now and then there is a transformer between the two.

Each transformer serves a local area containing a number of homes. Domestic roof top solar generators can share their excess electricity with their neighbours but nowhere else because the transformer will not permit a reverse flow when there is a higher voltage on the supply side. A transformer is like a one-way valve between a large diameter high pressure pipe and a small diameter lower pressure pipe.

Coal-fired generators were built in the coal fields to avoid costly transport of coal and many major energy-intensive industries such as aluminium smelters were built close to generators in case high voltage lines were damaged or shut down further along the line. In Australia, some 15% of electricity is used for the production of aluminium which results in Australia having a high *per capita* energy use.

We have prospered with 24/7 power use for computers, charging smart phones, illumination, food preservation, heating and cooling depending upon need and for keeping businesses running day and night. Previously we stocked our houses with candles, kept food cool in iceboxes, used wood for heating and cooking and opened windows for air conditioning. I remember those times.

The building of wind and solar electricity generating plants out of sight in remote parts of the country introduces a few problems. The one-way grid was not designed to take power from the end-of-the-line low-capacity generators and send it the opposite direction to consumers. The wind and solar generated electricity is fed into low-capacity lines before it can reach high-capacity lines. This is the limit of growth for renewable energy unless a whole new grid is built.

By far the most efficient method of generating renewable energy is to generate it as close as possible to the consumer.

A collection of 200 metre-high wind turbines in every city park and public space, playgrounds, schools and beachfront would solve the one-way problem of the grid. Unless green activists are hypocrites, they should be loudly demonstrating to have wind turbines erected as close as possible to end users of electricity.

Wind energy is very erratic. Too much wind one minute, not enough the next. It may produce 35% of peak capacity, it often produces peak power at night when demand is low and may produce no electricity whatsoever for a few days. A sudden wind burst sends an electricity surge to the grid and then falls to zero as the wind dies. A wind-battery system would need installed wind capacity of triple the expected demand.

There are often times when neither wind nor solar energy can do the job and these are commonly at peak load demand times at 6 to 9 am and 5 to 8 pm. Base load electricity is required in the middle of the night in order to keep society operating with its necessary lights, computers, refrigerators and communications.

Look up the Aneroid Energy site for any day, week, month or year. The combined notional capacity is 7,728 MW. Collapses of over 3,000 MW over a couple of hours are routine. This maniacal reliance on chaotically intermittent wind and solar poses an existential threat to south-eastern Australia's power grid. Such threats are nothing to do with the grid. It is because there are hundreds of occasions each year when the grid's entire wind fleet struggles to deliver more than a tiny fraction of its combined capacity.[1094]

Wind and solar surges of electricity destroy the economic viability of reliable cheap coal-fired power stations that cannot ramp up quickly but yet are expected to keep running for when there is the inevitable green energy drought. Cheap reliable coal-fired electricity has been priced out of the market by green subsidies and mismanagement of green electricity. Why should the consumer subsidise a second-rate product?

The largest and longest outages are when wind turbines are becalmed. There are many more short, sharp and very sudden failures in high wind conditions where wind turbines cut out. The sharpest power cuts are happening in between the high-pressure weather cells. As the wind picks up, production maximises, only to crash as turbines hit their safety cut off points and drop out of production suddenly.

About 50 times a year generation across the entire Australian wind turbine grid falls by 500 MW or more within one hour or less. High pressure weather systems becalm large regions of Australia for about a week. There were 64 periods of sustained power loss of 9 hours or greater of 1,500 MW to 3,700 MW. The longer and large losses are due to large high-pressure cells sitting over the Eastern states and refusing to move. The intermittency problem is getting worse, there are more occurrences of power losses and those losses in power generation are becoming larger.

This is green activist sabotage of an electricity generating and grid system paid for by their forbears.

Solar

In 2016, a 2.3 tonne solar-powered aeroplane with a wingspan wider than a Boeing 747 and covered with 17,000 solar cells circumnavigated the planet in 505 days carrying only a pilot and one passenger.[1095] A fossil-fuelled aeroplane can circumnavigate the planet in under 50 hours with only three fuel stops and a payload of at least 150 tonnes.

Why do we carry passengers and freight to the four corners of the globe using fossil fuel-driven aircraft? Because solar aircraft don't have the payload and depend on the vagaries of the weather.

Australia had cheap reliable electricity at $30 per MWh for decades. Since then, two million solar panels have been added and electricity prices have risen substantially resulting in business closure, unemployment and shipping of jobs to China. If renewables are so cheap, why have electricity costs risen? When the Sun shines, solar energy floods the network and prices collapse.

At best, solar delivers eight hours of energy a day at maximum efficiency if there are no clouds, smoke or dust and the panels are spotlessly clean. A solar facility needs three times nameplate capacity just to cover darkness, sunrise and sunset. There is a good reason not to subsidise solar energy for an industrialised and urban society unless you can make the sun shine 24/7.[1096]

One poor lonely cloud got in the way between the Sun and the Uterne solar complex (Alice Springs, Northern Territory [NT]) solar station. On 13th October 2019 on a 38°C-day, there was a relatively constant 3.3 MW of power and, in the blink of an eye, it went to 0.5 MW. Many residences use rooftop solar and were also affected by this small cloud.

Many businesses and residences were using grid air conditioning and the uptake of diesel and gas power from the Ron Goodin and Owen Springs Power Stations was not possible because they were not functioning at the time.[1097] A solar-charged battery storage system failed because it was on the wrong settings.

Some 12,000 residences and businesses, including a hospital, had no power for 10 hours and yet the NT insists on continuing down the path to 50% renewables after having 100 years of stable reliable electricity. Even remote communities 230 km away had no electricity, telephone systems collapsed, automatic teller machines (ATM) didn't work, air conditioning shut down, refrigerated food was spoiled and not even a cup of tea could be made. It took remote communities two weeks to get back to normal. They could not even pump fuel for a food run into town or collect money from an ATM for shopping.

The 2017 report called *"Roadmap to Renewables"* supported the NT going to 50% renewables by 2030 and also commented that the NT Government should improve its planning and knowledge of the whole power system. This is bureaucratic speak for saying that the NT Government is flying blind and has no idea about electricity systems.

At the time of the October 2019 blackout, the fossil-fuel fired Ron Goodin power station was in the process of decommissioning. The Electrical Trades Union[1098] warned in April 2019 that without that power station, power supply for 29,000 people would be compromised. A cable fault in Easter 2019 meant no solar power could reach Alice Springs, there was a blackout and it was the Ron Gooding power station that saved the day.

The government-owned company Territory Renewables was broke, wanted to push to renewables which would make electricity supply matters only worse and wanted to increase prices in one of the most impoverished parts of Australia.[1099] The NT government had been repeatedly warned from 2014 that there were problems yet blamed the October 2019 blackout on technical and human error. Welcome to the renewables future run by green know-all fools.

Maybe the Australian Broadcasting Corporation (ABC) in South Australia should only be powered by solar power. The taxpayer-funded ABC gushed that for the first *"phenomenal time"* in the world, SA managed to produce enough solar power for the whole state. This was not put in perspective. It was for just an hour on Sunday 11th October 2020 on one of the lowest demand days of the year.[1100]

The ABC did not mention that it took $2 billion of taxpayers money to achieve this incredible feat in an electricity market that only uses 5% of power from the National Electricity Market. An hour of power for $2 billion is a steal for an underpopulated state with little money and little value adding industry. On that wonderful warm fuzzy day, SA gas power plants were exporting electricity to Victoria because SA was very cleverly only using electricity from solar.

Why didn't the ABC ask questions that a real journalist would ask? Just one would be enough. Who can believe a word that the ABC broadcasts with such misleading and deceptive nonsense touted as news? On a quiet dark windless night, you can hear Chinese laughter at our pagan stupidity.

Australian government subsidies for rooftop solar cost taxpayers more than $1.7 billion annually. The real beneficiaries are the Chinese solar panel manufacturers. Those with solar panels on their roof blend avarice with virtue signalling yet don't thank their neighbours for subsidising them. Grid managers have absolutely no way of controlling the surges and collapses of output and new rules are being created whereby households with solar panels can have their output shut off remotely.

Sunshine is free but solar power is unreliable and more than double the cost of fossil fuels.[1101]

Killing the environment to save the planet

Solar power is ethical and moral. Or so we are told. There are numerous Civil War battle sites which are sacred for Americans. The Civil War gave rise to a great nation that emerged from an aggregate of disparate, squabbling, parochial states.

There were 750,000 soldiers who died. Hundreds of Union and Confederate soldiers died in a battle in 1862 at Savage's Station, Virginia. The solar company Northam has covered the unmarked graves at Savage's Station with solar panels.[1102]

In 2016, a study showed that hundreds of solar facilities around the US may kill nearly 140,000 birds annually. As very little solar radiation is converted to electricity and most is converted to heat, solar facilities have a heat island effect and raise local temperature by 3 to 4°C.

The modern irrational green worship of wind turbines and solar panels merrily ignores the wave of environmental destruction because virtuous energy is being occasionally generated. At times the positioning of wind and solar facilities creates environmental problems that are easily solved.

In Portugal, a solar electricity company enlisted the local lads with their high-powered rifles to massacre a few hundred deer that threatened to derail the clean green environmentally friendly project.[1103]

The boom in renewable energy has created unintended consequences with the release of large quantities of the world's most potent greenhouse gas into the atmosphere. It is sulphur hexafluoride which is 25,000 times more potent than carbon dioxide, is very difficult to break down in the atmosphere and lasts in it for 1,000 years.[1104]   In the atmosphere, carbon dioxide lasts for five to seven years.

Sulphur hexafluoride is u sed t o m ake w ind t urbines, s olar p anels and electrical switching gear. Leakage of this gas across Europe is equivalent to having an extra 1.4 million cars on the roads. The UK has measured increasing sulphur hexafluoride in the atmosphere as a consequence of the green energy boom.

Of the 73 bills related to the solar industry that were introduced in the California Legislature during 2007 and 2008, not one of them addressed the end-of-life or manufacturing environmental hazards. Most of the bills focussed on installation targets and tax incentive rebates for solar panel adoption.[1105]

Solar energy peaks around midday and falls to zero from dusk to dawn and is much reduced by clouds, dust and smoke. Over a year it may produce about 16% of name-plate capacity hence a solar-battery would need installed solar capacity six times the demand. However, solar panels degrade twice as fast as had been predicted and financial and operational models are challenged.[1106] Don't worry. There are subsidies that rescue the solar business.

These solar industrial complexes are very land-hungry per unit of usable energy and sterilise large areas of food-producing land. Those not on arable land are so far away that there is a massive voltage drop and hence power loss during the transmission to users.

After a decade or so of use, solar panels are crushed and used as landfill. The toxic chemicals will only take years to get into groundwater and by then it's someone else's problem.

In remote outback locations such as homesteads, shearing sheds, rail signals, public telephone boxes etc, low load solar power does make sense when supported by lead-acid batteries or backed up by diesel generators when required.

At *Fossil Downs* in the Kimberley (WA), an attempt was made to go solar. Millions was spent on solar pumps, solar hot water systems, solar-powered station buildings and electric vehicles charged from a solar hybrid power station with a lithium battery backup pack. Although diesel use was slashed, the solar outback station power generation was uneconomic.[1107] If Australia wants to go solar, there is just not enough money in the kitty to give to farmers to change to subsidised solar.

The Integrated System Plan of the Australian Energy Marketing Organisation plans to sterilise 600,000 acres of farming land with solar panels. When the panels are retired after 2030, there will be 100,000 tonnes of toxic waste for landfill. Similarly, the large-scale batteries required have little salvage value and a life limited to 15 years. This costly and environmentally damaging destructive energy transition attracts $2.4 billion in subsidies and is being implemented solely because of the IPCC claim that human emissions of carbon dioxide are warming the planet.

With hype about solar, it's a case of don't mention the waste. In the US, there are no mandates for recycling wind and solar wastes yet there are for the mining, petrochemical and the nuclear industries. Contrary to green activist hype, sunshine is not free fuel. It's not clean. It's not renewable. It's not sustainable. It's delusional.

To accommodate a 2,000 MW coal, gas or nuclear power generating facility, the area of two 18-hole golf courses is required. The area required for the generation of 2,000 MW of renewable energy is the size of Belgium.[1108] God has stopped creating land and what we have must be used efficiently.

No government mandates that solar companies set aside money to dispose of, store or recycle wastes generated during manufacturing or after massive solar installations have ceased functioning and been torn down. Solar panels have never seemed like a good idea. At any time.

Millions upon millions of solar panels have reached their use-by date. They can't be recycled and they face eternal life in the local dump.[1109] If the local landfill dump will not accept toxic turbine blades, then they are just illegally dumped.[1110] Each solar panel is a carcinogenic chemical cocktail of gallium arsenide, tellurium, silver, crystallised silicon, lead, cadmium and a long list of other toxic heavy metals just waiting to contaminate water and soils and to kill humans.

Because mines and nuclear plants generate waste, funds are required to be set aside for site remediation and waste disposal for decades. Not so for the solar industry. When solar panels are in sunlight, they add 3 to 4°C to local temperature because most solar energy is converted to heat and not electricity. If you want to stop alleged global warming, then solar panels will not do the job.

The dark side to solar is that panels produce 300 times more toxic waste than high-level nuclear waste[1111.] Solar and wind customers are not charged for waste cleanup, disposal or reuse and or for recycling costs. This, plus the massive subsidies distort and hide the true costs of solar power. In the US the cost of recycling exceeds the revenue by an order of magnitude and, because of leaching of toxins, landfill is not an option.

Solar panels, their installation and the transmission of power, require aggregate and crushed stone (for concrete), albite (solar cells), arsenic (gallium arsenide semiconductor chips), bauxite (aluminium), boron (glass, cadmium (thin film solar cells, clay and shale (for cement, coal (steel, cement and component manufacture), copper (wiring), gallium (solar cells), gypsum (cement), indium (solar cells), iron ore (steel), lead (batteries), limestone (concrete), lithium (batteries), manganese (steel), molybdenum (steel), oil (plastics, paints, transport, maintenance), phosphate rock (phosphorus), selenium (solar cells), silica (solar cells), silver (solar cells), soda (solar cells), tellurium (solar cells) and zinc (galvanising).

In the EU, the Waste Electrical and Electronic Equipment Directive does not include PV solar waste. A simple solution is employed. Shift solar panel waste to developing nations where environmental protections are weak and panels can be reused or dumped as toxic landfill. According to the International Renewable Energy Agency, in 2016 there was 250,000 tonnes of solar panel waste worldwide and in 2050 it will be 78 million tonnes.[1112]

Mad dogs and Englishmen
Even the midday Sun cannot make solar energy viable for an industrialised society.

A 1,000 MW electricity demand requires 3,000 MW of installed capacity at a very generous capacity factor of 60% plus another 600 MW to make up for losses. Solar panels need to be replaced every decade or so leaving hundreds of thousands of hectares of glass for disposal.

The same scammers who gave us wind and solar are now queuing up to give themselves another massive source of income: batteries. With wind and solar, five days of battery storage are necessary at a cost of $200 per kWh. There are often 20 to 25% losses associated with battery storage as they have a self-discharge rate of 10% per month.[1113]

France hands out €2 billion to solar investors who produce less than 1% of France's electricity but consumes a third of public spending on renewables. The French government plans retroactive cuts to generous solar subsidies it granted between 2006 and 2010. More subsidy cuts are expected. This is the price the green lobby is paying for claiming that renewable energy is now dirt cheap. Not surprisingly, the green lobby is up in arms as the bone has been taken from the dog.[1114,1115]

How well is money spent on renewables in the US? For $1 million worth of utility-scale solar panels, 40 million kWh will be generated over a very optimistic operating period of 30 years but more like 10-15 years. For $1 million, a shale gas well will produce 320 million kWh over a 30-year operating period.[1116]

Germany is held up as the world's wind and solar green exemplar. In the winter of 2020-2021, millions of Germans had no electricity because there was no wind and solar panels were blanketed by thick snow.[1117] Wind and solar are fashion accessories while the weather is sunny with a mild zephyr.

At all other times, society relies on coal, nuclear and gas. Germany is now demolishing wind turbines faster than it can build them. During 2020, it added a paltry 200 turbines to its fleet of 30,000.[1118] Subsidies for 5,700 turbines ran out in 2020 and Germany is looking to hide its turbine blade waste disposal in deepest darkest Africa.[1119]

Its renewable energy transition is in free fall as are its ersatz virtue signalling clean green credentials. The renewables transition has now given Germany the joy of being in the top two of the world's highest power price countries coupled with power rationing, grid instability and gas supply risks.[1120] German politicians with green tinges must live in a parallel universe.

Germany attempted to have a renewable energy transition and forced the shutdown of 11 coal-fired power stations. A change in the weather meant that eight days later the coal-fired power stations had to re-open to prevent country-wide blackouts.[1121] Consumers, of course, would bear the cost.

It is the ever-reliable coal that keeps the lights on. The German experience gives us a glimpse into a non-coal Net Zero future.

In the first half of 2021, the Statistische Bundesamt reported that coal was the No 1 energy supplier, wind power dropped by 21% because of a lack of wind and natural gas consumption increased by 14% despite record high gas future contracts at the Dutch TTF hub, a rise of 250% in that six months. Benchmark power contracts in Germany doubled in 2021. Germany's attempt to finally close coal-fired power is a bit like St Augustine's prayer asking the Lord to make him chaste, but not yet.

It happened again. In mid-September 2021 when gas and electricity markets were surging, energy in Europe hit record prices because the wind stopped blowing.[1122,1123] Was this unprecedented?

The UK power prices tripled and their power prices are now the highest in Europe which is because of low wind, dependency on gas when prices are rising and a reduction in nuclear- and coal-fired electricity.[1124] Yet again, the consumer pays and the poor get poorer and colder.

The summer 2021 energy crisis in the UK is a warning that companies could collapse[1125], taxpayers will be paying billions to bail out energy companies at the same time electricity bills go through the roof and supermarket shelves empty[1126] resulting in a rise in inflation.[1127] The UK is now exposed to gas price and supply blackmail by Russia's Gazprom[1128] and is exposed to security and financial risks.[1129]

None of this would have happened if the UK had embraced fracking for gas.[1130] Some UK factories are looking at taking control of their destiny and using their own diesel generators.[1131]

When will it be realised that a modern industrial economy cannot possibly survive on sunshine and sea breezes?

Solar panels have never seemed like a good idea. At any time. Millions upon millions of solar panels have reached their use-by date.

When will green activists take responsibility for the mess they created?

The answer's not blowing in the wind

It's an ill wind

Global Wind Day is 15th June. This is a feel-good orchestrated media event celebrated by rent seekers with their snouts in the trough, getting subsidies and raising electricity prices in a guaranteed market. Landowners receive compensation indirectly from the consumers. Green left environmental advocates promoting various UN agendas at the expense of sovereignty, freedom, the environment and financial common sense.

Global Wind Day is not celebrated by birds and bats; nor by nearby residents who suffer decreased property values, humans proximal to wind turbines suffering health problems from thumping blades, nor by consumers and industry hurting by paying higher electricity prices.

Although aesthetics is subjective, most folk see no scenic enhancement in hilltops covered by wind turbines. If giant wind turbines were attractive, why don't green activists lobby for them in their own back yard, in public parks, along beaches and headlands rather than supporting their building in areas well away from their residences?

On Global Wind Day in the UK in 2021, 2% of the electricity was generated from wind turbines.

Wind history

Why don't shipping companies carry freight and passengers in sailing ships? Wind is free. What a beautiful sight to behold. An environmentally friendly wind freighter carrying 10,000 40ft containers or 250,000 tonnes of iron ore. The average speed for the fastest clipper was 8 knots. Why don't we continue to use sailing ships for international travel and trade?

Is it something to do with the unreliability of the wind, the doldrums, speed of transport and costs? Imagine if you had purchased 250,000 tonnes of Australian iron ore and when the boat berthed in Asia 150 days later the price had halved.

A billion tonnes of freight is now transported around the world each year in containers and this is not a job for wind, especially if the container carries perishable food and needs to be kept refrigerated using electricity generated from fossil fuels. This freight is transported 24/7 at an average speed of 21 knots by 200,000 tonne ships burning fossil fuels.[1132]

Wind power reached its zenith 400 years ago. Centuries ago, wind was used for pumping water and grinding grain and these processes did not have to be undertaken all the time. Since then, the increasing energy requirements, and wind's poor energy density, inefficiency and unreliability have made that power source more and more expensive.

Wind power was replaced by the steam engine in the industrial revolution. The steam engine is now old technology. It will only be replaced when new technology is cheaper and more efficient and, at this stage, there is no new technology on the horizon. Today we use steam power to make electricity in nuclear and coal-fired power stations.

Wind, solar and hydrogen are often touted as new technology and greens advocate that we should abandon old technology. The electric motor predates the internal combustion engine by five decades.

The wheel, a technology invented at least 5,500 years ago, is the oldest technology we use. It was not discovered at all by some cultures. Should the wheel and electric motor also be abandoned because they are old technology? Greens use glib throwaway lines which, upon investigation, are shown to be just that.

When the wind stops, the disasters start.

When the wind dies

In the UK on Wednesday 13th January 2021, the wind died. Most of Britain's fleet of thousands of wind turbines stopped turning. Temperatures and wind speeds had been low all week, were forecast to fall further and so the lack of wind was no surprise. A 'margin notice', a warning to generators that extra capacity was going to be needed, was issued. Prices rocketed from their normal £40 per MWh to over £100 per MWh. Some generators held off taking this remarkable price, and when the grid still couldn't balance supply and demand in the evening peak, they were rewarded with even higher prices.[1133]

On Thursday 14th January 2021 when wind speeds were even lower, prices for balancing during the evening peak rose again reaching £1,400 per MWh. By peak time on Friday 15th January, that figure had risen to a record-breaking £4,000 per MWh. Such price spikes were never seen before 2016 when Britain's electricity system was not dependent upon wind. Many have warned that decarbonising the economy destabilises the electricity grid.

For 20 years, governments have given renewables generous subsidies and preferential access to the grid. Renewables companies neither constructed nor paid for the grid. The economics of coal- and gas-fired power stations have been wrecked. It is hardly surprising then that these generators, the backbone of the UK's grid, have simply left the market. Big power users are financially persuaded to switch off during such times, the result is a surge in electricity prices and those on 'time of use' contracts are hurting.

The middle classes will probably be happy enough to pay extra for power in the winter. They can cope with the hit to their cash flow for a few weeks, or even fork out £6,000 for a battery pack allowing them to take advantage of lower prices in times of surplus power. As always, the poor are left high and dry and they must tighten their belts even more.

That's the good news. In January 2021, the UK announced it would close a coal-fired power station and the Hunterston and Hinkley B nuclear power stations. EVs charging at night will drain even more power from the grid.

The reality of the windless summer weather, gas shortages and an electricity price hike has now caused the UK to reopen coal-fired power stations.

In the 2020-2021 European winter, polar cold waves, low wind speeds and snow covering of solar panels coincident with the concurrent shut down of 13 French nuclear power stations strained electricity systems towards a near catastrophic blackout. Germany was on the brink of disaster and 30,000 wind turbines took a holiday. Solar panels were covered with snow and just refused to operate as clean green electricity providers.[1134,1135,1136]

Fossil fuels and hydroelectricity came to the rescue. It is the same fossil fuel and hydroelectric plants that are massively fought against by green activists. Power rationing was the only thing that prevented the total collapse of the grid on 8th January 2021. If a hitch in the power system in any EU country causes a frequency drop, machines across Europe switch themselves off to protect themselves. Green energy-obsessed despots, rather than power engineers, now control electricity in Europe's greatest industrial economy. Power problems arose and persisted in Europe when cheap reliable electricity was given the boot a few decades ago.

Bureaucrat lies, bungles, policy failures

The home of green activist bureaucrats, the ACT, prides itself on its green credentials and claims that it is powered 100% by renewables. Data shows a different story. In the 2020 winter, renewables generated just 31% of the utilised power for June 2020, renewables failed to cover 59 of 60 peak consumption periods for the month and could not maintain a consistent stable minimum base load of 250 MW for it.

The ACT sourced more than 90% of its power from the NSW grid comprising seven days equivalent of electricity.[1137] And how was most of the back-up power generated? Coal.

The ACT generates no electricity. It games the system by buying renewable energy, mainly from Victoria and South Australia, sells it back to those states, and then claims it uses 100% renewables. People, government and businesses in the ACT rely on coal-fired power stations in NSW. The Emperor has no clothes.

Governments are now dictating household power use because they cannot solve self-inflicted problems. Western Australia is doubling power prices at the peak evening times and reducing them during the day when there is far less power consumption.[1138,1139] This is to incentivise consumers to move to use the power they need to the middle of the day.

From now on, the WA government wants you to cook the evening meal and watch evening TV in the middle of the day. This is an attempt to contain the damage that government-subsidised rooftop solar energy has done to the grid. Rooftop solar is expensive, this can only be afforded by the wealthy and is a transfer of wealth from the poor to the rich.

Ten million jobs are threatened by the UK government's net zero commitment. The UK has staked its household and industrial future on wind energy and will expend £160 billion a year over the next three decades to achieve net zero.[1140] And who will suffer? The average taxpayer will and not the elite bureaucrats who cannot be sued and who are first in line for taxpayers' money. If you think government's policies over COVID-19 were confused, catastrophic and contradictory, try energy. If you really want to lose sleep at night, imagine how these same folk would handle a global financial crisis or war.

Green activists have always touted that green energy creates employment which sounds too good to be true. If it's too good to be true then it probably isn't true.

Winter blackouts are now common in Europe. This is because of the weather. This did not happen before wind and solar partially replaced coal, gas and nuclear.[1141] European winters are becoming more savage recently, perhaps due to the lack of sunspots and the Grand Solar Minimum.

If the predicted 2020-2053 modern Grand Solar Minimum is anything like the Maunder Minimum 350 years ago[1142], then the energy-starved Europe and/or a depopulated world will be longing for global warming.

Care for the environment

Green activists use words like sustainable for their environmental policies. The only sustainability about wind and solar power is the massive subsidies. Use of the word sustainability is code for pillaging the taxpayer's pocket and has nothing to do with the environment.

Murderous green activists are easy to spot. They love industrial scale wind power and couldn't care a less about the environmental and human destruction it causes. They totally ignore the killing of birds and bats (many of which are protected), and of insects (food for birds), the horrendous human health costs, the toxic waste generated during construction and the poisonous landfill legacy that releases bisphenol A from plastics into the soils and waterways.

The renewables industry has special exemptions that allows it to kill and pollute but other industries do not have them. Green bureaucrats and governments negotiated these sweetheart deals with no care for the environment or the electors they claim to represent. One wonders whether they could be that stupid or whether palms were greased?

In a blow to Germany's ambitious wind power expansion plans, the European Court of Justice has strengthened protection aspects in turbine construction.[1143]

Rural law-abiding Germans are now in revolt against wind power roll out and farmers sabotage wind projects. Farmers blockaded turbine construction sites with their tractors and destroyed MET masts (used to measure wind speed). Never has such a simple issue created such division in communities.

The July 24th 2016 print edition of the national flagship *Die Welt* wrote a feature story on how German citizens are becoming fed up with the widespread crony capitalism of the wind energy business and are now mobilising a fierce rebellion. Governments are now no longer serving the interests of their people and are *"rolling out the red carpet for wind power companies"* and are *"no longer listening to the people and about the concerns of their everyday lives"*. The health issues for people living in the vicinity of a wind farm are horrendous.

Die Welt describes *Energiewende* (transition to renewable energies) that is *"dividing the people"* where those that live in big cities are not affected by the blight and are open to wind industrial complexes whereas those living in the countryside are fed up and are fiercely resisting them. People *"no longer view the Energiewende as a necessary national project but as a destructive force"*.

The living area preserved by generations is being ruined and destroyed for generations. Logic, respect for the individual and bulldozing of opinion by bureaucrats in league with corporations is no different in Germany than anywhere else.

Even in countries considered green such as Norway, there are growing protests about onshore and offshore wind turbines.[1144] Environmental protestors accuse some developers of building bigger structures than approved, despoiling landscapes and endangering birds.

It is not logically possible to be an environmentalist and concurrently support wind turbines which have been termed the modern-day green killing machine. Offshore wind turbines should have green activists out on the street wanting to save the whales. Where are the demonstrations? Or are green activists hypocrites? Offshore wind turbine noise and vibration have killed many whales in the UK because the whale's sonar guidance and communications systems are messed up.[1145] This is just another example of the wind industry doing its bit for the environment.

Rural Americans are sick and tired of wind development lies, trickery and deceit. They are fighting back.[1146] The wind industry has never had it so tough with litigation, evaporating subsidies and rural folk turning hostile.

Tenacious lobster fishermen in Maine did not believe the hype about a floating wind turbine platform and mounted an offshore protest against yet another environmentally unfriendly and community destroying project that would destroy their livelihood.[1147]

Locals erupted in anger with the proposal to disturb the quiet countryside of the Powys beauty spot, Wales. The proposal was to build wind turbines in national parks.[1148] The proposed turbines would have stopped right of way access, access along bridle ways, spoiled scenic views, hampered low-level RAF training flights and interfered with special protection areas. Wind turbine companies in India wanted to destroy forests and fertile farmlands, thousands of locals fought back in a battle for survival.[1149]

Developers of wind industrial complexes are a despised bunch whose penchant for bullying, lying and deceiving is commonly exposed at hearings on application to build a massive complex. If ever a rural community needs a trigger for unity, propose building a wind factory in their backyards.[1150]

Some 420,000 men comprising 39% of the male population between 18 and 44 from an infant nation of four million people enlisted in the Australian Infantry Forces (AIF) and 295,000 were sent to France and Belgium to Hell-holes like the Somme, Messines, Passchendale and Villiers-Bretonneux. At least 46,000 lost their lives and 11,000 have no known grave. About 132,000 were wounded.

The renewable company Energie planned to build turbines at the site of the April and May 1917 battles of Bullencourt (France), a significant site on the World War I Western Front. It is also the final resting place for thousands of Australian, British, Canadian and German soldiers killed in the Battle of Cambrai between 20th November and 6th December 1917. Some 2,300 Australians who died at Bullencourt have no known grave.[1151]

Pressure has been applied by the Australian government and the French people are up in arms about this officially sanctioned desecration.[1152]

The battle of Villiers-Bretonneux took place on 25th April 1918. This battle changed the course of World War I. There is a war memorial at Villiers-Bretonneux in memory of the 18,000 AIF soldiers who died in France, the local school was rebuilt by AIF soldiers awaiting transport back home and rebuilding was funded by Australian school children[1153.] The school has an annual memorial service on the day of the battle and a permanent sign "*Do not forget Australia*".

The plan at Villiers-Bretonneux by the French international energy company Energie was that each turbine base on the battlefield needed to be excavated to a depth of 15 metres to house 45 tonnes of steel reinforcing and 500 cubic metres of concrete. To add to the unsightly desecration, trenches carrying cables will criss-cross the battlefield littered with human remains.[1154] Energie, which also operates in Australia, knows no bounds of moral decadence or shame.

On the Energie website, they claim that[1155] "*For us, ethics are an essential part of our strategy, management style, and individual professional practices. Ethics helps drive performance*". Why didn't the ethics of Energie lead them to let the thousands rest where they lay?

Why were green activists not up in arms, demonstrating and raiding the share register in order to attend the May 2021 annual general meeting of Energie and to question directors on their morals?

The silence of the green activists shows that they live in the same moral vacuum.

The quickest way to devalue a property is to have wind turbines nearby. Why should people in rural areas suffer a property devaluation to appease city-based green activists?

Wind turbines produce light shadows and flickers[1156] and infrasound that can have a profound effect on some people. Wind companies suggest planting trees to avoid light shadows and flickers. I'm sure we can all wait around for 20 years until the trees grow big enough to protect us from an adjacent wind turbine.

If you are an epileptic, steer clear of wind turbines. Infrasound, low frequency sound below the human threshold of hearing, is used as a military weapon because the inner ear pressure pulses are disorienting.[1157,1158]

The disastrous effects on human health from wind turbines is well documented.[1159] If this noise came from a factory, farm, mine or a construction site, they would be shut down immediately and fined. Why are those who live near wind turbines in rural areas sacrificial lambs to the god of green renewables?

If such turbines are safe and beautiful to view, why are they not in every city park and along city beaches close to the consumers of electricity in order to be more efficient and to not suffer voltage drop by transporting electricity over large distances?

These medical effects of noise from wind turbines at this stage are difficult to quantify at present, medical research is at an early stage[1160], further work is needed; but reports from residents living near wind turbines suggest serious unresolved problems resulting from low frequency infrasound.[1161]

Why don't green activists invoke their precautionary principle for wind turbines? Green left activists are very vocal about what can't be seen (e.g. radiation, carbon dioxide) but are hypocritically silent about what can't be heard. Inaudible sound maybe can't be heard but it can be felt.

Time and time again in court hearings, the expert useful idiots paid by the wind industry claim that effects of turbines on humans are in the silly discombobulated heads of affected residents.

Wind turbine neighbours have incessant turbine-generated low-frequency noise and infrasound that creates sleep deprivation and serious health effects. Some turbines emit screeching noises, rumbles, thumps and grinding noises from gears and generators. Scientific surveys show that affected people are driven out of their homes and many leave their houses to sleep well away in their cars.[1162] This is the ugly face of green-activist supported renewable electricity.

Meanwhile in France, 400 cows died due to low frequency turbine noise and vibration. In Poland, geese living near wind turbines had increased cortisol levels indicative of stress from exposure to incessant turbine-generated low frequency noise and infrasound. In Kansas, the rare Greater Prairie Chicken had abandoned their long-established nesting sites within eight km of wind turbines. In the UK, badgers living less than a kilometre from wind turbines had a 264% higher cortisol level than badgers that lived more than 10 km from wind turbines.[1163]

The Locke Foundation cites University of Kansas studies confirming that wind turbines create unsafe flying conditions. At high latitudes in winter, ice builds up on turbine blades which is later thrown as projectiles. This danger to people and property has been completely ignored by green activists.

While green activists are making noise about potential fatalities from nuclear power generators, they are silent about actual fatalities in the wind industry. In the UK, the wind industry, which is barely two decades old, had some 220 fatalities. Wind industry and direct support workers account for 125 fatalities and 93 were public fatalities.[1164] Are greens really concerned about the environment, safety and their fellow humans?

The land area used by wind industrial complexes is orders of magnitude greater than the land area used for gas generators, coal-fired power stations and nuclear power stations. A single unsubsidised shale gas pad of two hectares would produce as much energy in 25 years as 87 giant wind turbines covering 15 square kilometres over the same time period (with turbines visible from 30 km away).[1165]

Furthermore, shale gas used to drive turbines to make electricity is neither intermittent nor unreliable. To replace the one hectare of a gas-fired power station, we would need 73 hectares of solar panels, 239 hectares of onshore wind turbines or 6,000 hectares of biomass farming.

The nuclear power plant at Borsella, The Netherlands produces 3.5 billion kWh of electricity 24/7 in a year. It occupies about 16 square kilometres whereas the offshore Germini wind factory occupies 68 square kilometres to produce an intermittent 2.6 billion kWh of electricity a year. The Borsella nuclear plant produces 570 times more electricity per unit area than the Gemini wind factory and 370 times more electricity per unit area than the Sunport Delfzijl solar park.[1166]

Contrary to the mantra that wind turbines are super safe, clean and green, these giant industrial wind turbines are the perfect incendiary device. What do wind turbines do best? It's not generating electricity. It's spectacularly throwing their blades to the four winds, spontaneously combusting, causing pollution from leaking and burning fossil fuels, causing bushfires and collapsing in a catastrophic fashion. Wind turbines are a huge incendiary device just waiting to burn out grasslands, crops and forests.[1167]

Around the world, hundreds of wind turbines have exploded into palls of toxic smoke and balls of flame. In the process, molten metal, 1,000 litres of burning gear and hydraulic oils and burning plastic falls to the ground.[1168] For example, the Vestas V112 3MW turbine contains anti-freeze, over 1.2 tonnes of gearbox oil, hydraulic oils, grease and chemical cleaning agents. When a turbine spontaneously bursts into flames, everybody within cooee knows about it.

A typical wind turbine has a tower, a nacelle and three blades. The foundation is made of concrete, the tower of steel or concrete, the nacelle from steel and copper and the blades from composite materials.[1169,1170] Spectacular fires in nacelles fill the air with black toxic smoke from burning plastics, oils, resins and metals. Wind power companies state that such polluting fires are rare. The data shows the exact opposite.

The nacelle fires are let burn because fire-fighting equipment cannot reach such heights. We can only wait and see the long-term environmental and human health effects of these burning toxins.

The wind turbine company Vestas claim that these are extremely rare events. This was also said for Fenner (New York, USA) and Leystad (The Netherlands) in 2009; Whitelee (Scotland) in 2010; Donegal (Ireland) and Ocotillo (California, USA) in 2013; Kansas (USA), Germany, Starfish Hill (South Australia), Devon (UK), Mill Run (Pennsylvania, USA), Nebraska (USA) and Brazil in 2014; Tyrone (Ireland), Otsee (Germany), Kerry (Ireland), Hamburg (Germany), Menil-la-Horgne (France) and Sweden in 2015; Sigel (Michigan, USA), Fenner (New York, USA) and Pontesco (Spain) in 2016; Leisnig (Germany), Cape Breton (Nova Scotia, Canada) and Kilgallioch (Scotland) in 2017; Chatham-Kent (Ontario, Canada) and Grant County (Oklahoma, USA) in 2018; and Pernambuco (Brazil) in 2019.[1171] Fires do not occur so frequently in coal-fired power stations.

If a car company produced a vehicle with such frequent catastrophic failures, the vehicle would have been recalled and the company would be sued until bankrupt. Why is there one rule for one industry and not the same for the wind industry? Green activists are burning the planet to save it.

In Australia the wind industry has been forced to concede that at least four bushfires were started by turbine fires. Fires were at Ten Mile Lagoon (WA), Millicent (SA), Port Lincoln (SA) and Starfish Hill (SA). Every time such regular fires occur, the company responsible claims in the media that the event is unprecedented or rare. There is no transparent public investigation of such fires and pollution.

Some of us have an eye for aesthetics. We are the true environmentalists. Wind turbines do not get built in cities where the wind power is needed. They are built in rural areas where there is not a large population that can flood the regulatory system with objections. They are built all along ridges and on hills and can be seen for miles. Some of us who have an eye for Nature and who have spent a lot of time in the bush enjoy areas where the human footprint is small.

Turbines, roads, cables, power lines and other infrastructure destroy what we consider is beautiful. Large areas of land are cleared yet this would not be allowed in a mining or agricultural enterprise. If a mine was to make such a visual impact in a rural area, it would be closed down by green activists. Not so for wind turbines because, we are told, they are sacred and help save the planet.

Wind turbines will remain as a memorial to our stupidity well after their useful life and when the wind power scam has been throttled.

A faux environmentalist is easy to spot. They love industrial scale wind power and couldn't care a less about the resultant environmental damage.

The level of cynicism and hypocrisy among wind power advocates is breathtaking. The wind industry is slaughtering millions of birds and bats every year and yet there's not a squeak from those who moralise about saving the planet[1172]. Clear-felling mile after mile of forests and destroying natural habitats is par for the course for the industry.

Clean green killing machines with turbine blade tips racing at 350 km/hr kill apex predators like eagles, kites, hawks and falcons. Millions of tonnes of insects which are bird and bat food gets splattered each year, as do millions of birds and bats. Some birds are now endangered, thanks to clean green environmentally-friendly wind turbine blades.

The Dutch breeding population of sea eagles is small. With an increasing number of offshore wind turbines in the Netherlands and Germany a disturbing number of sea eagles are being sliced and diced by turbine blades.[1173] Greens are killing the planet in order to save it.

Endless subsidies for wind turbines are said to be justified by the carbon dioxide emissions they save. One fifth of global soil carbon is stored in peat bogs. Europe's peat bogs are natural carbon dioxide sinks.

They are ripped up, drained and destroyed to make way for wind turbines sitting on 30 metre-deep pits filled with over 500 tonnes of concrete. This only adds to the atmospheric carbon dioxide emissions which are already in negative territory with the construction and maintenance of wind turbines. The cement used in making concrete is made by heating limestone which contains 44% by weight of carbon dioxide and which is released to the air during heating.

Regulations can be applied in common sense even-handed ways or can be used against politically and environmentally disfavoured industries and activities. In the Obama era in the USA, the Department of Interior re-interpreted the Migratory Bird Treaty Act. This Act was signed by President Woodrow Wilson in 1918.

Obama bureaucrats tried to apply migratory bird regulations against oil and gas companies whose activities occasionally inadvertently killed birds yet countless more birds die from wind turbine blades. The Obama administration Interior Department Solicitor General wrote Opinion M-37041 which radically changed the way Federal officials were to enforce the migratory bird law.

It was issued 10 days before the Obama administration ended and was suspended less than a month later pending review by the Trump administration. The normal suspects such as the *Washington Post* went ballistic.

A 2011 helicopter survey over 45 days found 28 dead birds in the North Dakota oilfields. There was public uproar. Migratory birds were used as a weapon against coal and metals mining, oil and gas exploration and production, fossil fuel electricity generation, farming and other industries despised by green activists. Meanwhile, officials could grant waivers to their friends in the wind and solar industries.

Wind turbines are now the leading cause of multiple mortality events in bats with three to five million killed every year. Bats are our primary defence in keeping mosquito and crop-damaging insect populations in check. One bat can eat between 500 and 1,000 mosquitos and other insects in just one hour.

If farmers shot that many bats destroying crops there would be an outcry by environmentalists, political action, fines and/or incarceration and yet more regulations. If just one bat is killed at a mine site, there is an uproar from environmentalists. Are bats part of a necessary sacred sacrifice to the god of environmentalism?

In a study of migratory bats in North America, there are some sobering statistics.[1174] *"Large numbers of migratory bats are killed every year at wind energy facilities... Using expert elicitation and population projection* *models, we show that mortality from wind turbines may drastically reduce population size and increase the risk of extinction. For example, the hoary bat population could decline by as much as 90% in the next 50 years if the initial population is near 2.5 million bats and annual population growth rate is similar to rates estimated for other bats species ($\lambda = 1.01$). Our results suggest that wind energy development may pose a substantial risk to migratory bats in North America".*

This slaughter of birds, bats and insects leads to a loss of the natural ecological balance. The Audubon Society states that *"Wind turbines kill and average of 140,000 to 328,000 birds each year in North America, making it the most threatening form of green energy".*

In the US, wind turbines are perfectly designed to slaughter airborne endangered predators like eagles, hawks and kites with the outer tip of a 60 metre-long blade travelling at 350 km/hr.[1175] Bird and bat carnage ends up unseen in landfill together with the turbine blades that spent a lifetime killing them.

Paul Cryan (research biologist, US Geological Survey) wrote[1176] *"Unprecedented numbers of migratory bats are found dead beneath industrial-scale wind turbines in late summer and autumn in both North America and Europe"* and *"There are no other well-documented threats to populations of migratory tree bats that cause mortality of similar magnitude to that observed at wind turbines".*

Why is there no outcry from green activists? Why do green bureaucrats accept such environmental damage? Why is there no Extinction Rebellion demonstration against the wholesale slaughter of birds and bats by wind turbines?

Why do Extinction Rebellion salivate about their projected death of a billion people from warming yet care nothing about humans killing wildlife?

The kittiwake joins the red list of UK birds facing extinction because of offshore wind facilities such as Hornsea 3.[1177] W hy d oesn't D avid Attenborough, the British nature narrator and journalist, use his considerable public influence to stop extinction by human actions?

In Scotland, some 17,283 acres comprising 14 million trees have been clear felled to make way for wind turbines that will save the planet.[1178] There is now no wildlife left for the turbines to kill.

In India, the pressure for generating electricity for 1.3 billion people has resulted in an increase in the number of wind turbines. In 2013, the Bombay Natural History Society expressed concern about bird mortality and a wind energy company was fined $1 million in a criminal case for the deaths of protected birds such as raptors and critically endangered birds such as the great Indian bustard.[1179]

Nothing is farmed at a wind or solar "farm". These are wind and solar industrial complexes. Land is sterilised, destroyed and contaminated with pollutants and food growing cannot take place.

Wind turbine blades last up to 20 years and then are tossed into land fill. This is, of course, if they have not catastrophically separated from the turbine, broken up or burned as so many do. Blades snap, crackle and drop with alarming frequency. Blades are not recyclable and 43 million tonnes of blade waste will be added to landfill sites over the next few years.[1180]

China will have 40% of the waste, Europe 25%, USA 16% and the rest of the world 19%.[1181] There are no established industrial recycling routes for end-of-life concrete and composites[1182] and there is an unresolved nightmare of the health risks of blade disposal.[1183]

The wind turbine graveyard stretches hundreds of metres at the North Platte River at Casper (Wyoming) in the remote Rocky Mountains. Between September 2019 and March 2020, over 1,100 turbine blades were dumped after blades had exceeded their working life at three wind facilities. It cost the taxpayer $200,000 per blade to decommission, transport, cut into three pieces, stack and dump them in a remote part of the USA.[1184]

Why should the taxpayer fork out $200 million after paying extra for subsidised unreliable electricity in what has been called the great wind power fraud? This is green environmentalism at its best.

These were blades from the 1990s. There are far more blades now in service. Just wait until we have to dispose of the millions of blades and millions of toxic explosive batteries from future electric cars.

In many jurisdictions, there is no requirement to dismantle and clean up the mess. These green fashion accessories soon turn into an ugly nightmare. There are now 25,000 derelict wind turbines in the USA.[1185]

Thousands of Germany's wind turbines have reached the end of their economic lives. Replacing them is not simple as the costs are huge and more modern turbines are so large that they exceed the height limits imposed by planning rules.[1186]

The rush into a wind- and solar-powered future was undertaken in such haste that the consequences were not considered. Germany only produces 2.5% of annual global carbon dioxide emissions. Some 30,000 gigantic wind turbines have spread across Germany over the last 20 years, many are now at the end of their life and there is no plan to dispose of the 90,000 turbine blades which cannot be recycled. Some of these are the length of a soccer field.

Whole forests have been clear-felled in a country with a modern and ancient culture where people are at one with their forests[1187]. The Baden-Württemburg state government in south-western Germany will clear fell state forests to create a permanently-disfiguring industrial landscape with 1,000 wind turbines.[1188]

This plan to sacrifice the environment, people's health and prosperity was by the eco-socialist coalition of Die Grünen[1189] and the conservative Christian Democrats.

What will Germany do with its 1.4 million tonnes[1190] of toxic waste from wind turbines? In former times, West Germany exported its toxic waste to communist East Germany which was paid hard currency for burial of toxic waste. Problem solved.

After reunification this was no longer possible and Germany got its toxins back. There are also no plans in Germany to dismantle the tens of millions of tonnes of massive steel reinforced turbine foundations. Do they get covered by dirt and forgotten?

Wind turbine blades are a toxic amalgam of unique composites, fibreglass, epoxy, polyvinyl chloride foam, polyethylene terephthalate foam, balsa wood and polyurethane. The plastic-composite-epoxy can't be recycled. Landfill is one of the few options. In the EU, some used blades are cut up and burned in kilns or in power plants. Incineration produces toxic gases and soot.

Blade disposal is not a trivial problem. In 2012, there were more than 70,000 turbine blades deployed globally. To achieve the US's 20% wind production goal by 2030, more blades will be manufactured and blades up to 100 metres long are now being produced. The designed life span of twenty years is because of physical degradation or damage beyond repair.

Blade damage from striking wildlife is minor. It is fatal for birds, bats and insects. The constant development of more efficient blades with higher power generation capacity is resulting in blade replacement well before the 20-year life span. Each year some 3,800 blades break off, are thrown to the four winds and end up in land fill.

When this 20-year lifespan is achieved, there will be disposal of 330,000 to 418,000 tonnes of blade composites per year. This is equivalent to the waste generated by four million Americans. Blades will not decompose in landfill because they are designed with a high resistance to heat, sunlight and moisture. Blades dumped in landfill will deliver a toxic cocktail into aquifers and water supplies for centuries.

Worldwide, there are 249,365 tonnes of epoxy resin containing highly toxic bisphenol A in wind turbine blades[1191] just waiting to enter aquifers and life as part of the environmentalists' dream to have a wind-powered clean-green world.

We are told blades have a 20-year life. Turbine operations have shown that this is not true. The life of the whole turbine unit is a little more than a decade. Blades commonly fail. For example, at AGL's Hallett 1 (Brown Hill) wind industrial complex south of Jamestown (South Australia), each and every one of its 45 turbines failed within their first year of operation requiring their wholesale replacement.

The 2.1 MW Indian-built turbines commenced operations in April 2008 and soon after stress fractures appeared in the 44 metre-long blades. And what happened to the failed blades? They were ground up and mixed with concrete and used in the bases of turbines erected later.

Did AGL tell the local community that the blades and now the concrete bases contain bisphenol A, a chemical so toxic that it has been banned by the EU[1192] and Canada[1193]? Did AGL tell the community that bisphenol A will leach from concrete?

AGL clearly knew what they were doing and knew that they were adding toxins to the environment. If a farmer, heavy industry or coal-fired power station deliberately and knowingly allowed bisphenol A to be leached into the environment, their operations would be closed down by the authorities and there would be massive fines.

Governments have been conned by the likes of AGL who can get away with releasing long-lived toxins into the environment over long periods of time. We end up with an environmental mess because advocates of wind power have closed their eyes to the environmental and industrial realities of 'clean and green' energy.

In the US, some 8,000 blades comprising 32,000 truckloads need to be removed to landfill sites. Turbine disposal costs are $400,000 apiece which equates to $24 billion to dispose of the 60,000 turbines currently in use in the US.

Over the next 20 years, the US alone could have to dispose of 720,000 tonnes of waste blades.[1194] A 2018 report predicted a 15% drop in US landfill capacity in 2021 with only 15% capacity remaining. That's the tip of the iceberg as there will be mountains of solar and battery waste. The problem is solved by sending the waste to Third World countries.

The First World is cynically using the Third World as the dumping ground for millions of toxic wind turbine blades in landfill[1195] and concurrently hectoring the Third World to embrace renewables.

The toxic cocktail of bisphenol A, gallium arsenide, tellurium, silver, crystallised silicon, lead, cadmium and a host more toxic heavy metals is now a Third World problem thanks to green activists pushing for coal-fired electricity to be replaced by wind generation.

Spent fuel, erroneously called nuclear waste, is very tightly contained and monitored whereas waste from wind turbines and solar panels is just dumped. Another evolving nightmare is the waste from lithium batteries. More about this later.

Green energy is a dirty business and fossil fuels and nuclear are relatively clean because of strict and long-term regulation.

Mining and wind

A wind turbine requires 900 tonnes of steel, 2,500 tonnes of concrete, metals and 45 tonnes of plastic. The energy to source to the components for a wind turbine is more than will ever be produced in the turbine's working life.

Per MW, besides steel and concrete natural gas requires 1,100 kg copper and 48.3 kg chromium. Per MW for offshore wind 8,000 kg copper, 5,500 kg zinc, 790 kg manganese, 535 kg chromium, 240 kg nickel, 239 kg rare earths and 109 kg molybdenum is required.[1196]

Exploration geologists have not yet found such quantities of these commodities and the normal delay between discovery and production is about 20 years. It will be a long time during a period of escalating prices before the thirst for these metals is satisfied.

A transition to wind and solar has already started another mining boom. To produce one tonne of copper, about 100 tonnes of rock needs to be mined, 97 tonnes of this rock ends up as tailings and two tonnes end up as smelter residues. To access this copper ore, 500 to 1,000 tonnes of waste rock needs to be removed. Most of this is sulphide-bearing and needs to be sterilised against acid mine drainage. Is it any wonder that the copper price keeps climbing northwards?

The irony is that green activists are anti-mining and yet their policies create a mining boom and an increase in fossil fuel use to create the copper needed for their so called renewable energy.

Because green activists have all but prevented new mines in many Western jurisdictions where there are strictly-enforced regulations on occupational health and safety, tailings, air and water quality and post-mining rehabilitation, the commodities that green activists want for the clean green dream will come from undeveloped countries.

Materials derive from China and from Chinese-controlled operations in Africa, Asia and Latin America where mining, air, soil and water pollution, wildlife preservation, workplace safety, fair wages, child labour, land rehabilitation, and other laws and standards are appalling.

If these operations were in the Western world, the companies would be unmasked, vilified, sued, fined and bankrupted and directors would be gaoled. This certainly does not sound like the environmentalist's creed of think globally, act locally. It is in reality, kill globally where operations are out of sight and virtue-signal locally in a wealthy Western country.

Wind turbines require 200 times more raw materials per megawatt of power than modern combined gas cycle turbines. By supporting wind and solar power generation, greens support murderous destructive operations in impoverished countries.

In 2019, a Dutch government-sponsored study concluded that if Holland wanted to fulfil its green ambitions, a major proportion of the world's minerals would be consumed. These minerals will not be mined in Europe or the US and would come from mines in jurisdictions that do not have the employment, safety and environmental regulations of the Western world.

According to the International Energy Agency, world demand for rare earth elements will rise by 300 to 1,000% to meet the demands for the Paris Accord. Rare earth elements are used in wind turbines, smart phones, computers, aircraft engines, magnets and electric cars.

The biggest producer of rare earth elements is China's Bayan Obo mine in Inner Mongolia. At present, China produces 81% of the world's rare earths thereby exposing the world's electronic industry to the whims of China. Not only is Bayan Obo the biggest producer it has created one of the biggest toxic environmental disasters on Earth.

The mantra chanted by greens is that wind power is non-polluting. Not so.[1197] Processing of the rare earth minerals mined at Bayan Obo in China has left a huge toxic, radioactive waste pile that gives the locals cancers and respiratory problems but is unseen in the neodymium-samarium magnets in a wind turbine. Manufacturing of turbine blades is energy-intensive and uses many toxic chemicals.

There are a few small rare earth element deposits elsewhere in the world but at this stage there are not enough reserves of rare earth elements at present in the world to meet demand. If you are clean and green and support wind energy or drive an electric car, you are responsible for toxic pollution and killing humans.

Steel requires a long chain of processes[1198] of exploration, mining, blending, processing, transport, smelting and fabrication. All require large quantities fossil fuels. Mining involves high-density energy such as diesel fuel. About two litres of diesel fuel is used for every tonne of rock moved, whether it be ore or waste.

Transporting iron ore from the mine to the steel mill requires diesel and ship bunker fuel. Many ship fuels are dirty and emit large quantities of sulphurous gases into the atmosphere. To convert iron ore into steel, coking coal or rarely natural gas is used and carbon dioxide is vented into the atmosphere. The fossil fuels for steel manufacture are for both energy and chemical reduction of an oxide to a metal. Chemical reduction cannot be done with wind, solar, hydro or nuclear power.

Concrete is composed of aggregate, sand and cement. The precursor to cement is limestone and shale, these are heated and carbon dioxide is vented to the atmosphere. For every 100 tonnes of limestone cooked to make cement, 44 tonnes of carbon dioxide are released to the atmosphere. To quarry, transport and crush gravel, sand, limestone and shale requires diesel fuel. The energy to heat limestone to convert it to cement requires coal or natural gas.

It doesn't end there. Large wind turbines need to extract energy from the grid to start and when the turbine is not spinning it still requires energy for the controls, lights, communications, sensors, metering, data collection, oil heating, pumps, coolers and gearbox filtering systems. This comes from the grid and is provided by burning coal. The bottom line is that wind turbines cannot be built, operated or maintained without using fossil fuels.

The theory is that wind-generated electricity decreases emissions of carbon dioxide. This does not happen. They actually increase emissions. Wind turbines produce intermittent and unreliable electricity hence require 24/7 backup by a coal-fired power station which cannot be just turned off and on. Wind generating facilities cannot exist without fossil fuels. The energy density of wind is very low. To put this into perspective, if all of the electricity requirements of the USA were to be from wind, then an area the size of Italy or twice the size of California would be required.[1199] The resources that a wind turbine uses just don't seem to get mentioned by the green left promoters.[1200]

There are a few uncomfortable realities.[1201] Building wind turbines and solar panels to generate electricity, plus batteries to fuel electric cars, requires more than 10 times the quantity of materials compared with building machines using fossil fuels to deliver the same amount of energy.

Wind industrial complexes clearly demonstrate that environmentalism is not about the environment. It is about skinning us alive financially and control of every minute aspect of our lives by unelected green activist thugs.

7

FREAKONOMICS

If city-based green activists want their clean, green, safe, renewable energy, wind and solar generators, they should be placed in cities to minimise voltage losses and costly grid rebuilding.

Streetwise renewables companies have conned naïve green bureaucrats and politicians with little life experience into long-term subsidies for inefficient, variable and horrendously expensive energy production based on the weather.

Renewables companies are in the business of making money, not saving the environment. Wind and solar energy are not free, they can only survive on generous subsidies and the increased energy costs cripple families, businesses and the economy

While backed up 24/7 by fossil fuel-generated electricity.

Electricity prices have tripled as a result of renewables. Money from the poor has been transferred to the rich. Productive energy-intensive industries have closed and unemployment has increased.

Germany, UK, Texas, California and South Australia had catastrophic lethal and financially-crippling blackouts when the weather changed. No modern industrial economy can survive on renewables.

The dash to renewables is the biggest financial scam in history. China is laughing all the way to the bank.

From 1600, the Dutch had the highest *per capita* income in the world. Their huge amount of money allowed them to waste money on fads and fashions. And they did. On tulip bulbs. In the height of the Dutch tulip mania, bulbs were selling for as much as 10 times the annual wage of tradesmen. It ended in tears in February 1637 when the bubble burst.

In the West, we are so wealthy that we are giving bucket loads of money to scamsters who claim they can stop the mythical effects of one molecule in 85,000 in the Earth's atmosphere. History will repeat itself and it will all end in tears.

It has already ended in tears for many workers in rich Western countries. More than 100 nations have committed to Net Zero by 2050. If this is the case, then why are emissions rising? Because of an accounting sleight of hand.

Workers jobs are sacrificed in a Western country thereby lowering emissions, the manufacture is outsourced to developing countries and China whose emissions are not part of the Paris Accord stitch-up and everybody is happy. China grows, the Western countries sacrifice jobs, look good internationally and can claim that they are responsible by reducing emissions. Those put out of work may have a different opinion.

As a result of Western outsourcing, China's share of global manufacturing output has climbed from 8.7% in 2004 to 28.4% in 2018. Over the same time period, the manufacturing component of Britain's GDP has decreased from 25% to 11%. Goods imported into Western countries have embedded energy and the carbon dioxide emissions to make goods is released by developing countries and China.[1202]

As long as Western countries put workers in the manufacturing, cement, steel, energy and smelting industries out of work and import manufactured goods, then can play the emissions game, look virtuous and have their leaders strut around at taxpayer-funded international talk fests breast beating and showing their peacock feathers in front of their co-conspirators.

Workers are being deceived and sacrificed at the altar of Net Zero.

Sloshing in a swill of subsidies

Veteran international investor Warren Buffet does not mince his words *"We get a tax credit if we build a lot of wind farms. That's the only reason we build them. They don't make sense without the tax credit"*.

In 1983, the American Wind Industry Association claimed that wind and solar would be *"competitive and self-supporting on a national level by the end of the decade if assisted by tax credits and augmented by Federally-sponsored R&D"*. For 40 years there has been a bottomless pit of subsidies. Time to stop the rorts.[1203]

Investment in the renewables sector is booming because subsidies are locked in for 20 years, the useful working life of a turbine or solar panel.

The investment has had a fabulous return and, after 20 years, the mess is left for someone else to clean up. What a deal.

When electricity is produced, the money flows. When the electricity is not produced because the wind dies or blows too fast, even more money flows. If there is too much solar, the turbine companies are paid by the regulator not to put electricity in the grid. What business is paid not to produce?

Investors don't care if the poor become poorer, if businesses close or if the nation is economically or strategically weakened. Saving the planet or tackling the alleged global warming does not enter the equation. Investors only care about the return on investment and have no interest in the environment or the environmental damage that their operations produce.

Crafty wind power generators have out-foxed naïve green-tinged bureaucrats and ideological politicians, both of whom have no business experience. Monopoly money is printed at the expense of the consumer, electricity prices skyrocket, businesses close and create unemployment, fuel poverty increases and more and more people can't afford electricity in a wealthy 21st Century Western country.

China is playing the green card, encouraging naïve Western countries to go green and even more green. China is talking its own book. It produces and sells to the West wind turbines and solar panels made from cheap coal and produced at a far lower cost than the West could achieve.

China has also ended support for the building of coal-fired generators outside China, thus potentially reducing the pressure of demand on coal and improving China's economy which is currently struggling with high import coal prices.[1204] This is a wonderful wedge that China has put between themselves and the mugs in Western countries.

Even after 30 years and endless subsidies, no country is powering itself with wind or solar electricity. No industrialised country ever will. The idea of trying to run a modern industrialised society on sea breezes and sunbeams is complete nonsense.

Weather-dependent sources are simply no substitute for coal, gas and nuclear which are available 24/7 on demand. No amounts of battery storage or pumped hydro will create cheap reliable power 24/7. That doesn't stop the political elites berating us if we even dare to question the logic of throwing endless trillions of dollars of taxpayers' money at something that already has proven to be a total failure.

Wind and solar need a spinning reserve, normally from a coal-fired generator. If the wind stops or a cloud shields the Sun, the lost power must be replaced in a fraction of a second. If not, the electrical demand will overwhelm the grid.

The only way to avoid a blackout is to have back-up fossil fuel power units running at 90-95% of rated power because such plants can take up to 24 hours to steam up and come on-line. The spinning reserve burns coal, emits carbon dioxide to the atmosphere and creates no electricity.

Wind and solar do not lower emissions of carbon dioxide into the atmosphere by replacing burning of coal. Wind and solar companies are wasting coal and capital for unreliable expensive electricity. And who pays? The consumer. And who suffers the most? The poor. Green projects make the poor poorer.

The world has collectively spent more than $2 trillion for green energy over the last decade. During this time, the share of the world's energy from fossil fuels has dropped from 86% to 84%.[1205] Electricity generation has been captured by the idealogues and the engineers who built the system have been sidelined. What would engineers know? The inevitable disaster has occurred.

Subsidised wind and solar is a monumental rip-off from start to finish. Power companies charge massive premiums for what they claim is "green" energy when in fact the customer is still getting electrons generated by coal, gas or nuclear plants, especially when it is dark or the wind is not blowing.[1206] Some call this fraud.

The weather now determines whether we have electricity or not in an industrialised society. No economy can operate on the vagaries of the wind yet some valiantly try. Renewables drain bank accounts. After 30 years of subsidies the wind and solar industries can hardly be called infant or emerging industries. Why do they still need to be fed subsidies?

The only sustainability with wind and solar power is the massive subsidies. Use of the word sustainability is code for pillaging the taxpayer's pocket.[1207] Another cost passed on to the taxpayer is inflation. In 2021, German inflation was the highest since 2008 driven by energy costs. The higher energy costs are passed on to businesses and the consumer and the rise in consumer prices of 3.4% in August 2021 was significantly higher than the European Central Bank aimed to achieve for the Euro zone.[1208]

The dream of an all wind- and solar-powered future is a recurring nightmare: chaotically intermittent, dependent upon massive subsidies and environmentally destructive.[1209] How will millions of turbines and seas of solar panels benefit the planet is never explained. The wind industry will never grow up and stand on its own feet and perpetually wants to be financially supported.

Electricity that can't be delivered when power consumers need it is of no commercial or social value. On a cold still winter morning, neither wind nor solar can deliver the peak load power for people to have a hot meal and hot drink. The same applies at sunset. The great reset based on green energy is just not possible until we can ring someone to make the Sun shine and the wind blow at dawn and dusk.[1210]

No one wants to delay having a hot meal until mid-morning after the Sun is shining and the wind is blowing. Until then, it is a fantasy with scam subsidies supported by scared politicians. The economy is not like a computer than can be turned off and on at will. It is like the human body, once it is dead then it is dead.

The real costs of wind and solar are obscured by subsidised prices, renewable energy credits, production tax credits, green bond discounts, accelerated depreciation, property tax exemptions and tax credits. Wind and solar companies have not had to build grids. These were paid for by coal-fired generators. The real costs of wind and solar energy also include the investments needed to integrate them into a national power system that must reliably meet demand.

Vast grids were designed to operate on the basis of a fossil fuel supply and electricity generation for at least half a century and to provide non-stop 24/7 electricity. They will need to be completely rebuilt if 100% renewable energy is to be used in a Net Zero scenario.[1211] It is an impossible dream.

Green Europe is trying to keep the lights on with wind and solar. Electricity in the EU and the UK is at the mercy of the weather.[1212] The system is now so disrupted that winter gas shortages and an energy crisis are the norm, especially when high pressure systems are static for days. A lack of North Sea wind sends gas prices through the roof and, because of market manipulation, gas prices are pushed even higher.[1213] Gas is used to back up the wind and solar and the plans to decarbonise economies to Net Zero means near-record prices are paid for carbon permits.

The cost of natural gas and electricity is surging across Europe reaching records in many countries. Rising carbon taxes, renewable energy subsidies and fracking bans are hitting home. As a result of the gas shortages in Europe and the UK, a major fertiliser maker in the UK has closed its two northern England factories.[1214] Fertiliser factories have also closed in Europe.

A by-product of manufacturing fertiliser is carbon dioxide which is used to make beer and fizzy drinks and is also used for the transport of a wide range of products from bread to meat. The head of one of the British food chains claimed that this was a "black swan event".[1215] No. Actions have consequences and you reap what you sow. The UK and Europe now face a food and drink shortage due to a shortage of carbon dioxide.

The UK government has realised that the natural gas crisis will bankrupt many energy companies; factories and businesses will go to the wall and this will exacerbate the energy crisis.[1216] The gas shortage and insufficient gas storage facilities in Europe are such that if it continues, Europe will turn to use more coal despite its ambitious green energy target.[1217] It's already happening. After a blackout summer, bleak winters with blackouts are on the cards due to the surging gas and electricity prices coupled with a gas shortage.[1218,1219]

The Nord Stream 2 pipeline provides gas to Germany and bypasses Poland and Ukraine who have been critical of Russia. In the past, Russia turned the tap off for gas to the Ukraine and both Poland and Ukraine are exposed to having the gas cut off if they continue to be critical of Russia.[1220] Currently 30% of the gas consumed in Europe comes from Russia and this will increase.[1221] The US has been critical of Europe using Russian gas which damages the US liquid natural gas exports to Europe.[1222]

Germany is exposed to the political whims of Russia. Once Germany is dependent upon Russian gas do you think the price will rise or fall? Do you think inflation will fall in Germany? Germany has strategically weakened itself by not having its own reliable cheap energy sources. Russia has heavily supported anti-fracking groups in Europe and has provided $95 million to NGOs campaigning against shale gas.[1223]

One wonders how many roubles ended up in German green activist and political hands? Upon retirement, German Chancellor Schröder was appointed to the chairman of the board of the Russian energy company Nord Stream AG.[1224] Why? Was this because Schröder was trained as a lawyer or because he was chairman of the board of the football club Hannover 96?

With thousands of wind turbines and solar facilities spread wide and far, their control systems offer the perfect back door for hackers determined to disrupt economies via vulnerable power grids.[1225] Government-sponsored operatives abroad could have a field day for military, commercial or just malevolent reasons against those who don't share their political purity.

The EU faces an energy crisis of its own making as it tries to phase out fossil fuels to meet the bloc's green energy agenda as part of its long-term strategy against climate change. Europe claims it will be carbon-neutral by 2050 and will enjoy electorally damaging personal and financial pain to reach this goal.[1226] A few cold winters with shortages of gas and electricity may provide a reality check and a change in strategy.

The EU cheats on trade, energy, carbon dioxide emissions and virtuous policies. It is hypocritical and runs the world's best anti-competition rackets with its Common Agricultural Policy and threats of carbon taxes on more efficient low-cost competitors. The EU plans to raise €10 billion annually and is confident that there will no retaliatory action.[1227]

This will mainly affect African countries trying to climb out of poverty. The EU looks after its elites at the expense of the average struggling taxpayer. There will be a carbon tax on freight and passenger jets, but private jets and pleasure flights will be exempt from the carbon tax on fossil fuels.[1228]

The European Commission now classifies investment in terms of 0%, 40% and 100% green content and rounds up the numbers to the next higher target. So, 1% becomes 40% and 41% becomes 100%. You can always rely on the EU leaders to have smoke and mirrors tricks to put appearances ahead of content.[1229,1230]

The EU's unilateral climate strategy is being used to transform Europe into a green fortress and to force green protectionism worldwide.[1231] The EU need to be in contact with reality. Europe and the UK have lost their economic power and their climate policies, tariffs and targets will further weaken them.

The UK, EU and other wealthy Western countries that have cold winters give their poor people freedom of choice.

The German disaster

No country went as hard and fast as Germany to spear their countryside with wind turbines. Germans were told that with more and more wind power, coal and nuclear power could be made redundant by the wonders of sea breezes and the Sun.

Reality and ideology are streets apart. German wind power has averaged only 17% of installed capacity since 1990, a fraction of its capacity.

A study of 18 European countries' wind power capacity by German power engineers[1232] concluded that 100% backup by coal/gas/nuclear is needed for 100% of the time and that the more wind capacity is installed, the greater the volatility of the grid. Wind power is an expensive flop.

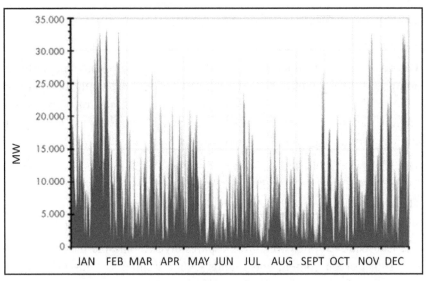

Figure 42: *Offshore and onshore wind energy generation in Germany. The enormous variability does not allow grid stability.*

Green activists commonly tout Germany as the benchmark for the inevitable transition to solar and wind energy. However, whether the wind is onshore or offshore, it is unreliable and continually stresses the grid. What green activists don't say is that the German energy system is a complete debacle. When renewables can't provide Germany with electricity, which is most of the time, Germany draws power from French nuclear reactors and Polish coal-fired generators and soon Russian gas will be added to the mix.

Germany has scored an own goal. Germany's shift to "renewable" energy has had a greater impact on traditional coal- and gas-fired power plants than originally planned. Some 57 of them traditional gas and coal plants will close as a consequence of Germany's *Energiewende*. It is now almost impossible to build a new modern plant.[1233]

In just one year, 330,000 households had their electricity disconnected, 6.2 million threats to disconnect power were made and 44,000 had their gas turned off because of the high cost of electricity.[1234] Costs have been driven up because of the legally mandated feeding in of wind and solar power. Its citizens have resorted to stealing trees from forests for heating and cooking because they have no electricity.[1235] This is the green dream. To push people back to the subsistence living conditions of 200 years ago.

The government plans to pay one third of consumers' bills. Even with this recipe for economic disaster, the price of electricity is so high that many poorer consumers cannot pay their bills and have been disconnected from the grid.[1236]

Subsidised renewable energy in Germany has reached dizzying heights with about €30 billion paid annually. The Renewable Energy Sources surcharge has increased to seven cents per kWh which is now about a quarter of the electricity price. Because of additional surcharges, the German electricity price is 43% above the average for EU countries. Consumers are suffering, manufacturers are looking at other jurisdictions and the government is feeling the political squeeze.[1237]

The costs of Germany's dash to green energy are rising and the number of green energy jobs are falling. The costs of Germany's subsidised wind and solar power will rise by €14 billion by 2025 when the costs will be €77 billion per year.[1238] Concurrently, China is banning wind power projects in six regions and building wind and solar equipment that is sold to Germany.

Germany relies heavily on Russian gas. Russia is not known to be a stable supplier and uses energy, especially in winter, for political blackmail. Russia and the Ukraine have pulled a three-card trick as a result of political horse-trading during UN negotiations. Most of the carbon credits created by them do not represent a cut in emissions because they involved emissions derived from fires on coal mine waste dumps and flaring of gas during oil production.[1239]

Bloomberg reported on 5th June 2020 that Germany's green power finance is becoming unaffordable. German electricity prices are three times that of the US. Closing fossil fuel and nuclear power generation has come at a horrendous price that is borne by the poor. The greens supposed solutions to an alleged climate crisis are delusional.

Germany has lost its solar industry to China and now is also losing its wind industry and jobs to China. The German labour union Nordmetall is not impressed.[1240] Maybe the wind and solar electricity generating plants should have a sticker "Made in China at the expense of jobs in Germany".

The Danish turbine company, Vestas, closed its blade factory in the town of Lauchhammer in the eastern German state of Brandenburg. Some 450 jobs were lost. Vestas also closed down other manufacturing facilities in Spain and Denmark and had moved its manufacturing base to Asia.[1241] The green New Deal becomes the green new dole.

Bloomberg reports that in 2020 Germany stole a record $38 billion in subsidies from consumers for renewables. Germany, which varies between the highest and second highest electricity costs in Europe, has a 46% green electricity supply share.[1242] Over the last eight years, some €200 billion has been paid in subsidies by Germans for unreliable green power that has resulted in job losses and industries moving elsewhere.[1243] And what has this achieved? An unreliable electricity system, higher costs and no change to emissions of carbon dioxide and no change to global temperature.

The wind and solar obsessed Germans reap what they sow. Over the last 20 years, the price paid by German industry for electricity has almost tripled and their credit rating is threatened.[1244,1245] The choice is now simple: go broke or go elsewhere. The Germans have found that calm weather and darkness cannot compete with nuclear, coal and gas.

In Germany over the last decade according to the German Trade Union Association, 'green' jobs in the German renewables sector collapsed from about 300,000 in 2011 to 150,000 in 2018.[1246] These jobs went to China. Green activists don't care that China has no interest whatsoever in green issues or 'clean' energy. Siemens Energy retrenched 7,800 people in its gas and power divisions to improve its long-term competitiveness.[1247]

Coupled with the export of jobs are threats of closure by energy-intensive heavy industries. The German Steel and Metalworking Association (WSM) asked member companies in early 2021 how they would react to carbon dioxide pricing and taxes. The answers ranged from "*immediate investment freeze*", "*downsizing*", "*relocating investments abroad*" to "*immediate company liquidation*" by a North Rhine-Westphalia hardening shop with 75 employees. Other industries severely affected are foundries, hot-dip galvanisers, ceramics, plastics processors and industrial textile manufacturers.

Germany is renowned for its high-tech industries and quality of products. The high electricity costs, taxes and levies are now driving German industry to Asia.[1248] Your future BMW may well be made in Vietnam with energy from Australian coal. China is interested in gaining a competitive advantage over the rest of the world while it powers full steam ahead with its fossil fuel-powered industries.

Germany once had the most stable and reliable power grids in the world when the business of power generation was in the hands of engineers. Electricity was cheap. However, the rise of climate alarmism in the 1990s empowered green activists, their political allies and bureaucratic fellow travellers to push for "green energy".

They insisted that fluctuating and intermittent electricity supply from weather-dependent energy could easily be managed. Germany is now looking to ration its electricity supply to stabilise its green power grid which, in turn, will threaten the stability of the larger EU grid.

German electricity is now controlled by the Renewable Energy Sources Act (EEG) that has given Germany almost the highest electricity prices in the industrialised world, has forced them to rely on imports and yet Germany consistently fails to meet its emission targets.[1249]

The German parliament addressed these problems by adding even more "green" electricity to the grid, ordered the closure of stable and productive coal and nuclear power plants, limited the amount of power consumed by employment-generating companies and individuals and, horror of horrors, will stop charging EVs during times of power shortages.

Germany's Renewable Energy Levy started in 2000 at 0.2 cents per kWh, the then environment minister, a Green MP, said that the cost to consumers would be no more than a *"scoop of ice cream"*. The Renewable Energy Levy has now risen over 3,000% and is still rising, despite renewables producing only 30% of the energy in Germany. Funds from the Levy are used to construct new renewables energy plants, only making matters worse.

The high proportion of renewables in the system has forced Germans to pay over €1 billion to stabilise major powerlines and the total cost for transforming to renewable energy would reach €520 billion by 2025 and will keep rising.[1250]

In January 2021, there was a serious cold spell and a lack of wind in Europe. Every kind of fossil fuel source was fired up to the maximum. The power demand surged across western Europe on January 8th and the continent's electricity network came close to a massive blackout because the grid struggled to keep a constant frequency. The problem originated in Croatia, then 200,000 households across Europe lost power and supplies to industries in Italy and France were cut.

The German Office for Civil Protection and Disaster Assistance advised Germans[1251] *"That in case of a prolonged power shortage, citizens should wear warm clothes and light a fires with a supply of coal or wood to make up for the lack of heating"* and *"It also advises to keep a stock of candles and flashlights, to prepare meals on a camping stove, and to have a sufficient reserve of cash in the house in case ATMs stop working due to the power future".*

European green energy is a great example of going back a couple of hundred years to reach the green nirvana called the future. Winter storms now threaten Germany's power system. This is the perfect storm with solar panels covered with thick snow, a lack of wind and an unstable grid.

The early 2021 US and German self-inflicted renewables disasters in the depth of a cold winter have driven demand for the ever-reliable nuclear power. This is the wake-up call. Wind turbines freeze, turbines catch alight and turbines only provide electricity at optimum wind speeds. The rest of the time the consumer sources reliable power from coal, nuclear and gas. No industrialised Western country has ever run itself entirely on wind and solar.

There was a climate consensus in Germany until the financial, reliability and electoral costs started to become too high. Cracks have appeared and legislation to increase renewable energy targets have stalled and Germany now plans to have wind turbine-free zones.[1252,1253] As Germany's post-COVID-19 economy rebounds, the rise in carbon dioxide emissions in 2021 is the biggest since 1990.[1254]

Germany industry has a simple choice. Go broke or go elsewhere.

France

Thousands of new wind turbines have been installed in France yet there has been no reduction in carbon dioxide emissions. However, the scenic vineyard slopes of Burgundy and the fortified medieval city of Carcassone are now polluted by wind turbines.[1255]

Thanks to nuclear energy, 90% of French electricity generation does not emit carbon dioxide with the energy mix being nuclear 71%, hydro 14%, coal 10%, wind 6% and solar 2%. France is spending €120 billion to increase wind and solar power generation to solve a problem it does not have because its nuclear power generation is not at capacity and much nuclear power is exported to neighbours when their renewable energy fails.

USA

The first renewable energy mandates in the US were imposed in 1983. Most states did not impose mandates until about 20 years later. Over 30 states have mandated voluntary renewable energy requirements. Utilities would have been required to provide 25% of the state's electricity from renewable sources by 2025.

Ohio was the first state in the US to freeze its renewable energy mandate and it halved its mandate level because of high costs. West Virginia repealed its renewable energy mandate and in New Mexico the renewable standards were frozen.

Kansas is well down the path to repealing its mandate which would save ratepayers $171 million (i.e. $4,367 for each household). States with a higher mandate had higher unemployment than states without a mandate.

The US Department of Energy has found that electricity prices have risen in states with renewable energy mandates twice as fast as those with no mandate. Electricity prices in states with mandates are 40% higher than those with no mandate.

A study by the University of California showed that the lower 60% of US households (by income) received about 10% of the amount of "green credits" whereas the upper 20% (i.e. with incomes above $75,000 *per annum)* extracted 60% of the amount. For electric vehicles, the upper 20% of income earners received 90% of the green credits.[1256]

In 2018, the US consumed 3.9 billion MWh of electricity. If only wind power were to be used, 14 million 1.8 MW wind turbines would be needed plus enough batteries to supply the grid on windless days.

Only one billion half-tonne Tesla batteries would be required and the manufacture of these batteries requires raw materials, hazardous chemicals and toxic metals. Batteries for 24 hours of energy for the US would cost $6.6 trillion. This is more than the entire budget of the US government.

The 14 million turbines would sprawl across three quarters of the lower 48 states and would require 15 billion tonnes of steel, concrete and other raw materials. They would wipe out almost every bird species in the USA.

Wind turbines don't last more than two decades and every wind turbine has 45 tonnes of non-recyclable plastic that must be disposed of as landfill. Decommissioning each wind turbine costs hundreds of thousands of dollars.

If the US were to be powered only by wind, 900,000 square kilometres would be required for turbines. That's twice the area of California. Generating just today's US electricity output with wind power could warm the surface of continental USA by 0.24°C. This is a tenth of the warming generated by solar photovoltaic systems.[1257] We have wind and solar power to stop the planet warming yet wind and solar facilities warm the planet.

How well is money spent on renewables? One million dollars worth of utility-scale solar panels will produce 40 million kWh over a very optimistic operating period of 30 years. For the same cost, a modern wind turbine produces 55 million kWh over the same optimistic operating period of 30 years. For $1 million, a shale gas well will produce 320 million kWh over a 30-year operating period.[1258]

If 100% renewable is achieved in the US, doubling the length of the high-voltage transmission system in the US would be required. Over 320,000 km of new high-voltage transmissions lines would need to be constructed as well as manufacture of hitherto unknown massive batteries equivalent to 788 million fully-charged Tesla S batteries.

The land area covered would require the US to import food and the minimum cost is an eye-watering $6.7 trillion. Battery storage would add only $49 trillion to the capital costs.[1259]

The alternative is that the US must get used to monumentally high costs for intermittent electricity, to power rationing and to high unemployment like never before with the associated civil unrest. All this is the price to avert the modelled climate apocalypse in a century when, if the models are correct, the temperature and sea level will be less that that 1,000 years ago. Tell them they're dreaming.

Doomsday climate activists in the US are telling people to give up their selfish demand for reliable electricity.[1260] Maybe the doomsdayers should do a poll of those who like to have electricity such that their children can have hot meals, hot drinks and do school assignments on computers. Maybe they should do a poll of people in hospitals on life-support machines.

Having a few hours of electricity each day is a feature of the Third World and this is where these doomsday climate activists want to take us. These killers should lead by example and immediately stop using electricity. A practice run would be to be stuck between floors in an elevator for a day or two with no lights, no air conditioning, no communication and no access to a bathroom.

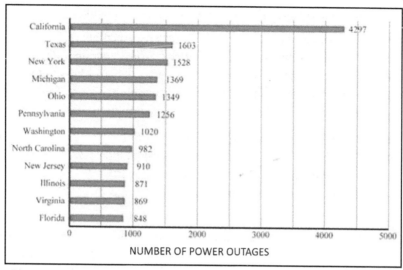

Figure 43: *Power outages in some US states with the biggest outages occurring in energy rich states with the highest proportion of wind and solar power.*

It takes time for simple things to dawn on many. If fossil fuels and nuclear power are eliminated from an industrialised economy, power shortages are inevitable.[1261] It's there for all to see in Germany and the UK. Once a jurisdiction relies on wind and solar for electricity and has anti-fossil fuel and anti-nuclear policies, it becomes a blackout zone.

The US Department of Energy's 2018 Wind Technologies Market Report showed that wind's capacity factor (i.e. percent of time actually generating electricity at full capacity) is only 35% compared to 57% for natural gas and 92% for nuclear. On the hottest and coldest days, it is often close to zero.[1262] This is why nuclear plants in the US produced 20% of the electricity in 2019, despite having only 9% of the nation's generating capacity.

Green activists use the word renewables as a throw-away line and clearly must know the killing and environmental catastrophe that wind turbines create. Are green activists happy to kill wildlife, use massive resources and pollute the landscape? If green activists support wind energy then they are knowingly committing green murder.

California has given those who embrace renewables a wake-up call, which will be undoubtedly ignored by green-tinted comfortable shiny bum bureaucrats. Climate zealots want clean green California to thrive on renewable energy. Apparently some believe that carbon dioxide emissions from the burning of those nasty fossil fuels and putting traces of a trace gas into the atmosphere can affect our giant planetary systems.

Sweeping blackouts in California show that its green energy fantasy has met reality and unreliable wind and solar power generators have forced Californians to rely on diesel generators to provide 450 MW of electricity for homes, businesses and hospitals.[1263] Unreliable power causes the grid to collapse[1264,1265] and wires to spark and cause wildfires.

California's energy policies impose disproportionate costs on the poor. Although California is the world's fifth biggest economy and is a place of unfathomable wealth, it has the US's highest poverty rate. The end result of expensive unreliable renewable energy and regulatory costs is now 40% of the cost of a new home.[1266]

Renewable energy advocates hate the poor and want to keep them that way while they make money using government command-and-control edicts and while renewable energy schemes transfer money from the poor to the wealthy. Green activists do not seem keen on cleaning up the environment and seem to be more interested in forcibly changing the way the average person lives.

California's wind and solar obsession leaves thousands without power. The Department of Energy responded to the Californian government's desperate pleas by declaring a state of emergency thereby permitting California to fire up thousands of fossil fuel-powered diesel generators and to even hook up to moored ships to get the benefit of their fossil fuel powered electricity.[1267] This shows the total failure of renewable energy. Green policies by deliberately sidelining gas and nuclear have given dark days and even darker nights.

California's green new normal fossil-free world has led to power outages despite its huge resources of gas and oil. Many nuclear plants in California are now closed and have been replaced by wind and solar with the chimera of green jobs.

Rolling blackouts affecting millions of Californians in August 2020 created shock, anger and financial losses.[1268,1269,1270,1271,1272,1273] California has put itself in a position where regular intentional rolling blackouts are inevitable as part of their all pain no gain green policies. California, the state of sunshine, is facing darkness.[1274]

Fifteen of the 19 high-efficiency low-carbon dioxide gas power plants along the Pacific coast were shut and not even kept in working order in case of a major hot weather power crisis. California imports electricity from adjacent states to cover wind and solar electricity shortfalls and even if there is one hot day with little wind, there are rolling blackouts. Pity the elderly with no air conditioning.

When there was a heat wave of a few days in the US Southwest, power to California was cut and there were numerous days of blackouts. It is no wonder that high-tech high electricity consuming businesses are moving to Arizona. If batteries were installed as the solution to California's self-inflicted wounds, the cost would be a trifling $8.125 trillion.[1275,1276]

Mandated blackouts using rotating power outages and a state of emergency were declared in August 2020.[1277] The Governor of California demanded an investigation. The answer is simple. Recommission California's nuclear and fossil fuel generation capacity because the drawbacks of wind and solar power are obvious, especially on hot days when air conditioning is used commonly just as the Sun starts to set. No costly blame-shifting investigation is necessary.[1278,1279,1280,1281,1282]

Over the last 20 years, the maniacal wind and solar obsession has cost California dearly. They suffer America's highest power costs and, in hot or very cold weather, they are lucky to have power. Energy poverty is now so entrenched, that for millions of poor and underprivileged, it is now part of daily life.

The solution: California reckons that electricity generated from the wind and Sun, when it's purportedly in excess, can be stored in giant batteries. It's never worked anywhere else but that doesn't seem to worry the Californian government and their green activist bureaucrats.

To achieve this, California needs to triple its grid capacity, expand its wind/solar generating capacity by 6 GW *per annum* for the next 25 years and build 54,000 GWh of battery storage at a bargain basement price of a bankrupting $10 trillion.[1283] Tell 'em they're dreamin'.

The August 2020 blackout have now forced California to build five gas-fired power stations. Green activist policies have a bottom line called reality. California, like Germany, the UK and Australia have decided that building no more gas- or coal-fired power stations is the way to avoid blackouts.[1284] The encroaching energy crisis is showing that this is bizarre nonsense.

The pro-wind and solar rent seekers have made every effort to conceal the cause of blackouts in California and elsewhere. The pliant media gives no indication to the average person what's really going on, and who has to grin and bear it. The growing burden of wind and solar on the power grid is at or near maximum levels in some states such as California and Texas.[1285]

The experience from California must surely be a loud and clear message for South Australia and Victoria. When the US rolls out its Green New Deal, California's chaos will be nationwide.

In February 2021, Texas enjoyed a deep green freeze and a huge demand for electricity. Five million people were left without electricity, some for four days, when 98% of its wind and solar capacity was shut down.[1286] Wind turbines froze in west Texas and energy-efficient techniques such as having fossil fuel burning helicopters pumping warm antifreeze onto turbines had to be employed.[1287,1288,1289]

During the big Texan freeze, thousands of wind turbines were frozen solid. The internal workings of wind turbines (gearboxes, generators, yaw and blade pitch controls) can be kept operable with on-board heating systems that chew up huge amounts of electricity, while the iced-up blades added nothing to the grid.[1290]

Texas modelled its energy systems on California and, although the Lone Star State is awash with gas, Texas reduced its use of gas and coal and concentrated on wind and solar.[1291] This was an own goal. Gas and electricity prices skyrocketed. In the past, gas, coal and oil had prevented past blackouts such as in December 2017- January 2018 with the Bomb Cyclone event. Ten years ago, 10% of electricity in Texas was from wind and solar.[1292] Now it is 25% which seemed like a good thing until an Arctic blast came along.[1293]

A coal-fired power station keeps 90 days of fuel on hand, a nuclear plant needs refuelling every 2 years and gas-fired power plants depend upon just-in-time fuel delivery hence pre-sold gas cannot be suddenly used.[1294]

Some green activists claim that the Arctic weather in Texas and resultant electricity blackouts were due to problems with gas, coal and nuclear plants and not to frozen wind turbines. Texas was a disaster waiting to happen, power outages fuel Texans' outrage and political leaders looking after their own jobs asked how such a disaster came about. Building more turbines and solar panels is unsustainable.

Texas might learn from the great achievements of South Australia. The Australian Energy Market Operator (AEMO) tabled that South Australia had served all of its electricity demand for more than an hour at low prices shortly after midday on October 11th 2020 through solar.[1295] Big deal. Look at all the other days in the year.

Normally, for about 75% of the time SA was importing coal-fired electricity, about 10% of the time the export and import of energy was balanced and for the rest of the time SA was exporting electricity.

The February 2021 blackouts in the Lone Star energy-rich state of Texas were not just poor policy, planning and preparation, bureaucrats had not thought to stress test the electricity system. Texas has more than enough gas for peak and base load electricity. Texas' over-reliance on gas turbines, which provide short-term peaking power, are not efficient for providing sustained power for hours or days. Combines cycle gas turbines, coal and nuclear are far more suitable.

Gas from wells needs to be compressed using electricity before it can be transported along pipelines to consumers. When electricity to the grid collapsed because of frozen wind turbines, there was no electricity to drive compressors to pump gas.[1296] In Texas, four million people were left freezing in the dark and many froze to death because of the failure of wind power.[1297,1298] The 200 deaths were the poor and vulnerable.

If wind and solar produced so little power during times of demand in Texas after an expenditure of $66 billion on renewable energy, then why bother? After the fatal disaster, blame was apportioned to Donald Trump, the moral and intellectual collapse of American conservatism and on fossil fuels rather than on analysis of the root cause of the problem.[1299] The renewables industry in Texas pays $250 million a year in tax whereas the oil and gas sectors pay 54 times that amount.[1300]

The Texas blackouts killed hundreds of people. They were caused by an epic government failure resulting from the embrace of green energy policies.[1301]

Washington State usually has the lowest cost electricity in the US. It imports most of its energy with local energy from gas (11%), coal (10%), hydro (8%), nuclear (5%) and wind (4%). In an attempt to decarbonise the state, the state's 2019 Clean Energy Transformation Energy Act requires all utilities to eliminate coal by 2025 and to provide carbon neutral electricity by 2030. Baseload brownouts and blackouts are predicted[1302]. The alleged climate crisis has morphed into an electricity reliability crisis.

Maybe legislators in Washington State should broaden their horizons and see what has repeatedly happened in California, Texas and Germany.

Subsidies are not only constrained to wind and solar power generation. In New York State, the New York Public Service Commission voted to impose a new Clean Energy Standard (CES) for the entire state. It requires that 50% of New York's energy must come from "carbon-neutral" sources by the year 2030.

The plan openly subsidises financially struggling nuclear power plants in upstate New York through Zero Emission Credits (ZECs). Other power utilities would be compelled to purchase ZECs from a government bureaucracy, which the government first obtains from the company operating the struggling upstate nuclear power plants.

This amounts to a wealth distribution from financially healthy power plants to financially struggling plants to satisfy carbon-free green energy regulations. This is little different from the wealth transfer from efficient coal-fired generators in Australia to inefficient wind and solar generators.

The costs of the ZECs, not surprisingly, will be paid by New York consumers, households and businesses. The scheme guarantees $1 billion to the struggling plants in the first two years alone with an estimated cost of $8 billion over the life of the CES until 2030.

Moody's warned investors that the costs for the duration of the program are $127.48 per MWh of production. Despite the subsidies for the nuclear generators, it estimates a 45% price increase. The subsidies will benefit a single company (Exelon) which controls the struggling nuclear plants of Ginna and Nine Mile Point[1303] that qualify for the subsidy.

Electricity prices have risen as a result. At best this is crony capitalism, at worst it is brown paper bag politics. This is essentially a tax to keep ageing nuclear plants operating, little different from the tax that keeps unreliable and inefficient wind and solar plants alive in Australia and elsewhere.

Households in New York State[1304] now pay $400 *per annum* more than the national average for electricity.[1305] Statewide this 53% extra cost over the national average is $3.2 billion a year. The 15 wind industrial complexes produced an output of 2.4 million MWh. The wind turbines need replacing every 10-13 years at a capital cost of $2 billion. The same amount of electricity could be generated by a small 450 MW gas-fired generating plant operating at 60% capacity and with a capital cost of 25% of the wind turbines and with an operating life of at least half a century.

New York was laid to waste during the COVID-19 crisis. The *New York Post* reported in August 2020 that the renewables-obsessed Governor

Cuomo planned to close safe, reliable affordable 2,069 MW nuclear power plants at Indian Point that have been providing New York with power since 1962.

Modern society, especially medical services, are vitally dependent upon reliable cheap electricity and without nuclear power, New York could not have kept its hospitals operating during the COVID-19 crisis. If New York were to survive on renewables, 1,300 times the land area of Indian Point would be needed for wind turbines.

The Rhode Island 30 MW capacity six-turbine offshore wind project, at best, would produce 10 MW. The Rhode Island Public Utility Commission rejected the project concluding that the sky-high prices it would charge would adversely affect consumers. But it all went ahead and, in 2016, the local utility paid $245 per MWh for the project's electricity and in 2019 the wholesale average price for electricity was $31 per MWh. And who pays for the decommissioning? With New York aiming to build 9,500 MW of offshore wind projects that would cost more than $2 billion for decommissioning.[1306]

Rhode Island is not an isolated example. The Japanese government has scrapped wind turbines in a failed $580 million offshore project.[1307] Despite these failures, the Biden administration wants to spear the Atlantic coastline with 10,000 offshore wind turbines at an astronomical cost and causing damage to the near shore environment and destruction of fishing grounds.[1308] This was sold as a *"bold set of actions"*.

If the wind was at optimal speed and turbines were spinning 24/7, these turbines would produce 75% of New York's summer power needs. Billions of tonnes of ore for steel, copper, lithium, cobalt, nickel and rare earth elements would need to be mined, crushed, ground, separated and transported to provide the materials for the turbines. Limestone would need to be burned with shale to make cement. All these mining, processing and transport operations of commodities requires the use of fossil fuels. Most of the turbines would be manufactured in China.

There is interference to shipping, submarines, aircraft and radar from the large turbines; submarine cables have shown to be prone to problems; turbine infrasound is disorienting and has been responsible for stranding whales and dolphins; and turbines slice and dice countless seabirds.[1309]

Of course, there are the obvious weaknesses. Where does power come from when the wind decides not to blow? How are the turbine blades disposed of in their shortened life in a salty wind?

The President is racking up huge power costs for the lower and middle classes of the US and exposing the US to strategic and energy risks. Biden's Green New Deal is lining the pockets of crony capitalists driving America's energy policy and keeping lobbying rent seekers happy with the squandering of hundreds of billions on utterly pointless wind and solar.[1310]

What does it take for another revolution in the US?

British hot air

In 2007, the UK government claimed that the country would be powered by wind energy by 2020.[1311] That did not happen. Under the current plans, it will only take 700 years for the UK to reach low-carbon heating.[1312] Which bureaucrats and politicians will be alive then to suffer the consequences of yet another failed pie-in-the-sky delusional policy?

In the history of British commerce, there's never been a financial scam that comes anywhere near rivalling subsidised wind and solar electricity. It is the greatest state-sanctioned wealth transfer in human history. Wealth has been transferred from the poor to the rich. Renewable investors have forced poor pensioners to freeze in the dark while racking up huge profits from subsidies.

Forget the pomposity and faux moral piety, the wind and solar industry is not about the environment. It is about persuading government to sign long-term contracts such that the industry can take lots and lots of other people's money. Most of this money is from the poor.

Wind generators in the UK are given "constraint payments" to stop them delivering power to the grid when the wind is blowing. Over the last decade, British wind power companies have been paid over £650 million not to generate electricity.[1313] Is this crazy, or is it crazy?

The UK National Grid warned that a lack of wind could plunge the UK into darkness. The National Grid's warning came a week after Prime Minister Boris Johnson said in 2019 that UK wind power could power every home in the UK within a decade.[1314] Don't wait up.

In the UK, more than half of the £50 billion wind investment in building offshore wind industrial complexes goes abroad and British companies lose jobs to foreign rivals. For the British to go green, they export jobs to China, have increases in electricity costs, have a grid straining to breaking point and regular blackouts.[1315] The British climate activists are trying very hard to turn the UK into another socialist Venezuela. Sounds like a normal green deal in Europe, Australia, North America and the UK.

The UK's Department of Business, Energy and Industry Strategy was written by consultancy firm Frontier Economics.[1316] The report had long been known to be in progress and supposedly was to address the "total system costs" of variable renewable energy generators. It was "delayed" for a year and released in 2017. The Global Warming Policy Foundation stated[1317] *"The study is not only very late, but contains no quantitative assessments of additional system costs"*.

Wasn't that the whole point of the exercise? In the released material, there are some peer review comments from which one can infer that quantitative estimates of those additional costs were in the drafts but had been deleted from the final. The Global Warming Policy Foundation's comment was entitled "Is the UK concealing very high renewables system cost estimates?"

Virtually all government reports on renewable energy are on the nose and try to conceal the real costs. The Global Warming Policy Foundation showed that the Frontier Economics report was no different. If you are starting to get the impression that you are being defrauded, you are right.

When will a government put out a remotely honest effort to calculate the real cost of the mostly wind-and-solar-generation system that they are busy trying to force onto people?

The UK's lowest cost offshore wind facility is Kentish Flats, owned by the nationalised Swedish conglomerate Vattenfall. It was commissioned in 2005, 20 years is the period for which operators can claim subsidies and it just so happens that the useful life of Kentish Flats is 20 years.

In 2017, the company made a pre-tax profit of £11 million, subsidies were £11.4 million, government grants were £0.5 million. Hence without those subsidies from UK taxpayers, this Swedish wind facility would have lost almost £1 million despite it being the lowest-cost offshore wind facility and despite it selling electricity far above market prices.[1318] Electricity generation by wind is wealth destruction.

UK's Hornsea 1 offshore wind facility will attract more than £500 million in subsidies each year and other large offshore wind facilities receive hundreds of millions of pounds a year in subsidies. Even larger offshore facilities are planned.[1319]

The subsidy rate for wind turbine owners could see them pocketing up to seven times more than the value of the amount of power they produce. The scheme was established to encourage homeowners to install a small windmill to supply their needs to feed into the electricity grid providing pots of cash for 20 years. The funds come from green levies on household and business utility bills. Each turbine reaps about £375,000 yet produces electricity worth £51,000.

In the UK, the 2010 environmental and social costs amounted to 4% of the average electricity bill. Today it is nearly 25%. Subsidies for renewables doubled over the last decade. Even if the UK reduces emissions, the amount of carbon dioxide entering the atmosphere is not reduced. The UK is now importing the high-energy produced goods that they once manufactured themselves. British jobs were lost and global emissions of carbon dioxide are not reduced. Looks like all pain and no gain.

In countries trying to look green such as the UK, governments are planning to cut off consumer's power when wind and solar output collapses without warning and without compensation thanks to smart meters.[1320] Imagine if your foundry business has pots full of molten metal and the electricity is suddenly shut off. Metal would freeze solid in the pots and the costs of retrieving the metal and salvaging the pots would be crippling. Why should the consumer and businesses suffer because of bureaucratic and government ideological green incompetence?

In the UK, they also heard far too often that wind is the cheapest method of generating electricity. That is correct only if subsidies and capital costs are ignored. Audited accounts show that the cost of wind power in the UK is not getting cheaper and is actually becoming more expensive.[1321]

China is the biggest winner from the UK's £50 billion of subsidies for wind power at the time the UK needs to be frugal and efficient to claw itself out of its self-inflicted recession. Most of the components for wind turbines are made outside the UK, especially in China.

There are two remaining blade factories at Hull and the Isle of Wight but most blades and other components are made abroad. The 260,000 tonnes of steel for the offshore Dogger Bank wind turbines will be made in Belgium and Poland post-Brexit. I'm sure British steelworkers are happy.[1322]

Government-mandated subsidies for wind and solar represent the greatest wealth transfer since England's King John sought to tax his enemies out of existence and gain ultimate power. Upon his death in 1216 AD, England was in economic chaos and poverty and disease were everywhere. And so too today. Some £10 billion a year of subsidies taken from householders goes to rent-seekers.[1323]

The UK energy regulator, Ofgen, has been criticised for misleading the public about additional rises in energy prices.[1324] They claimed the 9% rise was due to a rise in wholesale costs and that consumers should shop around for a cheaper electricity retailer. What a cop out.

Climate levies and system balancing costs account for most of the electricity bill, these apply to all suppliers hence there is little variability in costs between the various electricity retailers. Wholesale costs account for about 33% of electricity bills with the rest due to astronomical subsidies for renewable investors and other socialised costs.

For some years, the UK faces blackouts every time its wind drops a little or is too strong. The National Grid is constantly stressed.[1325,1326] This is the effect of green activism on a major G20 economy.[1327] How on Earth can this help poor people who the left have traditionally supported?

As green rent seekers infiltrate the bureaucratic and political halls of power even more, their schemes become even more fanciful and need to be supported by even more subsidies. One MW of intermittent offshore wind power capacity costs up to 17 times the cost of a thoroughly reliable MW of gas-fired power generation in the UK.[1328] The latter is available 24/7 whereas the former is when that fickle temperamental old cow Mother Nature deems fit.

The latest hare-brained scheme was to float wind turbines on pontoons and anchor them on the sea floor. Britain is following Germany's failed efforts there and is being blown away by wind turbine nameplate capacity, rather than the actual output. The only advantage of this is that they may stop an invasion by an armada.

In the UK, more than half of the £50 billion wind investment in building offshore wind industrial complexes goes abroad and British companies lose jobs to foreign rivals. For the British to go green, they export jobs to China, have increases in electricity costs, have a grid straining to breaking point and suffer regular blackouts. Sounds like a normal green deal in the Western World.

If wind and solar are meant to be the cure, then then they are worse than the disease. The allegedly cheap and reliable wind and solar energy in the UK, like everywhere else, is plagued by intermittency and electricity prices that yo-yo from one extreme to another. The UK grid needed to be balanced to cope with wind and solar and this only costs £1.5 billion a year.[1329] Who pays? You guessed it. The poor punter.

The UK is now facing a 'perfect storm' of food and beverage shortages, gas shortages, astronomically high electricity prices, 'greenflation' and price and tax hikes that will add £1,500 to household costs each year. This could break many families and businesses.[1330] Although the UK drive to wind energy has reduced carbon dioxide emissions, regular calm weather has resulted in a lack of electricity.

Britain has already turned coal-burning generators back on to fill the energy shortfall, but there's a gap emerging. By 2024, there will be no more coal-fired power stations left, and five of the UK's eight nuclear generating plants will be permanently closed.[1331] Political suicide is on the menu for winter meals in a poorer, meaner, slower, dearer UK.

The UK and Europe face difficult winters and a candle-lit future if gas and electricity prices continue to stay high.[1332] Russia continues to tighten gas supplies and outbreaks of seriously calm and cloudy weather are not uncommon rendering wind and solar useless. Successive UK governments had no long-term plan for keeping the UK heated and powered and surrendered to organised anti-fracking, anti-coal and anti-nuclear green activist zealots financially supported by Russia.[1333,1334]

The UK government will not bail out energy companies, many of whom face financial collapse.[1335] The steel industry is the biggest emitter of carbon dioxide in the UK. The Chinese-owned British Steel will suspend operations due to increases in power costs.[1336] The UK is paying the price for going green. Net Zero means zero electricity, zero jobs and zero hope.

Because of an energy crisis, the UK went to a three-day working week in the 1970s and because of green energy policies had power crises in August 2019, August 2020 and August 2021 hence self-inflicted wounds are not unprecedented. The UK is forced back to reliable coal to keep the lights on.[1337]

Meanwhile in Asia, it's a back to black post-COVID-19 recovery as Asian countries plan to build more than 600 coal-fired power stations[1338] at the expense of those countries that try to power industrial economies dependent on the weather. Politicians now have the awkward choice of admitting to deception or continuing with the charade.[1339] The UK Prime Minster is facing his bye-bye bonkers Boris moment.

It's the same Boris who lectured Australia's Prime Minister on a 14th May

2021 telephone call about emissions targets. When departing Australia in December 2018, the former British High Commissioner in Canberra referred to Australia's "visceral" connection with coal.[1340] He didn't mention that coal is Australia's second largest export, Asia is clamouring for Australia's high-energy, low-ash, low-sulphur coal and coal keeps the lights on in Australia, unlike the UK.

A sub-sea cable slung between the UK, Jersey and France took 12 months, 1,000 engineers and 3.2 million working hours to get it up and running. Through this 240 km-long cable Britain has enough electricity for 1.2% of demand. Except when the French use the supply to threaten the UK with differences over Brexit and French fishing in UK waters around Jersey. Britain imports more electricity than it exports through this interconnector.[1341]

Because of low wind speeds and a gas shortage, Ireland froze power exports to the UK to preserve domestic electricity as energy costs skyrocketed to preserve domestic electricity.[1342] To rely on neighbours for energy poses a national security risk.

Ten cables from Danish and British offshore wind turbines owned by Ørsted failed in the first quarter of 2021 resulting in a financial loss. Repair costs will be £350 million which, of course, will be passed on to the consumer.[1343]

The Basslink cable between Tasmania and mainland Australia has failed and now there are proposals from the rent-seekers to run a cable from Darwin to Singapore across ocean trenches, active earthquake zones and volcanoes to give Singaporeans the pleasure of Australian unreliable renewable energy.

Both the UK[1344] and Belgium[1345] have record high electricity prices and when the wind does not blow, rely on nuclear power from France. Britain is unfaithful in its love affair with wind. When the wind does not blow, it drains French nuclear electricity from France via the French interconnector and uses British coal and gas and burns more American forest timber.

The political game of Net Zero with resultant price increases to tackle the speculated climate emergency just does not wash. A survey shows that the British working-class public is far more concerned about day-to-day issues than climate change.[1346] Politicians beware.

Australia must learn from the UK, US and European energy disasters and build back blacker using coal.

The wind down under

Australia has used renewable energy for a very long time. Windmills have been used for pumping water to stock and many outback stations have homestead windmills charging lead-acid batteries for 32 volt power to the homestead and outbuildings. Some outback stations have total or back-up diesel electricity generation. Solar has long been used to charge batteries for isolated railway signals, public telephones, bores and tech-savvy travellers.

These renewables were not subsidised and cannot do the heavy lifting required by an industrialised society. City-based green activists could do themselves a favour and get into the red dirt country to see how the productive part of the economy operates. The outback, the rural and transport worlds run on diesel. Without diesel, no food could be produced and delivered to green-infested cities.

Wind turbines are increasingly littering rural Australia in an effort to satisfy the Federal Government's 2020 Renewable Energy Target. Every turbine is issued with 8,000 to 10,000 Renewable Energy Certificates each year worth $280,000 to $350,000. This is money for jam for the power generators but the consumers are forced to take electricity at up to four times the cost of hydro-, coal- or gas-fired power.

This is the Federal Government's tax on power consumers, retailers are forced to pay exorbitant prices for wind power; this is added to electricity bills and consumers are therefore forced to subsidise inefficient unreliable electricity. When the Renewable Energy Certificates rise to $90, wind power operators will be issued with $700,000 to $900,000 worth of Renewable Energy Certificates *gratis* until 2031 for every turbine and $52 billion worth of Renewable Energy Certificates will be issued before 2031.

Each wind turbine will earn $12 million just for being a blight on the horizon above and beyond the electricity generated that retailers must buy. You couldn't make this up if you tried.

Australia's Infigen spends $25 per MWh for operations and maintenance on wind turbines. This is contrary to what green propaganda tells us. How long do wind turbines last? Ten years or until the subsidies run out, whichever occurs first. The only thing that keeps them running is not the wind but an endless stream of taxpayers' and consumers' money.

The useful idiots have been telling us that renewables will result in a decrease in electricity prices. In a 2014 review of the Renewable Energy Target (RET), Jacobs (formerly SKM) and ACIL Allen predicted where Australia's energy market would be in 2017.[1347] Despite the bleeding obvious, they claimed that the electricity price would fall to around $50 per MWh. It didn't. It tripled. Jacobs claims that it is now carbon neutral and is now powered 100% by renewables.[1348] Does this mean they only operate a couple of hours a day? Unless they pick and choose which electron comes from a renewable source and which one comes from coal or gas, this verbal bilge is just a con and is supported by the normal suspects.[1349]

The average wholesale cost of electricity has doubled in the last 12 months. A year ago, the Australian Energy Market Operator (AEMO) was predicting a fall in electricity prices. This prediction was based on maintaining base-load coal-fired electricity, increasing renewable energy output and access to abundant cheap gas. However, the price of gas went through the roof.

Three years ago, the same AEMO predicted that scrapping the carbon tax, falling electricity demand and increased capacity would cause retail electricity prices to fall. They didn't. They doubled because the AEMO did not anticipate that rise in the gas price and the bans on coal seam and conventional onshore gas exploration. People are fed lies and think that renewables will be cheaper until they see their electricity bill.

Coal-fired electricity costs 3 to 4 cents per kWh. Most Australians have no idea that it is the cheapest electricity. The Tooth Fairy subsidies for renewables mean that some people with solar panels on their roof think that they are getting cheap electricity when in fact someone else in their neighbourhood is paying part of their bill. That someone else is usually someone who could not afford the capital cost of roof solar panels hence the poor are subsidising the rich.

If massive subsidies tacked onto your power bills that end up in the pockets of wind and solar companies are not enough, there are also Federal government slush funds channelling hundreds of millions of taxpayers funds into these companies.[1350] The Clean Energy Finance Corporation (CEFC) run by bureaucrats gives taxpayer's money to wind and solar companies struggling to raise the capital for a wind or solar facility. When questioned, the CEFC demonstrates it's short on numbers and accountability.[1351]

Melbourne and Sydney normally have 10-15 hot days each year when the wind does not blow. If dependent upon wind, then they are almost certain to have blackouts. Each summer. There is just not enough power for peak summer demand. Can ageing coal-fired generators and power distribution networks operate without breakdown? No.

Why are they ageing? Uncertainty in the investment environment because of renewables subsidies, hence reduced costly maintenance and no new facilities being built. If they fail, then gas-powered electricity generation is used and, as gas prices are increasing hence power costs will further increase.

Gas producers in Bass Strait and the Cooper Basin are selling gas to Gladstone LNG exporters who have long term contracts to sell more gas overseas than they can produce. These contracts were settled when gas was worth nothing and Australia used and needed very little gas.

The situation has now changed, the contracts cannot be broken and if gas is to be used for electricity generation, then electricity costs will rise even more than they have. Objections to onshore conventional gas exploration in Victoria and to onshore coal seam gas in NSW and Queensland have only exacerbated the situation. Power and gas bills will stress vast numbers of businesses and families. That's the grim picture facing eastern Australia.

Politicians and public servants don't really have any idea of what to do. This is typified by Victoria where the government announced it was erecting some 5,500 MW of renewable power generation (mostly wind) to offset the closure of the 1,600 MW Hazelwood brown coal generator.

Hazelwood was generating more electricity than all the Australian renewables combined and it provided 24/7 power and long-term cheap reliable electricity to Victoria's biggest consumer, the Portland aluminium smelter. Green groups gave them great praise for closing the Hazelwood generator and the overall green activist community felt good.

The Portland smelter was built when oil prices quadrupled and Australia had dependable, low-cost coal-based electricity. The transmission network is structured around transmitting electricity from the Latrobe Valley to Melbourne and Portland whereas wind turbines are in remote areas with no transmission infrastructure. Wind turbines are erected but do not have the transmission facilities or back up generation for when the wind does not blow.

Renewable energy costs made Portland uneconomic. The Victorian and Commonwealth governments saved the Portland smelter from closure by giving out more taxpayer's money which saved it from the hangman's noose that the governments themselves had made.[1352]

Tomago Aluminium, which uses 11% of the NSW power demand, is the biggest single electricity user in Australia. It has been forced to halt aluminium production a number of times a week because of skyrocketing electricity prices, unreliable electricity and because domestic consumers were using more electricity during cold or hot spells.[1353] Because Tomago stopped production, the lights were kept on in Sydney and political embarrassment was avoided. This is no way to run heavy industry in a highly competitive international market.

The war on gas is not about gas supply. It is to prop up wind and solar electricity which depend upon open cycle gas turbines that stop the grid collapsing when the wind does not blow and the Sun does not shine. Regular blackouts are political poison and the rorts may be stopped so stop-gap measures need to be created. The wind and solar output collapses on a daily basis. Open cycle gas turbines can be fired up in a few minutes and produce peak electricity at four to fives times the cost of coal-fired electricity. These turbines are rather like a jet engine and can run on gas, kerosene and diesel.

How much are people prepared to pay for electricity? Not very much. There is a limit to what people are prepared to pay for vanity signalling and moral posturing. There's a price for no power at all or unreliable power: freezing or boiling in the dark. When there are more blackouts or even deaths, politicians will panic. There will be short-term hare-brained schemes and the fundamental problem will not be addressed.

South Australia also blundered into generating large amounts of wind power without reliable backup and with a network that is not equipped for it. The backup and transmission economics alter the economics of wind and solar. Its blackouts were made worse by failure of poorly maintained equipment. Victoria and NSW have not learned and are going down the same track as South Australia. Queensland is planning major investments in wind.

Australia's hidden renewables taxes are buried in complexity, and complexity breeds corruption. In our renewables nirvana, we now pay companies to shut down in times of extreme demand when there are or could be blackouts. This is no way to run a country. The left-wing Australia Institute's energy apologist stated[1354] *"Big energy users like factories and farms will be able to earn money by saving electricity during heat waves and at other times when electricity prices are high"*. These blackouts are called *"wholesale demand response"*.

With all the governments, agencies, academics, enterprises, groups and individuals involved in the global climate change marketing campaign, and all of them with particular and often undeclared vested interests, the basic problem is conflicts of interest.

Even where the conflict of interest is recognised, there are multiple schemes devised to hide, obscure or overcome those problems, such as the disclaimers employed by CSIRO *et al.* and a user-friendly legal system for those with deep pockets.

It is no wonder that the average punters can't work their way through the complexity. The complexity is not because the system is complex. It is to make it very difficult for the average person to work out that they are being scammed. What do we want? Massively subsidised wind power or jobs, growth, economic stability and prosperity?

Whatever happens, the wind power companies make money. The government has created a business opportunity to make money out of air; wind generators have signed long term contracts with government to lock in cash flow for decades even if the wind does not blow and they sleep well to the gentle ring of the cash register.

Wind power, touted as renewable energy, has destroyed what was a cheap reliable electricity system. The message is loud and clear from South Australia which has 38% of its generating capacity from wind, all of which attracts the Renewable Energy Certificate tax, all of which is unreliable and all of which has placed South Australia in first place. The prize for being first place for unreliable and expensive electricity and continual drains on capital to keep the lights on.

Minimum wind velocity for power generation is 9.6 km/hr, full power output is at 34 km/hr and turbines are shut down at 90 km/hr to prevent damage. Wind power output is non-linear in relation to wind speed above 34 km/hr. If the wind is blowing at 70 km/hr there is no more power being generated that at 34 km/hr. Not only is wind power unreliable and too expensive, it is inefficient.

Wind power proponents should become wind watchers. On many days across the National Electricity Market, the wind gods regularly create a wind capacity of 1 or 2 % Over the last few months, the Australian wind power has been pathetically erratic. Collapses of over 3,000 MW that occur over the space of a few hours are normal. Any simple due diligence would show that wind power is not sustainable.

Unless the Australian Labor Party shows care for the poorer people in

Australia by at least halving electricity costs rather than trying to score pathetic short-term political points, there will never be cheap reliable nuclear power in Australia. Bipartisan support is needed.

In south-eastern Australia, on the 18th and 19th January 2019 were the days of the highest demand for the entire year. It was hot. At the point of maximum demand on those days, wind supplied 1.9% and 6.2% of total demand respectively. Where are the renewables when we most need them? If the planet savers want renewables to supply 50% of our electricity needs, then to meet maximum demand the wind capacity needs to be increased 26 times but electricity output would still be intermittent.

If the solution were to use batteries to store excess renewable power, then at 50% renewables there would only need to be about 2,700 of the 100 MWh ones to do the job. Easy. At around $100 million each, the costs would be a stunning $300 billion. Batteries only last about 10 years before they need replacing and it would be a juggle to have them fully charged before the next day. Never let a few financial and practical matters get in the way of the Greens Party or green activists.

The Hornsdale Power Reserve, South Australia's 150 MW Tesla battery, is owned and operated by the French energy company Neoen. Why are strategic assets in Australia owned by foreign companies? Neoen was paid by the Australian Energy Market Operator to be on standby to rescue the power grid in the event of a major power plant or transmission catastrophe.

During a major Queensland coal plant failure in 2019, the battery didn't do its job and Neoen is now being sued by Australian Energy Regulator (AER). AER had successfully sought financial penalties from various wind facilities in South Australia during a state-wide blackout.[1355]

Australia has one of the largest interconnected grids, covering Queensland, NSW, ACT, Victoria, Tasmania and South Australia. Wind turbines are plastered all over eastern Australia except for the ACT. What makes those who live comfortably off Australian taxpayers' risk-taking and hard work so special? Adding more wind and solar power to the grid replaces lower cost methods for generating electricity.

It costs $2,000 per kW to build a new wind turbine that requires expensive unreliable costly storage for backup for the small amounts of electricity generated whereas it costs $1,000 per kW to build a new modern coal fire power station that reliably operates 24/7, no matter what the weather.[1356]

Two centuries ago we replaced unreliable wind pumps to dewater mines with steam engines that could operate 24/7 and underground miners were not dependent upon the weather above. On average in Australia, every three days there is a wind power failure of 500 MW. This is the size of a small coal-fired electricity plant.[1357]

In the 2018-2019 financial year government subsidies to wind and solar industries were $3,087 million (Commonwealth regulators), $2,418 million (Commonwealth spending), $951 million (State regulators), and $457 million (State spending). Before governments were spending this $6,913 million on unreliable expensive weather-controlled sources of electricity, we had the world's cheapest reliable electricity which was owned and run by the states. Renewable subsidies add 70% to market costs.[1358]

Coal generators, which account for over 60% of supply, receive no subsidies and yet pay royalties to State governments and have funded the building of the electricity grids. We are at the end of the road and there is now no easy solution.

Some 70% of Australians drive to work and 30% spend almost an hour getting there. About 15% of Australia's emissions derive from road transport, mainly from heavy trucks. We were told that if we work from home, we would help to save the planet. Save from what? Working from home requires heating or air conditioning, this consumes additional electricity and puts further strain on unreliable grids using unreliable energy. In hot summers in Australia, air conditioning saves about 2,000 lives each year.[1359] Thanks to wind and solar, Australians have to to work from home to save the planet.

A typical example of a rort at the expense of consumers is from the Hepburn wind turbine co-operative (Victoria). Hepburn Wind[1360] gains 62% of its revenue from renewable energy certificate sales (priced at $74.54 per MW, forcibly purchased by energy retailers), the remainder (38%) from the actual sale of wind-generated electricity (at $42.73 per MW), probably to the same retailers. The net profit of this co-op was a tidy $213,961 (or 18% of gross revenue), for generating 9,872 MWh of non-dispatchable electricity over the year at a Capacity Factor of 27.5% of nameplate. No wonder there are so many snouts in this trough!

If you are starting to get the impression that you are being defrauded, you are right. When will a government undertake a remotely honest effort to calculate the real cost of the mostly wind-and-solar-generation system that they are busy trying to force onto people?

California and South Australia have committed energy suicide. Both have lost and continue to lose industry that is dependent upon cheap reliable electricity. They both created this problem by taxing, banning and demolishing reliable coal, gas, hydro and nuclear power generation while promoting and subsidising unreliable inefficient costly wind and solar generation. This all done to supposedly reduce carbon dioxide emissions yet it has never been shown that human emissions of carbon dioxide drive global warming anyway.

Wind power has a low energy density. Wind is old technology that was abandoned a long time ago because it is inefficient. There are good scientific reasons why more electricity cannot be extracted from turbine blades or silicon solar panels and wind and solar are now close to collecting the maximum amount of energy from a given land area. Furthermore, wind turbines generate no electricity from gentle breezes and gales.[1361]

A huge collection area is required when compared to a large centrally-located nuclear- or coal-fired generator with its strong walls, a roof and lightning protection. The infrastructure for wind and solar power generation is remote, at the wrong end of the grid and susceptible to damage from cyclones, hail, snow, lightning, bushfires, floods and sabotage.

The variability of wind makes the problem worse. Winds are unpredictable and input changes to the grid are very quick. At sunset when the winds die, fossil fuel energy is needed during this peak time of energy consumption. This can only be possible if coal-fired generators are wastefully running all the time on standby or gas/hydro generators are fired up quickly.

It is not economic for generating companies to have an intermittent unpredictable cash flow while competing against subsidised wind and solar rivals which have conned governments into providing subsidies and which get paid whether they deliver electricity or not. Coal-fired power has been legislated to be costly because bureaucracies have been infested with green ideologues and politicians are so frightened that they cannot make common sense decisions.

Coal and gas generators either lose money or shut down. No one is going to finance, build and operate a coal-fired power station that only runs intermittently and battles against subsidised competitors. Coal-fired generators are forced into the position of being mere backup to intermittent and unstable renewables. They must keep the turbines spinning 24/7 while burning coal and emitting carbon dioxide and suffering manpower and maintenance costs while waiting until they can despatch electricity immediately the wind drops or the Sun sets. The coal-fired electricity generating industry is not a sound financial investment because wind and solar competitors have massive taxpayer subsidies.

Coal miners pay State governments royalties, deposit a huge environmental bond of tens to hundreds of millions of dollars in case they go broke, and so these monies can be used for environmental rehabilitation; have mandated waste disposal plans and are enveloped in costly regulations. Wind and solar harvesters have no such constraints or stranded capital. If we want the lights to go on every time we flick the switch, this has nothing to do with the efficiency of wind and solar electricity generation. It is coal that keeps the lights on.

The end result is that the struggling power system is an expensive unreliable dog's breakfast that will get worse. The authorities have tried to solve the self-inflicted problem by making matters worse with rationing, rolling blackouts and terrorising customers with smart meters, which are an admission of failure.

Smart meters are designed by those who control their inefficient intermittent, expensive wind and solar electricity schemes so as to shut off supply of electricity to the consumer when renewable energy cannot deliver electricity.[1362] Generations of workers have bent their backs to make a better world for the next generation and in a few short decades, all this is lost because of the unsustainable electricity policies of the elites.

Chaotic wind and solar power, threats to shut coal mines and talk of closing thermal coal power stations have stimulated one of Australia's most powerful unions, the CFMEU, to suggest that governments ditch their virtue-signalling woke power policies, that they should support coal mining and coal-fired power generation and abandon Australia's infantile ban on nuclear power so we might to try to catch up with the rest of the nuclear-powered world.[1363] The subtext is that wind and solar power employ few people and hence attract very few union members.

The CFMEU and other unions donate to and control the Australian Labor Party (ALP); determine ALP pre-selections for parliamentary candidates for it, pay election costs and set the policies of it. Hence the marriage between the ALP and The Greens to try to attract inner city votes, and the divorce between the unions and The Greens in industrial areas have created even greater divisions in the ALP.

The unions have created a problem for themselves because union superannuation funds have become big businesses that have invested heavily in wind and solar rorts that extract money from the poor to give to the rich. They have pressurised their ALP members of parliament to support their protection rackets for the subsidised wind and solar industries. The myth that the unions and the ALP stand for the little man, the worker and equality is destroyed by their actions and lack of moral compass.

Australia once had clean, efficient, reliable and cheap power supplies. It was the envy of the world. Then wind and solar elbowed onto the scene, extracted green guilt money as massive subsidies to the tune of $7 billion a year and now Australia has unreliable power which is among the world's most expensive.

South Australia has excelled with the highest retail electricity prices in Australia, and they are up there with renewables-obsessed Germany and Denmark. The governing Labor Party in South Australia was warned by power engineers in 2006 that the grid would fail with the introduction of wind energy and again in 2009 that wind power would destabilise the grid.[1364] However, the green idealogues in the bureaucracy and government knew better than power engineers. Had reason prevailed, there would have been no electricity crisis in South Australia.

Twenty years ago, Australia (along with Poland and the US) enjoyed electricity prices that were less than half those of Japan and most of Europe. Australian electricity prices are now higher than Poland, USA, Japan and Europe (except for Denmark, Germany and Spain). Why have we committed economic suicide for no gain?

Australia can't have it both ways and ban nuclear as well as coal-fired electricity. In Australia, coal-fired power stations are the cheapest form of 24/7 electricity and cheaper than nuclear. Electricity is the key economic impact factor underlying economic growth hence cheap and abundant energy is economically beneficial. Australia has identified resources for thousands of years of coal and uranium production and, given cheap and abundant coal, it could be argued that nuclear is not necessary. However, energy from a diversity of sources is energy security.

The establishment of nuclear power stations accompanied by the necessary enrichment and reprocessing plants would provide the basis for a nuclear industry. South Australia's Olympic Dam copper-uranium-gold-silver mine exports around 4,000 tonnes *per annum* of uranium oxide. Concurrently, South Australia has an energy crisis from embracing renewables.

South Australia's renewables add carbon dioxide to the atmosphere. Renewables are like toys bought from a supermarket with the warning "Batteries not included". Australia's obsession with chaotically intermittent wind and solar has left it with a predictable power pricing and supply calamity, especially in South Australia. It believes it can run an industrial economy on sea breezes and sunshine yet blackouts and load shedding, business closures and unemployment show the government made a wrong and costly decision. Maybe the market should dictate rather than cynical inept politicians and their green bureaucrats.

There has been no due diligence on these renewable energy targets and wind turbines are insanely regarded by governments as perpetual motion machines with no costs. Politicians should be supporting systems that work, minimising red and green tape, and reducing costs rather than trying to destroy existing systems with their ideology.

The South Australian government allowed efficient cheap coal-fired powered stations to close on a matter of ideology. Alinta's Northern power station tried to negotiate a deal with the South Australian government for a $25 million subsidy. That $25 million to Alinta would have solved South Australia's electricity problem and would have produced cheap reliable base load power from the 20 years of reserves of Leigh Creek coal.

Some 70% of this would have been spent on the 250 km railway from the Leigh Creek coal mine to the 520 MW Northern power station at Port Augusta. The rest would have been spent on the Leigh Creek mine and its own town, airport and water supply. Coal mining royalty relief was also sought.[1365] This offer was kept secret by the South Australian government while they promoted the spending of $550 million on batteries, gas-fired generators and diesel generators to try to fix the problem it had created.

The South Australian government dynamited its coal-fired power stations to make sure that coal could not be used again for electricity generation. The Leigh Creek mine is now a mess needing costly rehabilitation, Leigh Creek town is abandoned and the Leigh Creek railway has been left to rot. This cynical act of vandalism was by a green activist Labor government and resulted in unreliable and expensive electricity.

The subsidised wind power in South Australia can't keep the lights on, elections come and go and governments have become creative in order to stay in power. South Australia now will use scores of diesel generators at the bargain basement price of $100 million.[1366] Each litre of the millions of litres of diesel consumed will create 2.7 kg of carbon dioxide per litre burned. Because of wind turbines, South Australia will now emit more carbon dioxide to the atmosphere.

That government will immediately throw $550 million of taxpayers' money at diesel generation; a gas-fired peak load generating plant, either open cycle gas turbines or perhaps gas fuelled piston-engine generators; and a $150 million 100 MW battery that will power South Australia for a whole four minutes when systems again fail. The fire risk of such batteries is very high, it is not known how many times they can be recharged and their working life is unknown. You couldn't make it up if you tried.

Australia is the No 2 seller of gas in the world yet suffers a gas shortage internally. The *Wall Street Journal* called the electricity disaster in South Australia as the energy shortage that no one saw coming.[1367] Total rubbish. It was a total stuff-up by governments and their bureaucrats who have opaque dealings with energy companies.

Many of us saw it coming but *The Wall Street Journal* was happy to believe its own warmist green ideology. To avoid a massive state-wide blackout again, during the January 2017 heat wave, regulators shut off electricity to 90,000 homes. It had happened before and was not unprecedented. It was predictable.

Next to the AGL Energy Torrens Island 1,280 MW gas-steam power plant in South Australia, 210 MW of reciprocating engines have been installed at Barker Inlet. In effect, these are giant ship's engines. These run on gas, diesel or bunker fuel, can reach peak load in five minutes and, of course, pump out carbon dioxide into the atmosphere.

AGL Energy spent $295 million on 12 Wartsila 50DF reciprocating engines, each with a capacity of 18MW.[1368] These engines only burn 18,000 litres of diesel per hour. Again, these pump carbon dioxide into the atmosphere, supposedly to save emissions of carbon dioxide into it. Almost all diesel is imported. What if the sea lanes are blocked? There will be no diesel to produce and transport food let alone produce electricity.

The cost of running a diesel generator, compared to an efficient coal-fired power plant on a dollar per MWh basis is eye watering. A modern diesel plant will, at its near optimal 60-70% loading, generate 3 kWh per litre of burnt fuel. If diesel is at $1.30 per litre and 333 litres is needed for 1 MWh, the cost is $433 per MWh.

Coal fired power in South Australia day-in day-out delivers electricity at $50 per MWh. South Australia further wants to generate 200-250 MW of diesel power backed up by a $150 million 100 MWh battery that would provide a squirt of power to SA for all of four minutes. It's wonderful to have grandiose schemes with other people's money.

If south-eastern Australia was to go to Net Zero, its electricity demand of 22,500 MW hence would need 150 South Australian-sized batteries at that $160 million cost for each, or a total $24 billion.[1369] If these costs are added to the massive COVID-19 debt that Australia has recently accrued, then we can expect generations of poverty except for the financial and political elites.

We have a choice. We can live in the 21st Century with electricity generated by the efficient reality of gas, coal and nuclear or in a pre-19th Century future with wind and solar. By the time we realise we are in the 21st Century, the rent seekers will have ripped off billions from the consumer.

The aim of subsidies for wind and solar is to knock reliable cheap energy systems out of the game for short-term profit. It is not to lower carbon dioxide emissions as backup power is horrendously expensive, produces carbon dioxide emissions and destabilises the grid. Wind- and solar-obsessed South Australia relies on backup from diesel fuelled ship engines and open cycle gas turbines that can also run on diesel. Renewable energy is gaming the system to financially wipe out more efficient and more reliable electricity generators. The renewable energy generators have shown that they have no concern about carbon dioxide emissions, the national economy or the environment.

A substation fire on Friday 12th March 2021 resulted in the shutdown of the 210 MW Barker Inlet power station; output at the larger Torrens Island gas power station was lowered because of safety concerns, supply from the Heywood interconnector with Victoria to bring coal-fired electricity to SA was reduced to half due to a planned week of maintenance and the combined SA wind and solar capacity fell to 2% of rated capacity.[1370]

Before the interruption the average cost was $26 per MWh and, during the interruption from 6 pm until midnight, wholesale electricity prices skyrocketed to their mandated maximum of $14,348 per MWh. The extra cost of more than $42 million was passed on to the poor consumer.

Rather than fix the problem at its source, the SA infrastructure minister claimed that these events showed why the 900 km NSW-SA power link, at a touted cost of $2.4 billion needs, to be built. And when did a government project ever come in on budget? However, the bulk of electricity in NSW is generated from burning coal. South Australians are learning to live without power when it is most needed.

In South Australia, Germany and the UK, large industrial power users are just bumped off the grid when the weather does not behave or the Sun is not at the optimum angle for power generation. No wonder SA is an economic backwater with the highest unemployment in Australia, yet it is still transitioning to 100% renewables. Why would any energy-consuming industry invest in South Australia?

To make matters worse, South Australia has now spent a fortune on the world's biggest battery, waffles on about hydrogen, has banished nuclear power and yet is a massive exporter of uranium from the world's biggest uranium deposit. The energy system is broken because green activist bureaucrats, government ministers and regulators rushed to renewables without appropriate due diligence.

Politicians and bureaucrats were outsmarted by renewable energy generating companies and these companies have long-term locked-in subsidies no matter the weather. Renewables were pushed as a matter of ideology without consultation with electrical power engineers and now, because renewables have failed, even more hare-brained schemes such as pumped hydro or hydrogen are being pushed.

When the wind blows for a few hours at a stretch in South Australia, its last remaining base load power station[1371] is unable to dispatch power to the grid and, accordingly, receives no revenue. However, it continues to incur fuel, maintenance, wage and other costs. It's that combination that destroyed the profitability of South Australia's last coal-fired power plant at Port Augusta and helped destroy the viability of Victoria's Hazelwood plant.

Subsidies paid in the form of Renewable Energy Certificates enable wind power plants to flood the market, either giving away power or even paying the grid manager to take it, leading to the closing of reliable base load power stations.

South Australia established an industrial base under the leadership of Sir Thomas Playford. The state generated its own electricity which was reliable and cheap and this attracted manufacturing businesses to South Australia. Since the advent of renewable energy, South Australia has now the highest household electricity prices in the world[1372] and relies on electricity from Victoria, NSW and Queensland to keep the lights on 24/7. Businesses are closing or moving. Who can blame them?

For example, a family-owned South Australian 38 year-old specialist plastics recycling company Plastic Granulating Services was forced to close, with a loss of 35 jobs, after its electricity bills soared from $80,000 per month to $180,000 per month over an 18 month period.[1373] In turn, that closure led to waste companies closing with further job losses and now there is no plastic recycling plant to service six local councils in South Australia.[1374]

This is green politics at its best. Create unreliable subsidised ideological energy based on intermittent sea breezes, close down tried and proven cheap electricity generating systems; and create expensive electricity that cripples the poor and bankrupts industry, creates unemployment, kills wildlife and destroys recycling, something supposedly dear to the heart of environmentalists.

In NSW, a November 2019 report to Parliament recommended a System Levelised Cost of Energy (SLCOE) whereby subsidies be cancelled for intermittent power generation; to add capacity market component to the National Electricity Market such that power can be delivered when needed by customers; and to remove the ban on nuclear power. Taxpayer's monies were spent on the report and the taxpayer-funded NSW parliament ignored the report.

In Tasmania, touted as Australia's green state, a proposal to build wind turbines each 240 metres tall along the main transmission corridor in the Central Highlands between Hobart and Launceston has met with a few problems. The Central Highlands is home to the Tasmanian wedged tailed eagle and the Tasmanian devil.

A 2015 Tasmanian Parliamentary Select Committee on wind turbines found this program would impose a total cost to the economy of between $30 and $50 million.[1375] To pretend to be green will kill wildlife and economically cripple Tasmania which has a population of just over 500,000 people and a shrinking industry base; and has a spaghetti entanglement of red, green and black tape that prevents new businesses.

Tasmania spent $64 million on leasing, site establishment and operational costs for 220 MW of diesel generation when the underwater Basslink cable, which carries electricity from the mainland, failed coincident with a drought which had depleted the hydroelectricity generation capacity. Environmentally pure Tasmania released huge amounts of carbon dioxide into the atmosphere by burning diesel. This plant food was gobbled up by the Tasmanian wilderness forests. Don't tell the green activists about reality.

Turning to Queensland, its government is destroying assets owned by Queenslanders. In a new report published by its Audit Office, it was shown that more than $1 billion was wiped off the value of Queensland government-owned coal and gas electricity generators.[1376] This is coupled with the increase in subsidies for wind and solar energy. Blind Freddy can see the choice is simple: we can have subsidised wind and solar power that employs few people and creates chaos for employers or we can have many well-paid jobs.[1377] We can't have both.

Anyone who objects to electricity from nuclear, coal or gas has never been in intensive care in hospital. Electricity in hospitals is a matter of life and death. If the grid fails, which is normal with wind and solar electricity, then each hospital has a diesel- or gas-fired generator that kicks in. If a green activist should have a cancer treatment using lutetium 177 which can only be manufactured in a nuclear reactor, will they reject treatment? Unless green activists reject all modern medicine and refuse to be treated in hospital, then they are total hypocrites.

Green activists have perfect solutions so a new government-conning growth industry has mushroomed. The same people who have conned governments to support renewables now realise that renewables cannot deliver what was promised and are now setting the scene for an even bigger con: the storage of electricity.

Pumped hydro, lithium batteries, compressed air, flywheels, hydrogen, capacitors and molten salt are possible batteries. They need to be able to provide charge for a few days when there is no wind or solar electricity hence they are huge and expensive. They also are net consumers of energy through their charge/discharge cycle.

Who is to say the half-tonne lithium-cobalt-lead batteries on top of today's popularity list are the best batteries for tomorrow? Maybe vanadium batteries will be better. It is not for governments to try to predict and subsidise unproven technology, especially as there is intense lobbying of uninformed governments by self-interested spivs.

The energy used for exploration, mining, processing, and transport for the commodities to make these batteries is monstrous. The energy to construct such batteries, charge them and absorb the charge/discharge cycle is huge. A massive amount of energy is also used to build battery warehouses and to recycle or bury worn out batteries that have a very short life anyway. These batteries have a habit of catching fire and such fires take a long time to be extinguished.

Each time green activist bureaucrats try to solve a problem they have created, the solution becomes far more costly, the spivs move in and persuade governments to mandate that more and more taxpayers money is transferred to them and the electricity system is far more exposed to unknown unknowns as the complexities increase.

Renewable energy rent-seekers have slowed down heavy industry investment in Australia because the grid can't take any more of the chaotically-generated wind and solar power. AEMO has introduced new rules about grid access for intermittent wind and solar electricity and, much to the horror of these industries, is actually enforcing them.[1378] Renewables are powered by politics, politicians don't seem to understand that the Sun goes down every day, that the wind often does not blow and that renewables are only possible with subsidies.

The Australian Federal government was not able to make the sensible decision and cancel subsidies for wind and solar rip-offs. Instead, it created subsidies in August 2021 for the coal-fired electricity generating industry to guarantee security of electricity supply. The end result is that the poor burdened electricity consumer just ends up paying more than before for a basic commodity for employers and households.

In the lucky country, there is now energy rationing in the land of plenty.

The oriental secret

China does not have such problems. China has some wind and solar power for remote areas but coal does the heavy lifting in cities and industrial centres. By contrast, Australia has fallen in love with wind and solar, is killing off coal and is suffering the inevitable high costs, unreliable electricity delivery and the strained grids just waiting to collapse.

China does not forget that coal pulled hundreds of millions of people out of agrarian poverty as it became an industrial giant and now it exports all manner of goods to the Western world that can only be manufactured using coal-derived power.[1379] China is selling wind and solar equipment to the West which is bringing the latter into poverty. China's success as an industrial power and manufacturing powerhouse is down to one policy: a focus on providing abundant reliable cheap supplies of electricity.[1380]

China's signature on the Paris Accord makes no difference to their increasing emissions yet they insist that other signatories lower their carbon dioxide emissions. This allows China to have cheap energy and to manufacture cheap products to be exported to those countries that signed the Paris Accord and that have expensive manufacturing. The West has been conned by China who have used the Paris Accord to create a commercial advantage. Why even bother? In 16 days, China emits as much carbon dioxide as Australia does in a year.

We are told that China is pushing towards a wind and solar economy. China's investment in new coal and nuclear plants totally dwarfs spending on intermittent wind and solar and its thirst for oil is insatiable.[1381] China's wind and solar is for show for the useful idiots in the West.

As a sop to the West, China has pledged carbon neutrality by 2060.[1382] Who believes this and who will be alive then? Australia has thousands of production years of coal and uranium sitting in the ground. Tourists might come to Australia from Asia to view the flora, fauna, scenery and vast open spaces. It may be a case of use it or lose it. If Australia does not use its in-ground resources, others wanting to use these resources may just come and harvest them.

China must be delighted that so many Western global competitors are sabotaging their electrical grids with erratic, unreliable expensive "renewable" energy using Chinese wind turbines and solar panels. If China ever wanted to invade Australia, all it would need to do is consult the weather forecast and invade on a cloudy windless day when infrastructure is impossibly strained.

Is the Chinese Communist Party using its useful idiots to mandate unreliable and expensive wind and solar in the US to wreck the power grid? Meanwhile, China is full steam ahead to use that black stuff to generate cheap reliable electricity to give Chinese businesses a huge competitive advantage over US businesses.

The West's obsession with wind and solar power, aided and abetted by China and Russia, is Western Civilisation's economic suicide pact.[1383] Rocketing power prices, power rationing and rolling blackouts are all part and parcel of the attempt to run an industrial economy on subsidised sunshine and sea breezes. Ask anyone in South Australia, California, Texas or Germany.

If I were President Xi, I would be funding Greenpeace, Friends of the Earth, Extinction Rebellion, the Greens and numerous green activist groups trying to destroy our economy and way of life. Who knows if he is? After all, the communist Soviet Union used to fund Western green and anti-nuclear movements.

Wind and solar can never replace nuclear, gas and coal-fired power. The recent wind and solar debacles in the US, UK and Germany have finally brought to bear and exposed the renewable energy idealogues as outright frauds.

It is speculated by market strategist Bill Blain that a green economy will trigger the next global financial crash.[1384] He speculates that if a couple of snowflakes roll down the hill and trigger an avalanche, the renewables sector will be the first to collapse. This will be exacerbated with the eye-watering debt that the West has incurred from COVID-19.

While the Western world fiddles with wind and solar, China is surging ahead with coal- and nuclear-fired electricity and Western countries that have embraced renewables have given themselves a huge national security risk.[1385] Society cannot function without 24/7 cheap and reliable energy.

Governments and their supine bureaucrats even control how we use the water that falls as rain on our properties. Farmers pay for building a dam to collect runoff water and government regulations control the use of that water. It took hundreds of years to devise a green- and government-backed scheme to tax air and tax the weather while running up costs of electricity, a basic commodity used by every business, consumer and household.

Households were once taxed for the number of windows in their residence. This resulted in the bricking up of windows. Newspaper publishers were once taxed for the number of pages which led to the shift from quarto- to broadsheet-sized newspapers. The cost of electricity has risen so much that businesses and residences have resorted to buying their own off-grid fossil fuel generators to reduce costs and keep the lights.

The self-induced chaos will not be solved by building more wind and solar. The future is already polluted. Too much electricity is produced when the wind blows and the Sun shines and not enough electricity is produced when it is needed.

The renewables energy industry is the greatest financial scam ever invented and, on a quiet night, Chinese laughter can be heard all over Australia.

8

FADS, FASHIONS, FOOLS, FRAUDS AND FINANCES

The manufacture and charging of electric vehicles (EVs) produces 11-28% more carbon dioxide emissions than for an equivalent diesel vehicle. EVs use far more metal than conventional cars.

EVs are a toy for rich virtue-signalling city residents who occasionally drive short urban trips. EVs survive on subsidies and subsidised infrastructure; overload the grid during charging; can't be used in grid-locked traffic or for long distances and regularly catch alight.

If countries change to 100% EVs, then grids are unable to cope because 70% of all electricity will be used for charging cars. The metals for 100% EVs have yet to be found.

EVs owners knowingly support black child slave labour in the Congo to produce cobalt, devastation of tropical rainforests to produce nickel and totalitarianism for lithium battery production in China.

When the weather stops renewable energy generation, extremely expensive mega-battery hubs will give a few minutes of electricity. Regular lithium car and mega-battery fires emit toxins such as lethal hydrofluoric acid and take days to extinguish.

Proposals to re-invent hydrogen power ignore the Laws of Thermodynamics. Hydrogen needs to be manufactured with at least 30% energy losses, extraordinarily high capital costs and operating costs and unthinkable safety risks. It's been tried before and failed.

The poster child for green activists is a poor little autistic, ill-educated, fragile Swedish child. She is abused by her parents, minders, manipulators and conferences of world leaders. Every speech is written by evil green activists who have stolen her childhood for their political purposes and used her as a cash cow.

Global temperature measurements have been altered, "homogenised" and ignored to make it appear that temperature has been increasing over time. Records are poor and inaccessible.

The keepers of the primary data have tampered with it so much that it's useless. Changes in temperature are within historical variability of ± 3.5°C hence there is no climate crisis or emergency. Carbon dioxide measurements could also be unreliable.

Climategate emails show that climate "science" and IPCC models are underpinned by long-term fraud. IPCC reports cherry pick scientific information, their models are unrelated to real world measurements and errors, rely on unvalidated assumptions and mistakes continue.

COP talk fests have nothing to do with climate or the environment but are UN-sponsored meetings to extract money from countries which do not protect their economy, freedoms and sovereign risk.

Net Zero, zero carbon and other uncosted clean green schemes are underpinned by the biggest scientific fraud in history. The hollow words by dreamers promote new business schemes to extract even more money from consumers via higher costs, subsidies, tariffs and penalties. There will be severe impacts upon already inefficient unreliable electricity systems that threaten food, industrial and military security.

Such ideology is supported by a scientifically-ignorant fawning media of fellow travellers who have corrupted free speech, free inquiry and democracy.

Electric vehicles (EVs)

Governments should not be predicting and subsidising untried and unproven new technologies. The UK government mandated for compact fluorescent lights and halogen globes to replace tungsten filament globes. This cost consumers £10 billion. This decision was made just as light emitting diodes (LEDs) were entering the market. LEDs swamped the market and, in 2021, the UK government banned halogen and fluorescent bulbs and tried to take the credit for the spontaneous market innovation of LEDs. This ban will cost consumers a further £2.5 billion. The ban is a stunt to cover up government failures.[1386]

Imagine if a government 120 years ago mandated to have steam-driven automobiles replace horse-drawn carriages rather than let the market sort out the best transport technology from horse-drawn carriages or steam, electric, gas, petrol or diesel vehicles.

Australia's chief scientist Alan Finkel made it very clear[1387] *"Maybe 20 or 30 years from now we'll have new kinds of batteries, vastly powerful, more extensive batteries and we can do it with batteries"*.

EVs are not transport vehicles. They are short-range carriers of large heavy batteries and metals.

Enormous hope rests on electric cars as the solution by the motor industry to climate change. This assumes that human emissions of carbon dioxide drives climate change and that the carbon dioxide emissions to build, maintain and operate an EV is less than that of a petrol or diesel car. Motor vehicles emit 15% of human emissions of carbon dioxide and EVs emit 11-28% more carbon dioxide than diesel vehicles when the emissions related to battery manufactures are considered.

Many tonnes of carbon dioxide are released even before the battery leaves the factory. Over the lifetime of a Tesla Model 3 car, 27.1 tonnes of carbon dioxide are emitted whereas a petrol-powered BMW 320i car releases 22.8 tonnes.[1388]

Here's the plot for a horror movie. In a moment of pious virtue signalling, you purchase an EV to run around your home city in Sweden, Germany, England, Canada or northern USA. In a mid-winter snow storm, you get stuck in an early evening peak-hour freeway gridlock and, after three hours, the car battery runs out of juice.

Once the battery is dead, the minimalist heating system cannot work, windscreen wipers are frozen stiff and the radio and global positioning system (GPS) with safety and rescue announcements are off air. Because you use the car as a city run-around, you have neither blankets nor food on board to keep you alive overnight.

Rescue and towing vehicles cannot navigate their way through grid-locked traffic, there is no way hundreds of cars can simultaneously be charged to clear the traffic and charged batteries are too heavy to be delivered by drones. Over to you. You can finish the plot of my horror movie. It certainly will not be pretty.

Maybe you could mercilessly steal my brilliant idea and write a variation based on summer travel in your EV on unsealed lonely roads between outback towns in Australia when the day temperature is 47°C in the waterbag.

The air conditioning needs to operate flat out, every now and then you need electricity for the windscreen washers and wipers to clean off the dead bugs so you can see where you are going and, once the battery is dead, you have no electricity for your GPS. In much of the outback, there is neither radio nor mobile phone reception. Both movie plots are not hypothetical. They are reality.

On our small street (approximately 25 homes), the electrical infrastructure would be unable to carry more than three houses with a single Tesla. For just half the homes to have electric vehicles, the system would be wildly over-loaded. Furthermore, houses cannot be insured if an EV is charged in a garage because of the high fire risk.

In the UK, electric car charging points in people's homes will be pre-set to switch off for nine hours each weekday at times of peak demand between 8 am to 11 am and 4 pm to 10 pm because of the fear of blackouts on the National Grid. The UK government will impose randomised delays to stop grid pressure when EVs owners all at the same time rush to charge their batteries.

Gushing articles record long drives in Australia in a Tesla Model 3.[1389] Nothing was written about the time taken for the drive, the number of recharge points and the difficulties in recharging. I do many long outback drives and have never seen an EV in my outback travels. I wonder why?

There is a different story when an EV is taken on a test drive by a motoring expert. Try this from the US.[1390] Eric Bolling test drove the Chevy Volt at the invitation of General Motors and writes *"For four days in a row, the fully charged battery lasted only 25 miles before the Volt switched to the reserve gasoline engine"*. Eric calculated the car got 30 mpg including the 25 miles it ran on the battery. So, the range including the 9-gallon gas tank and the 16 kWh battery is approximately 270 miles.

It will take you 4½ hours to drive 270 miles at 60 mph. Then add 10 hours to charge the battery and you have a total trip time of 14.5 hours. On a typical road trip your average speed (including charging time) would be 20 mph. *"According to General Motors, the Volt battery holds 16 kWh of electricity. It takes a full 10 hours to charge a drained battery. The cost for the electricity to charge the Volt is never mentioned so I looked up what I pay for electricity. I pay approximately (it varies with amount used and the*

seasons) $1.16 per kWh hence $18.56 to charge the battery which is $0.74
per mile to operate the Volt using the battery. Compare this to a similar
size car with a gasoline engine that gets only 32 mpg. The figure is $3.19
per gallon which is $0.10 per mile".

The gasoline powered car costs about $20,000 while the Volt costs at least
$46,000. So the American Government wants loyal Americans not to do
the math, but simply pay three times as much for a car, that costs more than
seven times as much to run, and takes three times longer to drive across
the country.

A Tesla on the flat has a 350 km range in summer and a 150 km range in
winter. The 70 kWh battery with a 75-minute charger requires 400 volts at
100 kW with about 40% of energy wasted as heat. Some 40 kg of coal is
required to charge a Tesla.

A Dutchman with a huge amount of time on his hands completed an epic
89,000 km publicity-seeking jaunt by electric car to Sydney from Holland
to prove the viability of such vehicles in tackling climate change.[1391] Did
it? He needs to first prove that human emissions of carbon dioxide drive
climate change.

His venture took three years and was funded by donations from the 33
countries he traversed. His trip took 36 lazy months. I can do the same trip
in 20 hours using fossil fuels and use the rest of my time constructively.
Did he use fossil fuels to get from south-east Asia to the island continent?
Did he use fossil fuels to get from mainland south-east Asia to Indonesia?
Will he fly the car back to The Netherlands?

His journey was not without problems. A tropical deluge in Surabaya
killed the battery pack rendering it completely unable to hold charge. To
get to Coober Pedy (South Australia) 260 km distant, he waited 12 hrs
for a tailwind and then trundled along at a power saving 60 km/hr while
being a dangerous pest for large trucks or road trains towing three of four
trailers. His car ran out of puff 15 km short of Coober Pedy and had to be
towed into town by a fossil fuel burning vehicle.

He claimed he saved 6,700 litres of fuel and in what was a totally
unproductive three years. He needed 12 hours for charging at a domestic
point and three hours at a commercial point. What did it prove? EVs are
totally useless for long distance travel.

We are told that EVs are emission free, climate friendly, socially and
ecologically responsible and more affordable each year. If so, why do

they need subsidies? If EVs are so great for the climate, why aren't green activists rushing out to buy them to save the planet. EV sales are very low. In the US, a 2021 $80,000 Tesla Model S Long Range can travel 412 miles (660 km) on a single time-consuming multi-hour charge. Range is reduced with battery age and depending on whether the heater or air conditioner is required. Elon Musk wants a carbon tax on all petrol and diesel vehicles to boost the sales of EVs.[1392] Maybe if EVs were cheap, didn't need battery replacements, had long range and didn't burst into flames then sales would increase.

Senator Chuck Schermer (Democrat, New York) wants $454 billion of the public's money to install 500,000 new EV charging stations, replace government vehicles with EVs and finance replacement of clunkers with EVs. The US tax code also allows buyers of an electric car to deduct $7,500 from federal taxes if the carmaker sells fewer than 200,000 units. Tesla is about to begin mass production and exceed 200,000 units and asking for the tax credit to be scrapped as they deem it to be unfair.[1393] I agree with Tesla. The taxpayers should not be funding people's purchase of a private car.

A *LA Times* article predicted the payoff for the public would take the form of significant pollution reductions, but only if solar panels and electric cars broke through as viable mass-market products.[1394] At that time both were considered niche products for mostly well-heeled customers. Nevada agreed to provide Tesla with $1.3 billion in incentives to help build a massive battery factory near Reno.[1395]

The California State Assembly passed a $3 billion subsidy program for electric vehicles, dwarfing the existing program. The bill is ostensibly an effort to put EV sales into high gear, but below the surface appears to be a Tesla bailout. Tesla will soon hit the limit of the federal tax rebates, which are good for the first 200,000 EVs sold in the US per manufacturer beginning in December 2009. In the second quarter after the manufacturer hits the limit, the subsidy gets cut in half, from $7,500 to $3,750; two quarters later, it gets cut to $1,875. Two quarters later, it goes to zero. Losing a $7,500 subsidy on a $35,000 car is a huge deal. The Tesla Model 3 would be tough to sell without the federal $7,500. But this new bill would push Californian taxpayers into filling the void. It would be a godsend for Tesla.[1396]

Tesla competes with a mature car industry that had seen massive federal bailouts for General Motors and Chrysler. As Tesla prepared to roll out mass production of its $35,000 "breakthrough" Model 3 vehicle, that is being put to the test. *The Wall Street Journal* stated that sales of Tesla cars in Hong Kong crashed after authorities removed a lucrative tax break, thereby increasing the cost of vehicles from $75,000 to $130,000.[1397] Not a single Tesla was sold in the country in April 2017. The local government slashed a tax break for electric vehicles on April 1, which resulted in no Model S or Model X deliveries during the whole month. Data from Hong Kong's Transportation Department also reveals only five privately owned electric vehicles were sold in May. The collapse reveals once again how sensitive the automaker's performance can be to government incentive programs.[1398]

If China sneezes, Tesla's coughing fit brings it to its knees. Tesla's share price dropped sharply after it released figures showing 22,000 cars were sold worldwide in the second quarter of 2017 as battery supply from China crimped output.[1399]

When Denmark's electric vehicle incentive program expired last year, new car registrations for electric vehicles of all brands fell 70 percent. Tesla's car sales in Denmark dropped by 94 percent.[1400]

A 2015 study found that the richest 20% of Americans received 90% of these generous EV subsidies. This reverse Robin Hood scheme also means that subsidies are financed by taxpayers. The poor, working class and minority families pay the most yet most will never be able to afford and EV, they will buy even more clunkers and the US traffic will look more like that of Cuba.

In 2020, the European car market was wobbling. Passenger car registrations in 2020 fell by about 27%. This decline was worse than the slump during the 2009 financial crisis.[1401] The green tinges were starting to turn brown. In 2021, Mercedes, BMW and Toyota all announced that the sales of their petrol vehicles have made a comeback.[1402] Fossil fuels are the big winner.

Big vehicles are becoming increasingly popular in Europe driving emissions higher and threatening manufacturers with fines. Over the last six years, there has been a 50% increase in the SUV market share in Spain, UK, Italy, France and Germany. Regulators want to cut car exhaust emissions yet consumers want SUVs that pump out more emissions. SUVs increase car sales and profitability and SUVs are now almost 50% of new vehicle sales.[1403] The European Environment Agency published that the carbon dioxide emissions from new cars in the EU, Nordic Countries and the UK increased in 2019 for the third consecutive year. Vans registered in 2019 emitted more carbon dioxide than in 2018. The EU should lead by example rather than trying to put pressure on non-EU countries.

In California, 20% of all electric vehicle owners in California switched back to fossil fuel-powered cars because charging an EV is such a hassle. In three minutes, you can fill the fuel tank of a Ford Mustang and have a 500 km range.

To charge the Mustang Mach-E, an overnight charge gives a 50 km range. This highlights the real problem of EVs. There is little infrastructure for charging EVs and, when it is found, it takes far too long. Tesla, the leading EV manufacturer, reversed its loss-making trend for one quarter in 2021 by selling energy credits and Bitcoin.[1404]

The European Automobile Manufacturers' Association (ACEA) states that if the EU targets of 30 million EVs are to be met, then EV numbers must rise by 5,000%. Just 0.25% (615,000 of the 243 million cars in the EU are EVs). At present, 200,000 charging points exist in the EU and the ACEA warns that another three million charge points will be required by 2030.[1405] A billion here, a billion there. Who cares? It's only taxpayers hard-earned cash.

A cordless electric motor such as an EV needs a battery. Batteries are heavy, bulky, contain large amounts of metals, are slow to charge and can explode. Recharging a battery creates emissions from coal-fired power stations that provide the electricity. Building an electric car generates considerably more carbon dioxide than a conventional petrol or diesel car will emit in construction and use.

In the UK, each year cars travel 250 billion kilometres. If the smallest electric car[1406] was used for these 250 billion kilometres, then that mileage would add an extra 16% demand on the existing electricity grid.[1407] Where will this extra electricity come from? If it was from wind, then there would need to be an extra 10,000 onshore and 5,000 offshore wind turbines requiring a subsidy of at least £2 billion *per annum*. What is not considered with electric vehicles is that 40% of road transport fuel is used by heavy vehicles that transport food, drinks and consumables.

In the US, there are 279 million fossil fuel vehicles and only one million EVs. If the US is to use only EVs, there will be up to an 8,000% increase in demand for a large number of critical commodities. The end result of having EVs with the Green New Deal will lead to an increase in fossil fuel use. Most of the supporters of the Green New Deal are anti-mining.[1408] What is the word for an own goal by a hypocrite? Maybe a gore?

California, the energy green dreamin' state, already has more than 70,000 public and private EV chargers with an additional 123,000 planned. The state has a 2025 goal of 250,000 EV chargers. Where will the energy come from to charge EVs in a State already challenged to keep the lights on? Where will the money come from for the charging stations? Clean green schemes are always uncosted and that means they are never financially reconciled. You can't manage what you can't measure.

In the UK, there is a 60% tax on petrol. If EVs replace petrol cars in the UK, where will the revenue come from? The poor who don't own cars will just have to cough up even more tax to fill the revenue shortfall. The obvious solution is that the UK government could cut costs. Pigs fly.

One in three UK motorists cannot afford even the cheapest electric car. Electric cars drive up energy costs.[1409,1410] Why do greens want to punish the poor and sacrifice the working class at the green altar? How many Green Party politicians drive an EV?

A January 2020 survey by the UK's *Auto Trader* of its 2,300 consumers showed 16% were planning to buy a battery-only car. A repeat survey of 2,700 consumers in August 2020 showed only 4%.[1411] The reason was that COVID-19 had contracted the economy, government subsidies for EVs were reduced and the personal finances of consumers were reduced. Green virtue signalling is the preserve of the wealthy with disposable cash.

If The Netherlands is to replace its eight million cars with EVs, 11% more electricity and 24% more capacity would need to be installed at a cost of $27 billion. In the UK, the numbers for 26 million cars are 36%, 50% and $140 billion. The German's 44 million cars would require 30%, 40% and $230 billion. If California gets its way and all 260 million cars will require 30%, 40% and $1.4 trillion.[1412] Grids will have to be rebuilt. Who will pay?

In the UK, a feasibility study commissioned by government ministers on ways to accelerate the uptake of electric vehicles has suggested making new petrol and diesel cars £1,500 more expensive to subsidise electric vehicles. This would penalise those who can't afford the high purchase price of electric cars or don't have the facilities for charging them. In effect, the rich are robbing the poor. The roads would become flooded with environmentally unfriendly old bangers as people could not afford to replace them with new cars.[1413]

Nearly half the British public say that they will never buy an electric car because of the lack of charging points and cost. EVs in the UK account for just 5% of new car sales[1414] and there are no long isolated outback drives on that small wet over-populated island. If the UK only had electric cars, only 0.25 % of UK cars could be charged by the UK's largest hydroelectric power station and 70 % of all the UK's electricity would be needed just to charge cars. The cabling and substations were sized and installed before electric cars were even thought of and if the UK were to have electric cars, the whole distribution system will need to be rebuilt.

Green activists claim that EVs are lovingly recharged by carbon-free sunshine and breezes. In south-eastern Australia, virtue signalling drivers of EVs have 85% of the charge coming from coal-fired electricity. If you want to sell your Nissan Leaf, buyers are aware that a battery only has a five-year warranty and a replacement battery costs $33,000. For a Mercedes it is $55,000. The resale value of an EV is low. What if you can't resell your electric car?

It is absurd for wealthy greens to tell themselves that eating less steak or driving an EV will rein in the alleged rise in temperature. EVs are touted as environmentally friendly but they require electricity generated from the burning of fossil fuels for charging. Construction of batteries involves emissions of far more carbon dioxide than saved until 60,000 km is driven. Manufacture of an electric car battery creates more carbon dioxide than a petrol car's emissions over eight years.

In 2018, EVs allegedly saved emissions of 40 million tonnes of carbon dioxide worldwide and, if carbon dioxide drives global warming, then this is sufficient to reduce global warming temperatures by a massive 0.000018°C. If you think that driving an electric car will change climate, then you are bonkers.

Don't even think that EV trucks can deliver food to the cities. The founder of the US electric truck maker Nikola, Trevor Milton, resigned after a demonstration of its new battery-powered electric truck. The truck was towed to the top of a hill and coasted downhill as a demonstration of a 1,000 hp zero emissions truck. The over-priced Nikola and over-hyped shares plunged after the fraud was exposed and investors lost their shirt.[1415] As a cost cutting measure, diesel-electric and electric trolley trucks have worked in mines for decades by generating electricity on the downhill run and using diesel and/or electricity on the fully-laden uphill run.

Australia has 913,000 km of public roads, of which 319,000 km are sealed.

These have been funded by a tax on fossil fuels which electric cars don't pay. Let's make the system fair. EVs will need to pay tolls to correct this imbalance and pay taxes for the grease, plastics made from fossil fuels, energy loss for long-distance high voltage power lines and, unless their food and consumables are delivered by electric trucks, will need to pay a surcharge for essentials produced and delivered using fossil fuels.

Let the elite green activists in the cities have an environment with no coal- or gas-generated electricity. Their wind turbines and solar panels can be peppered all over their city and they can drive their electric cars around so they can advertise their virtuousness.

Let those workers in rural and regional Australia have their farms, mines, smelters and drive around in fossil fuel-powered vehicles.

Politicians babble on about EVs. This is to keep city green activists happy who would never vote for them anyway. Our food production and transport to cities rely on diesel. We have only a few weeks of diesel and the supply of diesel to Australia could be stopped overnight by a hostile power.

The car market is singing from a very different hymn sheet from the green activist politicians and their lackeys. Politicians beware. Elections come with great regularity.

The plan was that virtuous owners of EVs return from a hard day locked away from reality in an office, plug in the car in a special power point in the garage and settle down for the evening. However, lithium battery bombs explode into toxic fireballs that burn for days. Parked and mobile cars can catch alight, houses burn down and there is a catastrophe waiting to happen.[1416] Because of insurance, some towns have now banned the parking of electric vehicles in underground car parks.

Over two years, 16 Kona EVs caught fire in Korea, Canada and Europe. As a result, Hyundai is expanding a recall to cover at least 74,000 top-selling EVs. Ford, BMW and Hyundai have recalled vehicles because of fire risk. The Chevy Volt EV had a habit of catching alight[1417] and General Motors recalled 69,000 older Bolt EVs and has recalled 73,000 2019-2022 models world-wide because they pose a fire risk, especially when fully charged. Total recall costs are now $1.8 billion.[1418]

GM is telling Chevvy Bolt EV owners not too charge vehicles overnight and not to park inside a garage.[1419] The National Highway Traffic Safety Administration has issued a safety alert recommending that vehicles should be parked outside and away from homes.[1420]   EV fires are intense, burn for a long time and are difficult to extinguish. President Biden needs EVs to reach his goal of cutting emissions in half by 2030.[1421]

Various types[1422] of lithium batteries in phones[1423], drones[1424] and piloted aeroplanes[1425] also catch alight to create a runaway fire with emissions of highly toxic fumes.[1426]

Our obsession with cutting carbon dioxide emissions has had terrible consequences. In the UK, 11 million people own diesel cars. A decade ago, the government advised that diesel cars would help the country lower its carbon dioxide emissions.

The push to diesel cars was driven hard by Professor Sir David King, Prime Minister Blair's personal scientific advisor.[1427] King was quoted as claiming that carbon dioxide was worse than terror. King has now stated that he was wrong. People purchased diesel cars.

Now the UK government is telling diesel car owners that their cars add particulates and nitrous oxides and they will be penalised.[1428] Car manufacturers are making a fortune out of the great green swindle.

The Chinese dominance of supplies for EVs makes supply chains vulnerable, especially as some of the critical minerals are tied up or stockpiled by China.[1429]

Green activists claim that EVs are lovingly recharged by carbon-free sunshine and breezes. Do the virtue-sodden allegedly ethical drivers of EVs ask where the cobalt, graphite and lithium in the battery come from? Or how it is mined? Have they looked at the life cycle of the EV construction materials, the energy used to produce these materials and how such materials are dumped or recycled?

Every petrol or diesel car need eight tonnes of mined material for its manufacture. An electric car needs eight tonnes for manufacture and a further eight tonnes for battery manufacture. A huge amount of carbon dioxide is released in the exploration, mining, processing and transport of mineral products.

If we drive electric cars to achieve Net Zero, what does Net Zero really mean? It means that commodity prices will go through the roof and EVs will become even more expensive. Does it mean that we emit carbon dioxide in Third World countries where mining is increasingly taking place? This has no effect on global carbon dioxide emissions.

A single electric car contains more cobalt than 1,000 smart phone batteries, the blades on a wind turbine have more plastic than five million smart phones and a solar array that can power one data centre uses more glass than 50 million smart phones.

Fossil fuels are needed to produce the concrete, steel and plastics to purify material to build green machines. This requires the equivalent of 100 barrels of oil to make a battery that can store the energy equivalent of one barrel of oil. A single electric car battery requires 200 tonnes of mined material. Averaged over a battery life, an EV consumes 4kg of mined material per kilometre driven whereas an internal combustion engine consumes 0.3 kg of liquids per kilometre.

EVs are a God-sent answer to the prayers of mining companies. According to the International Energy Agency[1430], the average petrol or diesel conventional car requires 2.3 kg copper and 11.2 kg manganese. The average EV requires 66.3 kg graphite, 53.2 kg of copper, 39.9 kg nickel, 24.9 kg manganese, 13.3 kg cobalt, 8.9 kg lithium and 0.5 kg of rare earth elements.[1431]

If the world is to change to EVs before 2040, The International Energy Agency calculated an increase in production of lithium, graphite, nickel and rare earth elements would rise by 4,200%, 2,500%, 1,900% and 700% respectively compared to 2020 production of these metals.[1432]

Such mineral deposits have not been found, we do not have enough geologists worldwide to undertake the exploration and mine geology and there are even fewer mining engineers and metallurgists worldwide to extract and treat the ores.

The world's lithium is produced from hard rock ores that contain about 1% lithium. There is considerable waste produced and diesel used in the mining and transport of the lithium mineral spodumene that contains 8% of lithium. To release it from this lithium aluminium silicate requires huge amounts of energy to break the strong atomic bonds.

The biggest importers of lithium are China (80% of world supply), Korea and Japan. China converts spodumene and lithium salts into lithium carbonate which is the precursor for making lithium batteries. Most of the world's lithium batteries are made in China because they control the world's lithium market. I'm sure that EV drivers are comforted that their lithium battery comes from China.

The top ten world graphite reserves are held by Turkey (90 Mt), China (73 Mt), Brazil (70 Mt), Madagascar (26 Mt), Mozambique (25 Mt), Tanzania

(17 Mt), India (8 Mt), Uzbekistan (7.6 Mt), Mexico (3.1 Mt), North Korea (2 Mt) and Norway (0.6 Mt).[1433] Only a small proportion of graphite in a deposit has the crystallinity to be used in a battery and synthetic graphite is horrendously expensive. Graphite is a strategic commodity and major Western countries such as the US, Japan and Germany have no domestic deposits.

Copper deposits occur worldwide with the biggest in the Andes, there are numerous deposits that today are uneconomic and with price rise, demand increase, capital and time, these could be producing the copper needed for EVs within 20 years. There is currently only one copper mine in the world that produces more than one million tonnes of copper *per annum*.

There are 200 copper mines that will close before 2035. Politicians and activists in California and elsewhere demand that a solution to "manmade climate crisis" is "clean, ethical, climate-friendly, sustainable" EVs but they don't say how they will do it.

Primary nickel deposits occur as small rich sulphide deposits at depth. Most of the world's nickel production comes from ancient and modern tropical soils developed on rocks below that contain traces of unextractable nickel. Mining requires the stripping of all vegetation and soils to reach the deep layer that contains nickel and traces of by-product cobalt. Many nickel mines in developing and Third World countries have left an environmental disaster with clear felled jungle, erosion, dumping of slurries and lack of land rehabilitation.

If green activists want to proudly parade in their EVs, they are directly responsible for desecration of the environment in tropical Third World countries. A quick search on the internet would show the EV owners where nickel comes from and how it is produced.

The critical component for an EV is cobalt. Most of the world's cobalt derives from the Congo with small amounts as a by-product from nickel mining elsewhere. Cobalt mining in the Congo is unregulated, children work in dangerous steep open pits and underground mines. Jungles are clear felled and waste is dumped where it is dumped. There is no technical supervision of pit stability, water ingress and underground safety and the mine fatality rate of children is scandalous. These children are slaves. If they were not slave miners, they would not eat. There is absolutely no chance that these children could get an education.

In an Amnesty International-Afrewatch report, the use of child slave labour for the metals for smart phones and electric car batteries is exposed.[1434] Traders buy cobalt produced by enslaved children and sell it to Congo Dongfang Mining, a wholly-owned subsidiary of the Chinese minerals giant Zhejiang Huayou Cobalt Ltd.

Mark Dummett of Amnesty International stated "*The glamourous shop displays and marketing of state of the art technologies are a stark contrast to the children carrying bags of rocks, and miners in narrow manmade tunnels risking permanent lung damage*". Will these slave labour children even live to see a wonderous world in 2060 of Net Zero?

If the car fleet in the UK only is replaced with electric cars, twice the annual cobalt production will be needed, three quarters of the annual lithium carbonate production will be needed, the entire annual world production of neodymium will be needed and half the world's annual production of copper will be used. There is not the resources and infrastructure required to deliver the electric cars and electric recharging required by 2050. Then there is the rest of the world to consider.

Buyers of EVs are knowingly supporting child slavery in Africa, massive environmental degradation in tropical Third World countries and despotic totalitarian regimes like China.

There is no excuse.

The energy policies of California are a mess because they have crashed and burned down the green path. They are committed to reducing carbon dioxide emissions, closed their main nuclear power generator, invested heavily in wind and solar energy which produce a little bit of electricity for a small amount of times and survives by importing electricity from other states.

Now they want EVs. Fewer than 2% of all cars in the US are EVs. It is a little higher in California which has enacted legislation to ban petrol-powered vehicles as from 2035.[1435] California's grid at present is so hopelessly inadequate that it can't charge the tiny number of EVs that try to stay mobile. How will they cope in 2035 without trillions spent on power and grid systems?

The CEO of the world's fifth biggest car maker said EVs would be far too expensive and would fail to reduce carbon emissions because the vehicles are far heavier than conventional cars.[1436] He said "*I can't imagine a democratic society where there is no freedom of mobility because it's only for wealthy people and all others will use public transport*".

Battery bonfires

Why do you think that airlines insist that lithium batteries not be carried in hold luggage? Lithium battery fires have brought down light aircraft.

Lithium is explosively reactive in air and water, fires can't be smothered with water or sand because it floats, melts at a low temperature (180°C) and burns at a high temperature (2,000°C).[1437]

Composite metals and aluminium are destroyed in a lithium fire and the fumes from burning lithium batteries are highly toxic and create brain damage. The best thing that can be done with a lithium battery fire is keep your distance, watch it burn and stay away for a few days until the area is safe. Lithium batteries are just waiting to explode, release toxic gas emissions[1438] and kill. Hydrofluoric acid is released during a fire. This acid is lethal and so potent that it dissolves metals, glass and rocks.

Methods of fighting lithium mega-battery fires have not been devised.[1439] In suburban Phoenix (Arizona) in April 2019, shorting of a battery led to flameless overheating hence fire suppression systems were not activated.[1440,1441] The system had a nameplate capacity of 2 MWh, there were 27 racks of batteries and 10,584 cells in a container. When the container door was opened, the battery exploded. Nine firefighters were injured, three seriously and one critically. More than 20 fires have been reported from South Korea as well as many others in Europe and Australia.[1442]

Two firefighters were killed on 16th April 2021 when a 25 MWh connected to a rooftop solar panel installation exploded into flames.[1443] Lithium batteries are especially dangerous in waste and recycling centres where there is other flammable material.[1444]

The fire in July 2021 of a 13-tonne Tesla Mega-battery in Victoria burned for three times longer than it had operated.[1445,1446,1447] Fire fighters were unable to control the blaze for days. In the toxic horror show, hundreds of fire fighters and many surrounding buildings were covered with a toxic metal plume.

Residents were advised to get themselves and their pets indoors, close windows and turn off heating so they didn't breathe in smoke. Nearby crops became enriched in heavy metals.[1448]

For a cool $160 million, South Australia's 100 MW Tesla battery requires about five million cells. Lithium cells are prone to spontaneous ignition from flaws in manufacture. If any of these cells has a meltdown its whole module will burn up and if adjacent modules are not sufficiently isolated they too will be ignited. Who would insure a firebomb? A further risk is also likely to exist in the form of short circuits igniting other modules if one fails. Lightning strikes on the battery farm itself or on nearby power lines are also a high risk. Furthermore, when there is a blackout, these batteries have enough charge to provide South Australia with five minutes of power and every time the battery is charged or discharged, its life is shortened.

Al Gore stated[1449] "*I have a lot of admiration for South Australia because it's now leading the world – the largest battery ever, and it will be the first of many*". That's the kiss of death.

As Matt Ridley wrote[1450], you would need only 160 million Tesla Powerwalls to cover one day's electricity consumption in just the UK or 3.3 billion for one week's consumption if all the UK's heating and transport was electrified.

The Minerals Council of Australia estimates the added cost of battery storage could push the cost of wind power from $92 per MWh to between $304 and $727 per MWh.[1451] Batteries store electricity yet some left wing activist groups seem to think that batteries generate electricity.[1452] We already have far better storage of energy than short-lived batteries. It is called coal, gas, oil and uranium.

The fast money will back subsidised solar, wind and batteries. This is at the expense of consumers. Not one country in the world relies on batteries for backup.

Hindenburg Han hydrogen hub

Renewable energy and hydrogen can be described in terms of the laws of thermodynamics. The First Law can be summarised as "You can't get something for nothing" and the Second Law as "You can't even break even". That's certainly the case with hydrogen.

You know when you are being conned when the normal suspects come out the woodwork and try to tell you about the greatest thing since sliced bread. In June 2020, we had an IT employee, a superannuation fund executive and a former Macquarie Bank executive telling Australians how we should make steel by using hydrogen and how great it would be if we installed cables from Australia to Singapore to transport electricity made from unreliable renewables. The word subsidy was never mentioned.

It took thousands of years for efficient steel making processes to evolve by trial and error, scientific research and engineering and what these publicity-seekers know about making steel could be written on a beetle's bum (and with space to spare). The press gave these suggestions an uncritical free ride because journalists are scientifically illiterate.

We've come a long way since hydrogen was discovered as a component of water in the 18th Century by Henry Cavendish. We now know hydrogen is the most abundant element in the universe. The idea of hydrogen fuel cells was developed in the early 1800s, the idea of a hydrogen economy was first raised 100 years ago and it has not eventuated for the obvious reasons. Technological research continues to create new hydrogen fuel cells. These fuel cells need cobalt. This is the problem that all geologists recognise. Where will the cobalt come from?

Before considering whether a future can be built on "green" hydrogen, we need to look at energy losses and gains. That is basic thermodynamics. Producing hydrogen from "clean green" chaotic intermittent wind and solar energy is a myth peddled by the renewable energy carpetbaggers and other rent-seeking crony capitalists.

Hydrogen has many industrial uses and is made by reacting natural gas with steam at high temperature ("blue hydrogen"). This process requires large amounts of energy and also produces poisonous carbon monoxide and carbon dioxide. Making "blue hydrogen" created 20% more carbon dioxide emissions than just burning fossil fuels.[1453]

Hydrogen can be made by simply passing an electric current (direct current) through water (i.e. electrolysis). If the electricity is supposedly from renewable sources, the product is called "green hydrogen". Nine litres of water are required to make one kilogram of hydrogen. Do we really have that much spare water in Australia? There are many experiments attempting to make hydrogen from water by different techniques.[1454] However, there is one important fundamental. Energy is needed to break the very strong hydrogen-oxygen bonding in water to produce hydrogen.

To make hydrogen requires massive capital investments, energy and increases risks of a catastrophic explosion. It is just not economic to use energy to make hydrogen to be burned to give out energy.[1455] Desperate policy makers are trying to reach Net Zero targets such as the premature adoption of hydrogen in order to save their agenda. Hydrogen is unaffordable and not feasible. If unsubsidised, the market will decide when the technology is feasible, not governments.

To convert electricity into a gas and then back to electricity is bizarre and wasteful. A hydrogen economy would be based on two electrolytic processes both associated with heavy energy losses: electrolysis and fuel cells. Furthermore, between the conversion of electricity into hydrogen by electrolysis and the re- conversion of hydrogen to electricity by fuel cells, the energy carrier gas has to be packaged and transported by compression or liquefaction.

Energy can neither be created nor destroyed. It can only be converted from one form to another. Some high-grade energy is lost with each conversion step. If hydrogen is to be manufactured with excess renewable energy, between the wind turbine generating electricity and use of hydrogen in your home, there will be highly significant energy losses. That's basic physics and chemistry

It has to be distributed by surface vehicles or pipelines, stored and transferred. No matter how hydrogen is ultimately used, in stationary, mobile or portable applications, the efficiency of the hydrogen chain between power plant and fuel cell output is hardly better than 30%.

Green hydrogen is four times as expensive as natural gas for heating and twice the price of petrol for driving your car.[1456] It's absurdly unaffordable. If wind and solar electricity generators are used to make hydrogen, then the carbon dioxide emissions for the manufacture, construction and maintenance of these generators are far higher than if coal were burned.

I guess when green activists are burning other people's money, costs don't enter the equation. Hydrogen made using solar energy costs $9.17 per MJ and hydrogen made using coal-fired electricity costs $5.42 per MJ whereas petrol costs $2.11 per MJ, diesel $2.36 per MJ and coal $0.34 per MJ.

School child science teaches us that hydrogen burns with a colourless flame hence can be dangerous. Chemicals are added to gas to make the flame coloured and ethyl mercaptan is added to gas such that a leak can be smelled.[1457] It has a horrible smell and is not very good for you.

Meaningless mumbo-jumbo like Net Zero, carbon neutrality and low-carbon hydrogen are used to befuddle the scientifically illiterate community and disguise the fact that the use of hydrogen is just another scheme to skin the consumer alive.

We were going to have electricity generated from hydrogen fusion that was too cheap to meter. Hydrogen cars (e.g. Toyota Mirai) and hydrogen-fuelled buses have been scrapped because hype and reality never meet.[1458] After 30 years of research and development on hydrogen fuel-cell cars, Daimler Mercedes-Benz has abandoned the program because hydrogen cars would be far too expensive. If Daimler Mercedes-Benz, who have been at the forefront of vehicle innovation for more than 120 years, can't succeed then it is doubtful that affordable hydrogen cars will become a reality any time soon. The market and not green ideology speaks.

A new hydrogen-powered fuel cell pickup the H2X Warrego with a fuel range of 500 km will be released in 2022 for a modest $200,000.[1459] Owners will initially need their own refuelling facilities until hydrogen vehicles are more prevalent and more refuelling stations are built. What happens if only a few Warrego vehicles are sold? Customers will be left high and dry and certainly could not travel long distances in a large state.

The hydrogen will be produced from solar and wind energy by electrolysis of water. The Federal Government has flagged hydrogen as a priority low-emission technology under its $18 billion Technology Investment Roadmap to lower Australia's carbon footprint. We only need to do this if it has been proven that human emissions of carbon dioxide drive global warming. The taxpayer is being ripped off. The Queensland government is adding five hydrogen-powered Hyundai NEXOs to its fleet of zero emission vehicles. The horrendous cost of yet another PR stunt is borne by the taxpayer.

Unlike iron, copper, coal and oil, there are no hydrogen deposits on Earth that can be mined. Hydrogen is the most common gas in the Solar System. It has been leaking out of our planet since the beginning of time. Pictures of Earth from space show a big cloud of hydrogen leaving our planet for space. If we are to use hydrogen as a fuel, we need to make it from liquids and solids that contain hydrogen which is held tightly in compounds.

The thought of piping pressurised hydrogen into houses, schools, hospitals, kindergartens, offices and factories or parking a car with a fuel tank full of hydrogen is insanity. The risk of a gigantic explosion is high because hydrogen is incredibly reactive. An exciting new future based on hydrogen will certainly be exciting. Hydrogen is highly explosive, leaks through solid metals which become brittle and is a gas not to be toyed with.

Hydrogen gas only occurs very deep on our planet and so to create hydrogen we must expend massive amounts of energy to free hydrogen from oxygen in water or fossil fuels.[1460] To make hydrogen has an energy loss of at least 30%.

Hydrogen is not a source of energy. Converting electricity to hydrogen and then burning hydrogen to make electricity is grossly inefficient and costly. Hydrogen is an expensive method of moving stored energy from one place to another.

Hydrogen is the smallest of all atoms and diffuses through solid steel tanks and escapes. Fuel tanks on cars need to be four times as large as current fuel tanks and liquid hydrogen presents a whole host of new problems. About 2 % of liquid hydrogen evaporates each day.

A massive amount of energy is consumed for compression to 700 times atmospheric pressure and storage below minus 252.87°C as liquid hydrogen. The coldest temperature possible in the Universe is minus 273.15°C so we have to expend huge amounts of energy to keep hydrogen 20°C above absolute zero.

The energy used to compress and keep liquid hydrogen cool could power a town, infrastructure would have to be rebuilt from scratch and the safety problems of having numerous public hydrogen filling stations are too frightening to contemplate.

All aeroplanes would have to be redesigned if hydrogen were to be used as fuel. They would be heavier and far more dangerous.

Fashionable fads are for fools who are incapable of undertaking a simple due diligence on technology that is not new but has been revisited time and time again as people struggle with the inefficiency of wind and solar electricity.

If "green" hydrogen is the latest cure for the inherent intermittency of subsidised wind and solar power, then this is just hot air. Why have a ludicrously expensive inefficient subsidised solution to an earlier expensive subsidised experiment with electricity generation totally dependent upon the weather?[1461] If an experiment fails, then it should be abandoned.

Even the left paper *The Guardian* agrees that, unlike fossil fuels, renewables cannot be relied upon to produce affordable hydrogen for the foreseeable future. They finally realise that using hydrogen for cars and home heating locks the UK into fossil fuel dependency.[1462]

I wonder whether it dawned on readers why hydrogen, known as a fuel for more than a century, has never been used in this innovative competitive world? Might it be due to that dirty little word money? Green schemes sound great until they are costed.

The perfect way to waste mountains of energy and money is to turn wind and solar power into hydrogen gas and then burn the gas. There is a 45% efficiency in converting electricity into hydrogen. Production of hydrogen is horrendously expensive.[1463] Renewable energy rent-seekers have seized on the concept of converting useless and unpredictable wind and solar electricity into something that can be used by consumers as and when it is required. Various politicians suffering from attention deficit disorder are touting a hydrogen-fuelled future.

The Australian Federal Government has expedited approval for a $53 billion project and the devastation of 78 square kilometres of desert wilderness to produce green hydrogen for a market that does not exist.[1464] The elephant in the room is the energy needed to liquefy the hydrogen and the difficulty in transporting liquid hydrogen large distances from the desert to the users. This is peak stupidity.

A proposal by the Australian Labor Party (ALP) for South Australia (SA) is to build a 204 MW hydrogen power station using "excess renewable energy". This is an attempt to seduce the electorate but does not mention that SA is a net importer of energy and does not have "excess renewable energy".

A 204 MW station would use 15.6 tonnes of hydrogen per hour and the energy for electrolysis to produce that hydrogen is 1,128 MWh giving an energy loss of 924 MWh. Using a $6 per kg cost of hydrogen, the cost of hydrogen is $460 per MWh.[1465] With hydrogen gas leakage and diffusion losses, some 300 tonnes of hydrogen will need to be delivered by road in tankers each day to power this 240 MW power station. Because of the damage that a hydrogen tanker could create if exploded, these tankers are a far more attractive terrorist target than those carrying chlorine, ANFO, acid or gasoline.

Germany wants a hydrogen economy but no one knows where the hydrogen will come from because there is no excess electricity for hydrogen production by electrolysis which is an energy intensive, uneconomic process. Germany already uses a large volume of industrial hydrogen sourced from natural gas.[1466]

California plans for 'zero carbon' electricity and by 2045 plan to have 40 GW of electricity produced from hydrogen cells. Currently there are no hydrogen fuel cells connected to the grid in California.[1467]

Renewables-obsessed governments around the world are now talking up hydrogen gas as if they've just discovered a perpetual motion machine. If you want to waste a large fortune, try turning wind and solar power into hydrogen gas.

Hydrogen fuel is the latest fad for fools and frauds. We are now told that hydrogen will cost $2/kg. There is no technology that can make hydrogen anywhere as cheaply as $2/kg and I fear rent-seekers will be looking to extract massive subsidies again.

In terms of the energy costs per gigajoule, this equates as $14.08 per GJ for hydrogen compared to natural gas ($8-10/GJ) and brown coal from the Latrobe Valley ($0.11/GJ). Why is expensive energy which will need to be subsidised even being considered? We are being conned. Again.

The energy density in terms of volume[1468] is diesel > petrol > jet fuel > liquid methane > ethanol > methanol > high pressure methane > liquid hydrogen > high pressure hydrogen. I choose to be really green and will stick to the winning high energy per volume liquids and keep driving diesel and petrol cars, thank you very much.

Another hare-brained scheme to save the world is "green" steel. Only green activists could promote such a scheme. Conversion of iron oxide ore requires the removal of oxygen from iron ore as carbon dioxide to yield carbon-bearing pig iron. In the second stage of steel making, oxygen is blasted through molten pig iron to remove small amounts of excess carbon in the iron as carbon dioxide. The resulting steel is a mixture of iron with traces of carbon.[1469]

Coking coal needs to be physically strong enough to support a column of reactants. It burns to provide gas to burn for melting the charge in a blast furnace, provides carbon to remove oxygen from the ore and provides waste components for the molten slag that floats on the molten iron. Making steel is a tried and proven technology resulting from more than 2,000 years of experiments and research. Using hydrogen has been tried in the past and failed due to costs.

Hydrogen can remove oxygen from the iron ores as steam but there would have to be substantial design changes such that the charge gravitates to produce a low viscosity slag and carbon would need to be added to make steel. Smelters would have to be recapitalised and rebuilt.

"Green" steel using hydrogen is six times more expensive than steel made using coal. And if the greens looked at how hydrogen is made, there would be further problems. Manufacturing costs, storage and transport make hydrogen an expensive and dangerous fuel. As per usual, green schemes to save the world do not look at costs, technological difficulties and the history of the evolution of technology.

Words like hydrogen hub or hydrogen economy flow off the tongue easily. A cost analysis shows that the "all in" cost of hydrogen produced by electrolysis is £137 per MWh whereas the cost for gas as a fuel is £13.60 per MWh.[1470] End of story. Why even bother thinking about hydrogen?

Attacks on farmers

Australian has a landmass of 7,692,024 square kilometres with a sparse inland population and greenhouse gas-emitting livestock. Combined with the transport of livestock, food and mined products long distances to cities and ports and the export of ores, coal, metals and food for 80 million people results in high *per capita* carbon dioxide emissions. Australia's exports of food, coal, iron ore and gas contribute to increasing the standard of living, longevity and health of billions of people in Asia.

The forestry, mining and smelting industries have been under constant attack by green activists who are happy to put thousands out of work and destroy the economy. They have now trained their sights on the cheapest and most reliable form of electricity and want it replaced by wind and solar which are underpinned by mining and fossil fuel power generation. The next target is food-producing farmers. They, like the forestry, mining and farming industries have nowhere to go if destroyed by green activists.

Australia has far greater economic priorities than to change a whole economy, increase energy costs, destroy successful efficient farming and decrease employment and decrease international competitiveness because of one poor lonely molecule of plant food emitted by Australia in 6.6 million other atmospheric molecules.

It is a very long bow to argue that this one molecule of plant food in 6.6 million other atmospheric molecules has any measurable effect whatsoever on global climate. Furthermore, it has yet to be shown that human emissions of carbon dioxide drive global warming. Why even bother with carbon theft schemes?

The green activist attacks under the guise of climate change are a method of controlling every minute aspects of our lives, producing unemployment and food dependency, impoverishing those in the developing world that import Australia's coal and energy and gaining total control of every activity in Australia.

Annual Australian *per capita* carbon dioxide emissions are in the order of 20 tonnes per person. There are 30 hectares of forest and 74 hectares of grassland for every Australian and each hectare annually sequesters about one tonne of carbon dioxide by photosynthesis. On the continental Australian landmass, Australians are removing by natural sequestration more than three times the amount of carbon dioxide they emit.[1471]

Crops remove even more carbon dioxide from the atmosphere. Australia's net contribution to atmospheric carbon dioxide is negative and this is confirmed by the net carbon dioxide flux estimates from the IBUKI satellite carbon dioxide data set.[1472]

Australia's continental shelf is 2,500,000 square kilometres in area. Carbon dioxide dissolves in ocean water and the cooler the water, the more carbon dioxide is dissolved in it. Living organisms extract dissolved carbon dioxide[1473] and calcium is constantly added to the oceans from rivers.

Australia has a huge coastline with currents bringing warm water from the north to the cooler southern waters (East Australian Current, Leeuwin Current) and through the Great Australian Bight (Zeehan Current). Corals and shells are built from carbon dioxide and calcium in seawater. This natural marine sequestration locks away even more Australian emissions of carbon dioxide and adds further to the negative contribution of atmospheric carbon dioxide made by Australia.

Using the thinking of the IPCC, UN and activist green groups, Australia should be very generously financially rewarded with money from populous, desert and landlocked countries for removing from the atmosphere its own emitted carbon dioxide and the carbon dioxide emissions from many other nations. By this method, wealthy Australia can take money from poor countries.

Australia has 440 million hectares of grasslands, 147 million hectares of native forests, 1.82 million hectares of plantations and 4% of the world's global forest estate. Australia has the world's sixth largest forest area and the fourth-largest area of forest in nature conservation reserves.[1474]

If trees absorb about 6.4 tonnes of carbon dioxide per year, then Australia forests adsorb 940 Mt of carbon dioxide per year compared to Australia's domestic and industrial emissions of 417 Mt *per annum*.[1475] Add to that the absorption of carbon dioxide into the huge area of grasslands, crops and continental shelf waters and Australia does more than its share of the heavy lifting of sequestration of carbon dioxide.

None of these calculations involve the fixing of biological carbon compounds and atmospheric carbon dioxide into soils. Soils contain two or three times as much carbon dioxide as the atmosphere, soil carbon increases fertility and water retention and reduces farming costs. New ways of more efficient and more productive farming are constantly being assessed. Natural sequestration in Australia locks away carbon dioxide and to lock it away carbon dioxide by industrial sequestration in deep drill holes is a fashionable way of wasting large amounts of taxpayer's money.

Green activists are keen to have more and more trees planted. They should be aware that trees use carbon dioxide as plant food and some of us see great benefits from having an atmosphere with increased carbon dioxide. Green activists just ignore the process of photosynthesis which flows through to grass growth and livestock farming.

All vegetation uses solar energy, carbon dioxide, rain and soils to grow. The EU and UK claim that harvesting trees for wood-fired power stations is a net zero process because burning wood puts carbon dioxide back into the atmosphere.

So does burning coal or grazing cattle. Grass uses solar energy, carbon dioxide, rain and soil nutrients. Grazing cattle convert grass into energy, protein, fat, and milk for human consumption.

Some of the carbon compounds consumed by cattle are quickly returned to the atmosphere by cattle farts and burps. The milk and meat consumed by humans is sequestered into human bodies and is eventually returned to the atmosphere by exhalation, defaecation and finally body burial or cremation. This is a net zero process that is quicker than the net zero process from tree harvesting.

Australia has about 26 million cattle for meat, milk and cream. China has 97 million cattle and is increasing herd size commensurate with becoming wealthier. Brazil has 230 million cattle and South America's beef consumption exceeds that of the USA and Australia in total and *per capita*. Meat consumption increases with increasing wealth.

Without cattle, grasslands would be fuel for monstrous bushfires, fuel for termites or would rot and put methane into the atmosphere. Our teeth and gut flora have evolved for eating and digesting meat. The digestion process is more efficient and healthier if the meat is cooked but not incinerated.

Real red meat is green, it sequesters carbon dioxide. The consumption of meat, milk and cheese has nothing to do with climate. It is a ploy by unbalanced vegan terrorists climbing onto the climate bandwagon trying to force consumers to live their misguided way of life. In our democracy, we tolerate them with amusement. In their world, we are on their hit list.

Some green activists are becoming hysterical about meat eating as part of their attack on farmers. Synthetic or cultured "meat" can be made in a laboratory by growing muscle cells in a nutrient serum.[1476] This makes a difficult decision for GM food haters, vegans and vegetarians because synthetic meat involves no animal killing. Another process removes starch from wheat to give a protein-rich material that tastes like chicken[1477] and yet is called "meat". The elephant in the room is the amount of energy and land required to make synthetic "meat" and the misleading and deceptive use of the word meat.

Some researchers, in the name of saving the planet from climate change, are advocating that we should eat kelp, maggots and algae instead of wheat, maize and rice. Other future foods they claim will be insect larvae, fungi protein and sugar kelp, currently used in sushi.[1478] Until they integrate these brilliant ideas with an understanding of how the digestive systems works, I'll stick to my tried and proven diet.

Methane is released from wetlands, termites, oceans, methane hydrates and domestic and wild animals. Inland Australia is the termite centre of the world and we take a bit global hit for the amount of methane released from gut bacteria in termites. The geographical distribution of domestic animal density shows no relationship to the global distribution of methane concentrations. Greens want to reduce methane losses from wetlands. This means that they want to starve every human dependent upon rice paddies to provide a staple diet.

The increase of methane in the atmosphere stabilised in the 1990s yet the world cattle population increased by more than 100 million between 1995 and 2005. Why did it stabilise? Because the former Soviet Union gas pipelines leaked like a sieve and, rather than having methane leak into the air, Russia sealed the pipelines to make more money from gas sold to Europe.

Climate change is not a scientific or environmental issue. It is a political movement. The green activist movement is the worst thing that has happened to the environment in Australia over the last 50 years and has driven total mismanagement of water, drought, vegetation and bushfires.

More than 90% of Australian farms are family-owned hence almost all the food we purchase provides income for rural and regional families who often do it tough. Farming families are the best environmentalists in the country because they are custodians of their own land that they normally pass on to the next generation. Australian farmers feed 80 million people outside Australia. What is the green activist agenda? It certainly looks like they want to send Australian farmers broke and deny food to their fellow humans.

If emission of carbon dioxide is a worry to activists, then Australia does more than its fair share to reduce carbon dioxide from the atmosphere.

Furthermore, the main population centres of Australia are in the south-east and the forests and grasslands in south-east Australia fix about 30% of all of Australia's carbon dioxide emissions. As per normal, it is rural Australia that carries the population centres of Australia.

Media networks are only too happy to broadcast that Australia is the worst polluter in the world but don't mention that carbon dioxide is not a pollutant despite some scientifically misguided activist US Environmental Protection Agency lawyer makes laws to call carbon dioxide a pollutant. Australia is therefore adsorbing and fixing carbon dioxide "pollution" from elsewhere on the planet.

However, *"worst polluters"* is a great headline whereas the facts are boring. For those that espouse that human emissions of carbon dioxide is the major cause of climate change, then Australia should be viewed as the hero and not the villain.

Socialist green activists could easily be put back in their boxes if those dreadful diesel burning trucks refused to deliver farmer's food to the cities for a month. Do it. This is the only way to save family farms.

Facts tend to pour cold water on hysteria.

Child abuse

Hitler Youth, fascists, communists and socialists have all used children as a front because it is far harder to criticise an innocent child, especially one with a mild form of autism. Poor little Greta Thunberg, she has had her childhood and education stolen and has been turned into a ball of fear.

A glitch in Facebook's software on 9th January 2020 showed the edit history of Greta Thunberg's posts.[1479] Her activist father and actor Svante Thunberg and Indian climate activist Adarsh Prathap, a delegate to the UN on climate change, posted on Facebook in her name. Mr Claptrap also serves as a delegate to the UN on climate change.

She has claimed that she organised a one-girl school strike at the Swedish parliament on 20th August 2018. It looks like Greta's actions were not the sole actions of a child and were cooked up by climate activist, PR guru and founder of the green activist social network start-up *We Don't Have Time* Ingmar Rentzhog who wanted a fresh new face to exploit for political activism.[1480] Without a prepared script and handlers, Greta is incapable of answering basic questions as various video clips show yet she wallows in being the centre of attention. She is the sad pawn of child abusers.

Greta was diagnosed as a child with obsessive-compulsive disorder and Asperger's Syndrome, just like her younger sister Beata. She should have been cared for rather than been flitted here, there and everywhere under a blaze of disorienting publicity. Her mother, opera singer Malena Ernman, writes that Greta[1481] "... *can see carbon dioxide with the naked eye. She sees how it flows out of chimneys and changes the atmosphere in a landfill*".

Carbon dioxide does not reflect or absorb any light within the spectrum that humans can see. It is the odourless, colourless, tasteless gas of life. Her activism is bizarre, people seem to want not to question her abilities and cognitive understanding. If her mother's claim is correct, hundreds of years of chemistry needs to be immediately discarded because it is wrong.

The Greta child Madonna is innocent but her parents are guilty of child abuse using their sensitive daughter as a high-profile cash cow. They should be looking after their child and have her at school rather than use her as a money-making machine to further their green activism. What has the world come to when an innocent abused victim with autism difficulties is treated as a saint with knowledge, insight and a cult leader beyond criticism? We've seen it before with brain-washed children as the front-line storm troopers and even as child soldiers.

The Greta phenomenon has also involved green lobbyists, PR hustlers, eco-academics and a think tank founded by a wealthy former minister in Sweden's Social Democratic government with links to the country's energy companies. These companies are preparing for the biggest bonanza of government contracts in history.[1482] Climate change should be a concern for Swedes. The country where Greta resides has continued to record nearly the coldest temperatures of the last 9,000 years[1483] and in the last glaciation Sweden was covered by 5 km of ice.

More than 60% of emissions since Greta Thunberg was born are attributable to China. She appeared at high profile functions in New York and Davos to chastise the West on their emissions but has never been seen in Beijing or Delhi, the capitals of the biggest emitting countries on the planet. The official mouthpiece of China[1484] stated *"Greta Thunberg, who at age 15 started skipping school on Fridays for her climate protests, is merely 18 years old this year. She is short of sufficient academic knowledge, study and lack of sound self-judgement capability"*.

When she has the stage, she delivers a tirade of emotive short sharp phrases about the effects of climate change, probably written by her actor father. Some of her gems are *"Entire ecosystems are collapsing"*. Ecosystems and species numbers are expanding. There is increased atmospheric carbon dioxide with increased plant growth and crop productivity because carbon dioxide is plant food. Increased plant life feeds more animals, all of which expand ecosystems.

She claimed that *"People are dying"*. The opposite is true. The world's population is at an all-time high, average lifespan continues to increase and the calorie intake continues to rise. The activities of climate change activists in limiting energy consumption is responsible for the deaths of six million people each year. Each year $400 billion is spent on climate change and world hunger could be solved for $30 billion a year.

The claim that *"People are suffering"* is also wrong. If this poor child went to school and read history, she would be aware that the worldwide quality of life has never been better. Over the last 25 years, one billion people have been lifted out of poverty, mainly due to the expansion of cheap reliable coal-based energy.

In 2019, the wealthy healthy spoilt Swedish child with freedoms that previous generations have never enjoyed stated *"You have stolen my dreams and my childhood"*. Does she know that African slave children work in illegal mines to provide the commodities that are needed for her travel, broadcasts and lifestyle?

These children also have dreams. One of these dreams would be the opportunity to go to school. Greta is so special that she can afford to skip school. Does she know that millions of young people are buried in war graves in Europe? They lost their lives before they reached adulthood enabling wealthy young Europeans like manipulated Greta to have the freedoms they enjoy today. Greta's scriptwriters are long on emotion and short on reality.

At Davos in 2019 she said *"I don't want you to be hopeful. I want you to panic. I want you to feel the fear I feel every day. And I want you to act"*. The audience of billionaires, politicians, celebrities and marauding NGOs in attendance all lapped it up, perhaps because adult society loves nothing more than having its own fears and confusions obediently parroted back to it by teenagers. She speaks to the deep self-loathing of the 21st Century elites. The world is not in good shape when those who claim they are leaders applaud the hysteria of a brain-washed child rather than listen to the considered opinion of experts.

She is from a privileged background in one of the most affluent countries in the world. She should have sat on the knee of her grandmother and asked her about the times of previous generations. Life expectancy of women in Sweden is 84 years[1485], the level of PM2.5 is almost half that of OECD countries and 96% of Swedes say they are satisfied with the quality of their life.

She has every reason to be proud of how her forbears created a healthy environment for her. Like the equally privileged young climate depressives of the UK, US and Australia, Thunberg believes she belongs to a generation doomed by the reckless behaviour of her forbears. We live longer, thanks to the affluence that previous generations have created, and the price of living longer in the Western world may be increased cancer, dementia, diabetes and coronary failure.

Humans are very prone to superstitions, panics, delusions, manias and hysteria. We have gone through thousands of them and many are still underway such as non-mainstream whacky religions, alchemy, witchcraft, astrology, phrenology, eugenics, pandemic panics, much traditional medicine and homeopathy. We can all be easily fooled, magicians do it all the time. Children, like Greta Thunberg, need to have a body of knowledge, think critically and analytically, have great joy from learning, promote opinions that they thoroughly understand and be willing to change. The antidote to fear is education and knowledge.

Greta Thunberg rejects all ideas of the enlightenment because she is now living in the best times ever to be a child on planet Earth. Would she prefer to live in the worst of times when there was panic and suffering coupled with neither hope nor education?

The worst years to live over the last 2,000 years were 535-550 AD because of massive volcanic eruptions (perhaps Kamchatka or Alaska in 535-536 AD and Ilopango, El Salvador in 539-540 AD)[1486] filled the Northern Hemisphere atmosphere with dust and acid sulphate clouds.[1487] These volcanic eruptions were coincidental with extraterrestrial impacts in March 536 AD in the Gulf of Carpentaria (Australia)[1488] and elsewhere in August 536 AD.[1489] To make matters worse, these were at the time of a Solar Minimum.

The Sun was dimmed for 18 months, a white sulphuric acid aerosol cloud enveloped Europe[1490], global temperature dropped by 1.5 to 2.5°C producing worldwide crop failures, famine and death by starvation. There was migration, political turmoil and the collapse of empires. Tree rings show almost no growth for a few years.[1491] Icelandic eruptions around 540-541 AD also lowered summer temperatures by 1.4 to 2.7°C in Europe.[1492] The weakened population was hit by the Justinian bubonic plague 541-543 AD killing 35-55% of the people through the Mediterranean and speeding the collapse of the Roman Empire.[1493]

Surges in atmospheric lead recorded as particles in Swiss glacial ice at 640 and 660 AD indicate an economic revival with lead mining and smelting in France to obtain silver co-product for coinage.[1494] At that time, coins changed from gold to silver. Bits of wood stuck onto pumice discovered in the GISP2 Greenland ice core came from numerous volcanic eruptions at Rabaul (Papua New Guinea) around 667 to 699 AD which led to another bleak few years.

Even though the eruption of Tambora (Indonesia) in 1815 gave "The Year Without a Summer" with poor harvests and malnutrition[1495], it was nothing compared to the multiple volcanic eruptions and at least two extraterrestrial impacts in 535-536 AD. The 1815 Tambora eruption demonstrated the capability of humans to adapt and help others who were worse off.

How will the climate moaners survive the next Grand Solar Minimum (2020 to 2053)[1496] if there are coincidental natural events such as large volcanic eruptions and small extraterrestrial impacts? Will some events of 536 AD repeat themselves? There are some highly speculative ideas floating around in the peer-reviewed literature concerning the relationship between epidemics and Grand Solar Minima.[1497]

The year 1349 AD was pretty grim when the Black Death wiped out half of the European population. The Spanish Flu from 1918 to 1920 wiped out more than 50 million people from a population of 1.8 billion and these were mostly young fit adults. To date, the COVID-19 pandemic has killed more than five million of the 7.7 billion global population and is a mere sniffle compared to the Spanish flu.

The world has many remarkable children. For example, Alma Deutscher was born on February 2005 in Basingstoke (England). She composed her first piano sonata at five, a short opera *The Sweeper of Dreams* at seven, a concerto for violin and orchestra at nine and, at ten, a full-length opera *Cinderella*. At twelve, she premiered her first piano concerto at Carnegie Hall. She has written many piano, orchestral, vocal and chamber music pieces.[1498]

Composer, violinist and pianist Alma Deutscher will be remembered for generations. Greta Thunberg will not. Deutscher's world is positive and constructive. That of Thunberg is negative and destructive for her and can only end badly.

I feel so sorry for the abused, manipulated, ignorant Greta. She is a window into the morality of green activists who use even their own children as pawns in their evil green activist political games.

I leave the last word to broadcaster Alan Jones.[1499] He always gets it anyway. *"To all the school kids going on strike for climate change, you're the first generation who have required air conditioning in every class room. You want TV in every room, and your classes are all computerised. You spend all day and night on electronic devices.*

More than ever, you don't walk or ride bikes to school but you arrive in caravans of private cars that choke suburban roads and worsen rush hour traffic. You're the biggest consumers of manufactured goods ever, and update perfectly good expensive luxury items to stay trendy. Your entertainment comes from electronic devices.

Furthermore, the people driving your protests are the same people who insist on artificially inflating the population growth through immigration which increases the need for energy, manufacturing and transport. The more people we have, the more forest and bushland we clear, the more of the environment that's destroyed.

How about this, tell your teachers to switch off the air-con, walk or ride to school, switch off your devices and read a book, make a sandwich instead of buying manufactured fast food.

No, none of this will happen because you're selfish, badly educated, virtue signalling little turds inspired by the adults around you who crave a feeling of having a noble cause while they indulge themselves in Western luxury and unprecedented quality of life. Wake up, grow up and shut up until you're sure of the facts before protesting".

Cooking the books

Land temperature measurement

Something really stinks. On Friday 19th April 2019, there were temperature records set at Albany (Western Australia). Snow fell. The live half hour observations recorded at Albany Airport show that the maximum that day was 10.4°C at 11 am. This was the lowest maximum ever recorded at Albany Airport.[1500]

Raw data disappeared and the Bureau of Meteorology (BoM) later estimated that the temperature on that day in the nearby town of Albany was 25.1°C. This estimate raises the average temperature for April 2019 at Albany Airport by 0.7°C. We are being told 0.7°C is the amount of global warming over the last 70 years yet by fiddling the data, it occurred in just one day at Albany. There was no reason given by the BoM for adjusting the temperature upwards by 15°C on a day that it snowed in the area.

In my field, all exploration and mine data undergoes a quality assurance (QA) and quality control (QC) process. QA/QC is a normal due diligence process when dealing with data in case there are errors, fraud or a fatal flaw. Regulation ensures that any data corrections have to be explicitly stated, explained and transparent. Qualified scientists must sign off on the QA/QC in order to protect investors from false data, leading to poor decision making.

There is no QA/QC process in climate "science" which allows the fraudulent "homogenisation" of data with no personal consequences. The taxpayer has invested billions of dollars into climate "science" yet the keepers of the data are neither accountable nor release data for independent checking.

The BoM states that temperature adjustments are secret.[1501] If they are secret, then the taxpayer should cease funding the BoM to the tune of $1 million a day. The data is owned by the taxpayer. If adjustments are made, then the tabulated data should show measured data, adjusted data, adjustment calculation and the reasons for an adjustment. This would happen if a QA/QC process was operative.

The taxpayer-funded ABC was telling their followers that the taxpayer-funded BoM needs more climate scientists to avoid expensive mistakes.[1502]

The taxpayer-funded Australian Academy of Sciences, lobbied for 77 extra positions in climate research because accurate climate modelling can potentially avoid costly, unnecessary investments.

What planet do these people inhabit? Before there can be accurate climate modelling, the data used for the modelling must be accurate. It is not. The investment in the BoM is already costly and the historical results are tainted with fraud. Maybe the original historical climate data should be preserved and not destroyed rather than using "homogenised" data to continue models that long ago were shown to be wrong.

Whenever climate "science" is investigated, be it Great Barrier Reef studies, measurements of air and sea temperature, sea level and atmospheric carbon dioxide, it can be quickly shown that the data is shoddy, the data manipulation leaves a lot to be desired and would not survive a basic QA/QC test that we use for data in the exploration and mining industries.

The financial implications of a lack of QA/QC for climate "science" are never considered. Because of shoddy, concocted and fraudulent data, decisions that cost the planet trillions of dollars are being made. If my data is wrong, I could go to gaol and my career would be in tatters. If a climate "scientist" gets it wrong, they stay in their job and there are no personal consequences.

In an independent QA/QC of the temperature data set, the normal scientific questions to be asked by a QA/QC team would be:

> When did measurements start?
> How were measurements made?
> Was there an overlap between instruments making historical measurements and modern measurements at the same site?
> Where were the measurements made?
> Who made the measurements?
> Has the measuring station moved?
> Has urbanisation encroached on the measuring station over time?
> What is the order of accuracy of measurements?
> What are the possible errors of measurement?
> Were the measurements independently validated?
> Are temperature measurements in accord with other validated data?
> Were measurements adjusted ("homogenised")?
> How is temperature data processed?
> How can a global trend be calculated when most measuring stations are in northern Hemisphere cities, mainly in the USA?

In addressing these questions, we can only conclude that the global temperature data set is shambolic, unreliable, can't be validated, is not in accord with validated past observations and measurements, covers a very small part of the planet, is biased and has been tampered with to achieve the desired result. The climate "science" coterie admit this themselves. Even the more reliable satellite measurements since 1980 have been adjusted. So don't even bother to read what I have written below. Go and make yourself a good cup of coffee in the knowledge that the whole global warming scam in underpinned by dodgy data.

Data is sacrosanct and the primary data must be preserved for ever. In my field, if I publish scientific papers on the geochemistry of rocks, the samples I used must be deposited in a secure repository and any reader of my scientific publications has access to these samples for checking or further research. This basic principle does not operate in the climate world.

When climate activists fill organisations entrusted with keeping official primary data and change the primary data, we can be very confident that green political activism has replaced science and that we can't trust climate "science". Instead of global warming, we have global fraud.

One of the first examples of these "homogenisations" was exposed in 2007 by the statistician Steve McIntyre, when he discovered a paper published in 1987 by James Hansen, the climate "scientist" who for many years ran the Global Institute for Space Studies (GISS). Hansen's original graph showed temperatures in the Arctic as having been much higher around 1940 than at any time since. But as Homewood reveals in his blog post[1592] "*Temperature adjustments transform Arctic history*". GISS has turned this upside down. Arctic temperatures from that time have been lowered so much that that they are now dwarfed by those of the past 20 years.

By extracting old data from papers of James Hansen and comparing them with data downloaded from NASA's GISS site, a progressive global warming appears. This is certainly man-made warming, created by men making adjustments to the data, blending badly sited urban data with correctly-sited rural data and then, in 2007, removing the urban adjustment for US data sites. The frequency and direction of the National Oceanic and Atmospheric Administration (NOAA) adjustments to US data were increased in 2007 at the same time as inarguable satellite data showing global cooling became public knowledge.

NOAA's "homogenisation process" has been shown to significantly alter the trends in many stations where the location suggests the data is unreliable. In fact, adjustments account for virtually all the warming trend. Unadjusted data for the most reliable sites (i.e. rural sites) show cyclical multi-decadal variations and no long-term warming trend.

Both NOAA and NASA resisted Freedom of Information requests for the release of all the unadjusted data and documentation for all the adjustments made. The US Data Quality Act requires that published data must be able to be replicated by independent audits. That is currently not possible given the resistance posed, despite promises of transparency.

After the raw temperature measurements are collected, warts and all, further adjustments are made. Curiously, each adjustment produces more warming. MIT meteorologist Richard Lindzen commented[1503] *"When data conflicts with models, a small coterie of scientists can be counted on to modify the data to agree with models' projections"*.

The term global warming is based on an increasing trend in global average temperature over time. This is based on measurements. Or is it? The IPCC reported in 2007 in Chapter 3 that *"Global mean surface temperatures have risen by 0.74°C ± 0.18°C when estimated by a linear trend over the last 100 years (1906–2005)"*.

This is total garbage because 100 years ago, thermometers did not have such a high order of accuracy. The order of accuracy of a thermometer then was, at best, ±0.5°C and in many cases ±1°C showing that the global mean surface temperature is within the order of accuracy of measurement hence is meaningless. It is based on poor massaged data that gives the desired result. This happens time and time again.

The August 2021 IPCC AR6 Report has removed the Little Ice Age and Medieval Warming. With one stroke of the pen, history has been deleted. The role of the Sun as a driver of climate was again ignored despite numerous landmark papers showing the Sun drives climate change and not carbon dioxide.

We are very much aware that this is a process of communist countries who want to whitewash the bad bits of history. In previous IPCC reports the Little Ice Age and Medieval Warming were present, they then disappeared with Mann's 2001 "hockey stick", reappeared and now have disappeared in the latest report. It is these reports that journalists use to scare people yet these same journalists do not look at past reports.

The temperature record is not a good record and it is really only the long-term rural stations that provide meaningful raw data. Because of the closure of so many stations, by 2017 there were more measuring stations in the US than in the rest of the world. Since 2000, NASA has further "cleaned" the historical record by making adjustments for area, time of observation, equipment change, station history adjustment, filling missing data and for urban warming adjustment. The rationale for the temperature station adjustments was to make the data more realistic for identifying temperature trends.

Urban areas are warmer than surrounding rural areas. This is especially the case at night. Airports, originally on the outskirts of urban areas, have seen cities grow around them and temperatures rise. A very large number of measurements for global temperature studies are from airports bathed in hot exhaust fumes.

Oke created an equation for urban heat island warming.[1504,1505] A hamlet with a population of 10 has a warming bias of plus 0.73°C, a village with 100 people has a warming bias of plus 1.46°C, a town with a population of 1,000 has a warming bias of plus 2.2°C and a city of a million people has a warming bias of plus 4.4°C. This can only be an approximation as some cities grow upwards and others grow laterally into what were agricultural areas. Heated urban areas increase plant growth.[1506]

These biased measurements account for more than 50% of the measured warming since 1880. NOAA initially denied that it was an issue and then asked the government for $100 million to upgrade and correct the locations of 1,000 climate stations.

The end result of all the adjustments is that additional apparent warming is created and one wonders whether this was accidental or deliberate. For example, the closest rural station to San Francisco (Davis) and the closest rural station to Seattle (Snoqualmie) have both had the older part of the record adjusted downwards.

The end result of this is to give a trend that looks as if there has been more than a century of warming in rural areas when the raw data shows a very different story.

Homewood[1507] checked a swathe of South American weather stations. He found the "homogenisations" for each station gave a warming trend. These "homogenisations" were made by the US government's Global Historical Climate Network (GHCN). There were 90 weather stations used of which 25 showed cooling and the rest measured warming. After "homogenisation", all stations showed warming. They were then amplified by two of the main official surface records, GISS and the National Climate Data Center (NCDC), which use the warming trends to estimate temperatures across the vast regions of the Earth where no measurements are taken.

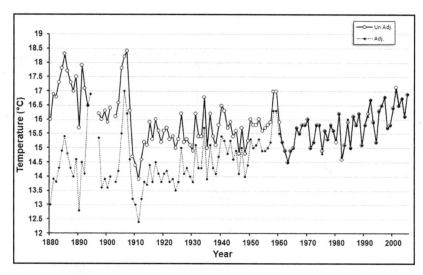

Figure 44: *Adjusted (thin line, dots) and raw (thick line, circles) for urban measuring station, Davis, California, USA.*[1508] *Note that adjustment gives apparent warming.*

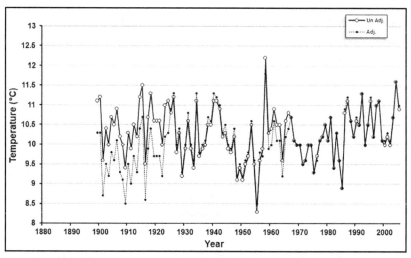

Figure 45: *Adjusted (thin line, dots) and raw (thick line, circles) for rural measuring station, Snoqualme Falls, Washington State, USA*[1600]. *Note that adjustment gives apparent warming.*

These are the very records on which scientists and politicians rely for their belief in "global warming" and which are used by the IPCC to scare the world to spend trillions on a concocted problem. Climate activists can easily do a literature search to show the unreliability of the data set, yet chose to ignore that data has been corrupted.

Homewood also looked at weather stations across much of the Arctic, between Canada (51°W) and the heart of Siberia (87°E). In nearly every case, the same upward "homogenisations" were made, to show warming up to 1°C or higher than was indicated by the data that was actually recorded.

This shocked Traust Jonsson, who was for a long time in charge of climate research for the Iceland meteorological office. Jonsson was amazed to see how the new version completely "disappears" Iceland's "sea ice years" around 1970, when a period of extreme cooling almost devastated his country's economy.[1509]

Homewood's interest in the Arctic is partly because the allegedly "vanishing" ice (and polar bears) which have become the poster-children for telling us we will fry-and-die. Homewood chose that particular stretch of the Arctic because it is where ice is affected by warmer water brought in by cyclical shifts in a major Atlantic current. The ice-melt is cyclical[1510] and is clearly unrelated to human emissions of carbon dioxide.

Of much more serious significance, however, is the way this wholesale manipulation of the official temperature record, for reasons GHCN and GISS have never plausibly explained, has become the real elephant in the room of the greatest and most costly scare the world has known.[1511] This really does begin to look like one of the greatest scientific frauds of all time.

It appears that those in charge of the records who promote human-induced global warming are climate "science" activists. Climate "scientists" have adjusted data and used poor measurements to change the past in an effort to control the present. This gives climate "scientists" the opportunity to bleat about a scary future while they are funded more and more.

The measured United States Historical Climate Network (USHCN)[1512] daily temperature data shows a decline in air temperature in the USA since the 1930s. Before this data is released to the public, it undergoes a number of "adjustments" which just happen to change the trend from cooling to warming.[1513] The difference between raw data and "adjusted" data sets shows that the past is cooled and the present is warmed. No matter what temperature is really doing, this shows exaggerated warming.

Such skullduggery is not restricted to the US. The UK's Met Office's Hadley Centre is responsible for compiling the HadCRUT global temperature datasets of land and sea surface temperatures with the Climatic Research Unit at the University of East Anglia.

There is a land temperature data set, CRUTEM, derived from air temperatures measured at weather stations near to the land surface across all continents of the Earth. The CRUTEM4 data set shows a 2.2°C temperature rise per century. When the data set was changed to CRUTEM4.3, the temperature rise had changed to 2.8°C per century. There was no explanation.

The HadCRUT 4.3 data set was changed to the HadCRUT 4.4 data set.[1514,1515] Without collecting new data there have been "adjustments" to the historical record wherein the past was cooled and the present was warmed[1516], the end result of which was to remove the 18 year pause in warming. It is a cardinal sin in science to adjust raw data. This data was paid for by the taxpayer and should remain as raw data for perpetuity.

Tampering with GISS air temperature data over a five-year period has suddenly retrospectively shown that instead of cooling, trends are now reversed.[1517,1518] Requests for explanations have gone unanswered.

Recent temperatures have been "adjusted" upwards by 0.2°C. Hence, since the warm 1930s and 1940s, the adjusted temperatures show an increase of 0.6°C. The adjustments account for at least one third of the post 1930 warming. With every adjustment, the warmest year on record shuffles up or down the ranking of the warmest year. This is fraud.

A pause on warming of 18 years lifted the lid on another angle of climate science fraud. Some of the climate "scientists", like the vociferous and energetic alarmist Ben Santer exposed in the Climategate emails, were previously telling us we would fry-and-die and are now giving us a different story.[1519] Santer previously denied there was a pause in warming and now claims that there was a pause.[1520] The pause in global warming could not previously be explained by modellers like Santer so he just simply dismissed the pause because it did not fit his narrative. He still can't explain the pause.

Cloud cover takes two thirds of the Earth's surface reflecting about 30% of solar energy back into space. A small change in cloud cover can easily cool or warm the planet. The IPCC stated that cloud feedbacks are the *"largest source of uncertainty"*. Changes in clouds can explain all the warming from 1986 to 2000 and explains the pause.[1521] We know that cloud cover is unrelated to carbon dioxide otherwise there would have been no pause during a time of rapidly-increasing human emissions of carbon dioxide. We don't know why cloud cover changed but it's odds on that the origin is extraterrestrial.[1522]

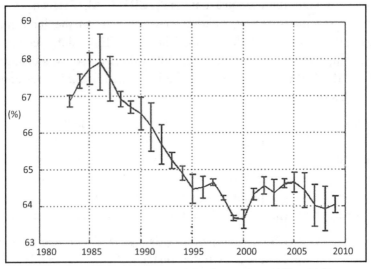

Figure 46: *Variation in the percentage of cloud cover from 1980 to 2008.*

Santer is a charming chap. In the Climategate emails, he wrote that he wanted to *"beat the crap"* out of climate sceptic Pat Michaels. In the draft Chapter 8 of the IPCC's second report, 28 authors wrote *"None of the studies cited above has shown clear evidence that we can attribute the observed changes to the specific cause of increases in greenhouse gases...No study to date has positively attributed all or part (of the climate change observed) to (man-made) causes...Any claims of positive detection and attribution of significant climate change are likely to remain controversial until uncertainties in the total natural variability of the climate system are reduced...When will an anthropogenic climate be identified? It is not surprising that the best answer to the question is 'We do not know '"*. This was a consensus opinion based on evidence by the 28 authors of the draft chapter.

After Santer's editing, none of these statements of caution appeared in the IPCC's bulky scientific Assessment Report and Santer included in the Summary for Policy Makers (the only part of the IPCC's AR2 report that people actually bother to read) the dogmatic and contrary statement *"the balance of evidence suggests that there is a discernible human influence on global climate"*. The IPCC reports cannot be trusted.

For years, GISS have been gradually "adjusting" historical records in order to increase the warming trend. Raw data is "adjusted" to show that the current year is the hottest one on record and there is an uncritical media fanfare. Raw data is "adjusted" to show that the 18 year pause in warming is not really a pause because the "adjusted" raw data now shows warming. Again, a big fanfare from the uncritical compliant media echo chamber.

No wonder the GISS data is now diverging from the more accurate satellite and balloon data.[1523] If this was in the commercial world, then there would be a prison cell waiting. It is very much a sign of how green left environmental activists have captured taxpayer-funded academia and feel free to conduct fraud to promote an ideology.

Data "adjustments" have occurred at Reykjavik (Iceland); De Bilt and Uccle (Netherlands)[1524]; Trier and Hannover (Germany); Alice Springs, Darwin, Bourke, Brisbane, Wagga, Deniliquin, Kerang and Rutherglen (Australia).[1525] Jennifer Marohasy has shown that the BoM just makes up an "adjustment".

They claim that moving a measuring station between paddocks near a small town in north-eastern Victoria (Rutherglen, Australia) required a change from a cooling trend of 0.35°C per century to a warming trend of 1.7°C per century.[1526] The trend was "adjusted" in order for it to be in accord with "adjusted" trends at neighbouring weather stations.

There may yet be a reason to "adjust" raw data but, despite requests, no explanations have been given. It appears that long-term temperature records have been "adjusted" to the point of unreliability. Taxpayers fund the BoM to keep accurate long-term records of raw data yet it appears that "adjustments" are made for ideological reasons. Much of the historical temperature and rainfall data in Australia has been collected by volunteers on isolated stations outback as part of their service to the nation. Why should the BoM deem that, a century later, these observations are wrong?

If the BoM "adjusts" raw data, then it is perfectly logical for governments to adjust the budget of the organisation. Such budgetary changes have been made elsewhere. The BBC sacked the UK's Met Office as the source of their weather information and has a New Zealand company do the job.[1527]

All "adjustments" always seem to go the same direction and exaggerate warming. Where is the "adjustment" that shows cooling? Surely with so many "adjustments" there would be just one lonely "adjustment" that was changed from warming to cooling. After decades collecting and interpreting raw data, this looks to me like systematic large-scale fraud by scientists who benefit from the catastrophist global warming gravy train.

The land-based air temperature measurement data is now not reliable and it would be reasonable to conclude that some data used for models is corrupted and altered beyond usefulness. The enigma of global warming has finally been solved. It is mainly due to fiddling the raw data. You heard it here first.

In 2015 there was a flurry to change the raw land-based temperature data (HadCRUT4, GISS, NOAA) to show exaggerated warming in the build-up for the Paris climate conference (whose true purpose is to establish an unelected and all-powerful global "governing body"). The satellite data (University of Alabama at Huntsville [UAH] and Remote Sensing Systems [RSS]) have not been amended yet and a new satellite system has been launched. Maybe the new satellite data will be ripe for the picking.

In Australia, there are a number of city-based measuring stations. Measurements are adjusted. Australia certainly has man-made global warming. It has nothing to do with carbon dioxide and occurs at the stroke of a pen. If all 328 stations are analysed from 1881 to the present, the raw data shows a statistically insignificant warming of 0.07°C.

Many historical measurements have an uncertainty greater than 0.5°C hence a warming of 0.07°C is well within the order or accuracy and is meaningless. To promote that this figure has any significance is fraud. Even if it were significant, if I stood up there would be a temperature increase of greater than 0.07°C. I don't think that we should have a nationwide restructuring of the economy because of a temperature rise of 0.07°C.

The analysis of the Rutherglen site in rural Victoria (Australia) is a well-documented example of some very persistent detective work by Jennifer Marohasy[1528] where measurements have been taken at the same site using the same equipment since 1912. There were no equipment failures or breaks in the data. Rutherglen should provide the best temperature data available because of the lack of the urban heat island effect and consistency of measurement from the one spot.

The data showed a cooling trend over the last century of 0.3°C, in accord with other measuring stations nearby.[1529] The official temperature record had been adjusted to create a warming trend of 1.7°C over the same period. This was because the pre-1974 minimum temperatures had been reduced by the BoM. This is fraud.

Surface air temperature measurements are administered by the taxpayer-funded BoM. There have been almost 2,000 sites and up to 112 of the measurements are weighted and combined to create various temperature averages for a whole continent. This forms the Australian historical temperature record and is fed into international datasets that are used to measure and predict global climate change.

When chased by Graeme Lloyd[1530], the environmental editor at *The Australian*, the BoM claimed that the change in recorded temperatures was because the site had moved and they needed more time to find the records. That was a lie. The BoM's own official Station Catalogue[1531] claimed there were *"no documented site moves during the site's history"*.

At Rutherglen, the BoM claimed that the Rutherglen trend needed to be consistent with trends at nearby stations. However, the trends from nearby station had been "homogenised" and the stations used at Wagga and Kerang are hundreds of kilometres away[1532] and can hardly be claimed as nearby.

The closest station to Rutherglen (Vic.) is Deniliquin (NSW) and its cooling trend over the course of a century of 0.7°C was "homogenised" to a warming trend of 1.0°C[1533]. In other areas of NSW, forty years of data at Bourke was just deleted, not even "homogenised". The 1.7°C cooling trend over a century at Bourke was "homogenised" to a slight warming trend and the site high of 51.7°C in 1909 was ignored, despite being recorded at nearby towns. Summer maximum temperatures at Bourke (NSW), Alice Springs (NT), Narrabri (NSW) and Hay (NSW) show a cooling trend.[1534] This is ignored by the BoM.

The records from the western NSW town of Cobar are used to "homogenise" temperatures at Alice Springs (NT), 1,460 km away as the crow flies. This is fraud. There is absolutely no way data from a town in a semi-arid vegetated area can be used to "homogenise" data from a far larger town in the arid dead centre of a continent. Why were closer towns such as Wilcannia (NSW), Broken Hill (NSW), Bourke (NSW), Port Augusta (SA), Roxby Downs (SA) or Birdsville (Qld) not used?

Data from Amberley in Brisbane in south-eastern Queensland was "homogenised" using data from the Coral Sea. After all, the climate on a terrestrial landmass will be exactly the same as that on distant tropical islands.[1535] Attempts to scrutinise the BoM by Lloyd[1598] met with delays then political obfuscation followed by inactivity. It is little wonder voters have lost confidence in politicians and institutions that were once proud of their independence.

Lloyd asked the BoM to justify its adjustments. The response was that's its methodology had been published in the peer-reviewed literature and that "homogenisation" was the world's best practice. This is not justification. It is obfuscation.

The BoM could not answer why all "homogenisations" gave a warming trend and why they changed cooling trends to warming trends in accord with "homogenised" warming trend of 1.0°C since 1910 elsewhere in the world. The climate science scare campaign is that temperature has increased by 1.0°C since 1910 and, if the data does not show such warming, then the primary data must be changed.

A recent study[1536] has shown that "homogenisation" of global surface temperature readings by scientists in recent years *are totally inconsistent with published and credible U.S. and other temperature data*. Climate "scientists" often "homogenise" primary data to account for what they perceive as biases in the data.

Each dataset pushed down the 1940s warming and pushed up the current warming, almost all "homogenisation" increases the warming trend and the paper authors[1537] state that *"nearly all the warming they are showing are in the adjustments"*. It is fraudulent to change data and, if for some reason data is corrected, then the primary data and amended data should be shown in the same table.

The raw data from Melbourne shows an increase in temperature from about 1950. This coincides with construction of large buildings thereby producing more back reflection, more vehicles and heating/cooling systems in office blocks. The raw Melbourne data shows the urban heat island whereas the adjusted data shows that Melbourne was warmer when there was little urban heat island effect and shows a warming from 1950.[1538]

The raw data from Darwin shows a slight cooling over the 150-year record. In the Second World War, the Darwin station was bombed and later moved from the city to the airport.[1539] The pre-1950 Darwin records have been adjusted downwards. The Darwin data looks like there has been significant warming over the last 70 years because the BoM has *"artificially shortened and cooled Darwin's history to make it consistent with the theory of human-caused global warming"*.[1607] Jennifer Marohasy is far too much of a lady to call it as it is. Fraud.

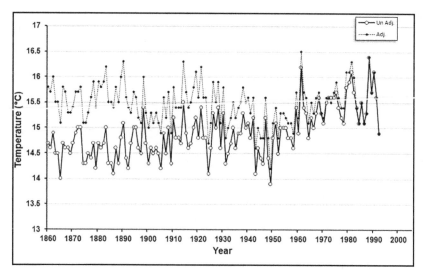

Figure 47: *Adjusted (thin line, dots) and raw (thick line, circles) for urban measuring station in Melbourne CBD.*[1631] *The station location suggests influence by post-1960 increased traffic and buildings giving apparent warming.*

It is this fraud that underpins Australia's decision to sign the Paris Accord, various COP agreements and to try to send the country broke with Net Zero. When politicians state that they follow the science, are they supporting fraud?

Exactly the same adjustments were made for Auckland with a rural station suffering substantial adjustments in the pre-1950 record such that the adjusted record looks as if there has been a sustained period of warming[1540]. The National Institute of Water and Atmosphere (NIWA), a New Zealand government research organisation, records average NZ temperature for the past 100 years. This shows a warming since 1900.

The world's longest recorded heatwave was at Marble Bar (WA). There were 160 days above 37.8°C (100°F) from 31st October 1923 to 7th April 1924. This record once had a whole page devoted to it on the BoM web site under "Climate Education". The web page disappeared and the 1923-1924 heatwave has been changed to make it cooler which make it look like there is a warming trend. This is fraud.

When the Federal Australian politician Craig Kelly tried to raise the changing of the temperature record by the BoM in the House of Representatives in the Federal Parliament, parliamentary procedure was used by the Australian Labor Party to prevent Mr Kelly from speaking. By not allowing Mr Kelly to expose climate fraud, the ALP is in essence supporting this fraud.

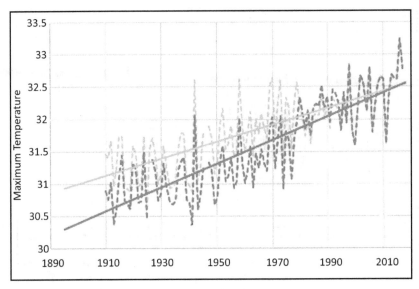

Figure 48: *Adjusted (red) and raw (yellow) for urban measuring station, Darwin, Australia.*[1632] *Note that adjustment gives apparent warming.*

Analysis in 2010 and 2020 of the Integrated Surface Database of temperature shows that the effects of urbanisation have not been removed from the official US land temperature datasets. The 'official' temperature datasets have been 'adjusted' several times. The analyses showed that there has been no global warming and that there is spurious warming in the thermometer data that has not been removed. This is in accord with an independent analysis in 2015.[1541]

In the US since 1985, temperature records over land have been steadily drifting higher than sea surface temperatures and are now about 1°C higher. This may be the result of both closing rural and often remote land stations, the urban heat island effect and the upward "homogenisation" of temperature trends.

The US annual temperature from NOAA showed warming maxima in the 1930s and cooling minima peaked in the 1960s and 1970s. The NOAA web site had James Hansen stating *"The US has warmed in the past century, but the warming hardly exceeds year-to-year variability. Indeed, in the US the warmest decade was in the 1930s and the warmest year was 1934"*.

There was constant friction as NOAA tried to infer warming from "homogenised" data. Indeed, by eliminating the urban heat island effect, warming since the 1930s suddenly appeared. Emails derived from Freedom of Information requests have revealed that David Easterling admitted

"One other fly in the ointment we have is a new adjustment scheme for USHCN(V2) that appears to adjust out some, if not most, of the 'local' trend that includes land use change and urban warming".

Clearly the new computer models are creating pre-ordained conclusions for use by scientific activists. This is contrary to the 2009 conclusions of Brian Stone[1542] who found in 2009 *"Across the US as a whole, approximately 50% of the warming that has occurred is due to land use changes (usually in the form of clearing forest for crops or cities) rather than the emissions of greenhouse gases"* and *"Most large US cities, including Atlanta, are warming more than twice the rate of the planet as a whole – a rate that is mostly attributable to land use change".*

Attempts to validate[1543] current surface temperature datasets managed by NASA, NOAA and the UK's Met Office failed. All these organisations make adjustments ("homogenisation") to raw temperature data, these adjustments dampen temperature cycles and increase the warming trend. Even with the doctored data, it is impossible to conclude from the datasets of NASA, NOAA and the UK's Met Office that recent years have been the warmest ever, despite claims of record setting warming. Global temperatures have dropped back to the pre-2016-2017 *El Niño* levels.

The claim that the Earth's atmosphere at the surface has warmed by 0.85°C over the last 100 years is unprecedented and is not in accord with the unadjusted evidence. History has a nasty habit of killing off scare stories.

In 1249 AD, it was so warm in England that people did not need winter clothes. There was no frost in England for the entire winter. It was so warm that people thought the seasons had changed. Was this due to the English burning fossil fuels and putting carbon dioxide in the atmosphere? Obviously not as coal was not used then. If such warmth was in the Medieval Warming, why are cooler times in the Modern Warming allegedly due to human emissions of carbon dioxide?

Are we really in a climate emergency? The longest temperature record is the UK Met Office's Central England Temperature Record that shows there have been warmings and coolings but no significant warming in the UK since 1659 AD.[1544,1545] However, there has been a significant increase in climate hysteria over the last 30 years. The Central England Temperature record shows that between 1693 and 1733 AD temperature rose at a rate of 4°C per century. During the Roman and Medieval times, it was warmer than now and the rate of warming at times was higher than in the 20th Century.[1546,1547]

Global temperatures have increased some 3 to 8°C since the zenith of the last glaciation, global temperatures have a variability of about ±3.5°C from the long-term mean[1548], in the 21st Century we are about 1°C above the long term mean but, before you go outside and slash your wrists, in the Medieval Warming the temperature was 2.5°C above the long-term mean. There is nothing special about the 20th and 21st Centuries temperature or temperature increase over the last 100 years. It is just the normal variability that planet Earth enjoys.

A large number of ground temperature measurements used for climate predictions have been made in cities and airports. Between 1950 and 2020, the world population tripled from 2.5 billion to 7.5 billion, the planet became far more urbanised and agricultural land around airports was covered by infrastructure and buildings.

The urban heat island effect is largely discounted by the IPCC. The only temperature measuring system that averages over the whole Earth is the satellite system which shows no global warming. The ground temperature record shows warming. This is probably a measure of population increase, urbanisation and positioning of measuring stations near population centres.[1549]

No one under 20 years old has experienced a trend of warming temperatures. Yet they are in the streets instead of classrooms demanding action to stop something they have never experienced in their lifetime. Even someone who was a teenager in the 1940s cannot show you whether it was warmer or colder then. Temperature changes over the last 100 years can be measured instrumentally but have little effect on humans. In my life, I have experienced four slight temperature trends and they had no effect on my life.

The cyclical pattern of climate reported earlier has very nearly been "adjusted out" of primary data taken from weather stations, buoys and ships. Almost all the surface temperature warming adjustments cool past temperatures and warm more recent measurements giving the appearance of a warming trend. Are those demonstrating about climate aware that everything they have been told is wrong and based on corrupted data?

How do we analyse and then integrate data from station to station? Some stations use hourly readings; some use a simple maximum and minimum temperatures; some massage and manipulate raw data for UHI and some don't; some use daily measurements, others use monthly or yearly measurements; some use windowing and running averages and some don't; some use all different means from area under curves to quadratic means, harmonic means, geometric means to plain old arithmetic means; some have lots of measuring stations and some don't; some are in cities surrounded by concrete, air conditioning and hot vehicle exhaust fumes and others are surrounded by trees, sheep and cows.

With all these different methods of collecting and treating data even before "homogenisation", the temperature record is a total dog's breakfast. It's not a case of not being able to compare apples with oranges, it's a case of trying to compare apples with bicycles. The whole concept of an average global temperature is nonsense.

If we start to plot global temperatures from the Holocene Optimum, Roman Warming or Medieval Warming, the global temperatures have decreased. If the starting point for a global temperature plot was from the Maunder Minimum 350 years or at the end of the Little Ice Age 170 years ago, then global temperatures have increased. It is so easy to cook the books.

In the 1990s, I had concerns about the mathematical treatment of the BoM data. One of my meteorologist staff members had purchased a block of BoM data for mathematical analysis. I asked him a few questions. How did he know the data was correct? Have you validated the data independently? Why don't you collect your own data? His answer was that he trusted the BoM data. That set off my alarm bells some 30 years ago about the lack of QA/QC used to predict human-induced global warming.

You heard it from the horse's mouth. If you cannot replicate the data, no matter how much it has been "homogenised", it is not science and is useless. No conclusions or predictions can validly be made, yet they are by climate "scientists", climate activists, the media and the IPCC.

The mixture of activist science with career survival by research grants dealing with allegedly pressing problems, politics, publicity, hero-worship publicity, the desire to influence public opinion and the takeover of institutions by the activist left has created the thread that ties the climate industry together. Cooking the books. Once science is politicised or fraudulent, climate "science" is no longer science.

What is unprecedented is that in the history of science, never has so much data has been tampered with by so few. The end result is that the present and historical climate record has been destroyed and any climate projections should be ignored. We are spending trillions to address climate change yet the basics underpinning this theory rely on compromised data promoted by frauds.

Primary data has been tampered with to produce pre-ordained results. The world temperature data record is already a hybrid of raw data, tampered data and estimates. There is now very little reliable temperature data for future generations. To change primary or raw data in science is a cardinal sin. It is almost a universal procedure in climate "science". That's how conclusions can be reached that are not in agreement with other areas of science.

The surface temperature record is so corrupted that it tells us nothing about global climate and temperature trends. It does tell us that climate "science" is underpinned by fraud and the premises for renewable energy, Net Zero and huge associated economic changes are unfounded.

Somewhere some place in the world right now, the weather is breaking a record. Time to hide under the bed in case the world ends.

Ocean temperature

The oceans play the dominant role in perpetuation and mediation of natural climate change as they contain far more heat than the atmosphere. Density variations linking the Northern and Southern Hemispheres of the Pacific and Atlantic Oceans via the Southern Ocean drive the ocean circulation system that controls hemispheric and global climate. Differences in temperature and salt concentrations produce these density variations. The oceans both moderate and intensify weather and decadal climate trends due to their great capacity to store heat. The process involves global currents, slow mixing, salt concentration variations, wind interactions and oscillations in heat distribution over very large volumes.

Currents are global heat conveyor belts.[1550] Ocean currents and cycles have a large decadal-scale effect on weather, regional climate and global climate.

In some parts of the global ocean heat conveyor, natural variations in heating, evaporation, freshwater input, atmospheric convection, surface winds and cloud cover can greatly influence ocean currents close to continents. In turn these modify carbon dioxide uptake or degassing, storms, tropical cyclone frequency, abundance of floating life that consumes carbon dioxide, droughts and sea level changes.

The ocean transfer of heat to the air is one of the major driving force of climate. As discussed earlier, we have no idea how much heat and carbon dioxide is added to the oceans by unseen submarine volcanicity. All we know is that it's a Hell of a lot. For climate "scientists" to suggest that it is only human emissions of carbon dioxide that drive global warming exposes their activist agenda.

The oceans occupy 70% of the Earth's surface. Sea surface temperature is measured. The Hadley Centre only trusts measurements from British merchant ships. These mainly use shipping routes in the Northern Hemisphere yet the Southern Hemisphere oceans occupy 80% of the ocean surface area. The larger land area in the Northern Hemisphere compared to the Southern Hemisphere means that the climates of each hemisphere are different. There is no such thing as global climate.

The change in methods over time for measurement of sea surface temperature from water collection in canvas buckets to later engine water intakes at various water depths has introduced measurement uncertainties. Ocean temperatures from ships, buoys and satellite also present opportunities for "adjustment", as the Climategate emails show.

Satellite measurements were removed from public scrutiny by NOAA in July 2009 after complaints of a cold bias in the Southern Hemisphere. Immediately after that, both ocean and global temperatures mysteriously increased. The 3,800 ARGO buoys have been measuring sea surface temperature since mid-2003.[1551] ARGO buoy data is not used in monthly assessments of global temperature. Why not?

The surface temperature records in the pre-satellite era (1850 to 1980) have been so widely, systematically and uni-directionally "homogenised" that it cannot be claimed that there is human-induced global warming in the 20th Century. These changes to the historical records mask cyclical changes that can be explained by oceanic and solar cycles. To increase the apparent trend of warming, earlier records of warming have been adjusted downwards.

It is probable that these temperature databases are now useless for determining long-term trends. To make matters worse, the Climate Research Unit of the University of East Anglia now denies public access to raw and "homogenised" HadCRUT4 data[1552,1553] lest data escapes for independent scrutiny.[1554] Data is collected and managed by taxpayer funded institutions who deny access to taxpayers. What have they got to hide?

More than three-quarters of the 6,000 global stations that once reported temperature are no longer being used in data trend analysis and some 40% of stations now report months when no data is available. These months where no data is reported required "infilling" which only adds to the uncertainty. The exaggeration of long-term warming by 30-50% is probably because of urbanisation, changes in land use, incorrect positioning and inadequately calibrated instrument upgrades.

Reliability of measurements

Five organisations publish global temperature data. These data centres perform some final adjustments to data before the final analysis. Adjustments are common and poorly documented. Data is not freely available for independent validation.

Attempts to validate[1555] current surface temperature datasets failed. All these organisations make adjustments to raw temperature data ("homogenisation"), adjustment of data is a cardinal sin of science, these adjustments dampen temperature cycles and increase the warming trend. Even with the doctored data, it is impossible to conclude from the datasets of NASA, NOAA and the UK's Met Office that recent years have been the warmest ever, despite claims of record warming. Global temperatures have dropped back to the pre-2016-2017 warm *El Niño* levels.

The World Meteorological Organisation (WMO) and NOAA supposedly have strict criteria for temperature measuring stations. Land-based measurement stations should be located on flat ground surrounded by a clear surface and more than 100 metres away from local heat sources, tall trees, artificial heating or reflecting surfaces such as buildings, parking areas, roads and concrete surfaces. Temperature sensors should be shaded from direct sunlight, ventilated by the wind 1.5 metres above mown grass no higher than 10 cm.

Anthony Watts found that 89% of the US measuring stations (i.e. more than 1,000 stations) did not meet the official WMO standards for temperature measurement regarding the distance between stations and adjacent heat sources.[1556] They still don't despite, the WMO and NOAA knowing that stations are poorly sited and will give anomalously warm measurements.[1557]

Many received back reflection from concrete and buildings, most received heat from hot car and aeroplane exhaust fumes and many were sited next to air conditioning units that emitted heat. Watts concludes *"The raw data produced by the stations are not sufficiently accurate to use in scientific studies or as a basis for public policy decisions"*.

Temperature measurements of the surface of the Earth are compiled into what is known as the HadCRUT data. The data derives from 5° latitude x 5° longitude grid cells covering the Earth's surface. This data gives a two-dimensional picture whereas balloon and satellite data give a three-dimensional temperature map of the atmosphere. There is a total of 2,592 cells. Cells that entirely cover land comprise just 21.7% of all cells and 18.1% of the total surface area and their data comes from observation stations.

Sea grid cells comprise 50.6 % of the total cells and 53.8 % of the total area, and their data comes from measurements of sea surface temperature. Cells that cover a mixture of land and sea account for 27.7 % of cells and 28.0 % of area, and their data might be from observation stations or sea surface temperatures, depending on what's available and what's regarded as most reliable.

The oceans cover 70% of the surface of the Earth. Both sea surface temperature and land temperature measurements have errors and the mixing data from two different measurement methods creates greater uncertainties. In any given month of measurements, coverage is around 80%, a slight fall from early 1980s levels. Data from the observation stations is commonly "homogenised" by the local meteorological authority. Data generally is not collected from more than 50% of the Earth's surface.

US climate research has received more than $73 billion in funding over the last two decades. And what do we have? Suspect data. In 1999, the NOAA administrator paved the way for the corruption of science by stating[1558] *"Urgent and unprecedented environmental and social changes challenge scientists to define a new social contract....a commitment on the part of all scientists to devote their energies and talents to the most pressing problems of the day, in proportion to their importance, in exchange for public funding".*

Since the 1960s, some 62% of temperature-measuring stations have been removed and most of these were in the colder remote, alpine, polar and rural regions where temperatures were lower. Most measuring stations are now near the sea or are at airports and cities which bias measurements because of urban and industrial heat. How can these sparse and biased stations tell us anything about the temperature of the planet?

According to the calculations of Lubos Motl[1559], 31% of the stations used by the UK's Hadley Centre Climate Research Unit (HadCRUT), the standard for surface temperatures, show that temperature fell since 1979. This agrees with the satellite temperature trends over the same period posted by John Christy and Roy Spencer (University of Alabama). The measurement of temperature is not simple and, with so many problems with data, the calculation of the meaningless global temperature is not possible.

Historical instrumentally-recorded temperatures exist only for 100 to 150 years in small areas of the world. From the 1950s to 1980s temperatures were measured in many more locations. Many measuring stations are no longer active. The main global surface temperature data set is managed by NOAA who state *"The period of record varies from station to station, with several thousand extending back to 1950 and several hundred being updated monthly"*.

There now have been so many stations shut down, especially in cooler high elevation, high latitude, rural and remote stations, that a significant warming bias has entered the overall record from land stations. Stations have been shut down because of costs and yet the patchy remaining data is used to make multi-trillion dollar decisions about the future of the planet. This is the main source of data for global studies, including the data reported by the IPCC.

Average surface air temperatures are calculated at a given station location. These are used to calculate the averages for the month which are then used to calculate the annual averages. This whole process is fraught with error and "homogenisation" can occur with any calculation.

The IPCC uses data processed by the Climatic Research Unit of the University of East Anglia. There is constant recalculation of a patchwork quilt of data until a half-reasonable result is obtained. This is not science. The CRU station data used by the IPCC is not publicly available. Why not? What are they trying to hide from the taxpayers who funded acquisition of this data?

Only the "homogenised" gridded data is available hence for global temperature data we must trust what the CRU at the University of East Anglia feeds us. Before Climategate I might have been nervous about their "homogenised" gridded data. But now that the keepers of the raw data have been shown to be frauds, we can only conclude that all the data used by the IPCC comes under a very big black cloud that just won't go away.

Years after the Climategate fraud, we see that the CRU has not changed its spots. The world's temperature data set is centralised at the CRU which controls what data will be used to construct climate models. In

December 2009, the former chief economic advisor to President Putin, Andrei Illarionov, stated that Russia had sent data to the CRU from 476 meteorological stations covering 20% of the globe's surface dating as far back as 1865.

The data from these 476 stations was ignored and some data from 121 station was cherry-picked from to show that temperatures between 1860 and 1965 were colder than they were which made temperatures from 1965 to 2005 artificially high.[1560] This gave the impression of warming. Is this fraud or is this fraud? If this were done in the financial world, a time behind bars would be the reward. Jobs were not lost at the CRU, no one went to gaol and the world continues to make trillion dollar decisions based on fraudulent incorrect data.

The NASA Goddard Institute for Space Studies (GISS) is a major provider of climatic data in the US (with NOAA as the source for GISS). Some 69% of the GISS data comes from latitudes between 30 and 60 degrees north and almost half of those stations are located in the United States.[1561] This certainly does not cover the globe and is totally unrepresentative for any global picture. If these stations are valid, the calculations for the US should be more reliable than for any other area or for the globe as a whole. However, many recording stations closed down in the late 1980s and early 1990s.

The locations of many stations were moved to escape urbanisation. Some stopped collecting data during periods of conflict or financial cutbacks. With the number of temperature measuring stations changing over time, the so-called "global" record is not really global. Land surface thermometers increased from coverage of 10% of the land in the 1880s to about 40% in the 1960s. Since then, the coverage has been decreasing.

Coverage has been redefined as the "*percent of hemispheric area located within 1,200 kilometres of a reporting station*". This means that in remote areas where there may be no stations in a 5° x 5° grid box, the temperature is estimated from the nearest station within 1,200 km. In my experience, there are considerable variations in temperature over 1,200 km distance.

Many areas of the world have no historical temperature measurements (e.g. Third World countries). It was claimed that these areas showed great warming. It appeared that Greenland was much warmer but most of the 5° x 5° grid boxes in Greenland have no measuring stations and most of the other grid boxes have only one station. The two hottest areas of Greenland shown by NOAA showed 5°C warming yet had no measuring station.

Would it be wrong to state that these numbers were just plucked out of the air and, to our great surprise, showed significant warming that could not be validated by measurement?

It is the same for Siberia. In the 1940s, Siberia had areas that showed differences in warming and cooling by a few degrees. The measuring stations have now been closed down. Siberia appears now to have warmed in 5° x 5° grids where there are no measuring stations. As the Siberian historical data only goes back to the 1940s, it is hard to calculate a trend of warming or cooling.

It appears that the Arctic of Canada is also sweltering. Almost all 5° x 5° grids have at least one measuring station. The only grid without a measuring station just happened to have warming of 5°C. Many stations are no longer maintained in the GHCN or CRUTEM3 databases yet NOAA claims that there has been a 4°C warming over the last 40 years. The record shows that similar warming occurred in the 1930s.

Temperature changes in the Canadian Arctic correlate with the *El Niño-*Southern Oscillation. The HadCRUT data shows 60-year cycles of temperature, probably related to the *El Niño*-Southern Oscillation. As temperature has been increasing for 350 years since the Maunder Minimum at the end of the Little Ice Age, the cycles are secondary to the main trend. These 60-year cycles have been measured as far back as 2,000 years.

The closure of temperature measuring stations was not completely random. There has been a decrease in the number of Russian measuring stations and the Hadley Centre ignores many continuous long-term records. In Canada, the number of stations has decreased from more than 600 to 50. NOAA used only 35 of these Canadian stations. The percentage of those at lower elevation (less than 100 metres above sea level) tripled and those above 1,000 metres reduced by half.

More southerly locations using population centres hugging the US border dominated over northerly areas. In fact, only one thermometer remains in Canada for everything north of the 65th parallel. This site is called Eureka[1562] because life is more abundant and the summers are warmer than elsewhere in the High Arctic thereby creating a temperature bias for the whole region. Hourly readings from Russia and Canada can be found on the internet yet these are not included in the global data set.

China had a great increase in measuring stations from 1950 (100 stations) to 1960 (400) and then a great decrease to only 35 in 1990.[1563] Recorded temperatures in China rose due to increased urbanisation during their current industrial revolution.

In Europe, high mountain stations were closed and there were more measurements from coastal cities, especially along the Mediterranean Sea. In northern Europe, the average results showed warming as the number of stations decreased. After 1990, Belgium showed warming yet the adjacent country (the Netherlands) showed no warming.

A similar story is apparent for South America where alpine stations disappeared and the number of coastal stations increased. There was a 50% decline in measuring stations.

African raw data also shows a warming bias.[1564] In North Africa, stations from the hotter Sahara were used in preference to those from the cooler Mediterranean and Atlantic coasts.

In Australia and New Zealand, some 84% of stations are at airports where there are hot fumes from planes and vehicles, concrete and buildings with air conditioning. Stations that closed were at cooler latitudes.

In the US, some 90% of all measuring stations did not appear in the global historical climate network GHCN Version 2 climate models. Most stations remaining were at airports and most of the high mountain measuring stations have closed. In California, only San Francisco, Santa Maria, Los Angeles and San Diego were used. The warming trend is still not significant despite this bias. Climate activist James Hansen stated *"The US has warmed during the past century, but the warming hardly exceeds year-to-year variability. Indeed, in the US the warmest decade was the 1930s and the warmest year was 1934"*.

Recent temperature analyses in China show that "global warming" in China is actually due to the heat emitted from scores of large cities in China. Most people can't name the 10 largest cities in China by population (Chongqing >30M, Shanghai 25.8M, Beijing 18.8M, Chengdu 14M, Tianjin 13M, Guangzhou 11.1M, Baoding 11M, Harbin 10.5M, Suzhou 10.4M, Shenzhen 10.3M). No matter what Australia does to reduce carbon dioxide emissions, Australia is dwarfed by China.

For a long time, various scientists have argued that almost all temperature measuring stations are poorly located and give artificially high results. NOAA attempted to disprove this, perhaps because there was too much press questioning NOAA. They established an experiment in an Oak Ridge (Tennessee) parking lot close to steel buildings to see if the claims had any veracity. NOAA looked at the impact of urban development on temperature with stations 4, 30, 50, 124 and 300 metres from a built up area[1565] and showed that stations up to 50 metres from a built-up area can have important impacts on the maxima and minima temperature.[1566,1567] As a result of this experiment, no admission of incorrect measurement has been made.

Models of future climate are based on very shoddy measurements and some pretty wild assumptions. Satellite measurements for 42 years show that warming is occurring much more slowly than the average climate model has predicted.[1568] This is exacerbated by "official" measurements in disagreement with "official" climate models. Climate models were also used to predict ocean warming. Oceans provide the best gauge of how fast extra energy is accumulating in the climate system.

Measurements of sea surface temperature since 1979 for 68 model simulations from 13 different models were compared with 42 years of satellite measurements. The measurements were 50% of climate model predictions. Deep ocean warming shows a very slight energy imbalance in the climate system requiring a very optimistic level of faith that is involved in the adjustments made to climate change models. This slight warming could be natural and the obvious source is heat from below.

Those who promote a doom-and-gloom future in return for research funds are the very same people who use and adjust data from poorly-sited weather stations and who were caught fiddling the data in the Climategate fraud.

No data shows us that there is a climate crisis or climate emergency. Any computer projections of global climate using land temperature measurements are deeply flawed and we should treat all claims, models, trends and predictions based on these with a very large pinch of salt. If the data is dodgy, why do we even bother with the projections and models developed from this data?

This is the science that we are being asked to follow. This is the science that underpins the signing of ruinous international agreements and expenditure of trillions to solve a non-problem. This is the science that underpins a complete structural change of the economy. No thanks.

Carbon dioxide measurements

There is a crowd out there trying to convince us that man-made carbon dioxide gas is "pollution" which, unless **YOU** stop emitting it, will result in the inevitable Armageddon where life on Earth perishes. They don't worry about naturally occurring carbon dioxide of course and the fact that in the past carbon dioxide was far higher than at present. What happens if the measurements of carbon dioxide are unreliable?

Attempts to experimentally replicate global warming by increased carbon dioxide in the atmosphere failed.[1569]

Because the atmosphere has greater than 100 parts per million (ppm) carbon dioxide, a doubling or quadrupling of human emissions of carbon dioxide will have very little effect on temperature unless atmospheric carbon dioxide residence times change to two orders of magnitude higher than past times. Leave carbon dioxide alone. It is only following what water vapour does which, in turn, is following the energy output of the Sun. Those westerners who would want to try to reduce plant food and hence contribute to starvation in the Third World need to consider their moral position.

The measurement of carbon dioxide in the atmosphere is fraught with difficulty. There is a 180-year record of atmospheric carbon dioxide measurement by the same method. There have been 90,000 measurements with an accuracy of 1–3% from 1812 until 1961 by a chemical method.[1570] These showed peaks in atmospheric carbon dioxide in 1825, 1857 and 1942. In 1942, the atmospheric carbon dioxide content was 440 ppm.[1571] At the time of writing, the "official" atmospheric carbon dioxide content is 412 ppm. For much of the 19th Century and from 1935 to 1950, the atmospheric carbon dioxide content was higher than at present and varied considerably. A variable carbon dioxide content is exactly as expected in Nature because all natural processes show variability. A smooth curve raises alarm bells about data collection, acceptance and rejection and data processing.

The Hawaiian station was established to measure *"the rise in atmospheric carbon dioxide resulting from fossil fuel combustion"*. The Hawaiian station was established to prove a pre-determined conclusion and not to test a hypothesis or acquire a body of data requiring explanation. There are other stations (e.g. South Pole, Cape Grim, La Jolla, Point Barrow, Shetland Islands). What the data shows is that the seasonal variations are far greater than the annual increase in atmospheric carbon dioxide and that there is a lag between an increase in temperature and the seasonal increase.

In 1959, the measurement method was changed to infra-red spectroscopy with the establishment of the Mauna Loa (Hawaii) station. It was a simple, cheap and quick method. The infra-red technique has never been validated against the earlier chemical method.

As with temperature data, the raw carbon dioxide data from Mauna Loa is "edited" by an operator who deletes what is considered poor data. Some 82% of the raw infra-red carbon dioxide measurement data is "edited" leaving just 18% of the raw data measurements for statistical analysis.[1572,1573] With such a savage editing of raw data, whatever trend one wants to show can be shown. In publications, large natural variations in carbon dioxide were removed from the data set by editing to make an upward-trending curve showing an increasing human contribution of carbon dioxide.

The early Mauna Loa and South Pole carbon dioxide measurements were considerably below measurements made at the same time in north-western Europe from 21 measuring stations using the chemical method.[1574] During the period these 21 stations were operating (1955–1960), there was no recorded increase in atmospheric carbon dioxide.[1575] There is a poor correlation between temperature and the greatly fluctuating atmospheric carbon dioxide content measured by the chemical method.

Chemical measurements in north-western Europe showed a variation between 270 and 440 ppm. There was no tendency for rising or falling carbon dioxide levels at any one of the measuring stations over the five-year period. Furthermore, these measurements were taken in industrial areas during post-World War II reconstruction and increasing atmospheric carbon dioxide would have been expected. While these measurements were being undertaken in north-western Europe, a measuring station was established on top Mauna Loa in order to be far away from carbon dioxide-emitting industrial areas.

However, the basalt volcano of Mauna Loa emits large quantities of carbon dioxide, as do other Hawaiian volcanoes.[1576] During a volcanic eruption, the observatory was evacuated for a few months and there was a gap in the data record which represented the period of no measurement. There are now no gaps in the Mauna Loa data set.[1577] Someone has taken a wild guess and just added a number rather than leaving a gap. This gives no confidence in the keepers of the data.

The annual mean carbon dioxide atmospheric content reported at Mauna Loa for 1959 was 315.93 ppm. This was 15 ppm lower than the 1959 measurements for measuring stations in north-western Europe. Measured carbon dioxide at Mauna Loa increased steadily[1578] and the 1989 value is the same as the European measurements 35 years earlier by the chemical method. This suggests problems with both the measurement methods and the statistical treatment of data.

When the historical chemical measurements are compared with the spectroscopic measurements of air trapped in ice and modern air, there is no correlation. Furthermore, measurement at Mauna Loa is by infra-red analysis[1579,1580] and some of the ice core measurements of carbon dioxide in trapped air were by gas chromatography.[1581] Comparing data using different measurement techniques is fraught with difficulty.

The Mauna Loa results change daily and seasonally. Night-time decomposition of plants and photosynthesis during sunlight change the data, as does traffic and industry. Downslope winds transport carbon dioxide from distant volcanoes and increase the carbon dioxide content. Upslope winds during afternoon hours record lower carbon dioxide because of photosynthetic depletion in sugar cane field and forests.

The raw data is an average of four samples from hour to hour. In 2004, there were a possible 8,784 measurements. Due to instrumental errors, 1,102 samples have no data, 1,085 were not used because of upslope winds, 655 had large variability within one hour but were used in the official figures and 866 had large hour-by-hour variability and were not used.[1582]

The Mauna Loa measurements show variations at sub-annual frequencies associated with variations in carbon sources, carbon sinks and atmospheric transport.[1583] The behaviour of carbon dioxide in the Northern Hemisphere is different from the Southern Hemisphere. Oceans dominate the Southern Hemisphere which has a lack of deciduous tree forests. Smaller land masses and fewer carbon dioxide emitting industry.

Air that arrives during the April–June period favours a lower carbon dioxide concentration. Seasonal changes derive from Northern Hemisphere deciduous plants that take up the food of life in spring and summer and release it in autumn and winter due to the decay of dead plant material. Every April, the Northern Hemisphere reduction shows that Nature reacts quickly to carbon dioxide in the atmosphere and can remove large amounts in a very short time. This is not news. For *millennia* farmers have called this time the growing season.

About 25 times as much carbon dioxide is emitted by humans in the Northern Hemisphere than in the Southern Hemisphere. These emissions occur essentially between the latitudes of 30° and 50° North. Does the Northern Hemisphere carbon dioxide mix with the Southern Hemisphere carbon dioxide to give a global atmospheric carbon dioxide content?

Clues come from the nuclear industry. The only source of the gas krypton 85 in the atmosphere is from nuclear reprocessing plants, reactors and bombs. Most reprocessing plants, reactors and bomb testing were in the Northern Hemisphere. A very distinct change in krypton 85 is evident from the Northern Hemisphere to the Southern Hemisphere showing that gases from each hemisphere do not mix very easily.

If krypton 85 does not mix quickly from one hemisphere to another, why would carbon dioxide from the industrialised Northern Hemisphere mix with the Southern Hemisphere atmosphere?

Measurements of increasing carbon dioxide from Hawaii (Northern Hemisphere) and the South Pole shows that there is no time lag difference between hemispheres in the increase of carbon dioxide in the atmosphere. This strongly suggests that ocean degassing in both hemispheres is dominating carbon dioxide increases. Furthermore, we estimate the amount of carbon dioxide emitted by human activities but we cannot accurately measure the amount of carbon dioxide emitted by natural processes.

The range of uncertainty is larger than the actual amount of carbon dioxide emitted by humans. Satellite measurement of carbon dioxide globally does not find sources concentrated where we would expect them in industry and population centres of Western Europe, USA and China. Instead, the carbon dioxide sources appear to be in places like the Amazon Basin, south-east Asia and tropical Africa, where human emissions are minimal.

The Mauna Loa measurements show annual change or increases in atmospheric carbon dioxide by up to three parts per million. Some years there were no changes. If emissions were from human activity, every year should show an increase correlating with human emissions. It does not. Murry Salby who has used carbon fingerprints to investigate this enigma and concludes that man-made emissions only have a small effect on global carbon dioxide levels[1584] and states that "*anyone who thinks the science is settled in this subject is in fantasia*".

Salby confirms what we have known from ice core measurements. An increase in temperature precedes an increase in atmospheric carbon dioxide. He was once an IPCC reviewer and comments that if this was known in 2007 then "*the IPCC could not have drawn the conclusions it did*". The IPCC knew in 2007 the uncertainty with measurements, emissions and conclusions.

There may be errors in sampling and analytical procedure.[1585] Measuring stations are now located around the world and in isolated coastal or island areas to measure carbon dioxide in air without contamination from life or industrial activity to establish the background carbon dioxide content of the atmosphere. The problem with these measurements is that land-derived air blowing across the sea loses about 10 ppm of carbon dioxide as it dissolves in the oceans. If the ocean is colder, more carbon dioxide is lost.

A greater problem is that the infra-red absorption spectrum of carbon dioxide overlaps with that for water vapour, ozone, methane, dinitrogen oxide and CFCs.[1586] Some infra-red equipment has a cold trap to remove water vapour. The NOAA Global Monitoring Laboratory at Mauna Loa (Hawaii) recorded the atmospheric carbon dioxide content 410 ppm and atmospheric water vapour as 10,000 to 40,000 ppm.

According to the 1979 Chaney Report, the main atmospheric greenhouse gas is water vapour yet NOAA expunged water vapour from their 2020 Annual Greenhouse Gas Index. However, carbon dioxide dissolves in cold water and some is also removed. These other gases are detected and measured as carbon dioxide. Gases such as CFCs, although at parts per billion in the atmosphere, have such a high infra-red absorption that they register as parts per million carbon dioxide. Unless all these other atmospheric gases are measured at the same time as it, then the analyses by infra-red techniques must be treated with great caution.

If the chemical method was used concurrently with infra-red measurement for validation, then there could be more confidence in the infra-red results. The infra-red carbon dioxide figures are now at the level recorded by the chemical method 70 years ago. Do we really have absolute proof that carbon dioxide has risen over the last 70 years? Caution needs to be exercised about numbers without the historical background as given above.

The IPCC's AR3 to AR6 reports argued that only infra-red carbon dioxide measurements can be relied upon and prior measurements can be disregarded.[1587] No scientific reasons were given. The atmospheric carbon dioxide measurements since 1812 do not show a steadily increasing atmospheric carbon dioxide content as shown by the Mauna Loa measurements.

The IPCC chose to ignore the 90,000 precise carbon dioxide measurements compiled despite the fact that there is an overlap in time between the chemical method and the infra-red method measurements at Mauna Loa.

If a large body of validated historical data is ignored, then a well-reasoned argument needs to be given. There was no explanation.

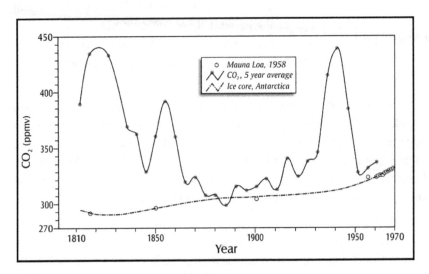

Figure 49: *Determinations of atmospheric carbon dioxide by the Pettenkofer method (solid line of 5 year averages) between 1812 and 1961, deductions of atmospheric carbon dioxide from Antarctic ice core (gas chromatography) and edited measurements of atmospheric carbon dioxide from Mauna Loa (infra-red spectroscopy, 1958 and onwards). One method of measurement shows great variability in atmospheric carbon dioxide yet another method does not. The high values of carbon dioxide by the Pettenkofer method have been rejected by the IPCC yet the lowest value is used by the IPCC as the baseline pre-industrial value for atmospheric carbon dioxide.*

A pre-IPCC paper used carefully selected data obtained by the chemical method. Any values more than 10% above or below a baseline of 270 ppm were rejected.[1588] The rejected data included a large number of the high values determined by chemical methods. The lowest figure measured since 1812, the 270 ppm figure, is taken as a pre-industrialisation yardstick that underpins all claims that human emissions are increasing the atmospheric carbon dioxide content.

The IPCC want it both ways. They are prepared to use the lowest determination by the chemical method as a yardstick yet do not acknowledge chemical method measurements showing chemical concentrations far higher than now many times since 1812. When the basics of the climate catastrophist claims are looked at, there is always an element of exaggeration, cooking the books or fraud.

Ocean degassing is a major source of atmospheric carbon dioxide. Cold seawater surface dissolves it; when the seawater is at the tropics it releases the gas. It is only the surface seawater that is releasing carbon dioxide. Degassing of the oceans is taking place in the tropical waters around Hawaii. Deep ocean water contains dissolved carbon dioxide much of which is from submarine volcanoes.

This increases the amount of water vapour and carbon dioxide in the atmosphere and reduces the uptake of it by the oceans. As a result, some of the carbon dioxide is misattributed to human activity.[1589,1590] As shown in Chapter 2, the problems of the chemical fingerprint of carbon and the lack of analysis of submarine volcanoes does not give confidence in the measurement of atmospheric carbon dioxide and any conclusions derived therefrom.

If each carbon dioxide molecule in the atmosphere has a short lifetime, it means that they will be removed fast from the atmosphere to be adsorbed in another reservoir. But how fast? Because atmospheric carbon dioxide is increasing, it was argued that it has not been dissolved in the sea and must have an atmospheric lifetime of several hundred years.[1591] The IPCC suggests that the lifetime is 50–200 years.[1592] The IPCC lifetime has been criticised because lifetime is not defined[1593] and because the IPCC has not factored in numerous known sinks of carbon dioxide.[1594,1595]

There is a considerable difference in the atmospheric carbon dioxide lifetime between the 37 independent measurements and calculations using six different methods and the IPCC computer model. This discrepancy has not been explained by the IPCC. Why is this important? If the carbon dioxide atmospheric lifetime were five years, then the amount of the total atmospheric carbon dioxide derived from fossil fuel burning would be 1.2%.[1596] In order to make the measurements of the atmospheric carbon dioxide lifetime agree with the IPCC assumption, it would be necessary to mix all the gas derived from the world's fossil fuel burning with a different reservoir that was five times larger than the atmosphere.[1597]

Maybe the IPCC's troubles derive from concerns that ice cores do not give reliable data for past atmospheres, including the pre-Industrial Revolution atmosphere. Maybe the IPCC's troubles derive from the fact that recent atmospheric carbon dioxide measurements were by a non-validated instrumental method[1598] where results were visually selected and "edited", deviating from unselected measurements of constant carbon dioxide levels by the highly accurate wet chemical methods at 19 stations in Europe.[1599]

The contribution of fossil fuel carbon dioxide to the atmosphere, and the lifetime of the gas in the atmosphere, are the cornerstone of the whole IPCC *raison d'être*. There is now no *raison d'être*. For the atmospheric carbon dioxide budget, marine adsorption and degassing, carbon dioxide from geological sources (volcanic degassing, cooking of rocks, mountain building, faulting; all totally ignored by the IPCC), carbon dioxide loss from weathering and carbon dioxide adsorption by ocean phytoplankton must be far more important than assumed by the IPCC.

There are many papers that use the same data set as the IPCC and derive a different conclusion. For example, studies in China concluded that global warming is not related only to carbon dioxide and that the effect of carbon dioxide on temperature is grossly exaggerated.[1600] An assertion by the IPCC is that global atmospheric temperature has risen by 0.7°C since 1850. What do you think would happen after the Little Ice Age? A temperature fall or a temperature rise?

The bulk of the global temperature rise was before massive industrialisation (1850 to 1940), then during the post-World War II economic boom temperature decreased while the industrial emissions of carbon dioxide greatly increased, and from 1976 to 1998 temperature increased. It has been static since 1998. We saw no change in the rate of carbon dioxide increase in both the Global Financial Crisis and COVID-19 crisis when industrial activity was greatly reduced.

The measurement of carbon dioxide by different methods is unresolved. The residency of carbon dioxide in the atmosphere is far less than that used by the IPCC and this greatly affects the estimation of the amount produced by humans. Not all sources and sinks of carbon dioxide are considered by the IPCC. The calculation of the transfer of carbon dioxide between the atmosphere and the oceans uses a body of incomplete data. By using various general circulation models, computer models are compared and for some strange reason agree with other computer models which used the same data set.[1601] There is clearly a lot more to learn about carbon dioxide.

A final word on measurement. The methods of measurement and data treatment for sea surface temperature, air temperature and carbon dioxide have errors, bias, lack of correlation between different methods, lack of validation, selective culling of data and uncertainties. If you want to measure something to obtain a predestined conclusion, it is easy.

While climate activists are getting into a lather about traces of a trace gas in the atmosphere, they ignore the main greenhouse gas. Water vapour.

Uncertainties around the role of clouds remains the biggest problem in the failed climate "science" models.

Massaging measurements: Climategate fraud

In November 2007, an international scandal erupted when 61 MB of emails internally circulated among the head honchos of the Climate Research Unit (CRU) of the UK's University of East Anglia were made public. The emails showed vast scales of international fraud was driven by the climate "science" community in officially-supported policy driven deception.

Phil Jones (Director of the CRU) demanded that data sets be ignored and massaged to justify the climate models that showed that human emissions of carbon dioxide drive global warming. It is these models that are still used by the UN's IPCC, underpin COP talk fests and fed into every school, university, bureaucracy, government and corporation to show that we will fry-and-die and it's all our fault.

Peter Thorne (UK Met Bureau) wrote in one of the leaked emails about the IPCC's 2007 report "*I also think the science is being manipulated to put a political spin on it which for all our sakes might not be too clever in the long run*". Too right. That's exactly what has transpired. Phil Jones (CRU) admits the political bias in the Summary for Policy Makers (SPM) which is written for journalists, bureaucrats and politicians. "*He says he'll read the IPCC Chapters! He hadn't as he said he thought they were politically biased. I assured him they were not. The SPM may be but not the chapters*".

An email from an IPCC coordinating lead author gave the game away "*The trick may be to decide on the main message and use that to decide what's included and what's left out*". The trick was used and there are many references showing that thousands of contrary valid scientific papers were left out by the IPCC (as I showed in my book *Heaven and Earth*). Is leaving out the contrary data genuine science or is it fraud?

Warmist Mike Hume also gave the game away in the Climategate emails "*...the debate around climate change is fundamentally about power and politics rather than the environment...There are not that many 'facts' about climate change which science can unequivocally reveal...*".

Giorgio Filippo, who contributed to all five IPCC reports, wrote "*I feel rather uncomfortable about using not only unpublished but also unreviewed material as the backbone of our conclusions (or any conclusions)...I feel that at this point there are very little rules and almost anything goes*".

True. Climategate showed how the climate industry is underpinned by fraud.

Email 5286 from Hans von Storch again gave the game away *"We should explain why we don't think the information robust yet. Climate research has become a postnormal science, with the intrusion of political demands and significant influence by activists driven by ideological (well meant) concerns"*.

If von Storch was honest, he would have resigned. Why not? Because, like his colleagues, he is unemployable outside a climate industry. Von Storch correctly wrote *"The concealment of dissent and uncertainty in favour of a politically good cause takes its toll on credibility, for the public is more intelligent than is usually assumed"*. The boys club having what they thought were secret discussions was exposed. They knew their data was dodgy but were quite happy to be politically manipulated.

As for IPCC climate "scientists" understanding the science of climate, the Climategate emails tell a different story. Richard Somerville write in 2004 *"We don't understand cloud feedbacks. We don't understand air-sea interactions. We don't understand aerosol indirect effects. The list is long"*. In effect, climate scientists don't understand the science of climate. For me that's nothing new because it's very complex, the data is almost useless and the past climate cycles are not integrated with the present.

Of interest in the Climategate emails, there is no mention of past climates and the fact that these were not related to atmospheric carbon dioxide. Warmist Kevin Trenberth wrote *"We are nowhere close to knowing where energy is going or whether clouds are changing to make the planet brighter"*. We still don't know the role of clouds in climate.

The infamous Michael Mann wrote in 2006 *"We certainly don't know the GLOBAL mean temperature anomaly very well, and nobody has ever claimed we do"*. The Climategate emails show us that climate scientists have kept the wheels of research grants and their own employment spinning knowing full well that there is huge uncertainty and they really know very little about what drives climate change. Human-induced global warming is the greatest scientific fraud in the history of time as shown by the Climategate writings of climate "scientists".

When two papers contrary to their consensus of scaring people witless were published by Chris de Freitas, CRU director Phil Jones and his mob pulled out all stops to get the editor sacked to prevent such papers being considered by the IPCC. Jones wrote (8th July 2004) "...*In can't see either of these papers being in the next IPCC report. Kevin and I will keep them out somehow, even if we have to re-define what the peer-review literature is*". Those green activists who prattle on about the peer review process do not know how it can be totally corrupted and how valid contrary scientific ideas are buried.

Michael Mann wrote on 3rd July 2003 "*It seems clear we have to go above...I think the community should ... as Mike H previously suggested in this eventuality, terminate its involvement with this journal at all levels – reviewing, editing, and submitting, and leave to wither way into oblivion and disrepute*". It's clear. If a scientific journal publishes papers against the party line, the climate mafia will destroy the journal. And this is the peer review system lauded by green activists.

According to Labor and the Greens Parties, climate change is fundamentally a moral issue. I agree. Non-scientific politicians, greens activists and ideologues are being led by the nose by fraudulent scientists and the followers don't have the ability, knowledge or inclination to ask simple searching questions or read contrary information.

One of Jones' colleagues, programmer Ian 'Harry' Harris who commented on "*[The] hopeless state of their (CRU) database. No uniform data integrity, it's just a catalogue of issues that continues to grow as they're found*" and "*This whole project is SUCH A MESS. Almost all the data we have in the CRU archive is exactly the same as in the GHCN [Global Historical Climate Network] archive used by the NOAA National Climatic Data Center*".

When interviewed by the BBC, Phil Jones of the Climate Research Unit stated[1602,1603] "*Surface temperature data are in such disarray they probably cannot be verified or replicated*".

The story in the UK is repeated in the US. In a series of e-mails obtained through a Freedom of Information Act request by the Competitive Enterprise Institute, additional doubt has surfaced over NASA's GISS temperature data.

Senior NASA scientist Reto Ruedy, in a response to *USA Today* reporter Doyle Rice in 2007, stated that it is clear that NASA GISS temperature data was flawed when he advised the reporter to "*continue using NCDC's (NOAA's National Climatic Data Center) for the U.S. means and Phil Jones' [HADCRU3] data for the global means. . .We are basically a modelling group and were forced into rudimentary analysis of global observed data in the 70s and early 80s... Now we happily combine NCDC's and Hadley Center data to ... evaluate our model results".*

There are serious concerns with what this e-mail exchange demonstrates, which is the use of the University of East Anglia's Climate Research Unit (CRU) data in what was thought to be an independent US data set. It appears that American data is partially derived from the corrupted data set that has been criticised as too political and unscientific as a result of the Climategate scandal. NASA's website states that NASA uses NOAA's GHCN.

Neither NASA nor the NOAA declare that their data is in such disarray and that data has disappeared yet many have expressed concern about how the data is collected, analysed and cherry picked.[1604] Some of the original criticism of NCDC's data, methods and analysis published by NOAA has been removed from the record[1605] but is recorded elsewhere.[1606]

Former NASA scientist Ed Long showed that after adjustment, the rural data trend agreed with the urban data trend when an artificial warming trend was introduced.[1607] Urban warming was allowed to remain in the urban data sets and a warming bias was artificially introduced to rural data sets that in their unadjusted state showed no warming.

If green activists, politicians or fellow travellers use the terms *"I believe in the science"*, *"respect the science"* or *"follow the science"* they clearly are not referring to the Climategate emails released in 2009, 2013 and 2015.

We're told that warming of 2°C above a pre-industrial level is some sort of tipping point after which we face doom. Phil Jones, Director of the Climate Research Unit at the University of East Anglia (UK) emailed on 6th September 2007 that the supposed 2°C limit was *"plucked out of thin air"*. As one who has followed this fraud for decades, it was obvious that 2°C and may other numbers are concocted.

It looks like Michael Mann's "hockey stick" was also plucked out of thin air and he never provided the primary data for other scientists to validate his "hockey stick". Mann's work has underpinned the UN's climate policy, the IPCC and various Democrat US presidents. He, one of the Climategate fraudsters and of "hockey stick" infamy, allowed no discussion or criticism of his science. Scientific criticism is part of the process but Mann was too high and mighty to have his work exposed to scrutiny. As soon as it was, Mann sued for defamation.

The defendant in the libel trial, Canadian climatologist Tim Ball, instructed his British Columbia attorneys to trigger mandatory punitive court sanctions, including a ruling that Mann acted with criminal intent when using public funds to commit climate data fraud. Penn State climate "scientist", Michael 'hockey stick' Mann committed contempt of court in his own defamation action against Ball by defying the judge and refusing to surrender data for open court examination.[1608] This was not very smart as his case was thrown out after years of delays attempting to wear down Ball, costs were awarded against him and punitive damages were sought. He also risks US authorities looking at how he spent taxpayer's money on his so-called research. Mann faces other litigation.[1609]

Mann's cherry-picked version of science makes the Medieval Warm Period (MWP) disappear and shows a pronounced upward 'tick' in the late 20th century (the blade of his 'hockey stick'). Below that, Ball's graph, using more reliable and widely available public data, shows a much warmer MWP, with temperatures hotter than today, and showing current temperatures well within natural variation. The IPCC used Mann's hockey stick in their 2001 report.

As Ball explains *"We believe he [Mann] withheld on the basis of a US court ruling that it was all his intellectual property. This ruling was made despite the fact the US taxpayer paid for the research and the research results were used as the basis of literally earth-shattering policies on energy and environment. The problem for him is that the Canadian court holds that you cannot withhold documents that are central to your charge of defamation regardless of the US ruling".*

It gets worse for the litigious Mann. Close behind Ball is celebrated writer Mark Steyn. Steyn also defends himself against another one of Mann's suits, this time in Washington DC. Steyn claims Mann *"has perverted the norms of science on an industrial scale"*. American climate scientist, Judith Curry, has submitted to the court an Amicus Curiae legal brief exposing Mann.

The world can now see that his six-year legal gambit to silence his most effective critics and chill scientific debate has spectacularly backfired. The perpetrator of the biggest criminal "assault on science" has now become clear: Mann, utterly damned by his contempt of the court order to show his dodgy data.

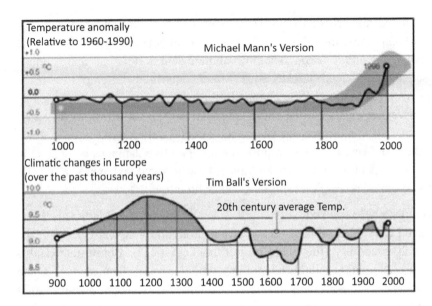

Figure 50: *The Michael Mann temperature history compared to the Ball temperature history which is in accord with proxies, solar data and history.*

When changes in data sets were introduced in 2009, there was a sudden warming. However, satellite temperature monitoring has given an alternative to land-based stations and this record is increasingly diverging from the land-based measurements consistent with adding a warming bias to the land records.

The Climategate emails show us that climate scientists have kept the wheels of research grants and their own employment spinning knowing full well that there is huge uncertainty and they really know very little about what drives climate change. Human-induced global warming is the greatest scientific fraud in the history of time as shown by the Climategate writings of climate scientists.

Net Zero

Why even bother when a carbon dioxide rise follows temperature rise. Wishful thinking is not sound public policy.

If we assume that carbon dioxide drives climate change, then a reduction of Australia's carbon dioxide emissions should reduce the global temperature. Right. Let's do some very simple calculations.

Reduction of emissions	Reduction of global temperature
0 %	0.0000 ºC
5 %	0.0008 ºC
10 %	0.0015 ºC
20 %	0.0031 ºC
50 %	0.0077 ºC
100 %	0.0154 ºC

Table 3: *Effect of emissions reduction by Australia on global temperature*

The ALP and the Greens Party are pushing for 100% reduction in emissions whereas the Liberal-National Coalition are pushing for a 50% reduction by 2050. All States, Territories and the Commonwealth have signed up for Net Zero. If I stand up or move from one room to another, the temperature change is far greater than 0.0154°C. Why bother? Why bankrupt a nation with a futile gesture? To make matters worse, no one has ever shown that human emissions of carbon dioxide drive global warming. Why even spend a second worrying about a 0.0154°C temperature rise?

Let's just ignore the annoying little fact that there has been no global warming since 1998 yet carbon dioxide emissions have been increasing. If every person on Earth stopped the planet warming by the predicted 0.24°C from 2011 to 2020 as a result of human carbon dioxide emissions, it would cost every man, woman and child on the planet $60,000. This is 60% of global GDP and 22 times the maximum estimates for doing nothing about possible global warming.

Are you prepared to pay $60,000 for less than a quarter of a degree temperature change? I don't think so. Especially as the planet has been cooling in the 21st century and there is no correlation between temperature changes and human emissions of carbon dioxide. Without correlation there can be no causation.

Now you can see why the climate alarmists, governments, industry and banks don't want their policies questioned because it could stop the flow of money from your pocket into theirs. It's called Net Zero. They've had a taste of it with the flight to wind and solar and they loved taking your money via subsidies.

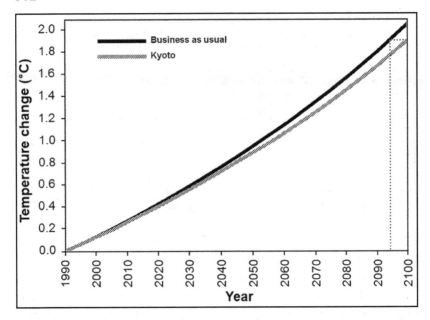

Figure 51: *The effect on temperature of 37 western countries reducing emissions by 5% from their 1990 emissions according to the Kyoto Protocol on the assumption that human emissions of carbon dioxide drive climate change. In effect, the 0.2°C temperature reduction ignores natural climate change and show the pointlessness of trying to legislate to change climate.*

It is the young people who will pay. It will stop them owning a house or having the same standard of living as their parents. Government action on human-induced global warming has nothing to do with climate or the environment. It is a method to take money out of our pockets, to keep the big banks happy and to take away some of our basic freedoms.

Calls for moves for clean energy are driven by those who will profit from its adoption. Why on Earth should we pay for expensive clean energy (that is not clean) when we already have cheap clean energy? In Australia, the black-coal power stations are really marvellous.

Burning coal heats water to make steam. This steam drives turbines that produce electricity. The boilers are 96% efficient, exhaust heat is captured and reheats water, the 4% heat loss is through the boiler wall, some moist air is condensed, precipitators remove 99.99% of small particles derived from coal burning and carbon dioxide escapes up the stack to the atmosphere.

And what does the carbon dioxide emitted from the stacks do? Fertilises plants. What can be wrong with this? Plants do not know whether the carbon dioxide they use is from human emissions or from natural emissions.

Coal-fired power stations generate electricity efficiently with some 8% of the generation cost being the cost of coal. Cheap electricity produces jobs. The world was once envious of Australia's cheap electricity which attracted investment. China is shutting down small old dirty inefficient generators and building a new large modern efficient coal power station every week. Australia did that years ago. If Australia had a 5% decrease of carbon dioxide emissions, the increase in Chinese emissions alone is expected to be over 100 times as large as Australia's reduction hence, whatever Australia does, it will have no effect on the planet.

Australia, like the USA, has coal-fired power stations because we have hundreds of years of coal reserves. It is crazy to think that what is sinful in Australia is good in China. The politics of the Green Party clearly are not pragmatic as they do not involve global solutions to a perceived global problem. The solution to perceived problems is nuclear energy. History shows that ideology and economics have never been good bedfellows.

What on Earth do the hundreds of public servants in Australia's Department of Climate Change and Energy Efficiency do each day? They certainly don't change climate but they certainly consume huge amounts of your money.

How can these public servants come home at night and proudly feel that they have earned their money? Maybe they think that dreaming up new ways to control people is actually work. What really is the point of a nation-bankrupting Net Zero that only decreases global temperature by 0.0154°C by 2050? Has it really been thought through?

What is Net Zero going to cost? We never get told. I don't sign blank cheques. Why should I accept Net Zero unless I have seen the costs, a financial due diligence and a financial QA/QC?

Across the world politicians are going out of their way to promise bank-breakingly expensive climate policies culminating in Net Zero. President Joe Biden has promised to spend $500 billion each year until 2050 on climate policies, about 13% of the entire Federal revenue. The EU has gone one better and is planning to spend 25% of its annual budget on climate.

Where will the money come from? What goods and services will be cut to pay for the climate policies? Who will be personally responsible if it all goes horribly wrong? Where are the annual targets and budgets up until the year 2050 so we can hold the decision-making politicians' feet to the fire?

In many wealthy countries, politicians have been bombarded by climate activists and bureaucrats who want us to achieve be carbon neutrality by mid-century. They just cave in. For Australia to go Net Zero carbon dioxide emissions by 2050 will cost about $1.13 trillion. This is money we don't have in a debt-ridden country. Few countries have costed the pain of going Net Zero. Even fewer have made such costings public. It is very much in the interest of woke banks to have massively expensive Net Zero schemes and they don't care one *iota* about whether ordinary people lose their jobs. Net Zero is about the transfer of money and has nothing to do with climate.

Not one of the nine governments in Australia has announced the number of people in heavy industry, smelting, manufacturing, transport, farming, fishing, shipping, coal and power generating areas who will be booted out of a job. Calculations[1610] can be made using published government statistics.[1611]

To achieve Net Zero will be a breeze. All fossil fuel burning equipment will have to be decommissioned such as tractors, harvesters, food processors, trucks, cars, ambulances, trams, trains, buses, boats and aeroplanes that currently annually use 1.1 million GWh and we will have to exchange these fossil fuel-burning engines with zero emissions engines.

Let's stress test the system now. Farmers and truckies should go on strike immediately for a month and not deliver food, medicines and essentials to the cities. Workers in coal-powered fire stations could also join in. There would be food and car fuel riots within a few days and a total breakdown of society within a fortnight. Where are the unions when they are really needed?

We are already seeing glimpses of the glorious Nirvana of Net Zero in the Northern Hemisphere autumn of 2021. Germany and the UK have snookered themselves by relying on weather-dependent energy. Net Zero will unleash a monster. We are now seeing massive increases in prices, scarcity of energy and food and increased unemployment. The next step along the road to disaster is the environmental devastation and the lack of land for growing food.[1612] The final step will be a hungry cold unemployed society with debt up to their ears. This imposes unthinkable sovereign risk.

Stationary high-pressure systems have created a collapse in European wind power[1613] and some UK energy companies are now facing bankruptcy.[1614] Because of a ban on fracking for gas, lack of gas storage facilities and of Asia signing long-term gas contracts that have pulled a large amount of gas off the spot market, the gas price has skyrocketed.[1615] The EU and the UK are now dependent upon Russia for gas.[1616] Europe was outsmarted by Putin[1617] and is now wrapped around his little finger.[1618] The UK and

Germany are resurrecting old coal-fired and nuclear power stations but these just can't be turned off and on at will.[1619] Germany does not have enough operating coal mines and is now trying to buy Russian coal to stockpile for winter.[1620]

Unemployment is increasing, supermarket shelves are being emptied and household costs, especially energy, have gone through the roof. The political backlash from the self-imposed energy crisis in the EU has already started. The UK would be awash with gas if fracking was allowed in its North Sea oilfields.[1621] Australia should learn from these disasters and keep well away from the path to disaster.

The proposed French eco-tax, billed by Macron's government as a means of facilitating an energy transition, was the last straw for those in the provinces who felt ignored by the city-based political elites No nationality can protest like the French and the continual Yellow Vest protests forced Macron to drop his environmental tax on heavy vehicles. Alternative methods of transport did not exist and the government showed that it was way out of contact with the people. People's anger did not subside and protests continued. The same sort of protests are possible in the UK, US and Australia because centralised government and the political elites have no idea how hard-working politically marginalised people survive. A green tax here, an environmental tax there and, before you know it, the government has been kicked out.

China has kicked an own goal. Its energy crisis was sown in the five-year national strategic plan that ran from 2016 to 2020 and, because of a hissy fit, it stopped importing top quality Australian thermal coal. More inferior coal comes long distances overland from Russia and Mongolia and many of its domestic coal mines have reopened. These coals are wet, high ash, produce smog and acid rain and don't have the bang for the buck that Australian coals have with 6,000 kcal/kg and only 15% moisture.[1622]

Power was rationed in China over the 2021 summer, factories closed or reduced production, street lights and domestic electricity were turned off, lifts in apartment complexes stopped and heavy energy users such as aluminium smelters and steel mills cut output[1623,1624] (hence the collapse in the Platts 62 iron ore price and the highest global prices for aluminium in more than a decade).[1625] Food security has become a concern. China has bitterly cold winters and 2021-2022 will be no exception. This will stress test the Chinese energy and food system and any civil unrest in China could have global implications. They have now started importing

Australian coal because of an energy crisis.

Food and energy are the basic commodities required by humans. They are inextricably linked. The UK, EU and Chinese examples above show that unless energy is abundant, cheap and reliable, food supply is threatened and civil unrest raises its head. There can be no better example than Australia. If the Net Zero path is followed, there would be no land for growing food.

To replace fossil fuel energy in Australia, an additional 119,000 wind turbines at an economy-destroying cost occupying an area of 60,000 square kilometres would be required, leaving little land for growing food. This area is the same as three million MCG stadiums. Construction of the turbines will use countless million tonnes of coal, 36 million tonnes of metals and 145 million tonnes of concrete. The energy for making the steel and concrete is enough to cripple any economy.

On top of this, energy would be required to make six million rooftop solar panels and solar industrial complexes leaving even less land for growing food. If fossil fuel machines could not be decommissioned, then under Net Zero carbon credits would need to be purchased. There would be 516,000 GWh of fossil fuel-burning for essential and emergency infrastructure that could not be changed so carbon credits would have to be purchased. The planting of 17 billion trees each year as carbon offsetting for a mere $238 billion and over a land area of 201 million hectares, which is about 50% of the current agricultural land in Australia at present, would be needed. Where will food be grown?

The task has not been achieved anywhere in the world. When Germany tried to increase its proportion of renewables, it led to chaos, unemployment, unreliable electricity, cost escalation and energy reliance on an unfriendly Russia.

If Australia is going to try to add to the existing 33,000 GWh of wind and solar by a factor of 337 to reach 1.1 million GWh, then you had better go bird watching now as none will be left in our Net Zero world. Take some food with you as it will become a precious resource.

The only beneficiary of a bankrupting Net Zero policy will be the international mining industry and China, the manufacturer of wind turbines and solar panels. All others will be losers.

There are just a few other little problems. By-products from fossil fuels provide us with over 5000 essentials of life. If biofuels rather than fossil fuels are used to make plastics then, I'm sorry, we have run out of land. No plastics can be made without fossil fuels. The cheapest and most efficient way of insulating a wire carrying electricity is with colour-coded plastic insulation. Imagine a world with no insulated electric wires. Batteries, electric cars and houses would be continually exploding and burning down and people would be zapped all-day-every-day going about their daily business.

Those advocating Net Zero must be passionate supporters of a cradle-to-grave nuclear industry in Australia. If not, then the green activist warriors are on a mission to destroy prosperity and demonstrate no concern for the environment or climate. Maybe we are at the dying stage of mass prosperity brought to the Western World by coal? There are no examples of low-energy societies providing a decent standard of living for their citizens.

If US climate activists have their way Americans must cut energy use by 90%, live in 640 square feet, fly once every three years, reduce calorie intake to 60% of the FAO's minimum for survival and have a clothing allowance of 4 kg a year (as long as clothing is washed only 20 times a year).[1626] Welcome to the Net Zero world. Maybe you wondered why I earlier described the green activists as a death cult.

Even if the West decides to bankrupt itself by virtue signalling, other countries will not follow. Climate policy has already brought created civil unrest[1627] and has brought down many political leaders.[1628]

Few people have ever warned about the financial and human costs of Net Zero. The harsh reality is that an energy transition, Net Zero or any other socialistic idealistic scheme creates higher prices for and shortages of everything. Industry struggles, cuts production and costs and puts people out of work. A reduction in food and freezing in the dark are a perfect recipe for governments to be thrown out at election time, civil unrest or a revolution.

I'll think I'll move to a more realistic country that has an economy using perpetual motion machines, fairy dust and unicorn pee.

UN's Intergovernmental Panel on Climate Change, Sixth Assessment Report (AR6)

All IPCC documents are political documents. The IPCC does not undertake any research on climate, they are not a scientific organisation and have a brief to promote the view that human emissions of carbon dioxide will result in dangerous warming. Various climate "scientists", green NGOs and green activists compile what they claim is a scientific report. The most illuminating part of the lengthy scientific reports is not what is reported but what was not reported. As the Climategate fraud shows, IPCC authors have an active policy of not citing any science that does not fit the party line.

The IPCC's AR6 2021 document has a bodybuilding weight at 3,949 pages. There is an introductory 41-page Summary for Policy Makers that gives grabs for journalists who do not read the scientific part of the reports. Terms such as "code red" are used by the IPCC authors in the Summary to whip the media into a frenzy.[1629,1630] Loud hysterical claims are not repeated validated evidence and the evidence presented for catastrophic climate change in the scientific body of the report is weak.[1631]

Summary for Policy Makers is generally released months before the main report (except for the 2021 AR6 Report). The Summary for Policy Makers is full of apocalyptic hype for the media who do not read the main report. In the body of the report, the true story of the planet's climate and associated uncertainties is described. For example *"In Climate Research and modelling, we should recognise that we are dealing with a coupled non-linear chaotic system, and therefore that the long-term prediction of future climate states is not possible".*[1632] They forgot to mention how many variables contribute to climate.

The AR6 report is true to form and has winnowed thousands of temperature proxies down to the chosen few which show that the hundreds of years of warmth in the Medieval Warming and the bitterly cold Little Ice Age did not exist. The fraudulent hockey stick of Mann reappeared after making its first appearance in 2001. The previous AR5 report stated: *"On a continental scale, temperature reconstructions of the medieval climate anomaly (years 950 to 1250) show with high confidence intervals of decades that were as warm in some regions as in the late 20th century".*

No new evidence was given to show why the position taken in the AR5 report was reversed. Hockey sticks don't die, they get resurrected with increasingly corrupt UN lackeys. The main IPCC brief assumes all warming is caused by human emissions. The 18-year pause shows that this is just nonsense.

The warming from 1850-1900 to 2011-2020 is given as 1.07°C. This is nonsense. No thermometer between 1850 to about 1970 could measure temperature to an accuracy of a hundredth of a degree. There is no explanation given for the cooling of 0.00024°C per year from April 2014 to June 2021. If carbon dioxide drives global warming, it is not possible to have a pause or cooling during a time of rapidly increasing human emissions of carbon dioxide.

IPCC climate models continue to fail and are not in accord with satellite and balloon measurements, are highly suspect and use outdated criteria. The models continue to show modelled warming is far greater than measured warming. This is well known.[1633] Hidden in the voluminous scientific part of the report is data that shows if fossil fuel consumption was stopped, then it would only make a 0.2°C change to global temperature yet some of the model speculations in the Summary screech that there will be a 5.7°C warming by 2081-2100.[1634]

The main scientific arguments used to support the AR6 claims were extreme rainfall, droughts, tropical cyclones and sea level rise. As shown earlier in this book, there is nothing alarming about these four natural events. The IPPC's statistical treatment of all data has consistently been challenged by statistician Pat Michaels, the "homogenised" temperature data to construct definitive models is invalid, models are claimed to be evidence and old failed models are warmed up and represented.[1635]

The scientific narrowness of climate "scientists" is such that they don't even know that the planet is losing heat and most of this heat is transferred from below into the deep oceans. Climate "scientists" don't have a clue about what will happen to clouds and temperatures if the planet warms.[1636]

The IPCC's AR6 report assumes the Sun has a negligible effect on temperature trends. This is wrong. A comprehensive compilation of 16 estimates of solar irradiance using five methods of estimating Northern Hemisphere temperatures shows that urban bias gives a much greater warming than previous estimates and that solar irradiance can account for temperature changes in the Northern Hemisphere.[1637] This compilation was published 4 months before the IPCC AR6 report was released. It was ignored.

The IPCC AR6 report wants to use a period in the geological past as an example of what could happen today. This time, the Palaeocene-Eocene Maximum (PETM) from 56.3 to 55.9 million years ago was when the average global surface temperature was up to 26°C compared to the 14.5°C today.[1638] The ocean pH dropped slightly, atmospheric carbon dioxide was higher than now and there was limited ocean extinction and increased mammal diversity.

There was no polar ice at that time, there was no Tibetan Plateau interfering with the jet stream, ocean currents were very different from today because Antarctica and South America were joined and sea level was higher. The origin of the PETM is unknown although candidate are a methane release from the oceans, volcanicity and oxygen depletion. There is no comparison between the PETM and the present yet the IPCC provide incomplete unrelated information as part of their scare campaign. The IPCC's attempt at geophysics is a fail.

They chose a warm time when the atmospheric carbon dioxide was higher than now and, by torturing the data as much as possible until it confessed, they could only show that half the carbon dioxide in the PETM could account for the warming.[1639] If there were another PETM, humans would thrive.

When IPCC reports are evaluated in the cool light of day after the media storm of impending catastrophes, a different story emerges. Despite the IPCC claiming it had increasing confidence in the role of carbon dioxide in global warming, its use of words like *"unequivocal"* is a warning sign that the IPCC know their methodology is flawed. A recent study showed that the IPCC continues to make fundamental mathematic errors in its statistical analysis.[1640]

Concurrent with this statement, in the Summary for Policy Makers there are climate models with long term predictions showing that we are going to fry-and-die. I don't trust the IPCC. Why should you?

What is not considered by the IPCC is important. The AR6 Report shows an absence of humanity and an overload of hypocrisy. There is no mention made of the 6,000 petroleum products that we use to make our lives better. There is no suggestion that anything suggested by the IPCC will change the fact that 11,000 children, mainly in Third World countries, are dying each day. There is no mention that climate-disaster deaths have reduced by 98% over the last century.

There is no mention that there are 20,000 private jets, 10,000 superyachts, 23,000 commercial airliners that carry four billion passengers annually, 56,000 fossil fuel burning merchant ships and 300 cruise liners. They fail to mention that there are more than six billion people living on less than $10 a day and billions with little to no electricity. Most importantly, they fail to mention that their decarbonising of the planet will bring humanity back to the gripping poverty of 500 years ago.[1641]

In my view, the shambolic data base, cooking the books, methodology, omissions and errors make the IPCC reports scientifically useless. No pollical policy can be based on such a weak document. In 2016, a Queensland Ph.D. student working from home was able to find errors in the key HadCRUT4 data which is the IPCC's favourite data set maintained by the UK's Met Office Hadley Centre at the University of East Anglia.

These errors had not been found by an army of the 2,100 employees at the £226 million *per annum* institute. The Hadley Centre did not disagree with any of the errors found. Some errors had been present for decades. Adjustments for site moves were not made and the older records were adjusted to give cooling to compensate for buildings. However, the buildings had not been constructed when the measurements were made.

The Summary for Policy Makers is used by the UN to persuade legislators to part with $100 billion a year to the UN Green Climate Fund. The IPCC is not about climate, people or the environment. It's about a massive wealth transfer that goes through the sticky fingers in the UN before trickling down to a few of the multitude of unfortunates.

Follow the money.

COP

There have now been 26 Committee of Parties (COP) meetings. Why didn't the alleged problems get solved after one or two meetings? Are these jaunts or problem-solving missions? Long ago they evolved from COP gabfests to FLOP meetings with an army of hangers-on, mostly funded by taxpayers. These conferences are not about climate or the environment. They are about the transfer of wealth.

Every four years some 11,000 athletes compete at the Olympics. There is a bigger hanger on and press contingent. The annual UN COP conferences attract even more. COP conferences give us an insight into eternal life.

The world had its last chance to change its filthy polluting ways of emissions plant food at the Copenhagen COP15 2009 conference. Again the world was presented with a last chance at Cancun 2010. The very last chance was again presented in Durban in 2011. No wait, the very very very last chance was presented to the world in Doha in 2012.

At COP19 in Warsaw 2013, an ultimatum was given and it was definitely the last chance. The last chance ultimatum was given again at COP20 in Lima in 2014. The high-profile Paris COP21 climate conference presented the world with its last chance. At the 2015 COP21 gabfest in Paris, 196 parties signed a non-binding agreement on 12th December 2015.[1642] Australia committed to reduce its greenhouse gas emissions by 26 to 28% below the 2005 levels before 2030. The devil is in the detail. Although the media make a lot of noise about the Paris Accord, each country sets its own targets based on its own self-interest and economy and the biggest emitter of carbon dioxide in the world signed nothing. Why does Australia even bother wasting taxpayers' money in sending delegates?

The UN Paris Climate Change Treaty became law in September 2016 committing the global community to arrest global temperature rise below 2° C. How is this law enforced? How does the climate elite think that by sitting around together sipping Bollinger and negotiating a treaty that they have the power to smugly twiddle the dials and keep global temperature rise to less than 2°C? This assumes that only one factor drives climate change (i.e. human emissions of carbon dioxide). The arrogance and ignorance are breathtaking.

The science was settled, the world was given its large chance yet again at COP22 in Marrakech in 2016. No one seemed to listen so at the COP23 talk fest in Bonn in 2017, the world was given its very very last chance.

No one thought this was serious and at Katowice COP24 2018 the world was given a very serious warning that this was definitely the last chance to change the ways of the world. No one was listening and, again the rhetoric was hysterical and the world was given its last chance at COP25 in Madrid in 2019.

COVID-19 travel restrictions has meant that the COP26 Glasgow gab fest was postponed until 2021. The host country, Scotland, set a legal target to cut greenhouse gas emissions to Net Zero by 2045 which is five years ahead of the UK governments aspirations. Scotland scored a hat-trick of own goals by missing its annual emissions target three years in a row.[1643]

It seems that every one of the thousands of green activists, politicians and their bureaucrats flying from one side of the planet to the other have now had 26 all-expenses-paid jaunts to tell us that we must stop flying and living the high life, normally at our expense. And who attends?

The Madrid COP25 2019 climate talks had 13,643 people representing specific parties from 198 countries, 9,987 from observer organisations (scientists, business groups and NGOs) and 3,076 journalists who will uncritically wrote what we know they will write. And a good time was had by all at someone else's expense. The delegate score sheet shows that countries that can least afford to travel around the world have the most delegates.

The number of delegates for the top 30 were Côte d'Ivoire (348), Democratic Republic of the Congo (293), Spain (172), Brazil (168), Congo (165), Indonesia (163), Guinea (159), Canada (145), Bangladesh (143), Japan (138), Morocco (137), Chile (136), EU (125), France (124), Sudan (121), Uganda (121), Benin (116), Senegal (108), Ghana (106), Burkino Faso (104), Germany (102), Dominican Republic (100), Zimbabwe (98), Kenya (95), United Arab Emirates (93), Honduras (83), Tunisia (82) and Turkey (81). In my eyes, both the US (78) and Australia (20) had excessively large delegations. Looks like a scheme of buying a COP vote by the UN in return for an all expenses paid European jaunt.

Europeans came represented by the EU and their home countries and were by far the biggest number of delegates. The biggest carbon dioxide emitters China (76) and India (35) were swamped by countries that have no coal-fired electricity. The Cape Verde Islands were depopulated when 37 delegates descended on Madrid. Many countries that sent a large number of delegates can't even feed their own people

The COP gabfests are not a climate conference. They are self-interest economic conferences trying to create new ways to redistribute the world's hard-earned wealth via the UN's sticky fingers.

In the months before the Glasgow COP26 talk fest, the UN Secretary-General Antonio Guterres speaking at the launch of the IPCC's AR6 Report claimed that the world was *"out of time"* to act on climate change and the consequences would be catastrophic unless there were immediate large-scale reductions in greenhouse gases.

He claimed that no country was safe from extreme weather and claimed that the 2021 floods in Germany, Hurricane Ida and a heatwave in the Pacific northwest of the US were examples.[1644] These weather events must be viewed in the context of historical weather events and, when they are, there is nothing extraordinary about weather events in 2021. Guterres was telling lies.

Applications by 15 nuclear bodies to attend and display exhibits at the COP26 gabfest in Glasgow were rejected by Mr Alok Sharma's COP26 Unit of the UK Cabinet Office.[1645] If COP26 is serious about reducing carbon dioxide emissions, a fundamental existing industry and technology that could help achieve this has to be nuclear. Clearly the IPCC and COP meetings have been captured by the wind and solar industry and has nothing to do with reducing carbon dioxide emissions, the environment or climate.

You can always tell that the next UN COP meeting is coming. The number of press releases increases, green activists and politicians increase their international air travel, the IPCC and other discredited groups make louder and even more apocalyptic predictions, the IPCC try to release reports such as AR6 to soften up the public and media in catastrophe overdrive yet they don't ask critical questions. For example, we were told that the Great Barrier Reef has greatly reduced in area since 2009. It did. What we were not told is that it grew back.

The 25,000 that travelled to the Glasgow COP26 talkfest were not required to have COVID-19 PCR swab tests upon arrival in the UK.[1646] The exceptions were introduced over the UK Bank Holiday Weekend when it was hoped no one would be watching. COP26 delegates trying to save the world from a confected climate crisis are clearly the elite who are too high and mighty to undertake simple public health testing.

Media scare campaigns always precede a COP gabfest. President Biden's climate czar said that the world is doomed unless 20 nations take climate action.[1647] These 20 countries[1648] are responsible for 80% of all global emissions. At President Biden's Climate Summit in April 2021, only 55% of global economies had committed to taking measures to achieve 1.5°C. Lord knows what this means and how it would be achieved.

Kerry went on a special trip in a fossil fuel-burning aeroplane to try to convince China that they should reduce emissions and they impolitely told him to get lost. The UK's Alok Sharma also went to China, panda-ed to them and hailed their efforts to tackle climate change.[1649] John Kerry went to India to persuade them that it should go Net Zero. He was told to take his bat and ball and go home.[1650] Did Kerry get into the slums and see how cheap coal-fired electricity could help the poor?

Just before COP26, various health journals tried to frame a joint editorial arguing that climate change is a health emergency. It certainly is. The anxiety and stress of constant bombardment with apocalyptic propaganda can create health problems. By far the greatest health problems arise from a contraction of economic development. The advances in medicine, diet and general welfare derive from economic material development which means we live longer healthier lives than any time in human history.[1651]

It must be right, I read it in the newspapers

Journalists are illiterate.

Their knowledge of science is the literary equivalent of not being able to read or write. In today's scientific world, they are therefore illiterate.

Give me a journalist and, after three questions, I will show you that they don't even know the basics of schoolchild science.

There is a slanted reporting of all matters climate and, because of scientific illiteracy and the need to sensationalise, journalists do not have the basic knowledge to know what questions to ask. A far greater breadth of information and opinion can be obtained with a simple search on the internet.

However, the internet is also the home for nutters to vent their spleens and background knowledge is required before any information can be deemed credible. Wikipedia is not a source of credible information. It even has my birthdate and career details wrong.

The BBC once was trusted. In 2006, the head of news Tony Hall authorised a seminar to determine how the BBC should report global warming and called on help from advisors. They described the 28 advisors as "*the best scientific experts*" and the advisors briefed about 30 BBC officials.

The BBC accepted the advice of "*the best scientific experts*" that the science was settled and dissenters were not allowed equal treatment. The seminar was one of many in a series that led to the BBC becoming an uncritical green activist propaganda network.

A lone pensioner in west Wales asked the BBC for the names of the 28 "experts". Using two barristers and four solicitors, the BBC used public money to successfully fight a freedom of information request to keep the names concealed.

Why did the BBC waste public money on fighting a freedom of information application if they had nothing to hide from the public? This indicates that they knew that they were presenting a biased commentary on climate.

In 2012, a blogger discovered the list of names on an obscure internet file. Two were climate scientists, one was another scientist from a different discipline and the remaining 25 "experts" were from Greenpeace, various business interests talking their own book and miscellaneous odd bods and sods with no scientific expertise. At times, the BBC has shown films made and supplied by green activist groups.[1652] Who can trust the BBC?

An annual style guide has been published by Associated Press. This is the Bible for many journalists. It favours left-wing bias over impartiality. It previously told journalists to refer to *"global warming"* as *"climate change"* and many journalists use these meaningless words.[1653]

Journalists' use of the words global warming inferring that all warming must be of human origin. This has not yet been shown. Using climate change, the journalist infers that whatever changes with climate or weather, it must be due to humans.

Journalists are clearly not aware that climate has changed for billions of years before humans were on Earth and there are cyclical warming and cooling periods. Is warming due to humans? Is cooling due to humans? Is any change whatsoever due to humans? What does the past show us about the present? The style promotes the use of propaganda rather than validated facts.

The guide now instructs lazy journalists to refer to climate sceptics who were previously labelled as *"climate deniers"* as *"doubters"*. Associated Press has made itself the science overlord and clearly does not want journalists to know that scepticism, doubt and criticism form the basis of science.

The journalists who slavishly follow the style guide have not checked whether catastrophic global warming predictions have come true. Why bother? A sensational scare story with goodies and baddies keeps bread and wine on the table whereas impartial reporting could threaten employment.

Serious journalists don't need a style guide and can think for themselves. Good journalism involves asking uncomfortable questions, being sceptical of everything, taking no position on any issue and putting the feet to the fire of those that live off the public purse. The great journalists I know contact me to validate a matter of science.

Five years after Charlie Hebdo journalists, editors and cartoonists were murdered in Paris by Islamic fundamentalists who considered them blasphemous, the Winnipeg Free Press (7th January 2020) had an editorial entitled *"Time to silence voices of denial"*. Climate change is apparently *"undoubtedly the most urgent problem of our time"*, *"many people still deny it exists"* and that *"scientists have reached a near-universal consensus on human-made climate change"*.

Many would have a different view about what may be the urgent problem of our time, no one denies that climate change takes place and there is certainly a consensus by those who live off taxpayers' money that this money should keep flowing so they can continue to scare us.

There was also a consensus against the Copernican theory of the Earth and Wegener's theory of continental drift. At times when we are required to be supportive of religious, ethnic, racial and sexual minorities, the editorial sought to silence a scientific minority by using twisted gaslighting to twice call the scientific minority *"climate change deniers"*. I have never denied climate change. A simple Google search will have me pegged as a "denier".

The Winnipeg Free Press editorial excelled by stating *"...in 2020, there's no longer room for debate about the existence of climate change...We need our leaders to make climate change a priority issue, but that can't happen until they, and we, stop wasting precious time with circular debates and denials while the world burns down around us"*.

This is exactly how totalitarian socialism works. The irony of this is that if there is future global cooling, Winnipeg will be one of the first places to be covered in thick ice. It's happened before and will happen again.

The Conversation is an academic website that has banned climate change sceptics. The online editor ex ABC journalist Misha Ketchell stated *"That's why the editorial team in Australia is implanting a zero-tolerance approach to moderating climate change deniers and sceptics. Not only will we be removing their comments, we'll be locking their accounts"*.

Who are these scientifically illiterate journalists to edit comments on the science of climate? This is how the battle for ideas and scientific debate is dealt within the media by those who do not understand the basics of science. This is what happens in totalitarian regimes like China.

The Conversation was founded with taxpayers' support and is funded by public-funded institutions such as the CSIRO, universities (Melbourne, Monash, RMIT, UTS, UWA, ANU, ACU, Canberra, CDU, CQU, Curtin, Deakin, ECU, Flinders, Griffith, JCU, La Trobe, Massey, Murdoch, Newcastle, Notre Dame, QUT, Swinburne, Sydney UniSA, SCU, USQ, UNE, UNSW, UQ, UTAS, UWS, VU and Wollongong) and strategic partners are the Commonwealth Bank, Corrs Chambers Westgarth and the Victorian State Government.

Taxpayers' money is used for the silencing of dissent and the deliberate shrinking and censoring of scientific, academic, environmental, economic and political debate. Those who preach the loudest about tolerance are those least likely to show it. University benefactors beware. Do you really know where your gift is going?

The Editor-in-Chief of the *Guardian News* and Media sent a circular note to all staff about the language used in covering the environment. They want to up the ante and rather than use the passive detached factual language of reporting, they want to use hysterical language.

She wrote "*Therefore we would like to change the terms we use as follows:*

Use climate emergency, crisis or breakdown instead of climate change
Use global heating instead of global warming
Use wildlife instead of biodiversity (where appropriate)
Use fish populations instead of fish stocks
Use climate science denier or climate denier instead of climate sceptic"

They are certainly not a news service and have proudly announced that they are an emotive agent of propaganda.

The government *Landesanstat für Median* in Nordrhein-Westfalen has attempted to silence the 19-year old Naomi Seibt by demanding she remove two of her YouTube videos on the grounds that that they are not "climate friendly". Her videos contain policy prescriptions and make reference to the Heartland Institute. YouTube is quite happy to publish porn but not those of dissenting informed opinion. When has it been illegal to speak the truth or express an opinion in Germany? Oh yes, now I remember, it was nearly 90 years ago.

Temperatures are always changing, as is climate, and the pattern of change tends to follow cycles. Most reporters don't understand this, including science reporters. As the Earth spins on its axis relative to the Sun creating day and night, temperature changes are in the order of 10°C. Temperatures change with the seasons because of the tilt of the Earth relative to its orbit around the Sun. The change in the Earth's orbit around the Sun gives three major cycles of climate. All temperature changes are essentially driven by the Earth's position and distance from the Sun which in turn has its own measurable and predictable cycles. Yet we are told that a trace gas has a more significant effect on temperatures than the Sun.

The message communicated is not about the scientific complexities of cycles but about frightening the public witless as a mechanism for controlling their lives and wallets and bullying and berating the public about lifestyle and all the energy systems that support the modern world.

The media could not operate without energy from fossil fuels. When a media network does not use coal-fired electricity for broadcasting, fossil fuels for transport of journalists from one scary sensationalist scene to the next and fossil-fuel derived printing ink, then they might be a credible source of information. All metals, lighting, paints and plastics in newsrooms are made using fossil fuels. Most of the media should be treated as ignorant hypocrites who bathe in envy, misinformation and a swill of propaganda. They should practise what they preach.

At greens-supported climate conferences, renewable energy conferences and sessions to train media spokespersons to communicate climate change, an engineer told me that there were no participants who had worked in power generation in an operational or maintenance role, none who had synchronised a machine on, no one who knew what the power system strength is and no one who understood losses, frequency management, interconnectors or effects of transients in an electricity grid. Time to give electricity supply back to engineers who know what they are doing. While we are about it, give forests back to foresters.[1654]

Most of the media is in the entertainment business and is driven by advertising revenue. With the availability of a mountain of on-line information, there is little need now for a public-funded broadcaster except in rural and remote Australia. Television uses tragedy porn with large fluffy animals, preferably with big eyes, exaggerations and the presenter or scientist holding back tears. Generally, if money is given, the tears evaporate. Until next time.

The media have a huge emotional investment in a beaten-up climate change crisis. Propaganda is made so much easier when the education system has been dumbed down for decades and many people now don't have the general knowledge and critical and analytical skills to dissect what is nonsense.

With every green activist scheme, the media promotes such schemes as if it is the solution to a problem that could only be solved by greens. They just don't have the basic skills to see nonsense, ask critical questions, cost out the schemes of dreamers and show that you can't get something for nothing (i.e. the First Law of Thermodynamics).

To expand the hydro systems on the planet's driest habitated continent may be feasible if new dams are built. However, new dams are not on the agenda for green activists. They champion beating the laws of thermodynamics by wanting to use excess electricity from wind and solar to make hydrogen or pump water back up the mountains for the Snowy 2 hydro system. Australia is exceptionally rich in gas but not in common sense and many jurisdictions have banned the exploration and exploitation of onshore gas yet the media are resoundingly silent on this political idiocy. Hot dry rock geothermal energy is still in the experimental stage and has already failed in Australia. Tidal power just does not have the head and energy density and is misty-eyed green ideology with a history of failure.

The language of fear and panic repeated *ad nauseum,* as noted by George Orwell more than 70 years ago, is used as one of the main instruments of political control. There is no climate crisis or emergency. Linguistic language manipulation has not given us repeatable validated measurements that show unprecedented global warming. The use of weasel words to deliberately replace repeatable facts is used to bamboozle the community and to push the green activist agenda.

Don't believe a word of what you read, hear or watch. This is what the media will not tell you. Green schemes are un-costed, subsidised and are championed by those bathed in self-interest. These folk have no practical experience and willing to spend other people's money. Wind- and solar-generated electricity are horrendously expensive and increase human emissions of carbon dioxide to the atmosphere. They are unreliable. To try to increase reliability using giant hazardous batteries filled with renewable energy or to try to manufacture, transport, store and burn hydrogen as a stop-gap measure creates expensive and unreliable giant incendiary bombs. Hydrogen cars are an inefficient highly expensive mobile bomb.

EVs are subsidised and are designed to run off subsidised electricity. This is financial madness. EVs, renewable energy, batteries and hydrogen take more energy and carbon dioxide to produce than they save and are kept operational by fossil fuel-generated electricity. Mobile incendiary bombs called EVs have very limited use by wealthy hypocritical green activist show-offs for hunting and gathering their lattes from one café to the next. They are owned and driven by those who support child slavery and destruction of the environment. Why is this not reported by the media?

All green solutions to the alleged problem of human emissions warming the planet such as Net Zero are expensive, unreliable, subsidised, explosive and achieve the opposite of the desired ideal. All green solutions championed by green activists produce a wealthy class who impoverish the workers even more. The free market created the most efficient and reliable methods of electricity production and transport and brought the poor out of poverty.

Get rid of subsidies and mandates and let the market do what it does best.

9

CERTA BONUM CERTAMEN

There has never been a better time to be a human on planet Earth. By every measure, life is far better than 200 years ago. There is still much to do. Green policies aim to bring us back to those times. Coal bought the West out of grinding poverty and China is on the same trajectory with India and many other Asian nations following.

Green activist policies against cheap reliable coal-fired electricity keep hundreds of millions of African people in poverty which results in environmental destruction and the deaths of millions of women and children from indoor cooking fires. The Greens are killing poor electricity-starved people in Africa.

The dumbing down of the education system over the last 40 years has produced a generation with many who cannot read, write, calculate, think, solve problems and look after themselves. They have no knowledge of the past, Western civilisation, science, critical thinking and the brutalities of communist and socialist regimes.

Green activists are totally intolerant of a contrary view yet it is the tolerance of a democracy that allows them to espouse their treachery. A democracy survives with open, respectful public debate. This process of finding the truth and even changing one's mind now no longer exists. For those who have recently been through schools and universities, debate is a black and white issue in a divided world. Democracy requires one to listen to unpleasant truths, to challenge nonsense, hypocrites, dishonesty and fraud and not to shut down debate by *ad hominem* attacks when no contrary argument can be presented.

There have been four decades of constant noisy attacks on employment-generating businesses that create new wealth such as mining, farming and forestry; on the businesses that add value such as smelting, refining and manufacturing; and on the principles that provide a moral foundation to society such as the family, property rights, human rights, civic pride, churches and charities. The attacks have now created a narcissistic selfish society devoid of problem-solving skills.

There is aggressive editorial censorship in scientific journals, newspapers, media networks and the social media sewer when an alternative opinion on climate change is raised. The evidence is not critically analysed and only the reputation of the bearer of contrary news is sullied. This occurs with many political topics such as climate change, COVID-19, the behaviour of dictatorships such as Russia and China, the cognitive decline of President Biden, the atrocities of the Palestinians and the great achievements of President Trump which are now for all to see with the very weak President Biden. This censorship is universal yet, because there are many other ways of acquiring information in today's world, silencing critics is not as successful as the despots would like.

When the Catholic Church was losing some of its flock to Protestantism, debate was banned, critics suffered eternal damnation and excommunication and anyone who wanted a logical discourse was deemed a heretic. Luther refused to recant for his 95 Theses[1655] and was excommunicated on 3rd January 1521. This led to more and more people following him and the establishment of the Protestant churches. And so too with the banning of discussion on climate change. Politicians beware. There is a groundswell of opinion amongst those who use common sense, whose livelihoods are threated and who are sick of the lies.

My concern is not what sort of planet we are leaving for our children. My concern is what sort of children we are leaving to care for the planet and how they believe wokeness will solve problems and create prosperity. No country has ever taxed, regulated or woked itself into prosperity.

Climate change and COVID-19 are state-sponsored crises that shift attention from other political failures, the lack of financial accountability, hypocritical behaviour and random mandating of mutually exclusive activities. This has been exacerbated by a politicised ill-educated public service filled with party hacks presenting to Ministers what they think he or she might want to hear. There are very few now in the public service who still give fearless, dispassionate and informed advice no matter which political party is in power.

Climate change relies on total ignorance in a zone where feelings outweigh facts. I am sure that I can ask six questions about climate change to school teachers, climate "scientists'. climate activists, journalists, public servants and politicians and get a 99% failure rate.

Out of hundreds of politicians and journalists I have met, I know of only five journalists and four politicians who could answer my six questions. It has never been shown that human emissions of carbon dioxide drive global warming. Only the opposite has been shown. In the past, increased emissions of carbon dioxide have followed natural temperature rises.

Ice has been on Earth for only 20 percent of time, the six great ice ages started when the atmospheric carbon dioxide content was far higher than at present showing that atmospheric carbon dioxide content does not drive global warming.

For most of time sea level was higher than at present. Sea level has changed by up to 600 metres, present sea level changes are within previously recorded variability and land level rises and falls can give the appearance of a sea level change.

At present, the global carbon dioxide content is increasing and temperature is neither rising nor falling. There have been two decades of a pause in temperature rise which, if increasing carbon dioxide drives warming, should not have occurred. The pause has never been explained by promoters of human emissions driving global warming. Modern temperature measurements are within the long-term natural variability.

Over the last 500 years, there have been 40 warmings in Greenland. Why is it that only the last warming is due to human activity? Warmings 7,000 to 4,000 years ago and in Minoan, Roman and Medieval times gave temperatures about 5°C higher than at present yet we are expected to believe that the slight Modern Warming is of human origin. Past warmings derive from orbital and solar cycles which have not changed.

The number of species on Earth continues to increase despite five major mass extinctions and more than 20 minor mass extinctions. We constantly undergo species turnover, a process that's been taking place for 500 million years and continues today.

Climate activists cannot provide a bullet-proof case to support their ideology, they do not argue with eminent scientists who have come to different conclusions and they resort to character assignation, denigration, abuse, cancellation and sacking of those who have a different view. To claim that the science is settled is the claim of those who have something to hide.

The colourless, odourless, tasteless, non-poisonous gas carbon dioxide is plant food and, without it, there would be no life on Earth. The recent slight increase in atmospheric carbon dioxide has led to a greening of the Earth, increase in forest cover and more productive agriculture.

Modern weather events claimed to be unprecedented have occurred with greater intensity and frequency in the past. The past shows us that forecasts of death, disaster, heat, floods, hurricanes and ice have all been wrong. Catastrophic lethal bushfires result from forest mismanagement because of green activist pressure and the combination of lightning, arson and a massive dry fuel load.

Mega-droughts occur during cool times, floods occur on flood plains, lethal hurricanes have decreased, reefs have come and gone for 3,500 million years and the 3,000 reefs along the 2,000 km-long Great Barrier Reef are healthy, dynamic and unaffected by farming, shipping or mining. Much of the doom-and-gloom science of the Great Barrier Reef, at best, is questionable.

The encroachment of industry and suburbia on historical weather stations has produced an upward temperature bias due to heat-emitting machinery, concrete, paved roads and exhaust fumes. A very large number of measuring stations in remote areas have been closed, others have had measuring devices changed without cross-calibration, some have been moved, the data from many measuring stations has been ignored and, where there are no measuring stations, the temperature is estimated by using data from stations that may be hundreds to thousands of kilometres distant.

There has been universal tampering of the global temperature record. Older measurements have been made cooler thereby producing graphs that show an increase in temperature over time. The keepers of the data, as exposed by the leaked Climategate emails, admit that the temperature data used and given to the IPCC and COP for major energy policy decisions is useless. They also admit that they use the scientific publication process to stop publication of contrary views.

The temperature record upon which major policy decisions have been made does not survive a simple due diligence. Fraud is the thread that unites all climate data and predictions. The long-term "homogenisation" of the global temperature record is the greatest scientific fraud that has ever taken place and the financial implications are horrendous.

Models to predict future temperature have failed. It's arrogant and preposterous to think that the future can be predicted. After 40 years of models, measurements show that models run far too hot. Carbon dioxide measurements from ice and Hawaii show a straight line increase and older measurements from Europe show carbon dioxide has been far lower and far higher than at present. Natural features show variability and straight lines exist in people's brains and suggest data has been manipulated.

Carbon dioxide emissions are not measured. They are estimated. One of the major sources of heat and carbon dioxide has been ignored. The long-term cooling and degassing of the planet is unseen and occurs deep in the oceans. Because of this, the claimed 3% of annual carbon dioxide emissions by humans is probably far lower. If human emissions drive global warming, then it must be shown that the 97% of carbon dioxide produced by natural emissions don't drive it. This has not been done.

Just because work in the peer-reviewed literature is published does not make it correct. Science has no consensus, it is impossible to "follow the science" when there is refutation of previous preciously-held ideas, science is married to evidence which is ever changing and belief is not a word used in science.

Two giant planetary experiments took place with the Global Financial Crisis and the COVID-19 epidemic. In both cases, the reduction in carbon dioxide emissions of about 7% due to the grounding of transport and the closure of major industries made no difference to the measured rise in atmospheric carbon dioxide. This suggests that human emissions are swamped by natural emissions of carbon dioxide and may have no effect at all.

We are told by green activists that Australia is the biggest *per capita* polluter in the world. This is wrong. We are a large carbon dioxide emitter and carbon dioxide is plant food and not a pollutant.[1656] The countries that emit the largest *per capita* carbon dioxide emissions are servicing the needs of other countries by smelting and refining aluminium and zinc, metals that have a massive amount of embedded energy, and by a large fossil fuel or petrochemicals industry.

The grasslands, forests, crops and continental shelf of Australia sequester far more carbon dioxide than Australia emits from all energy, transport, agriculture and mining sources.

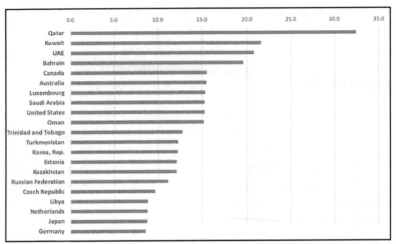

Figure 52: *The largest emitters of carbon dioxide in 2018[1747] in tonnes per capita.*

Climate change mantra is only possible because of a 40-year dumbing down of the education system resulting from the control of the school syllabus by Marxists. There is little knowledge of history and the murderous brutality of communism and socialism. Wokeness is the biggest threat to freedom since communism and is conducting war against the average person.

The march of the left through our institutions has people turning their backs on the great struggles to get us where we are now, what made the Western world great and the familial, religious, moral and community glue that holds unified society together. There is less unity; confected divisions are constantly being rammed home aided and abetted by an uncritical media.

The dumbed down alphabet soup of social causes does not create employment, generate wealth or make anyone's life better yet it has a prominent place in the media. The left was once concerned with material decadence but is now so wrapped in self-hate that it affects their judgement whereas ordinary people just want to get on with life, love and laughs.

Energy is not a product. It's what makes civilisation possible. Reduce the availability of electricity and increase the costs of energy and we go back to the miserable poverty of hundreds of years ago where we were ruled by unelected despots who controlled every aspect of our lives even to the point of if, who, when and where we could marry.

Why should a small minority of non-contributors drive a society wherein the majority work, create jobs, add value, have a family, pay tax, enjoy leisure and want a better world for their children and society? Politicians in their bubble listen to their "experts", inexperienced advisors and pollsters. Politicians should spend more time listening to conversations in the bleachers at the footy, the front bars of pubs and roadside fuel stops. Green activists do not change prime ministers. It is the disenfranchised average family trying to do the right thing that changes political leaders.

The concept of service to the community virtually does not exist as shown by the decreasing membership of people in volunteer organisations, clubs, churches and political parties. The incessant attacks on the Christian churches and Judaism, but not Islam, and the decrease in Christian principles has left a yawning spiritual hole in people's lives.

Many have embraced narcissistic greed and hedonism whereas others with no sense of service have embraced unelected power with neither regard to the consequences nor skin in the game. Climate activists have embraced their new religion of climate worship, a form of ideological decadent puritanism that divides society into good from bad, believers from non-believers and crusaders from heretics. They have the hallmarks of savage fundamentalist Christianity. The new religion has no history, architecture, scholarship, thinkers, philosophy, music, compassion, moral code, sense of service, charities, schools or universities. It is vacuous.

Green activists, politicians, bureaucrats and eco-terrorists exploit the hard-wired human survival mechanism which is the innate need to fear the unknown and unseen. This is especially the case if the doomsday object is unseen such as a distant comet or asteroid, a bacterium or virus, radiation or traces of a colourless, odourless, tasteless gas. To intertwine a prediction with sin, sacrifice, redemption, indulgences and salvation has been a common theme for 2,000 years and blends well with with scare-mongering.

Property rights are the central hallmark of Western civilisation. They are also fundamental for environmental stewardship, conservation, health of species, wildlife habitats, forests and other resources. The absence of enforced property rights in developing countries remains one of the largest barriers to improved prosperity and environmental well-being. In my travels as a geologist to some of the most remote areas of impoverished countries with no property rights, I have seen hopelessness, extreme poverty and trashing of the environment. There is a pride in ownership.

For those who have worked hard, taken risks and made sacrifices, there are rewards in a capitalist society such as acquiring property. However, there are a large number of people green with envy at successful and wealthy people and who are using the politics of eco-fascism to bring down the tall poppies and reduce everyone to the same level of misery.

Capitalism has won but is currently taking a backward step. For many, there is absolutely no interest in bringing our fellow humans in Africa, the Indian subcontinent, Asia or South America to our standard of living. For some, all that is left on the agenda is a choice of pronoun, learning the alphabet using genders and street demonstrations for dubious causes.

In the democratic Western countries, we have been striving for centuries for a better world for our citizens and those of other countries. Personal, financial and military struggles have come at a huge cost and created a trend towards fairness, equality and unity. The attacks on society by a very noisy few have created savage tribalism led by unelected elites. Time to fight back. Hard. *Certa Bonum Certamen.*[1657]

The world has changed. In Western countries, most people live in the cities and the rural population continues to decline as farming becomes larger and more efficient, people leave rural areas for employment and, as a result, city people have lost contact with rural Australia and don't know how their food, fibre, water, energy and metals are produced. These city folk have a yearning for space and think that by being green, they become custodians of the land they neither know nor understand.

Go to any country town and look at the World War I and World War II memorials in the main street. It was rural Australia that did the heavy lifting, provided most of the soldiers and suffered the most casualties. Some families lost all four sons in World War I. Rural people willingly volunteered and made huge sacrifices for the freedoms that every person in Australia enjoys today. These freedoms are being eroded and must be taken back.

Farmers who work for a living and are outnumbered and outvoted by those who vote for a living. This results in the biggest economic threat the country has faced. The country is being destroyed by those who lack a significant contribution to the nation. What gives them the right to destroy it when they have no skin in the game? The solution to the economic problems created by unelected city-based green activists is for farmers to pass the problem back to them.

Farmers have only one weapon and, if they don't use it, they will be faced with poverty and the occasional part-time job mowing the grass between solar panels on some rural solar facility. Bankruptcies, rural poverty, suicides and divorce rates will continue to rise. Green eco-media and eco-activists are heading down the path of telling farmers what they can produce, how they can produce and what job they should have.

This would take away the hard-earned freedoms of a farmer by an unelected activist. If activists feel they can completely turn a farmer's life upside down, then the farmers have the right to do the same. Give the cities a taste of Net Zero. Don't use fossil fuels for a month to deliver fresh food and produce to the markets or to go to sea to catch fish. With the just-in-time policy of the major food chains, there are only a few days of food stored in the cities.

Farmers are on a hiding to nothing. Slice-by-slice of their freedoms are lost by regulation each year and eventually family farms will be regulated out of existence. Without a fight, no one treats an adversary seriously. Muscles need to be flexed and farmers show that they are vital for the prosperity, feeding, social cohesion and spirit of Australia. Win, lose or draw they will be able to stand proud, be recognised for the value they create and again be a political force in Australia.

Australia runs on diesel heavy machinery. In a Net Zero world, truck drivers would be out of work and would have every reason to pass the message to cities that without trucks, Australia can't function. Truck drivers, farmers and fishermen have no political voice in a country where city-based non-productive eco-activists control energy and food policies. Get in first before the city eco-terrorists send you to the wall.

We are evolving towards a non-democratic authoritarian over-regulated capitalist society that is legislating for poverty with measures that will annihilate the middle class. Rather than spend taxation monies to build the nation, bureaucracies became bloated and inordinate amounts of money have been spent on totally useless ventures. If the country keeps going down the high taxation, monstrous debt, high regulation and green and red tape road, then we will face a recession, depression or loss of sovereignty. Measures such as personal carbon taxes are already being discussed, which will only speed up the impoverishment of Australia.[1658]

Where is the transparent risk analysis, financial due diligence and technical QA/QC of the wind and solar industries, electric vehicles, hydrogen and Net Zero? How much will the taxation base decrease and what services will have to be cut? How much will the endless subsidies increase? How many jobs, especially in rural Australia, will be lost? What will Net Zero cost each household, business and the GDP? Where is the plan if there is savage global cooling during the next Grand Solar Minimum when Australia will be colder, drier and windier? Why do we need subsidised electricity to charge subsidised electric cars and subsidised hydrogen when we are well aware of the known fires, explosions and energy losses?

There was once cheap reliable fossil fuel energy in Australia with minor peak load back up from hydro and gas. Once religious zeal by greedy carpetbaggers, green activists and left political parties was used to claim that Australia must reduce its emissions of the gas of life by closing down coal-fired power stations, then Australia started on the slippery slope towards economic hardship.

If green activists were really convinced that carbon dioxide was a pollutant, they should be actively campaigning to have nuclear power generation and rushing out to buy EVs. They don't. They also object to nuclear power despite the fact that we have the 20 MW Opal nuclear reactor at Lucas Heights for medical isotopes and the scientific and engineering skills to build and run reactors. This amounts to economic terrorism. Do greens get treated for cancer with radioisotopes? If so, they can only be huge supporters of building more nuclear reactors.

Australia has no energy policy and is heading for a financial meltdown. There is no reason why unreliable power generation based on the weather should be subsidised for 30 years. No Western country runs on sea breezes and sunshine. The UK and EU tried to increase the proportion of renewables concurrent with shutting down coal, gas and nuclear electricity generation. Only fools would put their country's economy in the hands of the weather. They did and there were inevitable consequences.

They now face economic hardship, loss of manufacturing businesses offshore, increasing unemployment, energy shortages, social turmoil and the energy survival of these countries is now in the hands of Mr Putin. Because of the flight to renewables, people in what were once wealthy countries have to make a decision: eat or heat.

The flight to renewables has consequences. Renewable energy creates unemployment, skilled people in heavy industries that use large amounts of electricity cannot be replaced and the few jobs in the renewables industry are low-paid menial jobs. The energy used to make wind turbines and solar cells is more than they will ever produce. The carbon dioxide emissions to make, maintain and provide standby power for wind and solar is far more than saved. A fossil fuel generator needs to be running on standby waiting to the inevitable drop in wind speed or passing cloud.

Forests are clear felled for wind turbine complexes and arable land is razed for solar panel complexes. Wind turbines change local weather, kill bird and bat populations, start bushfires, damage human health and don't produce electricity when needed. The short life of turbine blades and solar panels result in the dumping of them as landfill resulting in the long-term leaching of highly toxic chemicals into the soils and waterways. Renewable companies have no statutory requirement to clean up their mess at the end of equipment life. This is not environmentalism.

If any industry killed wildlife, destroyed forests and polluted the air, water and soil as much as the renewables industry, they would be closed down. Opaque sweetheart deals with bureaucrats and governments allow the subsidised renewables industry to continue unscathed.

Renewable energy is a war against ordinary people who live pay-to-pay. Green activist zealots are long on feelings, virtue and purity and short on facts and compassion. If city-based green activists are so keen on wind turbines and solar complexes, they should be placed in city parks, harbour foreshores and along beaches close to the user to reduce voltage losses.

Renewables are only possible because governments opened the door for streetwise businesses to con naïve bureaucrats and politicians who have little life experience and to create long-term subsidies for inefficient and horrendously expensive electricity. Wind companies are paid not to produce electricity as part of their perpetual price-gouging from the consumer.

Electricity prices have gone through the roof as a result of renewables, productive industries have closed, unemployment has increased and some jurisdictions have had crippling lethal blackouts when the weather changed. No modern economy can survive on the vagaries of the weather.

Renewables equipment is mainly supplied by China which has ignored populist renewables and is increasing the number of reliable, inexpensive coal, gas and nuclear generators during what is the biggest Industrial Revolution the world has seen. To date, hundreds of millions of Chinese have been dragged from poverty into the middle class.

The flight from fossil fuels has revived interest in technologies that failed a century ago. EVs produce 11-28 % more carbon dioxide than an equivalent petrol or diesel vehicle and electricity grids are not capable of charging a large number of cars without massive capital investment. They are subsidised, as is the electricity and charging system. The fuel range and charging times are such that their only use is as an over-priced inner city virtue-signalling toy for occasional short city trips.

EVs have a habit of catching alight, toxic fumes are emitted and fires cannot easily extinguished. If Western countries are to become Net Zero, the necessary minerals have not yet been discovered. The critical materials are processed and constructed in China. Those virtue signallers who choose to purchase EVs need to become aware that the cobalt for their vehicles is produced by black slave children and the nickel production requires the clear felling of tropical forests.

Renewables can't generate electricity when it is required. A brilliant solution to this was to spend hundreds of millions on mega-batteries that hold a few minutes of electricity. These batteries have a habit of catching fire, hydrofluoric acid and other toxins are released and the fires can take a week to extinguish.

When renewables generate electricity at the times when it can't be used, green governments have created yet another brilliant solution. Lose about 30% of this renewable energy and pump water uphill to holding dams to later generate hydroelectricity. The financial case is inspirational. Create energy losses and use subsidised renewable electricity to create subsidised hydroelectricity.

Another brilliant energy loss scheme is to use subsidised renewable electricity to electrocute nine litres of water to produce one litre of subsidised hydrogen followed by using huge amounts of electricity to liquify hydrogen for transport. Never mind the costs and the energy losses, a country littered with mobile hydrogen bombs will certainly not be boring.

Al Gore uses about 40 times as much energy as the average American and flits around the world in personal jets, yet tell us to reduce our carbon dioxide emissions. His predictions of a six metre sea level rise did not prevent him from buying a waterfront house. A UK court ruled that Gore's fantasy film was a crusade to make political points and contained numerous fundamental errors of fact. This did not stop the Gore gravy train promoting the same message which a court had ruled were false. Rules devised by green activists are for others, not for themselves. The green hypocrites are an elite group and are a law unto themselves.

Greens always want someone else investigated and constantly call for Royal Commissions. Why not start at home? Who funds the Greens Party? How much foreign money comes to the Greens by various diverse routes? Why are the Greens Party politicians continually having staff problems with bullying, sexual harassment and audits? Why is the Greens Party antisemitic? Why is the Greens Party anti-business yet the biggest ever political donation in Australia was to the Greens Party from a businessman? What is the relationship of the Greens Party to foreign economic terrorists such as socialists, communists and the Davos Forum leaders? What are the relationships between the Greens Party and Palestinian and other terrorist groups? Where are the independent cost analyses on various Greens Party proposals?

Don't let the Greens Party anywhere near money. Their fingers are sticky, costing projects is not in their lexicon and they have no economic training to vote on financial issues in the parliaments.

The whole basis of renewable energy, subsidies, carbon credits and offsets, taxes, electric vehicles, hydrogen, carbon capture, net zero, decarbonisation is based on the assumption that carbon dioxide is a pollutant. That's demonstrably wrong yet it underpins all Greens Party policy. The traitorous green activists supported by the Greens Party have weakened Australia. Compliant bureaucrats and politicians have followed and this weakness is being exploited by the UN, the EU and China.

This book shows that green activists occupy the low moral ground and green policies are expensive failures that destroy the environment and bleed the poor. Previous books dealt with the science that shows the gaping holes in the human-induced global warming theory.[1,2,3,4]

Poor autistic Greta Thunberg is an abused child who is the cash cow for her activist parents. Her speeches and internet posts are written by green activists, she needs to go to school to learn how lucky she is compared to previous generations and to learn some basic science. She needs care rather than being flitted around the world as the token oracle. No political party has ever won an election because they were aligned with a disturbed abused child who promised better weather if she was in charge and who abused international leaders.

We are dependent upon China for much of our iron ore and coal export revenue; we are dependent upon China for wind and solar facilities because we have moth-balled or destroyed our local reliable cheap energy systems; we are dependent upon China for lithium batteries for EVs, mega-batteries and almost all electronic consumer goods; major infrastructure, farms and mines are in Chinese hands and, without Chinese visitors, much of our local education, tourism and real estate businesses are uneconomic.

Australia's exports outstanding quality coal, gas and iron ore. Our food exports are free of toxins and meat is ethically killed. Our contractual and legal system makes us reliable traders. If China refuses to import some Australian goods, then our quality foods and coal can find other markets.

If China sneezes, we get something far worse than their COVID-19. Maybe we should not put all our eggs in the one basket and redevelop Australia's manufacturing industry. China has every right to grow into an economic superpower underpinned by coal, as did the UK, US and Europe earlier. They have done something that no nation has ever achieved. They have brought hundreds of millions out of poverty. No other nation can boast of such an amazing achievement.

Climate change has nothing to do with saving the planet, the environment or climate change. It is a mechanism whereby unelected elites control every single aspect of our lives. Government haste to appease green activists has resulted in increasingly expensive unreliable renewable energy, carbon capture, electric vehicles, "green" hydrogen and other schemes that lock in subsidies for decades. As a result, nimble clever businesses have outsmarted unelected green-tinged bureaucrats with the result that renewables companies get paid not to generate electricity.

The UN and its IPCC and COP are attempting centralised control of every aspect of our lives such as how we heat and cool our homes, where we live, what we eat, what vehicle we drive, what industries can operate, whether we stay employed in our chosen vocation and if we can travel.

The danger of unelected organisations controlling every aspect of our lives and our economy is that we lose national sovereignty and our natural competitive advantages. In the history of time, climate change has been the greatest transfer of wealth from the bulk of the population to the elites. Debt means that the lenders can control the climate, energy and environmental policies of countries which is not in the interest of taxpayers. This has already happened in Australia.

Political leaders will continue to be thrown out because of climate policies. My fears are that decision makers will be very slow to realise that climate change is not an existential problem. I fear that a major change in policy will only take place when the middle class has all but disappeared and our country has been ruined and is beholden to those we didn't elect.

It is my view that the only way a common sense economic policy on energy can be achieved is to have a few more Prime Ministers thrown out and for debt, inflation, interest rates, unemployment, cost of living and a decrease in the standard of living hit the electorate so hard that one-by-one voters realise that they've been led up the garden path. It is only then that change will take place. By then, the middle class would have contracted and the elites will have more power and more of our money.

We are all environmentalists and want a better world for ourselves and the next generations. This is exploited by the green activists, governments and their bureaucrats. The incessant doom and gloom of climate change propaganda has exploited the tendencies for humans to fear, panic, become hysterical and irrational, seek simple solutions to complex problems as salvation and to attack, censor and demonise those that differ.

For more than 50 years, green activists made false and failed predictions about the end of the world, famine, health, pollution, sea level rise, ice sheet melting, extinction, global cooling, global warming and population impact. They still do. If just one prediction were correct, we would not be here.

The biggest scientific fraud in the history of the planet has now been in progress for 40 years and trillion-dollar financial decisions and eye-watering subsidies based on this fraud will end poorly. This fraud underpins the IPCC and COP talkfests. The long march of the left through institutions such as schools, universities, churches, media, bureaucrats, politicians and some businesses has created two generations of people who lack basic earth history knowledge, independence of thought and critical and analytical thinking skills.

The incessant fact-free propaganda from the noisy media and the pornography from social media by climate activists prevent alternative views to be aired for consideration by rational thinkers. Untestable feelings have been replaced by facts, but not all ideas are of equal validity, and the truth can be determined by using validated repeatable evidence.

Journalists try to hold politicians responsible for their actions. The community needs to hold journalists responsible for their lies, exaggerations, concocted stories and bias. If company directors are legally required to tell the truth, present risks in a balanced fashion and not exaggerate, why don't journalists have the same constraints.

Australian has suffered the deepest cuts compared with any other G20 country yet more is demanded by our competitors. The IPCC is a political organisation. They condense thousands of scientific papers into a scientific report and a Summary for Policy Makers. The scientific report is measured, makes no outlandish claims about human-induced climate change and ignores all published science that presents a contrary view. The science in the IPCC reports uses the doctored and useless temperature record. The Summary for Policy Makers is written for journalists and politicians, is unrelated to the body of the report and is full of hysterical predictions about how the world will end and it's all our fault.

The UN's IPCC report underpins their regular COP talkfests which aim to cripple successful economies and transfer huge amounts of money through sticky fingers in an attempt to change climate. I am not aware that money, agreements or pledges can change a major planetary system. COP talkfests are used as a pile on to bully governments to impoverish their countries and yet key carbon dioxide emitting countries such as China and India are totally unaffected.

Net Zero requires communist-style central planning. We know how this went. It led to the deaths of hundreds of millions of people. Entrepreneurs are scrubbed out and these are just the people we need for economic prosperity. There is no point in using hollow words promoting a digital, clever or smart economy if there is no one willing to use their skills, invest and take risks.

The uncosted Net Zero raises alarm bells. Renewables were also uncosted resulting in unreliable electricity that tripled in cost and now the same old smooth talkers are giving their partial costings for Snowy 2 and hydrogen. The financial costs of Net Zero are unaffordable in a country with massive government and personal debt and a lack of manufacturing industry; mining, smelting and farming industries already suffering impossible energy and regulatory costs in a high cost country and these will only rise under Net Zero; and our export industry is based essentially on one customer. Governments need to reduce costs, deregulate, stop subsidies, step aside and let the market do what it does best.

Nothing in society could operate if there is Net Zero. Without abundant cheap diesel there will be no food. No water wells could be drilled. There would be no bus transport. No trucks. Where will the money come from if there are massive job losses and a universal basic unemployment wage? The rich will get richer, the poor will get poorer and the middle class will disappear. The seeds for civil unrest are being sown.

Talk of "green" hydrogen, "green" steel and "green" aluminium is the blabber of hucksters trying to look at making huge amounts of money from yet more subsidies. Basic science, engineering and economics kills off such ideas. The electricity needed to liquefy hydrogen would dim the lights of a big city. How is pig iron to be converted to steel without carbon?

Just tell me how long the Sun has to shine or the wind has to blow to melt and convert alumina to aluminium in hundreds of pots at a refinery and then pour high temperature molten metal into a mould? A scientist or engineer would check on the energy required for the Hall-Héroult process, the time it takes for alumina reduction, the melting point of aluminium, the latent heat of melting, the thermal conductivity, the melt viscosity and the cooling rate of liquid aluminium. As long as the verbiage of "green" aluminium sounds good, the basics don't need to be known.

We get a glimpse into Net Zero by looking at those places that have embraced renewables in what is a death clutch. The UK is suffering an energy crisis, pays massive prices for imported gas and yet has vast gas reserves in the North Sea. One legislative change and the UK's energy problems would be solved if subsidies were dropped and government got out of the way of people who actually achieved success. The same argument can be used for a cradle-to-grave nuclear industry in Australia. The EU, especially Germany, is suffering an energy crisis in summer. God knows what winter will be like. In California and Texas, failure of renewable systems led to blackouts, collapse of businesses and hundreds of deaths.

The banks love Net Zero, subsidised green schemes and steer clear of longer term more secure investments into coal, uranium and oil. It presents great opportunities for carbon trading, sucking up subsidies, debt and making money from insolvencies. When banks lobby on policy, it's not going to be good for customers or the health of the nation.

Green's promotion of renewable energy under the guise of human-induced climate change gives China complete dominance over our energy systems, infrastructure, food production, manufacturing and defence. We are exposed, have learned nothing from the 1930s and will have to rebuild the country. Green activists are essentially Chinese prostitutes. Many green movements are funded by Russia, China and totalitarians.

Many green leaders have an unhealthy obsession with death, killing, poverty, catastrophes, totalitarianism, restriction of freedoms and gaoling or killing those who have an alternative view.[1659,1660,1661,1662] This is no surprise considering the green movement arose from eugenics and totalitarianism.

Instead of mounting convincing arguments using knowledge and logic, green activists resort to abuse, threats, cancelling and violence, thus reflecting their totalitarian tendencies.

Virtue-signalling by greens attempting to demonstrate their moral superiority has failed. Green activists use fossil fuels for some or most of their energy and freely access the 5,000 products from petroleum, including medicines. By their actions, they are hypocrites.

A true environmentalist would move out of a city, get off the grid, use no electronics, buy land, plant trees, minimise intense forest fires, exterminate feral animals and introduced plants, live by hunting and gathering, remove plastics from the waterways, minimise waste and send messages written on bark by carrier pigeon concerning their imagined forthcoming catastrophe.

The green movement must be constantly brought to account to avoid lowering the quality of life and the population's standard of living; to avoid national bankruptcies resulting in a loss of sovereignty; to avoid loss of property and freedoms and to rebuild an education system that produces people who can think, create, solve problems, add value to society, love their country, have civic pride and look after themselves and others less fortunate.

The green policies increase infant mortality, shortened lives, prolong poverty, enhance pollution and cause starvation and allow otherwise curable fatal diseases to persist. Many greens state their wish to greatly reduce population and, although the number to be exterminated is often given, the method of extermination is unspecified.

The world's leading infection disease killer is tuberculosis (TB). It kills adults in their prime and leaves children without parents. TB is easily eradicated with cheap drugs. For only $6 billion annually, 1.6 million people could be saved annually from dying from easily curable diseases. Why do the greens ignore such immense but avoidable human tragedies and cheap solutions, but so stridently and unjustifiably aim to fund climate alarmism?

The World Health Organisation stated seven million people are killed each year by pollution[1663], these mainly being women and children exposed to indoor cooking fires because of the lack of domestic electricity. That's 13 deaths a second. Most were from Asia. These people are trapped in poverty and have no electricity.

About 45% of all people on the planet have a lower *per capita* annual electricity consumption than my refrigerator yet coal is cheap and widely available and able to provide cheap reliable electricity. Why do green activists object to coal when they themselves have their own coal-fired electricity and refrigerators full of food? Low cost and reliable fossil fuel-generated electricity reduces poverty.

There are more people killed by airborne sulphur and nitrogen oxide, particulates pollution and complex industrial chemicals than by malnutrition, TB, over-consumption and alcoholism. What are green activists doing to reduce poverty or pollution? Do the Greens run medical charities to medically assist pollution sufferers in their time of need? Why are the Greens happy to kill people, especially those from other races?

Some 44,000 people a month in medium to low-income countries die before their time because of air pollution and 97% of cities with more than 100,000 inhabitants in low to middle income countries do not meet minimum World Health Organisation air quality requirements. Are greens activists helping China, India, Mexico, Turkey, Egypt, Iran and other developing countries to reduce airborne and water pollution?

The World Bank estimates that particulates, sulphur and nitrogen oxides and chemical toxins in developing countries are attributable to 36% of deaths from lung cancer, 35% from chronic obstructive pulmonary diseases, 34% from strokes, and 27% from heart disease. Older people and children are most affected.

Half the deaths of children under five are due to acute infections of the lower respiratory tract from inhaling pollutants. Asthma is prevalent in 5% of the adult population and 12% of the infant population. Air pollution makes asthma worse.

Why do green activists make so much noise about carbon dioxide, the food of life, rather than showing concern about their fellow humans dying from real pollution? Do the green activists really care about poor people in developing countries who don't have enough money for food?

Air pollution degrades materials and coatings on buildings and decreases their useful life, increasing cleaning and repair and replacement costs. Many historical statues and memorials, especially those made of marble and other reactive porous stones, are decomposing in front of our eyes due to air pollution.

The World Bank estimates that air pollution costs the global economy $5 trillion in social assistance costs and $225 billion in lost income each year.

What have the greens done about this? The green activists should address the real issues, or is green activism just about control of Westerners' lives?

We all care for the environment, the planet is a far better place than it was decades, centuries and millennia ago. There is still much to do and environmental advances have only been achieved by wealthy countries using practical engineering solutions and not ideology.

Renewable energy requires more electricity than it will ever produce, sterilises farming land, destroys wildlife and human health, massively increases electricity prices thereby reducing employment and leaves mountains of untreatable toxic waste.

There are plans to build wind turbines over the First World War battlefields in France where thousands of Australian soldiers died and have no known gravesites. Green activists have been silent. Why? Is saving the planet more important than leaving our fallen in peace at their final resting place?

After nearly 40 years of subsidies, the green-supported renewables industry continues to bleed the taxpayer. Astronomical state and federal debt must be paid back. Cutting subsidies to that industry would be a good start. In Western countries, many live on struggle street and green policies make them poorer.

No environmentalist concerned about wildlife could possibly support the construction of wind turbines that slice and dice birds and bats. If any other industry operated like the wind and solar industries, they would be closed down by regulators. Why is the renewable industry exempt?

For every green job created, two and possibly four useful jobs were lost. In Scotland the VERSO study[1664] showed that for each green job created, 3.7 real jobs were lost.

Green leaders prove to be anti-environment, hypocritical, fraudulent and they use green politics to make greenbacks, despite what is preached by them about virtue. Unelected green activists try to control every minute detail of our personal life yet bear no responsibility. Climate activism is underpinned by hate speech.

Green policies support Chinese and Russian authoritarianism and provide a transfer of wealth from the poor in the West to China and Russia, as do some woke western corporations and governments. Schemes such as the Green New Deal result in sourcing what were goods manufactured in one country to "Made in China", with the consequent loss of employment in the West.

Many green technological solutions to climate change can only take place using slave labour in Africa and China. Green policies make Western countries even more economically dependent upon and subservient to China.

The tax deductibility of the numerous green activist organisations and Chinese fronts needs to be cancelled. Why should the taxpayer fund organisations that increase costs and unemployment and decrease the taxation base?

Green policies require an astronomical increase in mining of necessary mineral commodities and don't promote nuclear power, an industry that emits no carbon dioxide during power generation. Almost all greens object to mining and nuclear power yet can offer no viable solutions to the problems of their own making.

Apart from roads, buses, trains, trams, bikes, planes, buildings, glass, homes, bridges, water and sewerage pipes, concrete, cars, house heating and cooling, food, medicines, hospitals, dentists, schools, supermarkets.... what has mining ever done for us?[1665] Fears and hashtag activism don't solve problems and do not create food, medicines and necessary commodities.

Is there a single mine or cattle station in the world that has the wholehearted support of green activists? No. The greens claim they are concerned about humanity. What is the name of the nursing homes, shelters for homeless people, private schools or hospitals that are owned, financed and run by green activists? Maybe they moan from the comfort of their couch to save the planet?

According to green activists, climate change is a moral issue. Because it's a moral issue, they say there is no need to cost their policies. Do green activists have a household budget? Whether it's their own money or that of taxpayers, there has to be fiscal responsibility.

The first moral issue is that climate activists ignore the evidence of the totality of integrated Earth systems and of history, refuse to debate, do not analyse data, resort to fraud, do not consider the consequences of their policies and don't really care about the damage their policies create. The second real moral issue is their futility trying to engineer a climate within a narrow range having no understanding of the impact to the environment to the diversity of flora and fauna and to the damage of so many Western countries. The damage is treated as mere detritus on the path to salvation.

Taxpayer-funded climate "scientists" operate as green activists who admit they cheat, lie, commit fraud, change measurements, make false predictions, refuse normal scientific discussion and prevent the publishing of alternative data and conclusions. This is the seed of the whole corrupt climate change business.

Their antics are not expensive amusement but the destruction of the 2,500 year-old scientific method which is one way of finding a truth. Climate "scientists" remain unchallenged by a fawning scientifically illiterate politicised media.

For hundreds of thousands of years, humans and other organisms have adapted and flourished to great and rapid natural climate changes[1666]. Surely adaptation by an advanced technological society would be easier.

Future climate change predictions are based on computer models that have consistently failed over the last 40 years when compared with the reality of actual measurements. The end result of changing the whole energy system based on this nonsense is an energy crisis. During the time I was writing this book, the coal price rose 400% and LNG price 2,500%. Don't tell me that fossil fuels are on the path to extinction. They are needed for survival.

The new green religion is a doomsday cult where the naïve and gullible seek salvation through the veneration of cruciform wind turbines and "settled science" that has replaced Scripture.

Fire and brimstone preachers cancel, de-platform and abuse those who question green dogma with all the savagery that was used by the Spanish Inquisitors to root out heretics. What is touted today as "progressive" is in reality a regressive drift back to the Dark Ages with its misery, poverty and ignorance.

The Greens Party are not a soft cuddly feel-good political movement trying to make the planet a better place. Their green propaganda appeals to many of the young and well off. It is a communist-inspired destructive hypocritical Western movement that destroys democracy and cares not for the environment. By their actions and consequences, greens kill people. There is nothing virtuous or admirable about being green.

Today's green religion is like the children's game of musical chairs, a game designed to ensure that someone always misses out. With wind and solar electricity, someone always loses.[1667] These are the people on struggle street who cannot afford to pay the increasing electricity costs, especially when they face long-term unemployment as a result of the COVID-19 virus lockdowns. It is immoral, unnecessary, futile and anti-human.

There is a move now from European socialists, the UN and COP participants that the freedom losses experienced with COVID-19 lockdowns should be extended and that all citizens should have a Climate Passport where every atom of carbon used will be on a register to be monitored and taxed. This is communism. *Cui bono.*

Net Zero is a mindless meaningless slogan. It is undefined. What exactly is Net Zero? What do we use? What don't we use? Where is the due diligence? If it means no fossil fuel use then we go back two hundred years. We've been there. Endless mud, dung, shortened lives, disease, no heating, no cooling, nothing of the modern world and no hope. Some EU leaders are telling us that under Net Zero we will own nothing and be happy. I call this communist slavery. We've been there before and it all failed at great cost.

Net Zero has no costings and you can bet your bottom dollar, because that's all you'll have left, that your money will make the green billionaires even wealthier. Net Zero is a cooked-up recipe for economic and social chaos, class division and further destruction of the west and its values. The collapse in the values of Western Civilisation is a far greater threat than climate change or China. There is no need for a Chinese invasion or war, we are destroying and enslaving ourselves from within. In just one year, parts of the West have collapsed into an energy crisis of their own making. This is a window into Net Zero.

If this book has offended you, so be it. Take a daily teaspoon of cement to harden up, broaden your knowledge, reduce your biases and challenge your thinking. This book contains facts underpinned by repeatable validated evidence and the scientific method. No matter how offended you might be, feelings are not evidence. Green activists are able to get away with murder because they are unchallenged. *Certa Bonum Certamen.* Don't let them get away with it. Over to you.

In this book I charge the greens with murder. They murder humans who are kept in eternal poverty without coal-fired electricity. They support slavery and early deaths of black child miners. They murder forests and their wildlife by clear felling for mining and wind turbines. They murder forests and wildlife with their bushfire policies. They murder economies producing unemployment, hopelessness, collapse of communities, disrupted social cohesion and suicide.

They murder free speech and freedoms and their takeover of the education system has ended up in the murdering of the intellectual and economic future of young people. They terrify children into mental illness with their apocalyptic death cult lies and exaggerations. They try to divide a nation. They are hypocrites and such angry ignorant people should never touch other people's money.

The greens are guilty of murder. The sentence is life with no parole in a cave in the bush enjoying the benefits of Net Zero.

Notes

[1] Monaro, Marc, Arnold Schwarzenegger on global warming 'deniers': 'Strap some conservative-thinking people to a tailpipe for an hour and then they will agree it's a pollutant!'" *Climate Depot,* August 14th, 2013.

[2] www.breitbart.com/big-hollywood/2017/03/17/delingpole-climate-change-deniers-should-be-executed-gently-says-eric-idle/

[3] www.americanthinker.com/articles/2012/12/professor_calls_for_death_penalty_for_climate_change_deniers.html

[4] www.dailymail.co.uk/sciencetech/article-2566659?Are-global-warming-Nazi-People-label-sceptics-deniers-kill-MORE-people-Holocaust-claims-scientist.html

[5] Bailey, R. & Tupy, M. L. 2020: Ten global trends every smart person should know and many others you will find interesting. Cato Institute.

[6] Ronald Bailey & Marian Tupy, 2021: Ten global trends every smart person should know. *Cato Institute.*

[7] www.bas.ac.uk, 11th July 2011.

[8] Iverson, N. A. *et al.* 2017: The first physical evidence of subglacial volcanism under the West Antarctic Ice Sheet. *Scientific Reports* Article 11457.

[9] Van Wyk de Vries, M. *et al.* 2017: A new volcanic province: An inventory of subglacial volcanoes in West Antarctica. *Geol. Soc. Lond. Special Pubs* 461(1) doi.10.1144/SP461.7.

[10] www.ourworldindata.org.

[11] *The Wall Street Journal,* 9th June 2021, China rolls back climate efforts after climate officials prioritise growth.

[12] *Financial Times,* 13th August 2021, China puts growth ahead of climate with surge in coal-powered power stations and steel mills.

[13] www.eurasiareview.com, 27th October 2020.

[14] *S&P Global: E & E News,* 20th April 2021 US coal production set to rise, in blow to Biden's climate goals.

[15] www.iea.org, 20th April 2021, Global carbon dioxide emissions are set for their second-biggest increase in history.

[16] www.spc.int

[17] www.sdg.iisd.org, 11th February 2020.

[18] Kane-Berman, J. 31st January 2021: The Paris agreement: a costly and damaging failure? *Politics Web* 31.

[19] BP Annual Review of Energy 2019

[20] *The Times* 27th February 2021: World feigns climate concerns as most nations ignore Paris Agreement.

[21] *The Daily Telegraph,* 21st August 2021: The radical potential of nuclear fusion exposes the folly of our net zero deadline.

[22] *Time Magazine,* 20th August 2021: China to build 43 new coal power plants.

[23] www.eminetra.com.au, 27th April 2021, China doubles down on coal plants abroad despite carbon pledge at home.

[24] *Forbes,* 29th March 2021: China burned over half the world's coal last year, despite Xi Jinping's net-zero pledge.

[25] www.ela.gov

[26] www.iea.org

[27] www.iea.org, 6th December 2019.

[28] www.unicef.org, 18th February 2020.

[29] www.bp.com/content/dam/bp/pdf/Energy-economics/statistical-review-2020/BP-

statistical-review-of-world-energy-2020-primary-energy-section.pdf

[30]   www.ourworldindata.org, 11th November 2019.

[31]   Klümper, W. & Qaim, M. 2014: A meta-analysis of the impacts of genetically modified crops. *PLOS One* 3rd November 2014 doi.org/10.1371/journal.pone.0111629.

[32]   www.forbes.com, 5th February 2013.

[33]   www.skepticalscience.com

[34]   www.theguardian.com, 15th February 2020.

[35]   Bjorn Lomborg 2020: *False alarm: How climate change panic cost us trillions, hurts the poor and fails to fix the planet*. Basic Books.

[36]   Teleszewski, T. & Gladyszewska-Fiedoruk, K. 2019: The concentration of carbon dioxide in conference rooms: a simplified model and experimental verification. *Internat. Journ. Envir. Sci. Technol*. 16, 8-031-8040.

[37]   Rodeheffer, C. D. *et al.* 2018: Accurate exposure to low-to-moderate carbon dioxide levels and submarine decision making. *Aerosp. Med. Hum. Perform*. 89, 520-525.

[38]   Baldini, J. *et al.* 2006: Carbon dioxide sources, sinks and spatial variability in shallow temperate caves: Evidence from Ballynamintra Cave, Ireland. *Journ Cave Karst Studies* 68 (1), 4-11.

[39]   Giacomo, G. *et al.* 2014: Measurements of soil carbon dioxide emissions from two maize agroecosystems at harvest under different tillage conditions. *Scientif. World Journ* Article 141345, doi.org/10.1155/2014/141345.

[40]   www.co2science.org/education/reports/co2benefits/co2benefits.php

[41]   www.fao.org/worldfoodsituation/csdb/en/

[42]   www.edition.cnn, 13th August 2021, Get used to surging food prices: Extreme weather is here to stay.

[43]   Mendel, G. 1866: Versuche über Pflanzenhybriden. *Verhandlungen des naturforschenden Vereines in Brünn, Bd. IV für das Jahr 1865*, Abhandlungen, 3-47.

[44]   Yu, Q. *et al.* 2021: RNA demethylation increases the yield and biomass od rice and potato plants in field trials. *Nature Biotechnology* doi.org/10.1038/s41587-021-00982-9

[45]   www.healthyeating.sfgate.com/fortified-flour-1919.html

[46]   www.skepticalscience.com, 22nd May 2018.

[47]   Zaichun Zhu *et al.* 2016: Greening of the Earth and its drivers. *Nature Climate Change* 6, 791-795.

[48]   Donahue, R. J. *et al.* 2013: CO_2 fertilisation has increased maximum foliage cover across the globe's warm, arid environments. Geophysical Research Letters DOI: 10.1002/grl.50563.

[49]   www.judithcurry.com/2016/04/26/rise-in-co2-has-greened-planet-earth/#more-21465

[50]   Idso, S. B. & Kimball, B. A. 1993: Tree growth in carbon dioxide enriched air and its implications for global carbon cycling and maximum levels of atmospheric CO_2. *Global Biogeochemical Cycles* 7 (3) 537-555.

[51]   Saxe, H. *et al.* 1998: Tansley Review No. 98: Tree and forest functioning in an enriched CO_2 atmosphere. *The New Phytologist* 139 (3) 395-436.

[52]   www.co2science.org

[53]   www.try-db.org

[54]   www.garden.org

[55]   Huang, B. *et al., 2021:* Predominant regional biophysical cooling from recent land cover changes over Europe. *Nature Communications* 11, Article 1066.

[56]   Haverd, V. *et al.* 2020: Higher than expected CO_2 fertilization inferred from leaf to global observations. *Global Climate Change Biology* 26 (4), 2390-2402.

[57]   NASA Vegetation Index: Globe continues rapid greening trend, Sahara alone shrinks 700,000 sq km. *Pierre Gosselin No Tricks Zone* 24th February 2021.

[58]   Charles Rotter, 2020: Global change ecologist leads NASA satellite study of rapid greening across Arctic tundra. *Watts Up With That*, 23rd September 2020.

[59] Piao, S. *et al.* 2019: Characteristics, drivers and feedbacks of global greening. *Nature Reviews Earth & Environment* 1, 14-27.

[60] Myers-Smith, I. H. *et al.* 2020: Complexity revealed in the greening of the Arctic. *Nature Climate Change* 10, 106-117.

[61] www.carbonbrief.org, 2nd August 2012: Carbon uptake has doubled over the last 50 years – but where is it going?

[62] www.theguardian.com, 7th April 2015: Greenpeace activists board Arctic-bound oil rig

[63] www.arctictoday.com, 30th April 2019: Greenpeace activists target Norwegian Arctic drilling rig.

[64] www.skepticalscience.com, Positives and negatives of global warming.

[65] www.friendsofscience.org, Climate change science essay page 8.

[66] Gasparrini, A. *et al.* 2015: Mortality risk attributable to high and low ambient temperature: a multicountry observational study. *The Lancet* 386, 9991, 369-375

[67] Zhao, Q. Z. *et al.* 2021: Global, regional and national burden of mortality associated with non-optimal ambient temperatures from 2000 to 2019: a three-stage modelling study. *The Lancet* 5(7), E415-E425.

[68] Laschewski, G. & Jendritzky, G. 2002: Effects of the thermal environment on human health: an investigation of 30 years of daily mortality from SW Germany. *Climate Research* 21: 91-103.

[69] Guo , Y. *et al.* 2014: Global variation in the effects of ambient temperature on mortality: A systematic evaluation. *Epidemiology* 25: 781-789.

[70] Vardoulakis, S. *et al.* 2014: Comparative assessment of the effects of climate change on heat- and cold-related mortality in the United Kingdom and Australia. *Environ. Health. Perspect.* doi: 10.1289/eph.1307524.

[71] Falagas, M. E. *et al.* 2009: Seasonality of mortality: the September phenomenon in Mediterranean countries. *Canad. Med. Assoc. Jour.* 181: 484-486.

[72] Yi, W. & Chan, A. P. 2014: Effects of temperature on mortality in Hong Kong: a time series analysis. *Internat. Jour. Biomet.* 58: 1-10.

[73] Berko, J. *et al.* 2014: Deaths attributed to heat, cold, and other weather events in the United States, 2006-2010. *National Health Statistics Reports* 76: 1-16.

[74] Wu, W. *et al.* 2013: Temperature-mortality relationship in four subtropical Chinese cities: A time series study using a distributed lag non-linear mode. *Sci. Tot. Envir.* 449: 355-362

[75] Burkart, K. *et al.* 2011: Seasonal variations of all-cause and cause-specific mortality by age, gender, and socioeconomic condition in urban and rural areas of Bangladesh. *Int. Jour. Equity Health* 10: 32

[76] Egondi, T. *et al.* 2012: Time-series analysis of weather and mortality patterns in Nairobi's informal settlements. *Glob. Health Action* 5. Doi 10.3402/gha.v5i0

[77] Douglas, A. S. *et al.* 1991: Seasonality of disease in Kuwait. *Lancet* 337: 1393-1397

[78] www.databank.worldbank.org/data/databases.aspx

[79] Matt Ridley, 2010: *The rational optimist. How prosperity evolves.* Harper Collins.

[80] The Beer-Lambert Law.

[81] www.cfact.org/2020/08/01/watching-co2-feed-the-world/

[82] On 28th September 1893, a utopian communist settlement of 238 idealists was formed in Paraguay and called Colonia Nueva Australia. People trickled in from the UK and Australia, most returned disappointed to Australia and there are now about 2,000 descendants in Paraguay of the 8 families that remained.

[83] Most of the major UK cricket, rugby, football clubs formed at this time.

[84] Patricia Berger 1972: *The famine of 1682-1684 in France.* University of Chicago.

[85] *Power Engineering International,* 13th January 2013.

[86] Mutezo, G. & Mulopo, J. 2021: A review of Africa's transition from fossil fuels to renewable energy using circular economy principles. *Renew. Sust. Energy Reviews* 137, doi.org/10.1016/j.rser.2020.110609.

87   www.forbes.com, 11th January 2021
88   Alova, G. *et al.* 2021: A machine-learning approach to predicting Africa's electricity mix based on planned power plants and their chances of success. *Nature Energy* 6, 158-166
89   www.data.unicef.org
90   www.nationmaster.com
91   www.OilPrice.com, 19th April 2021, India is pushing for more coal capacity
92   www.OilPrice.com, 6th September 2021, India is running out of coal as energy demand sykrockets.
93   www.nsenergybusiness.com, 19th October 2020.
94   *Asia Times*, 14th August 2921, Skyrocketing coal prices defy climate goals
95   www.carbonbrief.org, 24th March 2020.
96   *VOA News, Reuters*, 3rd February 2021: Study: China's new coal power plant capacity in 2020 more than 3 times rest of the world's.
97   *The New York Post*, 15th March 2021: China's record of broken promises leaves no point in talking climate change.
98   *Financial Times*, 5th March 2021: Build back faster: China targets 6% growth after reining in coronavirus.
99   *Carbon Brief, 1st* March 2021: Build back faster: China's CO_2 emissions surged 4% in second half of 2020; Statistical Communiques on Economic and Social Development
100 *The Economic Times of India*, 12th February 2021: Coal projected to be India's largest source of power in 2040.
101 www.plattsinfo.platts.com
102 *AME Group*, 29th July 2021: Iron Ore Feature: Decarbonization drives China's pellet demand.
103 *Daily Express*, 13th March 2021: Net Zero destroys UK jobs and offshore problem to mega polluter China.
104 *Energy Voice*, 17th February 2021: India's coal use to surge as power demand is set to double.
105 *New Indian Express*, 9th March 2021: Coal India approves 32 coal mining projects.
106 *Bloomberg*, 16th March 2021: The world's three biggest coal users get ready to burn even more.
107 Tilak Doshi & C. S. Krishnadev, 25th August 2021: India's energy policies and the Paris Agreement commitments, *Real Clear Energy*
108 *The Guardian*, 6th May 2014: Climate change and poverty: Why Indira Gandhi's speech matters.
109 *Press Trust of India*, 9th February 2021.
110 www.coalindia.in, 15th June 2021.
111 www.wordometers.info
112 www.energy.economictimes.indiatimes.com, 5th May 2019.
113 *Nature Climate Change*, 19th October 2020.
114 *The Hindu*, 1st March 2021: India's percentage CO_2 emissions rose faster than the world average
115 *AFP*, 5th March 2021: China's 5-year coal plan: Build back blacker.
116 *Nikkei Asia*, 8th March 2021: China's addiction to coal clashes with carbon neutrality pledge.
117 www.ourworldindata.org, 28th January 2019.
118 www.statista.com, 15th June 2021.
119 www.theepochtimes.com/china-adding-new-coal-power-plants-equivalent-to-entire-european-union-capacity_3186439.html?fbclid=lwAR3CkkAFI56m1XV9qR086E6-JuMkiEOm2NA9ArV6gg8rlwxC20l-K4JCjJc.
120 *Reuters* 1st August 2020.
121 *Reuters* 3rd February 2021.
122 South African Development Community, www.sadc.int.
123 *The National, Business*, 25th April 2015.
124 Donn Dears, 14th July 2020: Don't ignore coal. *Power for USA.*

125 *Clean Energy Wire*, 29th May 2019.

126 www.ga.gov.au

127 www.world-nuclear.org

128 Plimer, I. R., 2000: *Milos – geologic history*. Koan.

129 Snoek, W. *et al.* 1999: Application of Pb isotope geochemistry to the study of the corrosion products of archaeological artefacts to constrain provenance. *Jour. Geochem. Explor.* 66, 421-425.

130 www.world-archaeology.com

131 www.worldhistory.org, 1st November 2013.

132 www.theguardian.com, 12th June 2012.

133 www.listverse.com

134 www.forbes.com, 12th May 2014.

135 Wei, T. *et al.* 2021: Keeping track of greenhouse gas emission reduction progress and targets in 167 cities worldwide. *Front. Sustain. Cities* doi/org/10.3389/frsc.2021.696381.

136 www.data.worldbank.org.

137 Particulate matter with a diameter of 2.5 micrometres or less.

138 Zhang, Q. *et al.*, 2019: Drivers of improved PM2.5 air quality in China from 2013 to 2017. *PNAS* doi.org/10.1073/pnas.1907956116.

139 www.scmp.com, 30th June 2021.

140 *S & P Global*, 23rd February 2021.

141 www.reuters.com, 13th January 2021

142 www.globaltimes.cn

143 www.tradingeconomics.com

144 www.globalconstructionreview.com

145 www.hydrocarbons-network.com

146 www.reuters.com, 3rd March 2021.

147 www.reuters.com, 16th April 2021.

148 www.tradingeconomics.com

149 www.feednavigator.com, 18th March 2021.

150 www.worldbank.org

151 www.cato.cato.org, 10th September 2020.

152 www.who.int

153 www.bloomberg.com, 22nd April 2021.

154 www.tradingeconomics.com.

155 www.wordometers.info.

156 www.nbcnews.com, 3rd September 2010.

157 www.governmentnews.com.au/high-cost-of-recycling-leads-to-more-landfill-waste/

158 Kevin Donnelly, 29th July 2021: Why our kids can't read: it's the ideology, stupid. *The Australian.*

159 Deidre Clary & Fiona Mueller, 2021: Writing matters: reversing a legacy of policy failure in Australian education. https://www.cis.org.au

160 Anthony Kronman, 2019: *The assault on American excellence*. Free Press.

161 www.theaustralian.com.au, 27th July 2021.

162 www.eminetra.com.au, 1st June 2021.

163 www.research-repository.griffith.edu.au

164 www.abc.net.au, 18th October 2008.

165 www.theaustralian,com.au, 30th October 2007

166 www.study.com/academy/lesson/what-are-greenhouse-gases-lesson-for-kids.html

167 www.abc.net.au, 9th April 2013.

168 www.jstor.org

169 www.washingtonpost.com, 14th May 2014.

170 www.abc.net.au, 3rd December 2019.

171   www.school-news.com.au, 4th April 2018.
172   www.thenewdaily.com.au, 11th March 2021.
173   www.startsat60.com, 5th December 2013.
174   www.smh.com.au, 3rd December 2019.
175   Alan Tudge, 22nd June 2021: Recruit, train teachers better for higher scores. *The Australian.*
176   www.pc.gov.au,
177   www.aeufederal.org.au, 9th September 2020.
178   www.abs.gov.au
179   www.abc.net.au, 20th March 2018.
180   www.abs.gov.au
181   The Manhattan Engineer District, 29th June 1946.
182   www.bbc.com, 9th August 2020.
183   www.forbes.com, 27th June 2019.
184   www.world-nuclear.org
185   www.ctbto.org
186   www.theconversation.com, 2nd November 2020.
187   www.history.com
188   www.americanscientist.org
189   www.cdc.gov
190   www.wsj.com, 6th November 2017.
191   www.cato.org, 28th October 2017.
192   www.diariocuba.com
193   *Miami Herald*, 12th July 2016.
194   www.averagesalary.com
195   *Washington Post*, 14th July 2021.
196   www.news.sky.com, 26th July 2021.
197   www.history.com
198   www.heritage.org, 2nd February 2010.
199   www.bbc.com, 23rd December 2017.
200   www.theguardian.com, 9th June 2017.
201   www.bbc.com, 8th February 2021.
202   www.news.un.org, 10th March 2021.
203   www.english.elpais.com, 5th March 2021.
204   www.bbc.com, 3rd December 2020.
205   www.opec.org
206   www.theholocaustexplained.com
207   www.washingtonpost.com, 5th February 2020.
208   www.nytimes.com, 3rd August 2006.
209   Greens Party Senator for South Australia.
210   *The Australian*, 21st January 2018, Sarah Hanson-Young pays own fare to mingle with world leaders at Davos.
211   www.smh.com.au/politics/federal/election-2016-greens-leader-fails-to-declare-home-pays-au-pairs-low-wage-20160519-goywxq.html
212   www.smh.com.au, 7th October 2019.
213   Examples of this took place at Agincourt, Chancellorsville in the American Civil War and at Rorke's Drift in the Zulu Wars.
214   www.independent.co.uk, 24th February 2014.
215   *Spectator Australia*, 5th August, 2021: Flat White Quick Shots. Why is Adam Bandt treading so carefully over the Julian Burnside ugliness?
216   www.cbc.ca, 2nd October 2020.
217   www.atag.org.

218 www.ourworldindata.org, 22nd October 2020.

219 www.theprovince.com.

220 www.usatoday.com, 24th March 2017.

221 www.news.com.au, 11th February 2019.

222 *Harry Zaremba,* 3rd July 2020: Oil Price China Plans To Dominate The Global Nuclear Energy Push.

223 *Donn Dears,* 16th October 2020: We need blackouts. *Power for USA.*

224 www.smh.com.au, 14th August 2019.

225 www.ourworldindata.org

226 www.heraldsun.com.au, 15th August 2019.

227 www.standard.net.au, 17th August 2019.

228 www.canberratimes.com.au, 20th August 2019.

229 www.thesun.co.uk, 5th May 2019.

230 *Currencylad.*

231 *Catallaxy Files,* 5th February 2021: The Only Choice.

232 www.foxnews.com/opinion/tucker-carlson-texas-green-new-deal-climate-catastrophe, 15th February 2021.

233 www.euronews.com, 2nd August 2019.

234 www.dailymail.co.uk, 2nd August 2019.

235 www.thetimes.co.uk, 20th August 2019.

236 www.vanityfair.com, 19th August 2019.

237 Whitmarsh, L. *et al.* 2020: Use of aviation by climate change researchers: Structural influences, personal attitudes, and information provision. *Global Environmental Change* 65, doi.org/10.1016/j.gloenvcha.2020.102184.

238 *The Times,* 20th October 2020 *Climate experts fly more often than other scientists.*

239 Douglas Woodwell, 2013: *Research foundations: How do we know what we know.* Sage Publications.

240 Horton, R.: What is medicine's 5 sigma? *The Lancet* 385 Offline: 1380.

241 Popper, K. 2005: *The logic of scientific discovery.* Taylor and Francis.

242 Israel, K. *et al.* 1931: 100 Autoren gegen Einstein. *Naturwissenschaften* 19: 254-256

243 Take no one's word for it.

244 www.judithcurry.com, 2014 and 2015: *Ethics of climate expertise* and *Conflicts of interest in climate science.*

245 In my view, the only committee that has done anything lasting and useful was the committee of scholars who translated the King James edition of the Bible into English from Greek, Hebrew, Aramaic and Latin. This committee's achievements are still being used on a daily basis by billions of people more than 400 years later.

246 *The Irish Times,* 12th March 2018: When government is the biggest funder of scientific endeavour.

247 David Gelernter, 2017: The closing of the scientific mind. *Science Matters.*

248 Ings, S. 2016: *Stalin and the scientists: A history of triumph and tragedy 1905-1953.* Faber & Faber.

249 *New Scientist,* 1st August 2007: The word: Vernalisation.

250 Robert Conquest, 1987: *Harvest of sorrow: Soviet collectivization and the Terror-famine.* Oxford University Press.

251 Large, R. R. *et al.* 2019: Atmospheric oxygen cycling through the Proterozoic and Phanerozoic. *Mineralium Deposita* 54, doi: 10.1007/s00126-019-00873-9.

252 Brink, H-J. 2014: Singnale der Milchstraße vorbogen in der Sedimentfüllung Zentraleuropäischen Beckensystems? *Z. Dt. Ges. Geowiss.* 166: 9-20.

253 Orbital eccentricity.

254 The combined effect of two precessions.

255 Axial tilt or obliquity.

[256] Pacific Decadal Oscillation, Indian Ocean Dipole and Atlantic Multidecadal Oscillation.

[257] ENSO = El Niño-Southern Oscillation.

[258] Sturtian Glaciation.

[259] Arkaroola Reef.

[260] Marinoan Glaciation.

[261] The chemistry of iron minerals over the history of time are a good way of measuring the cycles and amount of atmospheric oxygen in the past. See Large, R. R. *et al.* 2019: Atmospheric oxygen cycling through the Proterozoic and Phanerozoic. *Mineralium Deposita* 54, doi: 10.1007/s00126-019-00873-9.

[262] Ediacaran Fauna.

[263] www.geocraft.com

[264] Green, J. K. *et al.* 2020: Amazon rainforest photosynthesis increases in response to atmospheric dryness. *Science Advances* 6(47) doi:10.1126/sciadv.abb7232.

[265] Vazquez, M. & Montañes-Rodriguez, P. 2010: The Earth in Time *Researchgate* doi.10.1007/978-1-4419-1684-6.2.

[266] Eocene Warming.

[267] Van Wyk de Vries, M. *et al.* 2018: A new volcanic province: an inventory of subglacial volcanoes in West Antarctica. *Geol. Soc. Lond. Spec. Publ* 461.

[268] Dziadek, R. *et al.* 2021: High geothermal heat flow beneath Thwaites Glacier in West Antarctica inferred from aeromagnetic data. *Nature Communications* 2, Article Number 162.

[269] www.climatedata.info/proxies/ice-cores

[270] www.reddit.com

[271] Schmidt, M. *et al.* 2011: Abrupt climate change during the last ice age. *Nature Education Knowledge* 3 (10) 11.

[272] Salamatin, A. N. *et al.* 1998: Ice core age dating and paleothermometer calibration based on isotope and temperature profiles from deep boreholes at Vostok Station (East Antarctica). *Jour. Geophys. Research* doi.org/10.1029/97JD02253.

[273] Dansgaard-Oeschger cycles.

[274] Hawks, J. *et al.* 2000: Population bottlenecks and Pleistocene human evolution. *Molec. Biol. & Human Evol* 17 (1), 2-22.

[275] www.npr.org, 22nd October 2012: How human beings almost vanished from Earth in 70000 BC.

[276] Heinrich events.

[277] The latitude of Athens (Greece).

[278] EPICA Community Members, 2004: Eight glacial cycles from an Antarctic ice core. *Nature* 429, 623-628.

[279] Petit, J.-R. *et al.* 2004: Epica-Dome C ice core: Extending the dust record over the last 7 climatic cycles (740 kyr B.P.). *SCAR Open Science Conference* 25-1.07.2004, Bremen, Germany.

[280] Rorsch, A. *et al.* 2005: The interaction of climate change and the carbon dioxide cycle. *Energy & Environment* 16 (2), 217-238.

[281] Zwally, H. J. *et al.* 2021: Mass balance of the Antarctic ice sheet 1992-2016: reconciling results from GRACE gravimetry with ICESat, ERS1/2 and Envisat altimetry. *Glaciology* 67 (263) 533-559.

[282] Zhu, J. *et al.* 2021: An assessment of the ERA5 reanalysis for Antarctic near-surface air temperature. *Atmosphere* 12 (2) doi.org/10.3390/atmos12020217.

[283] Allerød.

[284] Younger Dryas.

[285] www.ocp.ideo.columbia.edu, Two examples of abrupt climate change – Columbia University.

[286] Holocene Optimum.

[287] Fleming K. *et al.* 1998: Refining the eustatic sea-level curve since the Last Glacial

Maximum using far- and intermediate sites. *Earth Planet. Sci. Letts* 163, 327-342.

[288] Alley, R. B. 2004: GISP-2 ice core and temperature accumulation data. IGBP PAGES/ World Data Center for Paleoclimatology Data Accumulation Series #2004-013. NOAA/ NGDC Paleoclimatology Program, Boulder, Co. USA.

[289] van Westen, R. M. & Dijkstra, H. A. 2021: Ocean eddies strongly affect global mean sea-level projections. *Science Advances* 7 (15) doi: 10.1126/sciadv.abf1674.

[290] Lang, P. & Gregory, K. 2019: Economic impact of energy consumption change caused by global warming. CAMA Workshop Paper No 55/2018, doi.org/10.2139/ssrn.3275803

[291] www.ncdc.noaa.gov, Glacial-Interglacial Cycles – National Climatic Data Center.

[292] Alley, R. B. 2000: The Younger Dryas cold interval as viewed from central Greenland. *Jour. Quat. Sci. Reviews* 19, 213-226.

[293] www.climate4you, Ole Humlum.

[294] Cosmogenic isotopes of Be^{10}, C^{14}, Al^{26}, Cl^{36}, Ca^{41}, Ti^{44} and I^{129}.

[295] Shaviv, N. J. *et al.* 2014: Is the Solar System's galactic motion imprinted in the Phanerozoic climate? *Nature Scientific Reports* Article 6150 doi.org/10.1038/srep06150 .

[296] Steinhilber, F. *et al.* 2012: 9,400 years of cosmic radiation and solar activity from ice cores and tree rings. *PNAS* 109 (16), doi.org/10.1073/pnas.1118965109.

[297] Christ, A. J. *et al.* 2021: A multimillion-year-old record of Greenland vegetation and glacial history preserved in sediment beneath 1.4 km of ice at Camp Century. *PNAS* 118 (13) doi.org/10.1073/pnas.2021442118.

[298] Champion, Rafe 2021: Thames Freezes. *Catallaxy Files* 14th February 2021.

[299] Nuessbaumer, S. U. & Zumbühl, H. J. 2012: The Little Ice Age history of the Glacier des Bossons (Mont Blanc Massif, France): a new high resolution glacier length curve based on historical documents. *Climate Change* 111, 301-334.

[300] www.jonova.s3.amazonaws.com/corruption/climate-corruption.pdf and Gervais, F. 2021: Climate sensitivity and carbon footprint. *Science of Climate Change* 1, 7-96.

[301] www.notrickszone.com, Pierre Gosselin, 28th August 2021, The most inconvenient region on the planet for global warming alarmists: Antarctica sees growing sea ice.

[302] www.nsidc.org, 9th September 2021, National Snow & Ice Data Center, Charctic interactive sea ice graph.

[303] *The Washington Post*, 1st October 2021: South Pole posts most severe cold season on record, an anomaly in a warming world.

[304] Misquote from H. L. Mencken in "The Divine Afflatus" (*New York Evening Mail*, 16th November 1917): For every complex problem there is a solution that is neat, simple and wrong.

[305] Holocene Epoch.

[306] Science and Environment Policy Project, 14th December 2020.

[307] Dixon, J. E. *et al.* 1996: An experimental study of water and carbon dioxide solubilities in mid-ocean ridge basaltic liquids. Part 1: Calibration and solubility methods. *Jour. Petrol.* 36, 1633-1646.

[308] Stolper, E. & Holloway, J. R. 1988: Experimental determination of the solubility of carbon dioxide in molten basalt at low pressure. *Earth Planet. Sci. Lett.* 87: 397-408

[309] Shilobreyeva, S. N. & Kadik, A. A. 1990: Solubility of CO_2 in magmatic melts at high temperatures and pressures. *Geochem. Internat.* 27: 31-41.

[310] Dasgupta, R. & Hirschmann, M. M. 2010: The deep carbon cycle and Earth's interior. *Earth. Planet. Sci. Letts* 298, 1-13.

[311] Carbonatite.

[312] Pêrez N. M. *et al.* 2011: Global CO_2 emission from volcanic lakes. *Geology* 39, 235-238

[313] Fytikas, M. 1989: Updating of the geological and geothermal research on Milos Island. *Geothermics* 18: 485-496.

[314] Plimer, I. R. 2000: *Milos – geologic history*. Koan.

[315] For example, Dr Claude Blot, a French volcanologist; unrecognised by his envious colleagues.

[316] Vail curves.

[317] Rampino, M. R. & Caldeira, K. 2020. A 32-million year cycle detected in sea-level fluctuations over the last 545 Myr. *Geoscience Frontiers* 11 (6) 2061-2065.

[318] Rothman, D. H. 2002: Atmospheric carbon dioxide levels for the last 500 million years. *PNAS* 99(7) 4167-4171.

[319] Large, R. R. *et al.* 2018: Atmosphere oxygen cycling through the Proterozoic and Phanerozoic. *Mineralium Deposita* 54, 485-506.

[320] Budyko, M I. 1977: Climate changes. *Amer. Geophys. Union* https://doi.org/10.1029/SO101.

[321] Kalderon-Asael, B. *et al.* 2021: A lithium isotope perspective on the evolution of carbon and silicon cycles. *Nature* 595, 394-398.

[322] www.journals.ametsoc.org/doi/abs/10.1175/2010JCL13682.1

[323] www.sciencedaily.com/releases/2009/05/090513130942.htm

[324] www.volcano.oregonstate.edu/submarine

[325] Crisp, J. A. 1984: Rates of magma emplacement and volcanic output. *Jour. Volcan. Geotherm. Res.* 20: 177-211.

[326] www.plateclimatology.com

[327] Dixon, J. E. & Stolper, E. M. 1995: An experimental study of water and carbon dioxide solubilities in mid-ocean ridge basaltic liquids. Part II: Applications to degassing. *Jour. Petrol.* 36, 6, 1633-1646.

[328] Marty, B. & Zimmermann, L. 1999: Volatiles (He, C, N, Ar) in mid-ocean ridge basalts: assessment of shallow-level fractionation and characterisation of source composition. *Geochim. Cosmochim. Acta* 63: 3619-3633.

[329] Bottinga, Y. & Javoy, M. 1989: MORB degassing: evolution of CO_2. *Earth Planet. Sci. Lett.* 95: 215-225.

[330] Kingsley, R. H. & Schilling, J.-G. 1995: Carbon in mid-Atlantic ridge basalt glasses from 28°N to 63°N: evidence for a carbon-enriched Azores mantle plume. *Earth Planet. Sci. Lett.* 129: 31-53.

[331] Des Marais, D. J. & Moore, J. G. 1984: Carbon and its isotopes in mid-oceanic basaltic glasses. *Earth Planet. Sci. Lett.* 69: 43-57.

[332] Dixon, J. E. & Stolper, E. M. 1995: An experimental study of water and carbon dioxide solubilities in mid-ocean ridge basaltic liquids. Part II: Applications to degassing. *Jour. Petrol.* 36: 1633-1646.

[333] Marty, B. & Tolstikhin, I. N. 1998: CO_2 fluxes from mid-ocean ridges, arcs, and plumes. *Chem. Geol.* 145: 233-248.

[334] East Pacific Rise, Mid Atlantic Ridge, Southeast Indian Ridge.

[335] Hauri, E. *et al.* 1993: Evidence for hot-spot related carbonatite metasomatism in the oceanic upper mantle. *Nature* 365: 221-227.

[336] Yim, W. 2019: Climate impacts of the SW Indian Ocean blob. *Imperial Engineer* Autumn 2019, 24-25.

[337] Yim, W. 2020: Volcanism generate ocean heat waves and biodiversity. Association for Geoconservation Hong Kong doi:org/10.13140/RG.2.2.32540.51844.

[338] Yim, W. 2019: Volcanic eruptions and the 2014-2016 ENSO. Climate Realists of Five Dock Conference doi:org/10.13149/RG.2.2.28206.97606.

[339] Leamon, R. J. *et al.* 2021: Termination of solar cycles and correlated tropospheric variability. *Earth Space Sci,* 8 (4) doi.org/10.1029/2020EA001223.

[340] Resing, J. A., *et al.* 2004: CO_2 and ^{3}He in hydrothermal plumes: implications for mid-ocean ridge CO_2 flux. *Earth Planet. Sci. Lett.* 226: 449-464.

[341] Mottl, M. J. & McConachy, T. F. 1990: Chemical processes in buoyant hydrothermal plumes on the East Pacific Rise near 21°N. *Geochim. Cosmochim. Acta* 54: 1911-1927

[342] Sansone, F. J. *et al.* 1998: CO_2-depleted fluids from mid-ocean ridge-flank hydrothermal springs. *Geochim. Cosmochim. Acta* 62: 2247-2252.

[343] LeQuéré, C. & Metzel, N. 2004: Chapter 12: Natural processes regulation the ocean uptake of CO_2. In: *SCOPE 62, The global carbon cycle: Integrating humans, climate, and the natural world* (Eds Field, C. B. and Raupach, M. R.). Island Press, 243-256.

[344] Jendrzejewski, N. *et al.* 1997: Carbon solubility in mid-ocean ridge basaltic melt at low pressures (250-1950 bar). *Chem. Geol.* 138: 81-92

[345] Pineau, F. & Javoy, M. 1994: Strong degassing at ridge crests: the behaviour of dissolved carbon and water in basaltic glasses at 14°N (M.A.R.). *Earth Planet. Sci. Lett.* 123: 179-198.

[346] Gerlach, T. M. 1989: Degassing of carbon dioxide from basaltic magma at spreading centers, II. Mid-ocean ridge basalts. *Jour. Volcan. Geoth. Res.* 39: 221-232.

[347] Jendrzejewski, N. *et al.* 1992: Water and carbon contents and isotopic compositions in Indian Ocean MORB. *EOS* 73: 352.

[348] Dixon, J. E. & Stolper, E. M. 1995: An experimental study of water and carbon dioxide solubilities in mid-ocean ridge basaltic liquids, Part II. Applications to degassing. *Jour. Petrol.* 36: 1633-1646.

[349] Weiss, R. F. 1974: Carbon dioxide in water and seawater: the solubility of a non-ideal gas. *Marine Chemistry* 2, 203-215.

[350] Teng, H. *et al.* 1996: Solubility of CO_2 in the ocean and its effect on CO_2 dissolution. *Energy Convers. Manag.* 37, 1029-1038.

[351] www.news.nationalgeographic.com.news/2006/08/060830-carbon-lakes.html

[352] Sohn, R. A. *et al.* 2008: Explosive volcanism on the ultraslow-spreading Gakkel ridge, Arctic Ocean. *Nature* 453: 1236-1238.

[353] Snow, J. *et al.* 2001: Magmatic and hydrothermal activity in the Lena Trough, Arctic Ocean. *Trans. Amer. Geophys. Union* 82: 193, 197-198.

[354] Helo, C. *et al.* 2011: Explosive eruptions at mid-ocean ridges driven by CO_2-rich magmas. *Nature Geoscience* 4 260-263.

[355] Edwards, M. H. *et al.* 2001: Evidence of recent volcanic activity on the ultraslow spreading Gakkel ridge. *Nature* 409, 808-812.

[356] Edmonds, H. N. *et al.* 2003: Discovery of abundant hydrothermal venting on the ultraslow-spreading Gakkel ridge in the Arctic Ocean. *Nature* 421, 252-256.

[357] Alexander, R. T. & Macdonald, C. 1996: Small off axis volcanoes on the East Pacific Rise. *Earth Planet. Sci. Letts* 139, 387-394.

[358] Reynolds, J. R & Langmuir, C. H. 2000: Identification and implications of off-axis lava flows around the East Pacific Rise. *Geochem. Geophys. Geosyst.* doi:org/10/1029/1999GC000033.

[359] Fialko, Y. 2001: On origin of near-axis volcanism and faulting at fast spreading mid-ocean ridges. *Earth Planet. Sci. Lett.* 190, 31-39.

[360] Fujii, M. & Okino, K. 2018: Near-seafloor magnetic mapping of off-axis lava flows near the Kairei and Yokoniwa hydrothermal vent fields in the Central Indian Ridge. *Earth Planets Space* 70, Article 188.

[361] Lupton, J. *et al.* 2006: Submarine venting of liquid carbon dioxide on a Mariana Arc volcano. *Geochem. Geophys. Geosys.* 7, doi:10.1029/2006GC001152.

[362] Huppert, K. L. *et al.* 2021: Hotspot swells and the lifespan of volcanic ocean islands. *Science Advances* 6, doi:org/10.1126/sciadv.aaw6909.

[363] Marty, B. *et al.* 1993: Geochemistry of gas emanations: A case study of the Réunion Hot Spot, Indian Ocean. *Applied Geochemisty* 8, 141-152.

[364] Dixon, J. E. *et al.* 1991: Degassing history of water, sulfur and carbon in submarine lavas from Kilauea Volcano, Hawaii. *Jour. Geology* 99, 371-394.

[365] Santana-Casiano, J. *et al.* 2016: Significant discharge of CO_2 from hydrothermalism associated with the submarine volcano of El Hierro Island, *Nature Scientific Reports* 6, Article number 25686.

[366] Batiza, R. 1969: Seamounts and seamount chains of the eastern Pacific. In *The Eastern Pacific Ocean and Hawaii* (eds Winterer E. L. *et al.*) *Geol. Soc. America* doi.org/10.1130/

DNAG-GNA-N.

[367] Such as the Hawaii Emperor seamount chain and Gulf of Alaska.

[368] Clague, D. A. *et al.* 2000: Near-ridge seamount chains in the northeastern Pacific Ocean. *Soli Earth* doi:org/10.1029/2000JB900082

[369] Guyots.

[370] Hillier, J. J. & Watts, A. B. 2007: Global distribution of seamounts from ship-track bathymetry data. *Geophys. Res. Letts* 34, L13304, doi:10.1029/2007GL029874.

[371] www.principa-scientific.org/volcanic-carbon-dioxide.html

[372] Fisher, A. T & Wheat, C. G. 2015: Seamounts as conduits for massive fluid, heat, and solute fluxes on ridge flanks. *Oceanography* 25(1) doi.org/10.5670/oceanog.2010.63.

[373] Hensen, C. *et al.* 2019: Marine transform faults and fracture zones: A joint perspective integrating seismicity, fluid flow and life. *Front Earth Sci.* doi.org/10.3389/feart.2019.0039.

[374] Hamling, I. J. *et al.* 2016: Off-axis magmatism along a subaerial back-arc rift: Observations from the Taupo Volcanic Zone, New Zealand. *Science Advances* 2, doi:org/10.1126/sciadv.1600288.

[375] www.citeseerx.psu.edu, Tom Quirk, Sources and sinks of carbon dioxide.

[376] Quirk, T., 2012: Did the global temperature trend change at the end of the 1990s? *Asian Pacific Journ. Atmos. Sciences* 48 (4), 339-344.

[377] Warwick, P. D. & Ruppert, L. F. 2016: Carbon and oxygen isotope composition of coal and carbon dioxide derived from laboratory coal consumption: A preliminary study. *Internat. Journ. Coal Geology* 166, 128-135.

[378] Karlsruhe Institute of Technology 2020: Corona-induced CO_2 emission reductions are not yet detectable in the atmosphere. *EurekaAlert,* News Release 21st October 2020.

[379] www.joannenova.com.au/2013/09/astounding-discovery-world-war-ii-had-low-carbon-footprint/print/

[380] Robert Youngson 1998: *Scientific blunders and a brief history of how wrong scientists can sometimes be.* Robinson.

[381] Karl Popper, 1963: *Conjectures and refutation: The growth of scientific knowledge.* Routledge.

[382] Randi, James, 1990: *The mask of Nostradamus. The prophesies of the world's most famous seer.* Charles Scribner's Sons.

[383] Randi, James 1995: *An encyclopedia of claims, frauds and hoaxes of the occult and supernatural.* St Martin's Press.

[384] Ian Wilson 2001: *Before the flood.* Orion.

[385] *Nuclear Energy and Fossil Fuels,* 7th March 1956.

[386] The Hubbard peak.

[387] Paul Ehrlich, 1968: *The population bomb.* Sierra Club/Ballantine Books.

[388] Thomas Malthaus, 1798: *An essay on the principle of population.*

[389] *Evolution and Ethics: Thomas Henry Huxley* (ed. Michael Ruse), 2009. Princeton University Press.

[390] Fairfield Osborn, 1948: *Our plundered planet.* Faber and Faber.

[391] Dixy Lee Ray, 1993: *Environmental overkill: Whatever happened to common sense?* Regnery Gateway.

[392] www.eugenicsarchive.ca

[393] Partington, J. S. 2003: H. G. Wells' eugenic thinking of the 1930s and 1940s. In: *Utopian Studies,* 74-81. Penn State University Press.

[394] www.diglib.amphilsoc.org.

[395] www.ncr.com.

[396] Paul Ehrlich, 1968: *The population bomb.* Sierra Club/Ballantine Books.

[397] William Vogt, 1948: *The road to survival.* William Sloan Associates.

[398] www.newscientist.com.

[399] Rachel Carson, 1962: *Silent spring.* Houghton Miffin.

[400] Paul Offit, 2017: *Pandora's lab: Seven stories of science gone wrong.* National Geographic

[401] Michael Crichton, 2004: *The state of fear.* Harper Collins.

[402] *World Economic Forum* 25th July 2019: Forests in Europe are expanding each year.

[403] *The Washington Post,* 9th February 1992: The ozone catastrophe: warning from the skies.

[404] William and Paul Paddock, 1967: *Famine 1975! America's decision: Who will survive?* Little, Brown and Co.

[405] Paul Ehrlich, 1968: *The population bomb.* Sierra Club/Ballantine Books.

[406] *The Moscow Times,* 14th June 2021: Russia's population decline more than doubles in 2020.

[407] *The New York Times,* 10th August 1969.

[408] www.aei.org, 21st April 2019: 18 spectacularly wrong predictions made around the time of first Earth Day in 1970, expect more this year.

[409] www.americanactionforum.org, 7th May 2010.

[410] *Audubon.* May 1970.

[411] www.climateandcapitalism.com, 10th May 2012.

[412] *Mademoiselle,* April 1970.

[413] *The Daily Telegraph,* 22nd April 2016.

[414] *The Redlands Daily Facts,* 6th October 1970.

[415] www.thegwpf.com, 22nd April 2020.

[416] *The Living Wilderness,* Spring 1970.

[417] www.thesmithsonianmag.cm, 27th April 2016.

[418] *The New York Times,* 19th November 1970.

[419] www.reason.com, 4th February 2004.

[420] 22nd May 2019.

[421] www.aei.org, 21st April 2019: 18 spectacularly wrong predictions made around the time of first Earth Day in 1970, expect more this year.

[422] *Time,* 2nd February 1970.

[423] www.groovyhistory.com, 3rd July 2018.

[424] Meadows, D. H. *et al.* 1972: *Limits to growth.* Universe Books, 1972.

[425] Meadows, D. H. *et al.* 2004: *Limits to growth: The 30-year update.* Chelsea Green Publishing.

[426] Bentley, R, W. 1972: Oil forecasts, past and present. *Energy Exploration and Exploitation* 20, 6, 481-492.

[427] M. J. Molina & F. S. Rowland 1974: Stratospheric sink for chlorofluoromethanes: chlorine-atomised destruction of ozone. *Nature* 249, 810-812.

[428] John Gribbin and Stephen Plagemann, 1974: *The Jupiter effect.* Vintage.

[429] www.aei.org, 21st April 2019: 18 spectacularly wrong predictions made around the time of first Earth Day in 1970, expect more this year.

[430] www.energy.gov

[431] Paul Sabin, 1980: *Paul Ehrlich, Julian Simon and our gamble over Earth's future.* Yale University Press.

[432] *The Guardian,* 28th September 2015.

[433] www.vineastrology.com

[434] Bardi, U. 2019: Peak oil, 20 years later: Failed prediction or useful insight? *Energy Research and Social* Science 48, 257-261.

[435] *Nobelsville Ledger,* 9th April 1980.

[436] *Associated Press,* 6th September 1990.

[437] *Forbes,* 4th December 2020.

[438] *The Guardian* 24th December 2002: Why vegans were right all along.

[439] Marte Gutierrez, October 2002, Colorado School of Mines.

[440] www.defenders.org, 18th April 2003.

[441] www.axwuotes.com, Top 20 quotes of Dave Forman.
[442] www.un.org, World Fertility Report 2009.
[443] www.politifact.com, 29th July 2009.
[444] www.goodreads.com, 10th October 2015.
[445] www.reddit.com, 19th February 2016.
[446] www.theeventchronicle.com, 30th March 2017: The unauthorised biography of David Rockefeller.
[447] *The Mirror*, 17th July 2017.
[448] www.dailystar.co.uk, 16th April 2020.
[449] www.quotefancy.com
[450] www.royalcentral.com.uk, 14th March 2020.
[451] Letter to the author, 30th April 2018.
[452] Brian Fagan, 2020: *The Little Ice Age: How climate made history.* Basic Books.
[453] www.who.int, Ambient air pollution map, 2018.
[454] www.unicef-irc.org
[455] wwwpopulation.un.org
[456] *World Grain News*, 5th March 2021.
[457] www.unicef-irc.org
[458] www.statista.com
[459] *Laudato si'*, Paragraph 161.
[460] *Laudato si'*, Paragraph 61.
[461] *Laudato si'*, Paragraph 67.
[462] *Laudato si'*, Paragraph 20.
[463] *Laudato si'*, Paragraphs 131, 132, 133 and 134
[479] Ridley, M. 2019: GM crops like golden rice will save the lives of hundreds of thousands of children *Quillette* 1st December 2019
[464] www.ourworldindata.org
[465] www.fao.org
[466] *The Wall Street Journal, Europe*, 23rd July 2015
[467] Bailkey, R. & Tupy, M. L. 2020: *Ten global trends every smart person should know and many others you will find interesting.* CATO institute.
[468] www.data.oecd.org
[469] www.jstor.org
[470] Husain, A. R. 2002: Life expectancy in developing countries: A cross sectional analysis. *The Bangladesh Development Studies* 28, 161-178
[471] Rao, V. 1988: Diet, mortality and life expectancy. *Journal of Population Economics* 1: 225-233.
[472] www.ourworldindata.org
[473] www.newscientist.com
[474] *Boston Globe*, 16th April 1970.
[475] www.gerardrennick.com.au
[476] www.earth.columbia.edu
[477] *The Washington Post*, 9th July 1971.
[478] www.johnlocke.org
[479] Zharkova, V. 2020: Modern Grand Solar Minimum will lead to terrestrial cooling. *Temperature* 7(3) 217-222.
[480] www.skepticalscience.com
[481] www.forums.tesla.com
[482] CIA 1974 National Security Threat: Global Cooling/Excess Arctic Ice Causing Extreme Weather.
[483] Center for Strategic and International Studies Report, December 21, 2012.
[484] *The Guardian*, 29th January 1974.

485 *Time*, 24th June 1974.

486 *The New York Times Book Review*, 16sh July 1976.

487 Rasool, S. I. & Schneider, S. H. 1971: Atmospheric carbon dioxide and aerosols. Effects of large increases on global climate. *Science* 173, 138-141.

488 www.climate.gov

489 www.theguardian.com, 21st February 2004.

490 *The Observer*, 22nd February 2004.

491 Brian Fagan, 2008: *The great warming: The rise and fall of civilisations*. Bloomsbury.

492 Green, K. C. & Armstrong, J. S. 2007a: Global warming: Forecasts by scientists versus scientific forecasts. *Energy and Environment* 18, 7+8, 995-1019.

493 Green, K. C. & Armstrong, J. S. 2007b: Structured analogues for forecasting. *Internat. Journ Forecast.* 23, 365-376.

494 *The Tuscaloosa News*, 18th May 1972.

495 Parry, M. L. *et al.* 1988: The climatology of droughts and drought prediction. In: Parry M. L. et al (eds). *The impact of climatic variations on agriculture*. Springer.

496 *Los Angeles Times,* 4th February 1989: 1988 was hottest year on record as global warming trend continues.

497 *The Miami News*, 24th June 1988.

498 John Steinbeck, 1939: *The grapes of wrath*. Viking Press.

499 *Associated Press*, 30th June 1989: If global warming is not revered by the next 30 years.

500 *The Canberra Times*, 26th September 1988: Threat to islands.

501 www.maldives-magazine.com

502 *The Canberra Times*, 26th September 1988.

503 Holdaway, A. *et al.* 2021: Global-scale changes in the area of atoll islands during the 21st Century. *Anthropocene* 33 doi:.org/10.1016/j.ancene.2021.100282.

504 Duvet, V. K. E. 2019: A global assessment of atoll island planform changes over the past decades. *WIREs Climate Change* 10 (1) doi.org/10.1002/wcc.557.

505 www.cei.org, 18th September 2019: Wrong again: 50 years of failed eco-pocalyptic predictions.

506 Kench, P. S. *et al.* 2018: Patterns of island change and persistence offer alternate adaptation pathways for atoll nations. *Nature Communications* 605, 9th February 2018.

507 Darwin, C. R. 1842: *The structure and distribution of coral reefs. Being the first part of the geology of the voyage of the Beagle under the command of Capt. Fitzroy, R. N. during the years 1832 to 1836*. Smith Elder and Co.

508 Lyell, C. 1830-1833: *Principles of geology, being an attempt to explain the former changes of the Earths surface by reference to causes now in operation*. John Murray.

509 *San Jose Mercury News*, 30th June 1989.

510 www.electroverse.net, 5th November 2020.

511 www.salon.com, 23rd October 2001.

512 *The New York Times*, 18th September 1995.

513 *The Guardian*, 29th July 1999.

514 *The Independent*, 17th January 2000: Children aren't going to know what snow is.

515 Climategate.

516 www.bbc.com, 6th December 2020.

517 Wei, M. *et al.* 2021: Could CMIP6 climate models reproduce the early-200s global warming slowdown? *Science China Earth Sciences* 64, 853-865.

518 *The Guardian*, 23rd December 2002.

519 *The Guardian* 29th January 1974.

520 *Sydney Morning Herald,* 19th May 2004: Sydney's future eaten: Flannery prophecy.

521 www.climatism.blog/2018/03/12/tim-flannery-professor-of-dud-predictions-and-climate-falsehoods/

522 *The Guardian*, 22nd February 2004: Now the Pentagon tells Bush: climate change will

destroy us.

523 *Sydney Morning Herald*, 19th May 2004: Sydney's future eaten: Flannery prophecy.
524 www.abc.news.com.au, 10th November 2015: Brisbane, Sydney among cities that will 'slip under the waves' with 2 degree Celsius global warming: study.
525 Tim Flannery, 2006: *The weathermakers*. Text.
526 *The Age*, 28th October 2006.
527 *The Daily Telegraph*, 29th March 2011.
528 *The Sydney Morning Herald*, 15th July 2007.
529 www.andev.project.org
530 www.wunderground.com, 24th July 2007.
531 www.bbc.com, 12th December 2007.
532 www.abc.net.au, *Landline*, 2nd November 2007.
533 *Sydney Morning Herald*, 20th April 2007: Global warming to fuel global drought.
534 www.phys.org
535 www.abc.net.au, *Landline*, 11th February 2007.
536 *Daily Telegraph*, 20th May 2009: Extreme and dangerous.
537 *Sydney Morning Herald*, 2nd February 2007: Temperature predictions conservative: Flannery.
538 www.abc.net.au, *Landline*, 11th February 2007.
539 www.cawcr.gov.au
540 www.watercorporation.com.au
541 www.news.com.au, 15th October 2015.
542 www.reddit.com
543 *Herald Sun*, 1st May 2015: Full dams but still no sorry from Flannery.
544 www.abc.net.au, 9th February 2009, ABC PM: Flannery backs geothermal energy.
545 www.arena.gov.au, Australian Renewable Energy Agency.
546 Scafetta, N. 2016: High resolution coherence analysis between planetary and climate oscillations. *Advances in Space Research* 57, 2121-2135.
547 Scafetta, N. 2017: Understanding climate change in terms of natural variability. In: *Climate change: The facts 2017*, 39-58. IPA (ed. Jennifer Marohasy).
548 Orlowski, A. 18th July 2011: CERN 'gags' physicists in cosmic ray experiment: What do these results mean? Not allowed to tell you. *The Register.*
549 www.economist.com/news/science-and-technology/21598610-slowdown-rising-temperatures-over-past-15-years-goes-being/
550 Kueter, Jeffrey, 29th September 2006: President of George C. Marshall Institute, letter to Congress.
551 O'Neill, Brendan, *The Weekend Australian* 2nd-3rd May 2015.
552 Mark Lynas, *Dagelijksestandard* 19th May 2006.
553 George Monbiot, *The Guardian* 23rd August 2008.
554 James Lovelock, *The Guardian* 29th May 2010.
555 Parncutt, Richard 25th October 2012: *Death penalty for global warming deniers? An objective argument...a conservative conclusion.* Internet text on website of Karl-Franzens-Universität Graz (Austria) until removed by order of university officials.
556 www.science.house.gov
557 www.al.com/news/huntsville/index.ssf/2015/04/7_questions_with_john_christy.html
558 www.docs.house.gov/meetings/SY/.../HHRG-114-SY00-Wstate-ChristyJ-20160202.pdf
559 www.al.com/news/huntsville/index.ssf/2017/04/shots_fired_at_office_building.html
560 www.drroyspencer.com, Shots fired into the Christy offices at UAH
561 Tebaldi, C. *et al.* 2021: Climate model projections from the Scenario Model Intercomparison Project (Scenario MIP) of CMIP6. *Earth Syst. Dynam.* 12, 253-293.
562 Voosen, P. 2021: U.N. climate panel confronts implausibly hot forecasts of future warming. *Science* 27th July 2021.

[563] Graham Lloyd, 2021: Climate change: Science magazine article blows the whistle on model future. *The Australian* 29th July 2021.

[564] www.scienceunderattack.com, 2nd November 2020, How clouds hold the key to global warming.

[565] Zelinka, M. D. *et al.* 2020: Causes of higher climate sensitivity in CMIP6 models. *Geophys. Res. Letts* doi.org/10.1029/2019GL085782.

[566] Myers, T. A *et al.* 2021: Observational constraints on low cloud feedback reduce uncertainty of climate sensitivity. *Nature Climate Change* 11, 501-507.

[567] Burgess, M. G. *et al.* 2020: IPCC baseline scenarios have over-projected CO_2 emissions and economic growth. *Environmental Research Letters* 16 (1) 014016.

[568] Flawed models are throwing off climate forecasts of rain and storms. *Science Magazine* 29th July 2020.

[569] Maher, N. *et al.* 2020: Quantifying the role of internal variability in the temperature we expect to observe in the coming decades. *Environmental Research Letters* 15 (5), 054014.

[570] Climate model failure, *Watts Up With That,* 2nd February 2021.

[571] Mitchell D. M. *et al.* 2020 The vertical profile of recent tropical temperature trends: Persistent model biases in the context of internal variability. *Environmental Research Letters* 15 (10) doi.10.1088/1748-9326/ab9af7.

[572] McKitrick, R. & Christy. J. R. 2020: Pervasive warming bias in CMIP6 tropospheric layers. *Earth and Space Science* 7(9) doi.org/10.1029/2020EA001821.

[573] www.scienceunderattack.com, 22nd February 2021 *Latest computer models run almost as hot as before.*

[574] Rode, D. & Fischbeck, P. 2021: Prophets of doom and the risk of communicating extreme climate forecasts. *Internat. Journ Global Warming* 23(2) doi:20.1504/IJGW.2021.112896

[575] Rode, D. & Fischbeck, P. 2021: Apocalypse now? Communicating extreme forecasts. *Internat. Journ Global Warming* 23(2) doi:20.1504/IJGW.2021.112896.

[576] The risks of communicating extreme climate forecasts. *Engineering & Public Policy,* 18th February 2021, Carnegie Mellon University.

[577] www.abc.net.au, ABC Saturday Extra, 9th March 1999: SA water crisis.

[578] *The Argus Press,* 24th June 2008.

[579] *New Scientist,* 25th April 2008: North Pole could be ice free in 2008.

[580] *The Guardian,* 1st August 2008.

[581] *The Daily Mail,* 30th July 2009.

[582] www.blog.metoffice.gov.uk, 28th October 2010.

[583] www.blog.metoffice.gov.uk, 29th March 2012.

[584] www.blog.metoffice.gov.uk, 5th November 2013.

[585] www.blog.metoffice.gov.uk, 29th November 2013.

[586] *The Herald Sun,* 4th December 2009.

[587] www.npr.org, December 15 2009: Al Gore slips on Arctic ice: Misstates scientist's forecast.

[588] www.cnn.com, 9th March 1999: Transcript: Vice President Gore on CNN's 'Late Edition'.

[589] www.m.economictimes.com, 9th July 2009.

[590] *The Independent,* 9th July 2009.

[591] *Watts Up With That?* 16th December 2018.

[592] *The Independent,* 20th October 2009.

[593] *USA Today,* 4th December 2009.

[594] *The Independent,* 20th October 2009.

[595] *USA Today* 4th December 2009.

[596] *The Huffington Post* 16th October 2009.

[597] *The Daily Telegraph,* 3rd November 2009.

[598] *The Sydney Morning Herald,* 19th February 2010.

[599] *The Weekend Australian* July 3rd 2012: Enjoy snow now…by 2020, it'll be gone.

[600] www.news.griffith.edu.au, 9th September 2012

[601] www.forbes.com, 12th September 2015

[602] Whiteman, G. *et al.* 2013: Comment: Vast costs of Arctic change. *Nature*, 25th July 2013

[603] *The Guardian* 25th July 2013: Ice-free Arctic in two years heralds methane catastrophe

[604] *ScienceDaily* 24th July 2013: Cost of Arctic methane release could be 'size of global economy', experts warn.

[605] www.businessinsider.com

[606] Liu *et al.* 2013: Reducing spread in climate model projections of a September-ice-free Arctic. *PNAS* doi.org//10.1073/pnas.1219716110.

[607] Wallman, K. *et al.* 2018: Gas hydrate dissociation off Svalbard induced by isostatic rebound rather than global warming, www.nature.com/articles/s41467-017-02550-9.

[608] *The Guardian* 10th December 2013: US Navy predicts summer ice free Arctic by 2016.

[609] www.stevengoddard.wordpress.com, 7th December 2013: Tim Flannery predicted and ice-free Arctic in 2013.

[610] *Fox Nation*, 23rd May 2014

[611] *The Washington Examiner*, 14th May 2014.

[612] www.econews.com.au

[613] *Herald Sun.* 3rd July 2021: Tim Flannery's latest prediction blown away.

[614] *Stuff News (NZ),* 27th November 2016.

[615] *The New York Post*, 1st October 2020.

[616] www.latimes.com/business/la-fi-travel-briefcase-climate-change-20170714-story.html

[617] Bombardier CRJ.

[618] Climate change and health. *World Health Organisation*, 1st February 2018.

[619] *The Guardian*, 8th October 2018.

[620] www.pv.magazine.com, 13th December 2018.

[621] www.bbc.com, 24th July 2019.

[622] www.nbc.news.com, Democratic debate transcript, 20th November 2019

[623] *The Guardian*, 20th November 2019.

[624] Thackeray, C. W. 2019: An emergent constraint on future Arctic sea ice albedo feedback. *Nature Climate Change* 9, 972-978.

[625] *NatureScot* 30th May 2019: Not too late to act for climate change.

[626] Jenouvrier, J. *et al.* 2019: The Paris Agreement objectives will likely halt future declines of Emperor penguins. G*lobal Change Biology* doi: 10.1111/gcb.14865.

[627] *Seattle Times,* 30th September 2019.

[628] Kulp, S. A. & Strauss, B. H. 2019: New elevation data triple estimates of global vulnerability to sea-level rise and coastal flooding. *Nature Communications* 10, article 4844.

[629] *The Global Warming Policy Foundation,* 8th August 2019: 2018 temperature prediction competition.

[630] www.3aw.com.au

[631] CNBC, 21st January, 2020.

[632] www.jennifermarohasy.com

[633] www.ipa.org

[634] www.thesun.co/uk/tech/15577539/moon-causes-flooding-wobble-orbit/

[635] www.droyspencer.com/global-warming-background-articles/the-pacific-decadal-oscillation/

[636] www.weather.plus

[637] www.psi.noaa.gov

[638] www.esa.int

[639] www.esa.int/Our_Activities/Observing_the_Earth/Cryosat/Arctic/sea-ice-up-from-record-low.

[640] www.data.giss.nasa.gov/gistemp/station_data_v2/; 2011-2015

[641] The eminently sensible Jeremy Clarkson was filmed drinking gin and tonic while driving across the sea ice. After complaints from temperance-type hand wringers, Clarkson claimed that he was over international waters and therefore sailing. The highest rating television show

in the world showed that the Arctic sea ice was not thinning. Greenpeace complained that *"Clarkson is a problem because he has represented some climate sceptic views. That's the true ecofascist face of Greenpeace, no one is allowed to have a different opinion or present data.*

[642] www.ukcop26.org

[643] Joughin, I. *et al.* 2004: Large fluctuations in speed on Greenland's Jakobshavn Isbrae Glacier. *Nature* 432, 608-610.

[644] Van der Veen, C. J. *et al.* 2007: Subglacial topography and geothermal heat flux: Potential interactions with drainage of the Greenland ice sheet, *Geophys. Res. Letts* 34 (12) doi. org/10.1029/2007GL030046.

[645] www.sky.com, 14th October 2020.

[646] www.maritimenews.net, 4th September 2019.

[647] A pedantic joke for chemists.

[648] www.nbcnews.com, 12th October 2007.

[649] www.wsj.com

[650] Gore made his first big money from an energy-intensive zinc business in Tennessee. Massive amounts of coal-fired electricity with the resultant carbon dioxide emissions are used to make zinc metal.

[651] www.reuters.com, 19th September 2014.

[652] www.abcnews.go.com, 27th February 2007.

[653] www.protectingtaxpayers.org, 4th August 2017.

[654] www.worldpropertyjournal.com, 13th May 2010.

[655] www.forbes.com, 3rd November 2013.

[656] www.ff.org, 29th January 2017.

[657] www.ballotpedia.org

[658] www.epw.senate.gov, 21st March 2007.

[659] www.vaticannews.va, 4th July 2018.

[660] *Laudato si.*

[661] www.filmink.com.au, 14th June 2017.

[662] www.abc.net.au, 6th June 2019.

[663] www.dfailymail.co.uk, 24th May 2019.

[664] Ian Plimer & John Ruddick, April 2021. REVEALED: Al Gore's real climate catastrophe. *Spectator Australia.*

[665] www.dailytelegraph.com.au, 16th July 2017.

[666] Stop the bullshit.

[667] Physicist found guilty of misconduct, *Nature*, 26th September 2002.

[668] Sokal, A. D. 1996a: Transgressing the boundaries – Toward a transformative hermeneutics of quantum gravity. *Social Text* 46/47: 217-252.

[669] Sokal, A. D. 1996b: A physicist experiments with cultural studies. *Lingua Franca* May/ June 1996: 62-64.

[670] www.linguafranca.mirror.theinfo.org/9607/mst.html.

[671] A quote by Sokal: *"The Einsteinian constant is not a constant, is not a center. It is the very concept of variability – it is, finally, the concept of the game. In other words, it is not the concept of something – of a center starting from which an observer could master the field – but the very concept of the game"*. I have read this many times with and without James Squire amber ale on board. It still means nothing to me. Another quote from the Sokal paper should have given the game away: *"the pi of Euclid and the G of Newton, formerly thought to be constant and universal, are now perceived in their ineluctable historicity"*. I have been an editor of a major scientific journal for years and have seen some dreadful papers submitted. Why didn't the editor at least insist that such great revelations were intelligible? An editorial tip: If something has to be read more than once to be understood, then it is poorly written or nonsense.

[672] Boghossian, P. & Lindsay, J. 2017: The conceptual penis as a social construct. *Cogent*

Social Sciences 2017, 3: 1330439.

[673] Alan Sokal 1996: Transgressing the boundaries: towards a transformational hermeneutics of quantum gravity. *Social Text* 46/47, 217-252.

[674] *New York Times*, 4th October 2018.

[675] Lindsay, J. A. *et al.* 2018: Academic grievance studies and the corruption of scholarship. *Areo*, 02/10/2018.

[676] Alan Sokal and Jean Bricmont, 1999: *Fashionable nonsense: Post modern intellectuals' abuse of science.* Picador.

[677] McNicol, J. H. 2017: Opinion: Why I published in a predatory journal. *The Scientist*, April 6, 2017.

[678] van Nostrand, M. 2017: Uromycitisis poisoning results in lower urinary tract infection and acute renal failure: Case report. *Urology and Nephrology Open Access Journal* 4, 3, 00132.

[679] Buckley, P. J. *et al.* 2017: Ten thousand voices on marine climate change in Europe: Different perceptions among demographic groups and nationalities. *Frontiers in Marine Science* 11, doi.org/10.3389/fmars.2017.00206 https://doi.org/10.3389/fmars.2017.0026

[680] *Global Warming Policy Forum*, 12th July, 2017

[681] Tranter, B. & Booth, K. 2015: Scepticism in a changing climate: A cross-national study. *Global Envir. Change* 33:154-164.

[682] Ian Plimer, 2009: *Heaven and Earth. Global warming: The missing science.* Connor Court

[683] www.blogs.scientificamerican.com, 17th November 2010.

[684] www.theaustralian.com.au, 7th November 2019.

[685] www.petitionprogram.org.

[686] www.rasmussenreports.com, 13th May 2021.

[687] Green, K. C. & Armstrong, J. S. 2007: Global warming forecasts by scientists versus scientific forecasts. *Energy and Environment* 18: 997-1021.

[688] Randi, James 1995: *An encyclopedia of claims, frauds and hoaxes of the occult and supernatural.* St Martin's Press.

[689] On the basis of biblical interpretations, Harold Camping publicly predicted the end of the world 12 times, the last date was 21st October 2011. His end of the world came on 15th December 2013. His critics had been right in stressing Matthew 24:36 ("of that day and hour knoweth no man").

[690] www.heritage.org, 13th December 2005.

[691] Andela, N. *et al.* 2017: A human-driven decline in global burned area. *Science 356,* 1352-1356.

[692] Doerr, S. H. & Santin, C. 2016: Global trends in wildfire and its impacts: perceptions versus realities in a changing world. *Phil. Trans. Roy. Soc. Lond. B. Biol. Sci.* doi:10.1098/rstb.2015.034.

[693] Pyne, S. 2010: The ecology of fire. *Nature Education Knowledge* 3(10): 30.

[694] Rimmer, S. M. *et al.* 2015: The rise of fire: Fossil charcoal in late Devonian marine shales as an indicator of expanding terrestrial ecosystems, fire, and atmospheric change. *Amer. Jour Sci.* 315 (8) doi.org/10.2475/08.2015.01.

[695] Fusain.

[696] Diessel, C. F. K. 1982: An appraisal of coal facies based on maceral characteristics. *Aust. Coal Geol.* 4, 474-483.

[697] Prince, T. J. *et al.* 2018: Postglacial reconstruction of fire history using sedimentary charcoal and pollen from a small lake in Southwest Yukon Territory, Canada. *Front. Ecol. Evol.* doi.org/10.3389/fevo.2018.00209.

[698] El Atfy, H. & Uhl, D. 2021: Palynology and palynofacies of sediments surrounding the *Edmontosaurus annectens* mummy at the Senckenberg Naturmuseum in Frankfurt/Main (Germany). *Z. Dt. Ges. Geowiss.* 172(2), 127-139.

[699] Burrows, G. E. 2013: Buds, bushfires and resprouting in the eucalypts *Aust. Journ Botany*

61(5) 331-349.

[700] Truswell, E. M. 1985: Preliminary palynology of deep sediments in the Lake George Basin. In: *Geology of the Lake George Basin* (ed Abell, R. S.) *BMR Geol Geophys Rec.* 1985/4, 52-57.

[701] Horn, S. P. & Underwood, C. A. 2014: Methods for the study of soil charcoal and forest history in the Appalachian region, U.S.A. In: *Proceedings Wildland Fire in the Appalachians: Discussions among managers and scientists* (ed Waldrop, T. A). *US Dept Agric Forest Service Tech Rpt* SRS-199.

[702] www.nasa.gov, 29th June 2017 NASA detects drop in global fires.

[703] Andela, N. *et al.* 2017: A human-driven decline in global burned area. *Science* 356, 1352-1356.

[704] www.italy24news.com/News/145873.html

[705] www.voanews.com/europe/arsonists-behind-more-half-italys-wildfires-officials-say

[706] www.wildfiretoday.com, 12th November 2014, 8th August 2016, 8th May 2017.

[707] Thompson, J. C. *et al.* 2021: Early human impacts and ecosystem reorganization in southern-central Africa. *Science Advances* 7 (19), doi:10.1126/sciadv.abf9776.

[708] Bill Gammage, 2011: *The biggest estate on Earth: How Aborigines made Australia.* Allen & Unwin.

[709] 1995 Yearbook Australia.

[710] Victor Steffensen, 2020: *Fire country: How indigenous fire management could help save Australia.* Hardie Grant.

[711] www.info.australia.gov.au

[712] Williams, N. J. *et al.* 2006: The vegetation history of the last glacial-interglacial cycle in eastern New South Wales, Australia. *Journ Quat. Sci* 21, 735-750.

[713] Harle, H. L. *et al.* 2004: Patterns of vegetation change in southwest Victoria (Australia) over the last two Glacial-Interglacial cycles: evidence from Lake Wangoom. https://handle.slv.vic.gov.au/10381/158713

[714] www.agriculture.gov.au, Australian forest and wood products statistics.

[715] Viv Forbes, 25th March 2021: Flood plains are for -gasp- floods. *Spectator Australia.*

[716] www.tlnews.com.au/kevin-tolhurst-a-life-devoted-to-forests-and-fires/

[717] Vic Jurskis, 24th March 2021: Of droughts and fires and floods. *Spectator Australia.*

[718] www.beefcentral.com/news/qld-landholder-hit-with-record-1m-penalty-for-making-fire-breaks-too-wide/

[719] www.smh.com.au, 2nd January 2020.

[720] www.kmc.nsw.gov.au/Your_Council/Organisation/News_and_media/Latest_news_media_releases/Catastrophic_fire_rating_declared_for_Kur-ing-gai_on_Saturday_21_December/

[721] www.thecanberratimes.com.au, 30th October 2014.

[722] www.aph.gov.au, House Select Committee on the recent Australian bushfires, Submission No 315.

[723] Giglio, L. *et al.* (2013), Analysis of daily, monthly, and annual burned area using the fourth-generation global fire emissions database (GFED4). *Jour. Geophys. Res. Biogeosci.,* 118, 317–328, doi:10.1002/jgrg.20042.

[724] www.agriculture.gov.au/abares/forestsaustralia/sofr

[725] www.globalfiredata.org/analysis.html

[726] *Herald Sun*, Monday 9th February 2009.

[727] www.bushfireseducation.vic.edu.au

[728] Cunningham, C. J. 1984: Recurring natural fire hazards: A case study of the Blue Mountains, Wew (sic) South Wales. *Applied Geography* 4(1), 5-27.

[729] Australian Institute for Disaster Resilience.

[730] www.romseyaustralia.com, Major bushfires in Queensland.

[731] www.theaustralian.com.au, 28th September 2019, False alarm: The great rainforest fire that

wasn't.

[732] www.dpac.tas.gov.au, 2013 Tasmanian Bushfires Enquiry.

[733] www.fire.tas.gov.au

[734] www.cfs.sa.gov.au

[735] R. H. Luke & A. G. McArthur 1978: *Bushfires in Australia.* AGPS.

[736] Cai, W. *et al.* (2009). Positive Indian Ocean dipole events precondition southeast Australia bushfires. *Geophysical Research Letters*, 36, L19710, doi.org/10.1029/2009GL039902.

[737] www.bnhcrc.com, Australian Seasonal Bushfire Outlook.

[738] www.theage.com.au, 12th March 2015, Bushfire scientist David Packham warns of huge blaze threat, urges increase in fuel reduction burns.

[739] Alan Moran, 2010: *Climate change: The facts.* Institute of Public Affairs.

[740] Keenan, R. J. *et al.* 2015: Dynamics of global forest area: Results from the FAO Global Forest Resources Assessment 2015. *Forest Ecol. Manag.* 352, 9-20.

[741] www.efi.int, 20th November 2020, EWWF Wildfire Conference.

[742] Tang, W. et al. 2021: Widespread phytoplankton blooms triggered by 2019-2020 Australian wildfires. *Nature* 597, 370-375.

[743] www.dailytelegraph.com.au, Let's tell the burning truth.

[744] www.nifc.gov/fireInfo/fireInfo_stats_totalFires.html.

[745] www.nifc.gov/PIO_bb/Policy/FederalWildlandFireManagementPolicy_2001.pdf.

[746] www.pbs.org, 30th April 2018, Decoding the weather machine

[747] Director Emeritus of San Francisco State's Sierra Nevada Field Campus and author of *Landscapes and cycles: An environmentalist's journey to climate skepticism.*

[748] Jim Steele, 14th September 2020: Gavin Newsom's exceedingly ignorant climate claim. *Watts Up With That?* .

[749] www.nifc.gov, National Fire News, National Interagency Fire Center.

[750] *Global Warming Policy Foundation*, 8th January 2020.

[751] Bjorn Lomborg, 14th September 2020: Sorry, solar panels won't stop California's fires. *New York Post.*

[752] Keeley, J. E. & Syphard, A. D. 2019: Twenty-first Century California, USA, wildfires: fuel-dominated vs. wind dominated fires. *Fire Ecology* 15, Article Number 243.

[753] Jim Steele, 14th September 2020: Gavin Newsom's exceedingly ignorant climate claim. *Watts Up With That?*

[754] Keeley, J. E. & Syphard, A. D. 2018: Historical patterns of wildfire ignition sources in California ecosystems. *Internat. Journ. Wildland Fire* 27(12), 781-799.

[755] www.turbinesonfire.org, Andrew Lees, 18th March 2021: Firefighters race to second turbine blaze in months at US wind farm. *Recharge News.*

[756] *The Spectator*, 22nd September, 2020.

[757] www.npr.org, 24th August 2020, Native American burning and California's wildfire strategy.

[758] www.calforestfoundation.org

[759] www.causeiq.com

[760] www.breitbart.com/national-security/2019/11/08/islamic-state-media-encourages-johadis-set-forest-fires-u-s-europe/

[761] www.abcnews.go.com/Blotter/al-qaeda-calls-massive-forest-firesmontana/story?id=16263981

[762] www.bbc.com/news/uk-england-leeds-49436957

[763] www.dailymail.co.uk/news/article-2138758/Unleash-Hell-New-Al-Qaeda-magazine-describes-start-huge-forest-fires-U-S-instructions-make-ember-bombs.html

[764] M. J. Matt, "Muslim Arsonist Targets Catholic University", *The Remnant*, 22nd. January, 2018.

[765] Paul Homewood, 27th September 2020: Canada wildfires at lowest level for decades. *Not a lot of people know that.*

[766] www.bbc.com, 19th February 2021: UK-owned pellet plant in US fined $2.5m over air quality breaches.

[767] Hazel Sheffield, 14th January 2021: 'Carbon neutrality is a fairy tail': How the race for neutrality is burning Europe's forests. *The Guardian*.

[768] www.sydney.edu.au, 8th January 2020.

[769] www.themoscowtimes.com, 7th September 2021.

[770] Sheffield, J. *et al.* 2012: Little change in global drought over the past 60 years. *Nature* 491, 435-438.

[771] Pachur, H-J. & Hoelzmann, P. 2000: Late Quaternary palaeoecology and palaeoclimates of the eastern Sahara *Journ. African Earth Sci.* 30, 929-939.

[772] Gasse, F. & van Campo, E. 1994: Abrupt post-glacial climate events in West Asia and North Africa monsoon domains. *Earth Planet. Sci. Letts* 126, 435-456.

[773] Dalfes, H. N. *et al.* 1994: Third Millennium Climate Change and Old World Collapse. In: NATO ASI Series 1: *Global Environmental Change* 49, Springer Verlag.

[774] Thompson, L. G. *et al.* 2000: Ice-core palaeoclimate records in tropical South America since the Last Glacial Maximum. *Journ Quat. Sci.* 15, 377-394.

[775] Petit, J. R. *et al.* 1999: Climate and atmospheric history of the past 420,000 years from the Vostok ice core, Antarctica. *Nature* 399, 429-436.

[776] Duce, R. A. & Tindale, N. W. 1999: Atmospheric transport of iron and its deposition in the ocean. *Limnol. Oceanogr.* 36, 1715-1726.

[777] de Baar, H. J. W. *et al.* 1999: Importance of iron for plankton blooms and carbon dioxide drawdown in the Southern Ocean. *Nature* 373, 412-415.

[778] McConnell, J. R., *et al.* 2007: 20th Century doubling in dust archived in an Antarctic Peninsula ice core parallels climate change and desertification in South America. *PNAS* 104, 5743-5748.

[779] Neff, J. C. *et al.* 2008: Increasing aeolian dust deposition in the western United States linked to human activity. *Nature Geoscience* 1, 189-195.

[780] Dalfes, H. *et al.* 1997: *Third Millennium BC Climate Change and Old World Collapse*. NATO ASI Series, Volume 49, Springer Verlag.

[781] Dumont, H. J. 1978: Neolithic hyperarid period preceded the present climate of the Central Sahel. Nature 274, 356-358.

[782] Nicoll, K. 2004: Recent environmental change and prehistoric activity in Egypt and Northern Sudan. *Quat. Sci. Revs* 23, 561-580.

[783] Wenxiang, W. & Tungsheng, L. 2004: Possible role of the "Holocene Event 3" on the collapse of Neolithic cultures around the Central Plain of China. *Quat. Geol.* 117, 153-166.

[784] Malek, J. 2000: The Old Kingdom (c.2686-2100 BC). *The Oxford History of Ancient Egypt*. Oxford University Press.

[785] Van Andel, T. H. *et al.* 1990: Land use and soil erosion in prehistoric and historical Greece. *Journ. Field Archaeol.* 17, 379-396.

[786] Jacobsen, T. & Adams, R. M. 1958: Salt and silt in ancient Mesopotamian agriculture: Progressive changes in soil salinity and sedimentation contributed to the breakup of past civilisations. *Science* 18, 1251-1258.

[787] Bell, B. 1975: Climate and the history of Egypt: the Middle Kingdom. *Amer. Journ. Archaeol.* 79, 223-269.

[788] Cullen, H. M. *et al.* 2000: Climate change and the collapse of the Akkadian empire: Evidence from the deep sea. *Geology* 28, 379-382.

[789] Shaowu, W. *et al.* 2004: Abrupt climate change around 4 ka BP: Role of the Thermohaline Circulation as indicated by a GCM experiment. *Advan. Atmos. Sci.* 21, 291-295.

[790] Williams, P. W. *et al.* 1999: Palaeoclimatic interpretation of stable isotope data from Holocene speleotherms of the Waitomo district, North Island, New Zealand. *The Holocene* 9, 649-657.

[791] Bell, B. 1975: Climate and the history of Egypt: the Middle Kingdom. *Amer. Journ. Archaeol.* 79, 223-269.

[792] Bell, B. 1971: The Dark Ages in ancient history. *Amer. Journ. Archaeol.* 75, 1-26.

793 Hsu, K. J. 2000: *Climate and people: A theory of history.* Orell Fussli.
794 de Menocal P. *et al.* 2000: Coherent high- and low-latitude climate variability during the Holocene Warm Period. *Science* 288, 2198-2202.
795 Weiss, H. *et al.* 1993: The genesis and collapse of third millennium North Mesopotamian civilisation. *Science* 261, 995-1004.
796 Smit, B. & Wandel, J. 2006: Adaptation, adaptive capacity and vulneribility. *Global Environ. Change* 16, 282-292.
797 Thompson, L. G. *et al.* 2002: Kilimanjaro ice core records: Evidence of Holocene climate change in tropical Africa. *Science* 298, 589-593.
798 Williams, A. N. *et al.* 2005: Interpreting trace-metal and stable isotopic results from a Holocene stalagmite from Buckeye Creek cave, West Virginia. *GSA Salt Lake City AGM*, October 16-19, 2005; paper 131-33.
799 de Menocal, P. 2001: Cultural responses to climate change during the Late Holocene. *Science* 292, 667-673.
800 Mandella, M. & Fuller, D. Q. 2006: Palaeoecology and the Harappan civilisation of South Asia: a reconsideration. *Quat. Sci. Reviews* 25, 1283-1301
801 Shaw, J. *et al.* 2007: Dates and pollen sequences from the Sanchi Dams. *Asian Perspect.* 46, 166-201
802 Rad, U. von, *et al.* 1999: A 5,000-years record of climate change from the oxygen minimum zone off Pakistan, Northeastern Arabian Sea. *Quat. Res.* 51, 39-53.
803 Chepstow-Lusty, A. J. *et al.* 1998: Tracing 4,000 years of environmental history in the Cuzco area, Peru, from the pollen record. *Mountain Res. Develop.* 18, 159-172.
804 Valero-Garces, B. L. *et al.* 2000: Paleohydrology of Andean saline lakes from sedimentological and isotopic records, Northwestern Argentina. *Journ. Paleolimnol.* 24, 343-359.
805 Iriondo, M. 1999: Climatic changes in the South American plains: Records of a continent-scale oscillation. *Quat. Internat.* 57, 93-112.
806 Holmgren, K. *et al.* 2001: A preliminary 3,000-year regional temperature reconstruction for South Africa: Research letter. *South African Journ. Sci.* 97, 49-51
807 Drysdale, R. *et al.* 2006: Late Holocene drought responsible for the collapse of Old World civilisations is recorded in an Italian cave flowstone. *Geology* 34, 101-104.
808 Hempel, L. 1987: The "Mediterraneanization" of the climate in the Mediterranean countries – a cause of unstable ecobudget. *GeoJournal* 14, 163-173.
809 Perry, C. A. & Hsu, K. J. 2000: Geophysical, archaeological, and historical evidence support a solar-output model for climate change. *PNAS* doi: 10.1073/pnas.230423297.
810 The Sub-Atlantic cooling.
811 Bell, B. 1971: The Dark Ages in ancient history. *Amer. Journ. Archaeol.* 75, 1-26.
812 Oliver, J. E. 1973: *Climate and man's environment.* Wiley, New York.
813 Laird, K. R. *et al.* 2003: Lake sediments record large-scale shifts in moisture regimes across the northern prairies of north America during the past two millennia. *PNAS* 100, 2483-2488.
814 Lebreiro, S. M. *et al.* 2006: Climate change and coastal hydrographic response along the Atlantic Iberian margin (Tagus Prodelta and Muros Ria) during the last two millennia. *The Holocene* 16, 1003-1015.
815 Haug, G. H. *et al.* 2003: Climate and collapse of Maya civilisation. *Science* 299, 1731-1735.
816 Fagan, B. 1999: *Floods, famines and emperors: El Niño and the fate of civilisations.* Basic Books, New York.
817 Hodell, D. A. *et al.* 1995: Possible role of climate in the collapse of Classic Maya civilization. *Nature* 375, 391-394.
818 Lachniet, M. S. *et al.* 2012: A 2400 yr Mesoamerican rainfall reconstruction links climate and cultural change. *Geology* 40(3), 259-262.
819 Nunn, P. D. 2000: Environmental catastrophe in the Pacific Islands around A.D. 1300. *Geoarchaeol.* 15, 715-740.
820 Touchan, R. *et al.* 2007: May-June precipitation reconstruction of southwestern Anatolia,

Turkey during the last 900 years from tree rings. *Quat. Resear.* 68, 196-202.

[821] Nunn, P. D. & Britton, J. M. R. 2001: Human-environment relationships in the Pacific Islands around A.D. 1300. *Environ. History* 7, 3-22.

[822] The Earth Institute, Columbia University, November 6, 2015.

[823] Mishra, V. 2019: Long-term (1870-2018) drought reconstruction in the context of surface water security in India. *Journ Hydrol.* 580(8) doi:10.1016/j.hydrol.2019.124228.

[824] www.rogerpielkejr.blogspot.com, Roger Pielke, Jr, Drought analysis 24th September 2012.

[825] Schwalm, C. R. *et al.* 2012: Reduction in carbon uptake during turn of the century drought in western North America. *Nature Geoscience* 5 (8), 551-556.

[826] Paul Dorian, 10th April 2017: *Vencore Weather.*

[827] Darryl Fears, 2nd March 2015: California's terrifying climate forecast: It could face droughts nearly every year. *The Washington Post.*

[828] Schwalm, C. R. *et al.* 2012: Reduction in carbon uptake during turn of the century drought in western N North America. *Nature Geoscience* 5 (8), 551-556

[829] Feng Zhang *et al.* 2017: Causality of the drought in southwestern United States based on observations. *Amer. Met. Soc.*, doi.org/10.1175/JCLI-D-16-0601.1

[830] Kalisa, W. *et al.* 2020: Spatio-temporal analysis of drought and return periods over the East African region using Standardized Precipitation Index from 1920-2016. *Agricult. Water Manag.* 237, 106195

[831] Vance, T. R. *et al.* 2014: Interdecadal Pacific variability and eastern Australian megadroughts over the last millennium. *Geophys. Res. Letts* 42, 129-137

[832] Donnell, A. J. *et al.* 2021: Megadroughts and pluvials in southwest Australia: 1350-2017 CE. *Climate Dynamics* https://doi.org/10.1007/s00382-021-05782-0

[833] *The Weekend Australian,* 12-13th September 2015

[834] www.abc.net.au/news/2018-12-07/indian-ocean-dipole-dominant-cause-drought/10571802

[835] Latin for meal or flour.

[836] Coughlan, M. J. 1986: *Drought in Australia. Natural Disasters in Australia.* Australian Academy of Technological Sciences and Engineering.

[837] Ashcroft, L. *et al.* 2019: Historical extreme rainfall events in southeastern Australia. *Weather & Climate* Extremes 25 doi.org/10.1016/j.wace.2019.100210.

[838] www.abs.gov.au/AUSSTATS/abs@.nsf/lookup/1301.0Features%20Article151988.

[839] www.bom.gov.au/climate/drought/

[840] www.australia.gov.au/about-australia/australian-story/natural-disasters

[841] Ashcroft, L. *et al.* 2019: Historical extreme rainfall events in southeastern Australia. *Weather & Climate* Extremes 25 doi.org/10.1016/j.wace.2019.100210.

[842] www.nla.gov.au/nla.news-article2553198

[843] www.nla.gov.au/nla.news-article255321

[844] *The Courier,* 8th December 1849.

[845] Hope, P. *et al.* 2017: *A synthesis of findings from the Victorian Climate Initiative (VicCI).* Bureau of Meteorology.

[846] www.climate.gov

[847] www.metoffice.gov.uk/research/climate/maps-and-data/uk-and-regional-series

[848] *Global Warming Policy Foundation,* 2014.

[849] www.theguardian.com.au, 13th November 2020.

[850] www.abc.net.au, 4th March 2014

[851] www.7news.com.au, 11th January 2020

[852] www.abc.net.au, 26th February 2021

[853] David Tanner, 24th March 2021: Don't blame Warragamba Dam for flood in the Hawkesbury. *The Australian.*

[854] www.waternsw.com.au

[855] www.giveadam.org.au

[856] www.efas.eu

857 www.sciencemag.org/news/2021/07/europe-s-deadly-floods-leave-scientists-stunned

858 www.wsj.com/articles/climate-change-natural-disasters-ayr-river-flood-germany-wildfire-risks-11628177742?mod=hp_opin_pos_5

859 www.floodlist.com

860 *Global Warming Policy Foundation*, Press release, 19th July 2021.

861 Mudelsee, M. *et al.* 2004: Extreme floods in Central Europe over the past 500 years: role of cyclone pathway "Zugstrasse Vb". *Journ. Geophys. Res.* 109, doi.10:1029/2004JD005034.

862 Tetzlaff, G. *et al.* 2002: Das Jahrtausendhochwasser von 1342 am Main aus meteorologisch–hydrologischer Sicht. *Wasser & Boden* 54, 41–49.

863 Mauch, F. 2012: The Great Flood of 1962 in Hamburg. *Arcadia* 6, https://doi.org/10.5282/rcc/3733.

864 Böhm, O. & Wetzel, K.-F. 2006: Flood history of the Danube tributaries Lech and Isar in the Alpine foreland of Germany. *Hydrolog. Sci.* 51(5), 774-798

865 Mudelsee, M. *et al.* 2003: No upward trends in the occurrence of extreme floods in Central Europe. *Nature* 425, 166-169.

866 Beurton, S. & Thieken. A. H. 2009: Seasonality of floods in Germany. *Hydrol. Sci.* 54(1), 62-76.

867 www.weather.gov, Mississippi River flood history 1543-present

868 Bjorn Lomborg, 5th August 2021: Climate change doesn't cause all disaster. Warming annually causes about 120,000 heat deaths but prevents nearly 200,000 cold deaths. *The Wall Street Journal*

869 World Economic Forum, Alex Gray, 8th June 2018

870 Press Release, Helmholtz Centre for Environmental Research, 17th October 2017

871 Steele, J. 2020: *Ocean health – Is there an "acidification" problem?* CO_2 Coalition of Climate Scientists

872 www.psml.org, Data and station information for Sydney, Fort Denison

873 Sloss, C. R. *et al.* 2007: Holocene sea-level change on the south east coast of Australia: A review. *The Holocene* doi.org/10.1177/0959683607084215

874 Dougherty, A. *et al.* 2019: Redating the earliest evidence of the mid-Holocene relative sea-level highstand in Australia and implications for global sea-level rise. *PLoS ONE* 14(7) doi.org/10.1371/journal.pone.0218430

875 Beaman, R. *et al.* 1994: New evidence for the Holocene sea-level high from the inner shelf, Great Barrier Reef, Australia. *Journ. Sedim. Res.* doi.org/10.1306?D4267EF1-2B26-11D-8648000102C1865D

876 www.portergeo.com.au, Ginkgo and Snapper

877 www.ga.gov.au, Mineral Sands

878 Miller, K. G. *et al.* 2005: The Phanerozoic record of global sea-level change. *Science* 310, 1293-1298

879 Chappell, J. 1974: Geology of coral terraces, Huon Peninsula, New Guinea: A study of Quaternary tectonic movements and sea-level changes. *Bull. Geol. Soc. Amer.* 85, 553-570

880 Solé, R. V. & Newman, M. 2002: Extinctions and biodiversity in the fossil record. In: *The Earth system: biological and ecological dimensions of global environmental change* (eds Mooney, H. A. & Canadell, J. G.), John Wiley, 297-301

881 Christie-Blick, N. *et al.* 1995: Sequence stratigraphy and the interpretation of Neoproterozoic earth history. *Precamb. Res.* 73, 3-26

882 Hoffman, P. F. & Schrag, D. P. 2002: The snowball Earth hypothesis: testing the limits of global change. *Terra Nova* 14, 129-155

883 Parker, A. & Ollier, C. 2019: Pacific sea levels rising very slowly and not accelerating. *Quaestiones Geographicae* 38 (1) 179-184

884 Cabanes, C. *et al.* 2001: sea level rise during the past 40 years determined from satellite and in situ observations. *Science* 294, 840-842

[885] Trupin, A. & Wahr, J. 1990: Spectroscopic analysis of global tide gauge sea level data. *Geophys. Journ. Internat.* 100(3), 441-453

[886] Douglas, P. C. 1992: Global sea level acceleration. *Journ. Geophys Res.* 97, 12699-12706

[887] Singer, S. F. 1997: *Hot talk, cold science. Global warming's unfinished debate.* The Independent Institute

[888] Belperio, A. P. 1993: Land subsidence and sea level rise in the Port Adelaide estuary: implications for monitoring the greenhouse effect. *Aust. Journ. Earth Sci.* 40, 359-368

[889] Vadivel, S. K. P *et al.* 2019: Sinking tide gauge revealed by space-borne InSAR: Implications for sea level acceleration at Pophang, South Korea. *Remote Sensing* 11 (277) doi:10.3390/rs11030277

[890] Kraft, J. C. *et al.* 1977: Paleogeographic reconstructions of coastal Aegean archaeological sites. *Science* 195, 941-947

[891] Kuniholm, P. I. 1990: Archaeological record: evidence and non-evidence for climate change. *Phil. Trans. Royal Soc. London* A330, 645-655.

[892] Acts 19: 1-7.

[893] Murphy-O'Connor, J. 2008: *St Paul's Ephesus: Texts and Archaeology.* Liturgical Press.

[894] Nunn, P. D. *et al.* 2002: Late Quaternary sea-level and tectonic changes in northeast Fiji. *Marine Geology* 187, 299-311.

[895] Lambeck, K. & Chappell, J. 2001: Sea level change through the last glacial cycle. *Science* 292, 679-686.

[896] Clemmensen, L. B. & Andersen, C. 1998: Late Holocene deflation of beach deposits, Skagen Odde, Denmark. *Geol. Soc. Den. Bull.* 44, 187-188.

[897] Frost, H. 2007: Some out-of-the-way European maritime museums and developments. *Internat. Journ. Naut. Archaeol.* 4, 143-13.

[898] Bloom, A. L. 1967: Pleistocene shorelines: A new test of isostasy. *Bull. Geol. Soc. Amer.* 78, 1477-1494.

[899] Walcott, R. I. 1972: Late Quaternary vertical movements in eastern America: quantitative evidence of glacio-isostatic rebound. *Rev. Geophys. Space Phys.* 10, 849-884.

[900] Chappell, J. *et al.* 1982: Hydro-isostasy and the sea-level isobase of 5500 B.P. in north Queensland, Australia. *Marine Geology* 49, 81-90.

[901] Gibb, J. G. 1986: A New Zealand regional Holocene eustatic sea-level curve and its application for determination of vertical tectonic movements. *Bull. Roy. Soc. NZ* 24, 377-395.

[902] Nakada, M. 1986: Holocene sea levels in ocean islands: implications for the rheological structure of the Earth's mantle. *Tectonophysics* 121, 263-276.

[903] Nakada, M. & Lambeck, K. 2008: Late Pleistocene and Holocene sea-level change in the Australian region and mantle rheology. *Geophys. Journ. Internat.* 96, 497-517.

[904] McCabe, A. M. *et al.* 2007: Relative sea-level changes from NE Ireland during the last glacial termination. *Journ. Geol. Soc., London* 164, 1059-1063.

[905] Lambeck, K. 1996: Late Devensian and Holocene shorelines of the British Isles and North Sea from models of glacio-hydrostatic rebound. *Journ. Geol. Soc., London* 153, 437-448.

[906] Shennan, I. *et al.* 2006: Relative sea-level changes, glacial isostatic modelling and ice sheet reconstructions from the British Isles since the last glacial maximum. *Journ. Quat. Sci.* 21, 585-599.

[907] Eitner, V. 1996: Geomorphological response of the East Frisian barrier islands to sea-level rise: an investigation of past and future evolution. *Geomorph.* 15, 57-65.

[908] Sharp, J. M. & Germiat, S. J. 1990: Risk assessment and causes of subsidence and inundation along the Texas Gulf Coast. In: *Greenhouse effect, sea level and drought* (eds Paepe R. *et al.*) 395-414, Kluwer.

[909] di Paola G. *et al.* 2018: Coastal subsidence detected by Synthetic Aperture Radar interferometry and its effect coupled with future sea-level rise: the case of the Sele Plain (Southern Italy). *Journ Flood Risk Manag.* 11, 191-206.

[910] Wang, H. *et al.* 2012: InSAR reveals coastal subsidence in the Pearl River Delta, China.

Geophys. Journ. Initernat. 131 (3), 119-1128.

[911] Blackwell, E. *et al.* 2020: Tracking California sinking coast from space: Implications for relative sea-level rise. *Science Advances* 6(31), 4551.

[912] Dodson, A. T. & Dines, J. S. 1929: Report on Thames floods and meteorological conditions associated with high tides in the Thames. *Geophys. Mem.* 47, 1-39.

[913] Dunning, F. W. *et al.* 1978: *Britain before man.* HMSO, 36.

[914] Clayton, K. M. 1990: Sea-level rise and coastal defences in the UK. *Quart. Journ. Engin. Geol.* 23, 283-287.

[915] Savage, A. 1995: *The Anglo-Saxon Chronicles.* Crescent Books..

[916] Matthew Paris, 1236: *Chronica Majora.*

[917] John Stow, 1580: *The Chronicles of England from Brute unto this Present Yeare of Christ,* London

[918] Samuel Pepys, 1666: *The diary of Samuel Pepys.* Cassell

[919] Poland, J. F. & Davis, G. H. 1969: Land subsidence due to withdrawals of fluids. *Rev. Engin. Geol.* 2, 187-269.

[920] Gilbert, S. & Horner, R. W. 1984: *The Thames Barrier.* Thomas Telford

[921] www.edition.cnn.com, 16th December 2019, London has spent billions but no one can escape climate change.

[922] *The New York Times* 3rd October 2020: Adaptation works, problem solved: Floodgates now protect Venice from flooding.

[923] Sheppard, T. 1912. *The lost towns of the Yorkshire coast and other chapters bearing upon the geography of the district.* A, Brown & Sons.

[924] Green, C. & Hutchinson, J. N. 1965: Relative land and sea levels at Great Yarmouth, Norfolk. *Geograph. Journ.* 131, 86-90.

[925] Green, C. 1961. East Anglian coastline levels since the Roman times. *Antiquity* 35: 21–28 & 155–156.

[926] Hansom, J. D. et al. 2003: The geomorphology of the coastal cliffs of Great Britain. *Project Coastal Geomorphology ofd Great Britain.*

[927] Anthony Watts, 26th February 2021: "Acceleration" in sea-level rise found to be false – an artefact of switching satellites. *Climate Realism*

[928] Arjen Luijendijk *et al.* 2018: The state of the world's beaches. *Nature Climate Change* Scientific Reports 8: 6641 doi: 10.1038/S41598-018-24630-6

[929] *The Times,* 27th October 2020: EU report about vanishing beaches was alarmist and wrong, scientists say.

[930] Douglas, B. C. & Peltier, W. R. 2002: The puzzle of global sea-level rise. *Physics Today* March 2002, 1-6.

[931] Shevenell, A. E. *et al.* 2004: Middle Miocene Southern Ocean cooling and Antarctic cryosphere expansion. *Science* 305, 1766-1770.

[932] Degens, E. T & Ross, D. A. 1969: *Hot brines and recent heavy metal deposits in the Red Sea.* Springer-Verlag.

[933] McGuire, W. J. *et al.* 1997: Correlation between the rate of sea-level change and frequency of explosive volcanism in the Mediterranean. *Nature* 399, 473-476.

[934] Satow, C. *et al.* 2021: Eruptive activity of the Santorini volcano controlled by sea level rise and fall. *Nature Geoscience* 14, 586-592.

[935] McGuire, W. J. 2008: Changing sea levels and erupting volcanoes: cause and effect? *Geology Today* 8, 141-144.

[936] Rampino, M. R. *et al.* 1979: Can rapid climate change cause volcanic eruptions? *Science* 206, 826-829.

[937] Peate, I. U. & Bryan, S. E. 2008: Re-evaluating plume-induced uplift in the Emeishan large igneous province. *Nature Geoscience* 1, 625-629.

[938] Peate, I. U. *et al.* 2003: The transition from sedimentation to flood volcanism in the Kangerlussuaq Basin, East Greenland: Basaltic pyroclastic volcanism during initial

Palaeogene continental break-up. *Journ. Geol. Soc. London* 160, 759-772.

[939] Meyers, G. 1996: Variation of Indonesian flowthrough and the El Niño:Southern Oscillation: Pacific low-latitude western boundary currents and the Indonesian flowthrough. *Journ. Geophys. Res.* 101, 12255-12264.

[940] Harrison, D. E. & Crane, M. A. 1984: Changes in the Pacific during the 1982-83 El Niño event. *Oceanus* 27, 21-28.

[941] Komar, P. D. 1986: El Niño and erosion on the coast of Oregon. *Shore and Beach* 54, 3-12..

[942] Muis, S. *et al.* 2018: Influence of El Niño-Southern Oscillation on global coastal flooding. *Amer. Geophy. Union* https://www.doi.org/10.1029/2018EF000909.

[943] Long, X. *et al.* 2020: Higher sea levels at Hawaii caused by strong El Niño and weak trade winds. *Journ Climate* 33 (8), 3037-3039.

[944] Kench, P. S. *et al.* 2018: Patterns of island change and persistence offer alternate adaptation pathways for atoll nations. *Nature Communications* 9, Article Number 605

[945] Nakamura, N. *et al.* 2020: Anthropogenic anoxic history of the Tuvalu Atoll recorded as annual black bands in coral. *Nature Scientific Reports* 10, Article Number 7338

[946] Alan Moran, September 2019: *Climate News.*

[947] Shepherd, A. & Wingham, D. 2007: Recent sea-level contributions of the Antarctic and Greenland ice sheets. *Science* 315, 1529-1532

[948] Vaughan, D. & Arthern, R. 2007: Why it is hard to predict the future of ice sheets. *Science* 315: 1503-1504

[949] www.theguardian.com, 12th August 2017: Scientists discover 91 volcanoes beneath Antarctic ice sheet.

[950] Mann, M. E. *et al.* 1999: Northern Hemisphere temperatures during the past millennium: Inferences, uncertainties, and limitations. *Geophys. Res. Letts* 26, 759-762

[951] Larsen, C. E. & Clark, I. 2006: A search for scale in sea level studies. *Journ. Coastal Res.* 22, 788-800

[952] Stone, J. *et al.* 2003: Holocene deglaciation of Marie Byrd Land, West Antarctica. *Science* 299, 99-102.

[953] Lambeck, K. *et al.* 2004: Sea level in Roman time in the Central Mediterranean and implications for recent change. *Earth Planet. Sci. Letts* 224, 563-575

[954] Müller, D. *et al.* 2008: Long-term sea-level fluctuations driven by ocean basin dynamics. *Science* 319, 1357-1362

[955] Kasting, J. F. *et al.* 2006: Paleoclimates, ocean depth, and the oxygen isotopic composition of seawater. *Earth Planet. Sci. Letts* 252, 82-93.

[956] Palaeozoic Era.

[957] Haq, B. U. & Schutter, S. R. 2008: A chronology of Paleozoic sea-level changes. *Science* 322, 64-68.

[958] Watts, A. B. & Thorne, J. A. 1984: Tectonics, global changes in sea level and their relationship to stratigraphic sequences at the US Atlantic continental margin. *Marine Petrol. Geol.* 1, 319-339.

[959] Haq, B. U. *et al.* 1987: Chronology of fluctuating sea levels since the Triassic. *Science* 235, 1156-1167.

[960] Vail curves.

[961] Mörner, N. A. 2004: Estimating future sea level changes from past records. *Global Planet. Change* 40, 49-54

[962] Otto-Bliesner, B. L. *et al.* 2006: Simulating Arctic climate warmth and icefield retreat in the last interglaciation. *Science* 311, 1751-1753.

[963] McCulloch, M. T. & Esat, T. 2000: The coral record of last interglacial sea levels and sea surface temperatures. *Chem. Geol.* 169, 107-129.

[964] Stirling, C. H. *et al.* 1998: Timing and duration of the Last Interglacial: Evidence for a restricted interval of widespread coral growth. *Earth Planet. Sci. Letts* 160, 745-762.

[965] Neumann, A. C. & Hearty, P. J. 1996: Rapid sea-level changes at the close of the last

interglacial period: [234]U-[230]Th data from fossil coral reefs in the Bahamas. *Bull. Geol. Soc. Amer.* 103, 82-97.

966 Rohling, E. J. *et al.* 2008: High rates of sea-level rise during the last interglacial period. *Nature Geoscience* 1, 38-42.

967 Sloss, C. R. *et al.* 2007: Holocene sea-level change on the southeast coast of Australia: a review. *The Holocene* 17, 999-1014.

968 Fairbridge, R. W. 1958: Dating the latest movements in the Quaternary sea level. *New York Acad. Sci, Trans* 20, 471-482.

969 Fairbridge, R. W. 1960: The changing level of the sea. *Scientific American* 202, 70-79.

970 Fairbridge, R. W. 1961: Eustatic changes in sea level. In: *Physics and Chemistry of the Earth, Vol 4* (eds Ahrens *et al.*) Pergamon Press, 99-185.

971 Baker R. G. V. *et al.* 2001a: Warmer or cooler late Holocene marine palaeoenvironments?: Interpreting Southeast Australian and Brazilian sea-level changes using fixed biological indicators and their $\partial^{18}O$ composition. *Palaeogeogr. Palaeoclimat. Palaeoecol.* 168, 249-272.

972 Baker, R. G. *et al.* 2001b: Inter-tidal fixed indicators of former Holocene sea levels in Australia: a summary of sites and a review of methods and models. *Quat. Internat.* 83-85, 257-273.

973 Baker, R. G. *et al.* 2005: An oscillating Holocene sea-level? Revisiting Rottnest Island, Western Australia and the Fairbridge eustatic hypothesis. *Journ Coast. Res.* 42, 3-14.

974 Holgate, S. J. 2007: On the decadal rates of sea level change during the twentieth century. *Geophys. Res. Letts* 34(1) doi.org/10.1029/2006GL028492.

975 Dietzel, A. *et al.* 2021: The population sizes and global extinction risk of reef-building coral species at biogographic scales. *Nature Ecology & Evolution* doi: 10.1038/s41559-021-01393-4

976 Webster, J. M. *et al.* 2018: Response of the Great Barrier Reef to sea-level and environmental changes over the past 30,000 years. *Nature Geoscience* 11, 426-432

977 Peter Ridd, 6th December 2020: It's the science that's rotten, not the Great Barrier Reef. *The Australian.*

978 Webster, J. M. *et al.* 2018: Response of the Great Barrier Reef to sea-level and environmental changes over the past 30,000 years. *Nature Geoscience* 11, 426-432.

979 Peter Ridd, 6th December 2020: It's the science that's rotten, not the Great Barrier Reef. *The Australian.*

980 Ryan, E. *et al.* 2015: Chronostratigraphy of Bramston Reef reveals a long-term record of fringing reef growth under muddy conditions in the central Great Barrier Reef. *Palaeogeogr. Palaeoclimatol. Palaeoecol.* Doi:10.1016/palaeo.2015.10.106

981 Hayne, M. & Chappell, J. 2001: Cyclone frequency during the last 5,000 years at Curacos Island, north Queensland, Australia. *Palaeogr. Palaeoclim. Palaeoecol.* 168, 207-219.

982 Nott, J. & Hayne, M. 2001: High frequency of 'super-cyclones' along the Great Barrier Reef over the past 5,000 years. *Nature* 413, 508-512.

983 *The Australian*, 20th August 2021: Bigger agenda in play to reduce Paris Agreement warming target.

984 Lewis S. E. *et al.* 2013: Post-glacial sea-level changes around the Australian margin: a review. *Quaternary Science Reviews* 74, 115-138.

985 Banha, T. N. S. *et al.* 2020: Low coral mortality during the most intense bleaching event ever recorded in subtropical Southwestern Atlantic reefs. *Coral Reefs* 39, 515-521.

986 Masselink, G. 2000: Coral reefs can accrete vertically in response to sea level rise. *Science Advances* DOI:10.1126/scladc.aay3656.

987 Peter Ridd, 2020: *Reef Heresy? Science, research and the Great Barrier Reef.* Connor Court.

988 *The Australian*, 24th 2021: When climate alarmism meets cancel culture.

989 Vecchi, G. A. *et al.* 2021: Changes in Atlantic major hurricane frequency since the late-19th century. *Nature Communications* doi:10.1038.s41467-021-24268-5

990 Bjorn Lomborg, July 20, 2021: Climate change and deaths from extreme heat and cold. *Financial Post.*

991 Brázdil, R. et al. 2021: Fatalities associated with the severe weather conditions in the Czech Republic, 2000-2019. *Nat. Hazards Earth Syst. Sci.* 21, 1355-1382.

992 Formetta, G. & Feyen, L. 2019: Empirical evidence of declining global vulnerability to climate-related hazards. *Global Environmental Change* 57, 101920.

993 www.theguardian.com/us-news/2020/aug/17/death-valley-temperature-rises-to-54.4c-possibly-the-hottest-ever-reliably-recorded

994 www.climod2.nrcc.cornell.edu/

995 *There was once a bowler called Clover,*
Who bowled sixteen wides in an over,
It had never been done
By a clergyman's son,
On a Tuesday, in August, at Dover.

996 Sturt, C. 1849: Narrative of an expedition into Central Australia, www.ebooks.adelaide.edu.au/s/sturt/charles/s93n/

997 Mitchell, T. L. 1846: Journal of an expedition into the interior of tropical Australia. www.gutenberg.net.au/ebooks/e00034.html

998 Ewart, J. N. 23rd January, 1861: Mr Stuart's party. *The Cornwall Chronicle.*

999 Nova, J. 2017: Mysterious revisions of Australia's long hot history. In: *Climate change: The facts 2017* (ed. J. Marohasy), IPA 117-13.

1000 Stewart, K. 2010: *The Australian temperature record – Part 9: An urban myth.* www.kenskingdom.wordpress.com

1001 www.CNSNews.com, October 2016, NOAA data

1002 Klotzbach, P. J. et al. 2020: The record-breaking 1933 Atlantic hurricane season. *Bull. Amer. Met. Soc.* Doi.org/10.1175/BAMS-D-19-0330.1

1003 NOAA/Geophysical Fluid Dynamics Laboratory. *Global warming and hurricanes – An overview of current research results. Has global warming affected hurricane or tropical cyclone activity.* Last revised March 2017

1004 www.climateatlas.com, Dr Ryan Maue, 2021 accumulated cyclone energy.

1005 www.notalotofpeopleknowthat.wordpress.com/2020/12/31

1006 www.climatlas.com/tropical/; https://www.spc.noaa.gov/wcm/#data

1007 Raup, D. 1990: *Extinction: Bad genes or bad luck.* Norton

1008 McGhee, G. R. 2006: Extinction: Late Devonian mass extinction. *Encyclopedia of Life Sciences,* John Wiley

1009 Elewa, A. M. T. 2008: *Mass extinction.* Springer

1010 Stanley, S. M. & Yang, X. 1994: A double mass extinction at the end of the Paleozoic era. *Science,* 266, 1340-1344

1011 Hallam, A. 1990: The end-Triassic mass extinction event. *Geol. Soc. Amer. Special Paper* 247, 577-583

1012 Sepkoski, J. 2002: Compendium of fossil marine animal genera (eds Jablonski, D. & Foote, M.), *Bull. Amer. Palaeont.* 363

1013 Bennet, K. D. 2013: Is the number of species on earth increasing or decreasing? Time, chaos and the origin of species. *Palaeontology* 54(6), 1305-1325

1014 Thomas, C. D. et al. 2004: Extinction risk from climate change. *Nature* 427, 145-148.

1015 Root, T. et al. 2003: Fingerprints of global warming on wild animals and plants. *Nature* 421, 57-60

1016 Parmersan, C. & Yohe, G. 2003: A globally-coherent fingerprint of climate change impacts across natural systems. *Nature* 421, 37-42

1017 Botkin, D. B. et al. 2007: Forecasting the effects of global warming on biodiversity. *BioScience* 57, 227-236.

1018 Pounds, J. A. & Schneider, S. H. 1999: *Present and future consequences of global warming for highland tropical forests exosystems: The case of Costa Rica.* U.S. Global Change Research Program Seminar, Washington D.C., 29th September 1999.

[1019] Lawton, R. O. *et al.* 2001: Climate impact of tropical lowland deforestation on nearby mountain cloud forests. *Science* 294, 584-587.

[1020] Schuster, P. F. *et al.* 2000: Chronological refinement of an ice core record at Upper Fremont Glacier in south central North America. *Journ. Geophys. Res.* 105, 4657-4666.

[1021] www.iucnredlist.org.

[1022] www.wattsupwiththat.com, 24th May 2019, Gregory Wrightstone: exposing the mass extinction lie

[1023] Loehle, C. 1998: Height growth rate tradeoffs determine northern and southern range limits for trees. *Journ. Biogeogr.* 25, 735-742

[1024] Gauslaa, Y. 1984: Heat resistance and energy budget in different Scandinavian plants. *Holarctic Ecol.* 7, 1-78

[1025] Levitt, J. 1980: *Responses of plants to environmental stresses.* Vol 1: *Chilling, freezing and high temperature stresses.* Academic Press

[1026] Kappen, L. 1981: Ecological significance of response to high temperature. In: *Physiological plant ecology. I. Response to the physical environment.* (eds O. L. Lange *et al.).* Springer-Verlag.

[1027] Axelrod, D. I. 1956: Mio-Pliocene floras from west-central Nevada. *Univ. Calif. Publics Geol. Sci.* 33, 1-316

[1028] Axelrod, D. I. 1987: The Late Oligiocene Creede Flora, Colorado. *Univ. Calif. Publics Geol. Sci.* 130, 1-235

[1029] Idso, S. *et al.* 2003: *The specter of species extinction.* The Marshall Institute, Washington, D.C., 1-39.

[1030] Idso, K. E. & Idso, S. B. 1994: Plant responses to atmospheric CO_2 enrichment in the face of environmental constraints: A review of the past 10 years' research. *Agric. Forest Meteorol.* 69, 153-203.

[1031] Cannell, M. G. R. & Thorley, H. H. M. 1998: Temperature and CO_2 responses of leaf and canopy photosynthesis. A clarification using the non-rectangular hyperbola model of photosynthesis. *Annal. Botany* 82, 883-892

[1032] Nemani, R. R. *et al.* 2003: Climate-driven increases in global terrestrial net primary production from 1982 to 1999. *Science* 300, 1560-1563.

[1033] Kimball, B. A. 1983: Carbon dioxide and agricultural yield: An assemblage and analysis of 430 prior observations. *Agron. Journ.* 75, 779-788.

[1034] Saxe, H. E. *et al.* 1998: Tree and forest fluctuating in an enriched CO_2 atmosphere. *New Phytol.* 139, 395-436.

[1035] Pauli, H. *et al.* 1996: Effects of climate change on mountain ecosystems – upward shifting of mountain plants. *World Resources Review* 8, 382-390.

[1036] Sobrino, E. *et al.* 2001: The expansion of thermophilic plants in the Iberian Peninsula as a sign of climate change. In: *"Fingerprints" of climate change: Adapted behaviour and shifted species ranges* (eds Walther, G. R. *et al.*) Plenum, New York, 163-184.

[1037] Sturm, M. *et al.* 2001: Increasing shrub abundance in the Arctic. *Nature* 411, 546-547.

[1038] Smith, R. I. L. 1994: Vascular plants as bioindicators of regional warming in Antarctica. *Oecologia* 99, 322-328.

[1039] Van Herk, C. M. *et al.* 2002: Long term monitoring in The Netherlands suggests that lichens respond to global warming. *Lichenologist* 34, 141-154.

[1040] Pollard, E. *et al.* 1995: Population trends of common British butterflies at monitored sites. *Journ Appl. Ecol.* 32, 9-16

[1041] Jackson, N. K. 1994: Pioneering and natural expansion of breeding distributions of western North American birds. *Studies Avian Biol.* 15, 27-44.

[1042] Southward, A. J. 1995: Seventy years' observation of changes in distribution and abundance of zooplankton and intertidal organisms in the Western English Channel in relation to rising sea temperatures. *Journ. Therm. Biol.* 20, 127-155.

[1043] Smith, R. C. *et al.* 1999: Marine ecosystem sensitivity to climate change. *BioScience* 49, 393-404

[1044] Sagarin, R. D. *et al.* 1999: Climate-related change in an intertidal community over short and long time scales. *Ecol. Monogr.* 69, 465-490.

[1045] Collins, Simon, 2004: Antarctic fish set to survive warmer seas. *New Zealand Herald*, 16th April 2004.

[1046] US Fisheries and Wildlife Service, Report 6th April 2006.

[1047] Amstrup, S. C. *et al.* 2004. Using satellite radio-telemetry data to delineate and manage wildlife populations. *Wildlife Soc. Bull.* 32, 661-679.

[1048] Crockford, S. J. 2020: State of the Polar Bear Report 2020 *Global Warming Policy Foundation.*

[1049] Haynes, G. 2002: The catastrophic extinction of North American mammoths and mastodonts. *World Archaeology* 33, 391-416

[1050] Barnosky, A. D. *et al.* 2004: Assessing the causes of Late Pleistocene extinctions on the continents. *Science* 306, 70-75

[1051] Haynes, V. *et al.* 2010: The Murray Springs Clovis site, Pleistocene extinction, and the question of extraterrestrial impact. *PNAS* 107(9), 4010-4015

[1052] Broughton, J. M. & Weitzel, E. M. 2018: Population reconstructions for humans and megafauna suggest mixed causes for North American Pleistocene extinctions. *Nature Communications* 9, Article 5441

[1053] www.iucn.org, IUCN red list of threatened species

[1054] www.un.org, 6th May 2019, UN Report Nature's dangerous decline 'unprecedented', species extinction rates accelerating

[1055] *The Federal*, 29th August 2021: Europe's tryst with biofuels destroyed 10% of world's orangutan habitats

[1056] www.transpoerenvironment.ord, 10 years of EU's failed biofuels policy has wiped out forests the size of the Netherlands - study

[1057] Jeswani, H. *et al.* 020: Environmental sustainability of biofuels: a review. *Proc. Roy. Soc. A,* doi.org/10.1098/rspa.2020.0351

[1058] James Conca, 20th April 2014: It's final—Corn ethanol is of no use. *Forbes*

[1059] www.polarbearsscience.com/2020/08/06/emperor-penguin-numbers-rise-as-biologists-petition-for-iucn-red-list-upgrade

[1060] *Daily Mail*, 7th September 2020: Free speech, fake science – and why we must take the fight to the climate zealots

[1061] *The Times*, 12th September 2020: Climate campaigners quit as Extinction Rebellion starts to split.

[1062] *The Sunday Telegraph*, 6th September 2020: Extinction Rebellion: Printworks protest 'completely unacceptable' says Boris Johnson

[1063] *The Guardian*, 6th September 2020: Extinction Rebellion calls move to class it as organised crime group 'ridiculous'

[1064] *Daily Telegraph*, 12th September 2021: Extinction Rebellion.

[1065] www.policyexchange.uk.org, Tim Wilson and Richard Walton *Extremism Rebellion: A review of ideology and tactics*

[1066] Kurt Zindulka, 25th August 2021: Climate hypocrites! Extinction rebellion is latest top green figure who drives a diesel car. *Breitbart*

[1067] Peter Foster, 21st October 2020: The IEA's solar spin cycle: Sorry folks, the world will still be overwhelmingly fossil-fuelled in 2030. *Financial Post*

[1068] *The Australian*, 6th September 2021: UN gives 10-year deadline to shutdown coal mining

[1069] www.abs.gov.au

[1070] www.world-nuclear.org/information-library/country-profiles/countries-a-f/appendices/australia-s-electricity.aspx

[1071] Bernard Weinstein, 5th April 2021: Until something better comes along, we need base-load power for grid reliability. *Real Clear Energy*

[1072] www.choice.com.au, 21st July 2021: Residential disconnections for gas and electricity on the rise again

[1073] www.energy.gov.au, Alan Finkel, 9th June 2017: *Review into the future security of the National Electricity Market*

[1074] www.malcolmrobertsqld.com.au, *Dr Alan Moran Report*

[1075] Alan Moran, 2nd December 2020: Subsidies drain power from electricity market. *Spectator Australia*

[1076] www.europeanscientist.com/en/energy/hydrogen-strategy-to-nowhere%E2%80%A8/utm_source-Energy+geopolitics&utm_campaign=fcbaa25021

[1077] www.wattsupwiththat.com/2017/07/05/monumental-unsustainable-environmental-impacts/

[1078] www.angora-energiewende.de, Making the most of offshore wind

[1079] Pierre Gosselin, 15th September 2020: New study shows German offshore wind turbines may cannibalize each other when improperly sited. *No Tricks Zone*

[1080] www.metoffice.gov.uk/news/releases/2017/study-shows-potential

[1081] *The Guardian*, 25th April 2021: UK solar panels manufactured by firms linked to Chinese slave labour

[1082] www.greenpeace.org, 10th December 2018: What does climate change have to do with human rights?

[1083] House of Lords debate by Lord Turnbull, 17th July 2017, on the Report from the Economic Affairs Committee *The Price of Power: Reforming the Electricity Market*

[1084] www.americaspower.org

[1085] www.instituteforenergyresearch.org

[1086] *Daily News*, 28th April 2021: The U.S. vs. Atlantic fisheries.

[1087] www.markey.senate.gov/GlobalWarming/index.html

[1088] www.finance.senate.gov.

[1089] *The Wall Street Journal*, 10th September 2021: Cost inflation is blowing through the wind turbine industry

[1090] *The Wall Street Journal*, 23rd August 2021: Wind-turbine makers struggle to profit from renewable-energy boom

[1091] *Daily Telegraph*, 2nd March 2011: Era of constant electricity at home is ending, says power chief.

[1092] www.europeanscientist.com/en/energy/hydrogen-strategy-to-nowhere%E2%80%A8/utm_source-Energy+geopolitics&utm_campaign=fcbaa25021

[1093] $v = I^2r$ where v = volts, I = current and r = resistance hence a higher voltage gives less resistance (i.e. losses as heat and sound).

[1094] Jo Nova, 10th October 2020: Extension cord to rescue renewable South Australia will now cost $2.4 billion. *Jo Nova Blog.*

[1095] Alan Taylor, 26th July 2016: Flying around the world in a solar powered plane. *The Atlantic.*

[1096] Centre for Environmental Research and Earth Sciences, 1st October 2020: Surprising science – There's no such thing as clean energy.

[1097] www.katherinetimes.com, 10th December 2019: Shine comes off solar from Alice Springs failure.

[1098] Letter to the Editor, Dave Hayes, 23rd April 2019: Without old power station we'd be 'stuffed': ETU. *Alice Springs News.*

[1099] www.abc.net.au, 12th December 2019: Can the Northern Territory cope with a transition to 50% renewables?

[1100] www.joannenova.com.au/2020/10/bargain-2-billion-in-solar-panels-powers-sa-for-whole-hour-on-sunday-in-spring/

[1101] David Middleton, 13th October 2020: Solar power costs 2-3 times as much as wind, fossil fuels and nuclear. *Watts Up With That?*

[1102] *The Washington Free Beacon*, 7th August 2021: Solar farms spark Civil War in Virginia.

[1103] Natasha Donn, 22nd December 2020: Over 500 animals slaughtered in a walled estate in Azumbuja purportedly making way for massive solar energy park. *Portugal Resident.*

[1104] Andrew Leonard, 2009: Solar's hidden poison. *The Australian Business Review.*

[1105] www.svtc.org, Silicon Valley Toxics Coalition 2009 report. *Toward a just and sustainable solar energy industry.*

[1106] www.instituteofenergyresearch.org/about/ier-site-manager/articles, *Degradation of operational solar units.*

[1107] www.theaustralian.com.au/business/mining-energy/mission-zero-gina-rinhart-backs-renewables-fears-farmers--and-australians-cant-afford-it/news:story/d5f3c77176c041c1a215fec09fd7a49b

[1108] Robert Bradley Jr, 1st April 2021: Land-intensive renewables: Three TW of wind and solar = 228 sq. miles, *Master Resource.*

[1109] Duggan Flanakin, 19th September 2020: Solar panels generate mountains of waste – They also heat the planet, blanket wildlife habitats and cause other ecological damage. *Watts up with that?*

[1110] *Iowa Capital Dispatch*, 7th July 2021: Company illegally storing hundreds of old wind turbine blades at three Iowa sites.

[1111] Michael Shellenberger, 22nd June 2021: Why everything you've heard about solar is wrong. *Climate Dispatch.*

[1112] Duggan Flanakin, 17th September 2020: Solar panels generate mountains of waste. *Townhall.*

[1113] www.reneweconomy.com.au/energy-insiders-podcast-batteries-in-the-street-the-new-face-of-storage/

[1114] *Global Warming Policy Foundation*, 13th November 2020: Green lobby up in arms as France plans to tear up solar subsidy contracts.

[1115] *Financial Times*, 12th November 2020: French solar investors up in arms over threat to renege on contracts.

[1116] Mark Mills, 2019: *Manhatten Institute.*

[1117] *Sky News Australia*, 14th February 2021.

[1118] Pierre Gosselin, 7th April 2021: Germany's windexit. *No Tricks Zone.*

[1119] Alex Reichmuth, 5th April 2018: Abbruchstimmung in Deutschland. *Baseler Zeitung Ausland.*

[1120] Marcus Wacket, 30th March 2021: Germany's energy drive criticised over expense, risks. *Reuters.*

[1121] Pierre Gossilin, 13th April 2021: 2021 German coal plant "Phaseout" lasted only 8 days… Put back online to stabilize shaky grid. *No Tricks Zone.*

[1122] *The Wall Street Journal*, 13th September 2021: Energy prices in Europe hit records after the wind stops blowing.

[1123] *Fortune*, 11th September 2021: Europe's ambitious Net Zero pledges hit home – with eye-watering energy bills.

[1124] *S&P Global*, 14th September 2021: UK electricity prices now most expensive in Europe.

[1125] *The Independent*, 20th September 2021: Energy crisis could last for months, Boris Johnson admits as companies warn of collapse.

[1126] *Daily Mail*, 20th September 2021: Taxpayers face multi-billion pound bill to bail out failing energy firms, soaring bills and empty supermarket shelves.

[1127] *Financial Times*, 19th September 2019: Energy prices will push up inflation across Europe, economists warn.

[1128] *The Daily Telegraph*, 18th September 2021: Environmental hubris has left Britain vulnerable to Putin's gas blackmail.

[1129] *The Daily Telegraph*, 19th September 2021: Suicidal energy policy is empowering Britain's enemies.

[1130] Matt Ridley, December 2020: How Britain's shale revolution was killed by green lies and Russian propaganda. *The Critic.*

[1131] *The Daily Telegraph*, 14th September 2021: UK factories may have to switch to diesel generators as energy prices rise.

[1132] www.riteon.org.au/netzero-casualties/#218

[1133] Andrew Montford, 2021: An ill wind for hard-up electricity users. *Conservative Woman.*

[1134] Henrik Paulitz, 21 Januar 2021: Knapp am Blackout vorbeigeschrammt, *Kalte Sonne, Täglicher Newsletter uber aktuelle Klimathemen.*

[1135] Pierre Gosselin, 9th January 2021: Power Supply fiasco: Green energy blackout hits Germany! Fossil fuels to the rescue. *No Tricks Zone.*

[1136] *Handelsblatt Energie-Gipfel*, 8 Januar 2021: Kurz vor Blackout: Europas Stromnetz Wäre im Januar fast zusammengebrochen.

[1137] www.catallaxyfiles.com/2020/07/06/city-of-sydney-and-act-claims-re-re/

[1138] Eric Worrall, 16th December 2020: WA "Solves" the Solar Energy Duck Curve by raising Evening Electricity Prices. *Watts Up With That?*

[1139] Daniel Mercer, 16th December 2020: Electricity prices would be slashed during the day doubled during peak under new Government trial. *ABC WA.*

[1140] Dominic Lawson, 10th January 2021: The rush to 'Net Zero' will most harm those Boris Johnson pledged to prioritise. *The Sunday Times.*

[1141] Rafe Champion, 7th February 2021: A perfect storm brewing in Europe. *Catallaxy Files.*

[1142] Zharkova, V. 2020: Modern Grand Solar Minimum will lead to terrestrial cooling. Temperature 7 (3), 217-222.

[1143] *Energy News Wire*, 5th March 2021.

[1144] GWPF *Report* 36, 19th June 2020: *Norway slows down onshore wind power developments.*

[1145] Andrew Montford, 1st April 2021: The RSPB, wind farms and a change of direction. *Conservative Woman.*

[1146] Lisa Linowes, 29th June 2020: Big wind's assault on the truth. *WindAction.*

[1147] *News Center Maine*, 21st March 2021: Dozens of lobster boats gather off Monhegan to protest floating wind turbine.

[1148] Jonny Drury, 13th April 2017: Group welcome u-turn over Powys solar and wind energy plans. *News North Wales.*

[1149] Shivani Gupta, 7th August 2021: Kachchh villagers on vigil to save Sangnara forest from a windmill project. *Gaon Connection*

[1150] *Gippsland Times*, 8th October 2020: Residents overturn wind farm planning permit at VCAI.

[1151] www.minister.defence.gov.au, 21st November 2017, Bullencourt wind farms.

[1152] *The Washington Newsday*, 3rd September 2021: France is up in arms over a proposed wind farm near WW1 memorial.

[1153] Andy Lines, 6th May 2020: Fury at plans to build wind turbines on graves of British First World War heroes. *Mirror.*

[1154] *Stop these things*, 8th May 2021: Let the Diggers rest in peace: French wind power outfit set to desecrate Australian war graves.

[1155] www.energie.com

[1156] Kevin Kilty, 14th August 2021: Shadows and flicker. *Watts Up With That?*

[1157] Jauchem, J. R & Cook, M. C. 2007: High-intensity acoustics for military nonlethal: a lack of useful systems. *Military Medicine* 172 (2) 182-189.

[1158] www.phr.org, 27th October 2020: Health impacts of crowd-control weapons: Acoustic weapons.

[1159] Sherri Lange and Steven Cooper, 30 July 2020: Health effects of industrial wind: The debate intensifies (update with Steven Cooper). *Master Resource.*

[1160] Salt, A. N. & Hullar, T. E. 2010: Responses of the ear to low frequency sounds, infrasound and wind turbines. *Hearing Research* 286, 12-21

[1161] Australian Senate Select Committee on Wind Turbines, 2015.

[1162] www.abc.net.au, ABC Gippsland, 9th September 2021: Wind farm trial in Supreme Court hears plaintiff slept in car to avoid 'roaring' of turbines

[1163] Pierre Gosselin, 16th January 2021: French Herder Believes Family Health Ailments, 400 Dead Cows "Clearly Linked" To Nearby Wind Farm. *No Tricks Zone*

[1164] www.caithnesswindfarms.co.uk, 16th December 2020 Crash involving semi hauling a wind turbine blade occurred near Sundance Wednesday. *County 10.*

[1165] Christopher Booker, 4th July 2015: Why are the greens so keen to destroy the world's wildlife? *Daily Telegraph.*

[1166] Jonathan Tennenbaum, 8th March 2021: Wind and solar reliance would black out the US. *Asia Times.*

[1167] www.turbinesonfire.org

[1168] www.turbinesonfire.org

[1169] Tremeac, B. & Meunier, F. 2009: Life cycle analysis of 4.5A and 250A W wind turbines. *Renew. Sustain. Energy Rev.* 13 (8) 2104-2110.

[1170] Guezuraga, B. *et al.* 2012: Life cycle assessment of two different 2 MW class wind turbines. *Renewable* Energy 37, 37-44.

[1171] Lars Paulsson *Bloomberg* 23rd November 2020 Giant Vestas wind turbine collapses in northern Sweden; https://stopthesethings.com/2020/12/09/turbines-tumble-another-230-metre-300-tonne-whirling-wonder-bites-the-dust/

[1172] Jessica Howard, 16th September 2020: Hundreds of bird, bat carcasses detected at Morton's Lane Wind Farm. *The Standard.*

[1173] *NL Times*, 28th February 2021: Tracked sea eagle dies in fatal collision with wind turbine

[1174] Frick, W. F. *et al.* 2017: Fatalities at wind turbines may threaten population viability of migratory bat. *Biol. Conserv.* 209, 172-177.

[1175] Michael Shellenberger, 10th August 2020: Do We Have to Destroy the Earth to Save It? *Prager University.*

[1176] Cryan, P. M. *et al.* 2014: Behaviour of bats at wind turbines. *PNAS* 111(42), 15126-15131.

[1177] *The Global Warming Policy Foundation*, Press Release, 19th January 2021: GWPF calls on Sir David Attenborough to save Red-listed Kittiwakes from giant wind turbine project.

[1178] *National Wind Watch, Scottish Forestry*, 16th January 2020: More than 13.9 million trees felled in Scotland for wind development, 2000-2019.

[1179] Andrew Montford, 2019: Green killing machines. The impact of renewable energy on wildlife and nature. *The Global Warming Policy Foundation*

[1180] Ortegon, K. *et al.* 2013: Preparing for end of service life for wind turbines. *Journ. Cleaner Prodn.* 39, 191-199.

[1181] Liu, P. & Barlow, C. Y. 2017: Wind turbine waste in 2050. *Waste Management* 62, 229-240.

[1182] Pimenta, S. & Pinho, S. T. 2011: Recycling carbon fibre reinforced polymers for structural applications: Technology review and market outlook. *Waste Management* 31 (2), 378-392.

[1183] Ramirez-Tejeda *et al.* 2017: Unsustainable wind-turbine blade disposal practices in the United States *SAGE Journals* 26 (4), doi.org/10.1177/1048291116676098

[1184] www.bloomberg.com, 6th February 2020, Wind turbine blades can't be recycled, so they're piling up

[1185] www.heartlanddailynews.com/2020/09/wind-turbines-generate-mountains-of-waste-no-more-green-than-solar/

[1186] Pierre Gosselin 9th September 2020 *No Tricks Zone* Green dream arrives in Germany! But repowering obstacles pose "Imminent Catastrophe" for wind power

[1187] www.vernunftkraft.de, Pierre Gosselin 1st December 2020: Germany's enviro-dystopia: Wind parks devastating rural regions at catastrophic proportions, *No Tricks Zone.*

[1188] Werner Köppen, August 2021.

[1189] The Greens.

[1190] Pierre Gosselin, 21st November 2020: 1.35 million tonnes of "hazardous materials", Germany admits no plan to recycle used wind blades. *No Tricks Zone*

[1191] www.epoxy-europe.eup/wp_content/uploads/2015/07/epoxy_erc_bpa_whitepapers_wind_energy-2.pdf

[1192] www.efsa.europa.eu/en/topics/topic/bsiphenol

[1193] www.canada.ca/en/health-canada/services/home-garden-safety/bisphenol-bpa.html

[1194] Duggan Flanakin, 26th September 2020: Wind turbines generate mountains of waste. *Watts up with That?*

[1195] *Climate Change Dispatch,* 27th August 2020.

[1196] www.iea.org, Sustainable Development Scenario – World Energy Model.

[1197] Song, N. *et al.* 2009: Extraction and separation of rare earths from chloride medium with mixtures of 2-ethylhexylphosponic acid mono-(-2-ethylhexyl) ester and *sec*-nonlyphenoxy acid. *Chem. Technol. Biotechn.* 84(12), 1798-1802.

[1198] Plimer, Ian 2014: *Not for greens.* Connor Court.

[1199] Bryce, Robert 2014: *Smaller, faster, lighter, denser, cheaper. How innovation keeps providing the catastrophists wrong.* Public Affairs.

[1200] Wind energy in the United States and materials required for the land-based turbine industry from 2010 through 2013: *US Geological Survey.*

[1201] Mark P Mills, 9th July 2020: Mines, minerals, and "green" energy: A reality check. *Manhatten Institute.*

[1202] www.theaustralian.com.au/commentary/no-virtue-in-rich-nations-outsourcing-their-emissions/news-story/de8981f7bb9327ba26cc5634091

[1203] Mark Mathis, 23rd March 2021: Wind and solar subsidies. *Clear Energy Alliance.*

[1204] *GWPF International,* 22nd September 2021: China plays its green card, proposing to sell the world its wind and solar projects produced by coal.

[1205] www.bp.com, Annual Statistical Review of World Energy

[1206] Bill Peacock, 2021: Spanish renewable giant Iberdrola enters Texas with a thud. *Master Resource.*

[1207] Robert Bradley Jr., 4th November 2020: A free market energy vision. *Master Resource.*

[1208] *Bloomberg,* 30th August 2021: German inflation climbs to highest since 2008 on energy costs.

[1209] Jan Smelik, 21st September 2020: Climate or environment. *YouTube*

[1210] Gail Tverberg, 17th July 2020: Why a great reset based on green energy isn't possible. *Our finite world.*

[1211] Gordon Hughes, University of Edinburgh, 2020: *Wind Power Economics – Rhetoric & Reality.*

[1212] *Bloomberg,* 17th September 2021: You don't say: Green push leaves UK energy supply at the mercy of weather.

[1213] *The Wall Street Journal,* 13th September 2021: Energy prices in Europe hit records after wind stops blowing Heavy reliance on wind power, coupled with a shortage of natural gas, has led to a spike in energy prices.

[1214] *Bloomberg,* 15th September 2021: Europe's energy crunch is forcing UK factories to shut down.

[1215] *GWPF Energy,* 18th September 2021, Britain faces food shortages as energy crisis shuts down factories.

[1216] *Financial Times,* 18th September 2021: UK energy groups in emergency talks with government over natural gas crisis.

[1217] *Yahoo News,* 10th September 2021: Europe may turn to more coal if gas crunch persists.

[1218] *Bloomberg,* 15th September 2021: Goldman warns of blackout risk for European industry this winter.

[1219] *Bloomberg,* 18th September 2021: Europe faces bleak winter energy crisis years in the making.

[1220] *The Global Warming Policy Foundation*, 1st September 2021: Green Europe faces gas shortages and energy crisis as winter looms.

[1221] www.oilprice.com, 12th April 2021: Russia tightens its grip on Europe's natural gas markets.

[1222] *Associated Press*, 27th April 2021: Merkel vs Biden: Europe's new gas wars.

[1223] Matt Ridley, December 2019: The plot against fracking: How cheap energy was killed by Green lies and Russian propaganda. *The Critic.*

[1224] *Deutsche Welle*, 8th September 2020: Politicians in Germany warn ex-Chancellor Schröder to quit Russian posts.

[1225] Bonner Cohen, 2nd April 2021 Cybersecurity of wind power a growing concern, *Epoch Times.*

[1226] *Daily Express,* 18th September 2021: EU energy crisis: Brussels faces 'tough' reality check over 'challenging' green policy.

[1227] www.ft.com/content/7a812f4d-ao93-4f1a-9a2f-877c41811486?

[1228] *Argus Media*, 6th July 2021: Green Europe: EU to exempt private jets and 'pleasure flights' from climate tax on jet fuel.

[1229] *The Street*, 14th April 2021: Revealed: The EU's stunning green hypocrisy.

[1230] Wolfgang Münchau, 13th March 2021: What do we mean by green? The EU puts too much emphasis on climate targets, and not enough on investments. *Euro Intelligence.*

[1231] Hans-Werner Sinn, 23rd July 2021: Europe's green unilateralism. *Project Syndicate.*

[1232] Linneman, T. & Vallana, G. S. 2017: Windenergie in Deutschland und Europa. *VGB Power Tech* 6: 63-73.

[1233] *Deutsche Welle*, 24th August 2015.

[1234] *DPA German Press Agency* ,2nd March 2017, t-online.de: Bundesnetzagentur

[1235] www.spiegel.de, 17th January 2013" Tree theft on the rise in Germany as heating costs rise.

[1236] *Bloomberg,* 1st February 2021: Europe just skirted blackout disaster.

[1237] Alex Reichmuth, 5th May 2021: Germany plans to hide the astronomical cost of renewable energy transition. *Schweizer Nebelspalte.*

[1238] Daniel Wetzel, 6th March 2017: Die Energiewende droht zum ökonomischen Desaster zu werden. *Die Welt.*

[1239] www.sei-international.org/mediamanager/documents/Publications/Climate/SEI-WP-2015-07-JI-lessons-for-carbon-mechs.pdf

[1240] *Welt an Sonntag*, 1st February 2021" First it lost its solar industry to China, now Germany's wind industry at risk of losing market to Chinese producers.

[1241] *Clean Energy Wire*, 21st September 2021: Vestas closes wind turbine factories in Europe, manufacturing moves to Asia.

[1242] Henrik Paulitz and Kalte Sonne, 17th January 2021: *Global Warming Policy Foundation*

[1243] *Bloomberg* 12th January 2021.

[1244] *Fitch Ratings,* 13th November 2020: Ratings agency warning: Germany's green energy transition reducing utilities' financial flexibility

[1245] Pierre Gosselin, 3rd March 2021: Germans spend 'More than ever before'…Consumer electricity costs reach record high in 2020. *No Tricks Zone.*

[1246] James Delingpole, 9th February 2021: Green jobs collapse in Germany and go to China instead. What a surprise! *Breitbart.*

[1247] *Yahoo Finance*, 2nd February 2021: Siemens slashed 7,800 jobs for green transition

[1248] *Handelsblatt*, 23rd February 2021: Chipmakers lament high taxes and green levies on electricity in Germany.

[1249] *Natural News*, 25th January 2021.

[1250] Dusseldorf's Institute for Competition Economics, 2016.

[1251] Pierre Gosselin, 6th February 2021: Winter storm threatens Germany's power…Freezing Hell threatens if already rickety grid collapses. *No Tricks Zone.*

[1252] *Bloomberg,* 21st November 2021: Germany's climate consensus cracks as costs mount.

[1253] *EurActiv,* 22nd November 2020: Germany plans 'turbine-free zones'

[1254] *The Guardian*, 15th August 2021: Germany 'set for biggest rise in greenhouse gases for

30 years'.

[1255] *Standpoint*, 28th December 2020: Thousands of new French windfarms – but no reduction in carbon emissions.

[1256] *The Australian*, 12th August 2015

[1257] Duggan Flanakin, 26th September 2020: Wind turbines generate mountains of waste. *Watts up with That?*

[1258] Mark Mills, 2019: Manhatten Institute.

[1259] Dan Shreve, *W*indpower 2019 conference.

[1260] Donn Dears, 16th October 2020: We need blackouts. *Power for USA.*

[1261] Ronald Bailey, 19th August 2020: Californian Blackout Fury: "It's not just the heat, it's also the anti-nuclear power stupidity". *Reason.*

[1262] Duggan Flanakin, 26th September 2020: Wind turbines generate mountains of waste. *Watts up with That?*

[1263] Charles Rotter, 13th June 2020: California will use diesel this summer to help keep the lights on. *Watts Up With That?*

[1264] Steve Goreham, 18th August 2020: Green California has the nation's worst power grid. *Washington Examiner.*

[1265] Nathan Solis, 13th January 2021: Report explains why California power grid couldn't take the heat in 2020. *Courthouse News.*

[1266] Steven Greenhut, 3rd July 2020: State energy policies help rich over the poor. *Orange County Register.*

[1267] Bronson Stocking, 6th September 2020: The Feds just threw a lifeline to California during energy crisis. *Townhall.*

[1268] Editorial, *The Wall Street Journal*, 20th August 2020: California's Green Blackouts.

[1269] *East Bay Times*, 17th August 2020: California blackouts expose problems in transition to green energy.

[1270] Katy Grimes, 15th August 2020: California's energy corruption and rolling blackouts. *California Globe.*

[1271] Michael Shellenberger, 15th August 2020: Why California's climate policies are causing electricity blackouts. *Forbes.*

[1272] Michael Shellenberger, 18th August 2020: Democrats say California is model for climate action but its blackouts say otherwise. *Forbes.*

[1273] Francis Menton, 16th August 2020: What is the cause of the recent power blackouts in California? *Manhattan Contrarian.*

[1274] Chris Horner, 18th August 2020: The year the lights went out in California. *The Pipeline.*

[1275] www.pubs.rsc.org/en/content/articlelanding/2018/cc/c7ee03029k#

[1276] www.windtaskforce.org/profiles/blogs/reality-check-regarding-utility-scale-battery-systems-during-a-one-day-wind-solar-lull/

[1277] *Institute of Energy Research*, 18th August 2020: Renewable mandates are leading to leading to electrical shortages and price spikes in California.

[1278] *Deadline*, 18th August 2020: California declares state of emergency as blackouts loom.

[1279] *Financial Times*, 18th August 2020: Californians face dark, hot summer as green energy is sapped.

[1280] *The Washington Times*, 18th August 2020: 'Forcing Americans in the Dark': Green energy push blamed in California's rolling blackouts.

[1281] *Bloomberg*, 19th August 2020: California may knock out power to 5 million people.

[1282] *The Daily Caller*, 17th August 2020: 'Gaps' in renewable energy led to blackouts for millions of Californians, Gov Newsom says.

[1283] Donn Dears, 24th August 2021: California orchestrates disaster. *Power for USA.*

[1284] *Bloomberg*, 9th September 2021: California declares energy emergency, seeks to avert blackouts by burning more fossil fuels.

[1285] Wayne Lubvardi, 24th May 2021: California summer blackouts? (Bureaucratic green doublespeak). *Master Resource.*

1286 *Bloomberg,* 16th February 2021: Almost 5 million across America plunged into darkness as energy crisis spreads in unprecedented deep freeze.

1287 *Bloomberg,* 16th February 2021: Almost 5 million across America plunged into darkness as energy crisis spreads in unprecedented deep freeze.

1288 *The New York Times,* 15th February 2021: Frozen turbines and surging demand prompt rolling blackouts in Texas.

1289 *Austin American Statesman,* 15th February 2021: Frozen wind turbines, surging demand triggers Texas blackouts.

1290 Hui Ha and Mike Krapil, 4th March 2021: Field studies show icing can cost wind turbines up to 80% of power production. *Iowa State University.*

1291 Editorial, *The Wall Street Journal,* 16th February 2021: A deep green freeze: An existential threat to America's future.

1292 *Forbes,* 15th February 2021: Sal Gilbertie Texas outages put reliability of renewable energy in the spotlight.

1293 *The Federalist,* 18th February 2021: Yes, green energy failures helped cause Texas blackout disaster.

1294 Francis Menton, 17th February 2021: Texas: Time to get rid of this ridiculous wind power. *Manhatten Contrarian.*

1295 Rafe Champion, 17th February 2021: The success of wind power in South Australia? *Catallaxy Files.*

1296 *Science and Environment Policy Project,* February 27th 2021: The week that was.

1297 *Dallas News,* 17th February 2021: Power outages fuel Texans' outrage, stir political leaders to ask how they came about.

1298 Judith Sloan, 8th March 2021: Texas is what you get when cowboys take over the grid. *The Australian.*

1299 Paul Krugman, 18th February 2021: Texas: Land of wind and lies. *New York Times.*

1300 Robert Bryce, 26th April 2021: After the Texas blackouts, follow the wind and solar money – all $66 billion of it. *Real Clear Energy.*

1301 *Dallas News,* 1st August 2021: The Texas blackouts were caused by an epic government failure.

1302 *Forbes,* 15th June 2021: Washington State's approaching energy crisis.

1303 Timothy Lee, 22nd December 2016: Cuomo's energy boondoggle triggers bipartisan rejection. *The Daily Caller.*

1304 www.nyiso.com

1305 www.newsmax.com/LarryBell/Climate-Change-Global-Warming/2015/008/03/id/665118/

1306 Jonathan Lesser, 30th July 2020: Offshore wind power vast boondoggle that New York can no longer afford. *New York Post.*

1307 *Kyodo The Japan Times,* 17th December 2020: ¥60 million wind power project off Fukushima to be dismantled.

1308 Robert Bryce, 9th April 2021: Lower- and middle-class Americans will pay a fortune for Biden's wind-power plan. *New York Post.*

1309 Craig Rucker, 6th May 2021: Joe Biden's offshore wind energy mirage. *Real Clear Energy.*

1310 *City Journal,* 20th March 2021: Green dreams.

1311 www.theguardian.com/environment/2007/dec/10/politics

1312 *The Guardian,* 8th October 2020: UK 'will take 700 years' to reach low-carbon heating under current plans.

1313 Caroline McMorran 11th June 2020: Sutherland windfarm operators paid a total of £630M to turn off turbines. *The Northern Times*

1314 *Daily Mail* 15th October 2029 National Grid warns lack of wind could plunge Britain into darkness.

[1315] *The Times,* 23rd November 2021: Boris' green jobs for China.

[1316] www.frontiereconomics.com, The economics of climate change.

[1317] www.thegwpf.org, Is the UK concealing very high renewables system cost estimates?

[1318] Andrew Montford, 14th October 2020 How is Britain's most efficient windfarm doing? *GWPF Energy.*

[1319] www.thecritic.co.uk/its-alright-for-some/

[1320] Paul Holmwood, 17th September 2020: New plans to switch your power off. *Not a Lot of People Know That.*

[1321] John Constable and Gordon Hughes, 21st September 2020: The costs of offshore wind power: Blindness and insight. *Briefings for Britain.*

[1322] *The Global Warming Policy Foundation,* 23rd November 2020: Boris Johnson's green jobs for China.

[1323] *Global Warming Policy Forum,* Press Release,28th January 2021: GWPF calls on Government to suspend £10 billion green levy on suffering households.

[1324] *Global Warming Policy Foundation,* Press Release, 5th February 2021: Ofgem condemned for misleading the public about energy price rises.

[1325] *Mail on Sunday,* 17th August 2019: Renewable energy is a blackout risk, warns National Grid after chaos during biggest outage in a decade.

[1326] *The Times,* 2nd May 2020: Blackout risk as low demand for power brings plea to switch off wind farms.

[1327] *The Times,* 4th November 2020: Blackouts fear forces power alarm at National Grid.

[1328] Andrew Montford, 15th October 2020: Floating windfarms to sink backers. *The Global Warming Policy Foundation.*

[1329] John Constable, June 2020 The brink of darkness: Britain's fragile power grid. *The Global Warming Policy Foundation.*

[1330] *Daily Mail,* 18th September 2021: Greenflation: Household bills to soar more than £1,500 a year, analysts warn.

[1331] *Business Daily,* 18th September 2021. Becalmed wind energy sector has UK turning to coal.

[1332] *The Independent,* 21st September 2021: Energy Minister admits 'very difficult winter' ahead amid fears of blackouts as gas prices soar.

[1333] *The Sun,* Editorial, 21st September 2021: We should have got cracking with fracking to keep Britain powered and not surrender to eco-protestors.

[1334] *Daily Mail,* 21st September 2021: Power mad: This energy crisis is a mere harbinger of the candle-lit future that awaits us if we do not change course.

[1335] *The Times,* 21st September 2021: Dozens of energy companies will be left to collapse.

[1336] *Financial Times,* 21st September 2021: British Steel warns of up to 50-fold increase in power prices.

[1337] Cap Allon, 7th September 2021: UK fires up coal power plant as European gas shortage worsens, *Electroverse.*

[1338] *AFP,* 30th June 2021: Asia's great reset: Asian nations to build more than 600 new coal power plants.

[1339] Andrew Montford, 3rd July 2020: Mr Net Zero and the great green smokescreen, *The Conservative Woman.*

[1340] *The Australian,* 26th June 2021: Great friends, sure, but don't lecture us on emissions targets.

[1341] *The Daily Telegraph,* 6th May 2021: Jersey crisis exposes UK's dangerous reliance on undersea power cables.

[1342] *The Daily Telegraph,* 10th September 2021: Ireland freezes power exports to UK as energy costs rocket tenfold.

[1343] *The Global Warming Policy Foundation,* 19th April, 2021: More cable failures at European and UK offshore wind farms.

[1344] *Bloomberg,* 6th December 2020: U.K. power prices jump after National Grid warns of supply risks.

[1345] *The Brussels Times,* 11th December 2020: Electricity prices reach record high in Belgium

[1346] *Guido Fawkes,* 18th May 2021: Working class Britons most worried about day-to-day issues, not climate change.

[1347] www.climatechangeauthority.gov.au, 2014 Renewable Energy Target Review.

[1348] www.jacobs.com, Jacobs Carbon Neutrality Commitment.

[1349] www.theguardian.com.au, 18th August 2014, Australia needs the renewable energy target

[1350] www.stopthesethings.com/2021/06/16/burning-cash-renewable-energy-slush-fund-wasting-billions-of-taxpayers-money/

[1351] Malcolm Roberts, 2021: Clean Energy Finance Corporation is going backwards – Senate Estimates. *One Nation.*

[1352] Alan Moran, 20th March 2021: Saving the Portland smelter: one problem solved, others created. *Catallaxy Files.*

[1353] www.theaustralian.com.au/business/mining-energy/tomago-aluminium-buckles-under-price-spikes-backs-snowy-gas/news-story/e12c49a2423f7e07ff1589599519c-379.

[1354] www.australiainstitute.org.au, 11th June 2020: Demand Response Rule Change: Electricity market competition will reduce prices, help with summer heat waves

[1355] www.abc.net.au/news/2021-09-23/sa-tesla-battery-sued-for-not-helping-during-qld-coal-failure/100484664.

[1356] Don Dears, 21st July 2020: It's time to abandon wind power. *Power for USA*

[1357] www.joannenova.com.au/2020/08/wind-power-failure-100-times-a-year-we-get-a-500mw-outage/

[1358] Alan Moran, 7th January 2021: Yes, the energy system is broken – but because of ministers, bureaucrats and regulators rush to renewables. *Spectator Australia.*

[1359] Jo Nova Blog 21st September 2020: Thanks to wind and solar power, Australians have to drive to work to save the planet.

[1360] www.hepburnwind.com.au/wp-content/uploads/2014/06/FY16_Hepburn-Wind-Annual-Report-.pdf

[1361] Viv Forbes, 1st September 2020: The Green Road to Blackouts. *Michael Smith News.*

[1362] Ross Clarke ,19th September 2020: The critics of 'smart meters' were right all along. *The Daily Telegraph.*

[1363] Geoff Dyke, July 2020: Nuclear power through the lens of an Australian trade union. *Public Policy Paper* 2/2020.

[1364] Sheradyn Holderhead, 20th March, 2017: Labor warned wind farms would destabilise energy grid in 2009. *The Advertiser.*

[1365] Michael Owen and Meredith Booth, 30th March 2017: Jay Weatherall rejected $25 million deal to save Northern power station. *The Australian.*

[1366] Michael Owen, 30th June, 2017: Interim power generation to cost SA taxpayers $100 M a year. *The Australian.*

[1367] *The Wall Street Journal,* 10th July 2017: How energy rich Australia exported its way into an energy crisis.

[1368] Terry McCrann, 18th September 2020: Power grab would leave us all in the dark. *The Australian.*

[1369] Viv Forbes, 1st September 2020, The Green Road to Blackouts, *Michael Smith News*

[1370] David Bistrup, 15th March 2021: The cost of a fragile grid; Power prices surge in SA after fore at AGL Energy's Barker Inlet power plant. *Catallaxy Files.*

[1371] AGL's Torrens Island, 1,280 MW gas-steam plant.

[1372] Michael Owen and Meredith Booth, 28th Jun, 2017: State leads the world – for power prices. *The Australian.*

[1373] Jade Gailberger, 27th June 2017: Recycling form to shut, 35 jobs lost, as SA government ignores plea for help over soaring power bills. *The Advertiser.*

[1374] www.joannenova.com.au/2017/06/in-sa-recycling-businesses-going-broke-due-to-electricity-cost.

1375 Rafe Champion, 17th July 2020 More windpower on the way from Tasmania, battery of the nation. *Catallaxy Files.*

1376 www.reneweconomy.com.au/more-than-1-billion-wiped-off-value-of-queensland-coal-and-gas-power-stations/

1377 Matt Canavan, 26th January 2021: The new reality is we need to protect our industries. *The Australian.*

1378 Angela Macdonald-Smith, 19th August 2020: No new renewable projects while the grid is in crisis. *Australian Financial Review.*

1379 David Wojick, 11th January 2021: China loves coal more than wind. *CFACT.*

1380 Jo Nova Blog, 1st September 2021: China poised to be the largest global nuclear power by 2030.

1381 John Constable, 8th August 2020: China's geostrategic priorities become clear. Oil not wind. *The Global Warming Policy Foundation.*

1382 *Reuters*, 21st January 2021.

1383 Tilak Doshi, 10th October 2020: The West intends energy suicide: Will it succeed? *Forbes.*

1384 *Real Clear Energy*, 11th January 2021.

1385 *The Global Warming Policy Foundation*, 20th August 2020: Boris Johnson's wind delusion poses national security risk.

1386 *Global Warming Policy Foundation*, Press Release, 9th June 2021: The truth behind the Government's latest light bulb ban.

1387 www.pv-magazine-australia.com/2020/05/28/chief-scientist-says-big-batteries-are-necessary-to-fast-track-energy-transition

1388 Johan Kristensson, 29th May 2017: New study: Large CO_2 emissions from batteries of electric cars. *New Technology.*

1389 www.thedriven.io, Tesla Model 3 completes round-Australia trip, and the photos are stunning.

1390 *Fox New Video,* 2nd February 2012: Eric Bolling test drives Chevy Volt.

1391 www.theguardian.com, 29th January 2109: Dutch man's epic 89,000 drive proves electric cars are viable in Australia.

1392 https://thewest.com.au/business/cnbc/elon-musk-endorses-federal-carbon-tax-downplays-concerns-over-methane-ng-b882032855z#:~:text=Tesla%20CEO%20Elon%20Musk%20once,annual%20shareholder%20meeting%20on%20Thursday.&text=A%20carbon%20tax%2C%20or%20tax,climate.

1393 www.latimes.com, 3rd July 2018: As Tesla tax credits disappear, will Model 3 deposit holders stick around?

1394 www.latimes.com, 30th May 2015: How it adds up: Three companies, 4.9 billion in government support.

1395 www.latimes.com, 24th June 2018: How Tesla uses cash from Nevada casinos to boost its bottom line.

1396 Wolf Richter, 17th July, 2017: *Business Insider.*

1397 www.wsj.com, 9th July 2017: Tesla sales fall to zero in Hong Kong after tax break is slashed.

1398 *Motor1 News*, 11 July 2017

1399 www.bloomberg.com, 3rd July 2017: Tesla quarterly sales falls as battery supply crimps output.

1400 www.cleantechnica.com, 1st May 2018: Denmark rethinks EV incentives after market collapses.

1401 *Nasdaq*, 24th November 2020: European 'green recovery' falters as car sales continue to tumble.

1402 *Bloomberg,* 29th April 2021, Petrol cars make a covid comeback, and that means burning more oil.

1403 *The Wall Street Journal*, 22nd September 2020: In Europe, regulators want to cut emissions, but consumers want SUVs.

[1404] *Business Insider*, 30th April 2021, 20% of electric vehicle owners in California switched back to fossil fuel because charging their cars is a hassle.

[1405] *Auto Express,* 10th December 2020: Car makers warn electric car plans are "far removed from reality".

[1406] Nissan Leaf.

[1407] Matt Ridley, 10th July 2017: How the electric car revolution could backfire. *The Times.*

[1408] *Mark P. Mills,* 27th October 2020: The Green New Deal can't break the laws of physics. *The Daily Caller.*.

[1409] *The Daily Telegraph,* 19th November 2019: Electric car push set to drive energy bills higher.

[1410] *Daily Mail,* 21st November 2020: One in three motorists cannot afford even the cheapest electric car, experts warn.

[1411] *Financial Times,* 3rd September 2020.

[1412] www.euanmearns.com, Energy Matters.

[1413] *Daily Mail,* 11th September 2020: Petrol and diesel cars could be made £1,500 more expensive to subsidise electric vehicles.

[1414] *The Daily Telegraph,* 8th September 2020.

[1415] www.nytimes.com, Nikola chairman Trevor Milton resigns amid fraud claims

[1416] *Business and Politics,* 24th August 2021, Gregory Wrightstone, Electric vehicle fire catastrophe: It is not a matter of if, but when.

[1417] *National Highway Traffic Safety Administration,* 2012: Chevrolet Volt battery incident overview. NHTSA DOT HS 811 573.

[1418] *Associated Press,* 21st August 2021, Growing fire risk of battery cars forces GM to recall all its electric vehicles.

[1419] www.electrek.co, 14th July 2021, GM asks Chevy Bolt EV owners not to charge overnight or park inside after 2 more fires.

[1420] www.nhtsa.gov, 14th July 2021, Consumer alert: Important Chevrolet Bolt recall for fire risk.

[1421] *Reuters,* 19th November 2020: From Hyundai to Tesla and BMW, battery fires turn the heat on electric cars.

[1422] $LiPF_6$, Li cobalt fluorophosphate, lithium titanate.

[1423] www.samsung.com/us/explore/committed-to-quality/?CID=van-brd-brd-0119-10000141

[1424] www.dailymail.co.uk/sciencetech/article-3883158/Nasapreveals-shocking-video-sercretevive-military-RoboSimian-EXPLODING-batteries-catch-fire.html

[1425] www.ntsb.gov/investigations/AccidentReports/AIR1401.pdf, 13th February 2017: Auxiliary power unit battery fire, Japan Airlines Boeing 787-8, JA829J, Boston, Massachusetts; NTSB/AIR-14-01.

[1426] Nedjalkov, A. *et al.* 2016: Toxic gas emissions from damaged lithium ion batteries – analysis and safety enhancement solution. *Batteries* 2, 5.

[1427] Victoria Allen, 5th April, 2017: Hoodwinked by a green zealot: The scientist behind the dash for diesel called CO2 "worse than terror". *Daily Mail.*

[1428] *The Times,* 26th April 2017.

[1429] *Financial Times*, 12th May 2021: Glencore boss warns of future China dominance in electric vehicles.

[1430] www.iea.org, Sustainable Development Scenario – World Report.

[1431] Tilak Doshi, 2nd August 2020: The dirty secrets of 'clean' electric vehicles. *Forbes.*

[1432] Mark Mills, 9th July 2020: Mines, minerals and "green" energy: A reality check. *Manhattan Institute.*

[1433] www.statista.com, World graphite reserves as of 2020, by country.

[1434] www.amnesty.org, 19th January 2018, Exposed: Child labour behind smart phone and electric car batteries.

[1435] *PowerLine*, 23rd June 2021: Electric cars on collision course with reality.

[1436] *The Daily Telegraph*, 1th May 021: Electric cars make driving too expensive for middle classes, warns Vauxhall chief.

[1437] Eric Worrall, 4th March 2021: How do you extinguish a lithium battery fire? *Watts Up With That?*

[1438] Larsson, F. *et al.* 2017: Toxic fluoride gas emissions from lithium-io battery fires. *Nature Scientific Reports* 7, Article number 10018

[1439] www.fireandemergency.nz, Lithium Batteries – What's the problem?

[1440] www.spectrum.ieee.org, 10th August 2020: Dispute erupts over what sparked an explosive Li-ion energy storage accident.

[1441] www.azcentral.com, 27th May 2020: Cause of APS battery explosion that injured 9 first responders detailed in new report.

[1442] www.spglobal.com, 24th May, 2019: Burning concern: Energy storage industry battles battery fires.

[1443] www.ctif.org, 25th May 2021: Accident analysis of the Beijing lithium battery explosion which killed two firefighters.

[1444] www.epa.gov, 10th July, 2021: An analysis of lithium-ion battery fires in waste management and recycling.

[1445] *Australian Financial Review*, 1st August 2021, Geelong Tesla big battery burns over weekend.

[1446] *The Age*, 1st August 2021, Blaze at Tesla Big Battery extinguished after three days.

[1447] *Business and Politics*, 24th August 2021, Gregory Wrightstone, Electric vehicle fire catastrophe: It is not a matter of if, but when.

[1448] *Australian Financial Review*, 1st August 2021, Geelong Tesla big battery burns over weekend.

[1449] www.thefix.nine.com.au/2017/07/11/08/22/al-gore-on-clive-palmer-south-australia-elon-musk

[1450] Matt Ridley, 21st October 2016: Is climate policy doing more harm than good? *The Australian.*

[1451] *The Australian* 17th June 2017: Energy giant flags 20 pc rise in power bills.

[1452] www.getup.org.au/sa-power?t=53t1t1OQ

[1453] Lindsay Clark, 3rd November 2020: H$_2$?Oh! New water-splitting technique pushes progress of green hydrogen. *The Register.*

[1454] Serra, J. M. *et al.* 2020: Hydrogen production via microwave-induced water splitting at low temperature. *Nature Energy* 5, 910-919.

[1455] *Global Warming Policy Foundation*, Press Release, 19th June 2020, Hydrogen: The once and future fuel.

[1456] *Global Warming Policy Foundation*, Press Release, 6th September 2021, The white elephant of green hydrogen.

[1457] www.abc.net.au/news/science/2021-01-23/green-hydrogen-renewable-energy-climate-emissions-explainer/13081872

[1458] Joe Romm, 2004: *The Hydrogen Hype.* Island Press.

[1459] *The Australian*, 12th September 2021: H2X Global's Warrego-branded hydrogen ute comes with $200k price tag.

[1460] www.transporextra.com, 19th October 2020: Wisdom of hydrogen called into question.

[1461] Viv Forbes, 8th July 2021: Hydrogen hype and hurdles. *Saltbush Club.*

[1462] *The Guardian,* 7th May 2021, Using hydrogen fuel risks locking in reliance on fossil fuels, researchers say.

[1463] www.ge.com, Paper GEA33861.

[1464] Eric Worrall, 14th November 2020: AU53 billion to service a green hydrogen market which does not exist. *Watts Up With That?*

[1465] David Bidstrup, 27th March 2021 Hydrogen hopes *Catallaxy Files.*

[1466] Eric Worrall, 15th June 2020: Germany's climate friendly hydrogen strategy. *Watts Up*

With That?
[1467] Francis Menton, 19th August 2021, Trying to see if California's energy plans add up. *Manhattan Contrarian.*
[1468] US Department of Energy, Energy density in terms of volume.
[1469] www.catallaxyfiles.com/2020/07/28/david-bidstrup-guest-post-green-steel/
[1470] Paul Holmwood, 18th March 2021: Hydrogen supply evidence base – BEIS, *Not a Lot of People Know That.*
[1471] www.spectator.com.au/2017/06/climate-notes/
[1472] www.global.jaxa.jp
[1473] As carbonate and bicarbonate
[1474] www/daf.gov.au/brs/pubprofiles
[1475] www.ourworldindata.org/co2/country/australia?country=~AUS#what-are-the-country's-annual-co2-emissions
[1476] *The Guardian*, 2nd December 2020: No-kill, lab-grown meat to go on sale for first time.
[1477] *The Conversation*, 29th November 2019: 'Cultured' meat could create more problems than it solves.
[1478] *The Times*, 14th May, 2021: How to turn almost everyone into a climate sceptic.
[1479] www.breitibart.com, 14th January 2020: Facebook bug undermines Greta Thunberg's Dad claiming she's not scripted.
[1480] www.news.com.au, 25th September 2019: The PR guru behind the rise of Greta Thunberg
[1481] Malena Ernman & Svante Thunberg, 2016: *Scener ur hjaertat.* Bokförlaget Polaris.
[1482] The *New York Times*,18th August 2019.
[1483] www.sverigesradio.se, 17th September 2019, Winter arrives early in Värmland.
[1484] *Global Times*, 9th May 2021: China says truant school girl Greta lacks knowledge and is full of herself.
[1485] www.statista.com, 18th March 2021: Sweden: average life expectancy by gender
[1486] Zielinski, S. 2015: Sixth-Century misery tied to not one, but two, volcanic eruptions. *Smithsonian Magazine.*
[1487] Larsen, L. B. *et al.* 2008: New ice core evidence for a volcanic cause of the A.D. 536 dust veil. *Geophys. Res.* Letts 35 (4) doi.org/10.1029/2007GL032450.
[1488] Subt, C. *et al.* 2010: Cosmic catastrophe in the Gulf of Carpentaria. AGU Fall Meeting Abstracts PP13A-1499.
[1489] Abbott, D. H. *et al.* 2008: Magnetite and silicate spherules from GISP2 core at the 536 A.D. horizon. *AGU Fall Meeting Abstracts* 41B-1454.
[1490] Stothers, R. S. 1984: Mystery cloud of AD 536. *Nature* 307, 344-345.
[1491] Baillie, M. 2008: Proposed re-dating of the European ice core chronology by seven years prior to the 7th Century AD. *Geophys. Res.* Letts 35 (15) doi:10.1029/2008gl034755.
[1492] www.nationalgeographic.com, 23rd August 2019.
[1493] Ann Gibbons, *Science*, 15th November 2018, Why 536 was 'the worst year to be alive'.
[1494] Loveluck, C. P. *et al.* 2018: Alpine ice-core evidence for the transformation of the European monetary system, AD 640-670. *Antiquity* 58 (366) 1571-1585.
[1495] Luterbacher, J. & Pfister, C. 2015: The year without a summer. *Nature Geoscience* 8, 246-248.
[1496] Zharkova, V. 2020: Modern Grand Solar Minimum will lead to terrestrial cooling. *Temperature* 7(3) 217-222.
[1497] Retejum, A. J. 2020: Epidemics during Grand Solar Minima. *Global Journal of Human-Social Science B. Geography, Geo-sciences, Environmental Science & Disaster Management* 20 (3) ISSN 2249-460X.
[1498] www.almadeutscher.com
[1499] Alan Jones, 20th September 2020: Quoting from a piece entitled "Growing Up" regarding the "school strike for climate." *Sky News Australia.*
[1500] https://joannenova.com.au/2019/05/albany-robbed-of-its-coldest-ever-april-day-bom-

adjusts-temp-up-15-degrees-c/

[1501] www.joannenova.com.au/2015/06/if-it-cant-be-replicated-it-isnt-%20

[1502] www.abc.net.au/news/2017-08-03/australia-needs-more-climate-scientists-review-urges/8767004?pfmredir=sm&WT.mc_id=-newsmail&WT.tsrc+Newsmail

[1503] Lindzen, R. S. 2011: A case against precipitous climate action. *Energy & Environment* 22 (6), 747-751.

[1504] Oke, T. R. 1988: The urban energy balance. *Prog. Phys. Geogr.* 12, 471-508.

[1505] Urban heat-island warming= 0.317lnP, where P=population.

[1506] Zipper, S. C. *et al.* 2016: Urban heat island impacts on plant phenology: Intra-urban variability and response to plant cover. *Envir. Research Letters* 11 (5), 054023

[1507] www.notalotofpeopleknowthat.wordpress.com

[1508] Don Easterbrook, 2016: *Evidence-based climate science.* Elsevier.

[1509] Holmwood, Paul 28th January, 2012: *Iceland's "Sea Ice Years" disappear in GHCN adjustments.* www.notalotofpeopleknowthat.wordpress.com

[1510] www.nside.org/arcticseaicenews/2016/03/

[1511] Christopher Booker, 7th February 2015: The fiddling with temperature data is the biggest science scandal ever: New data shows that the "vanishing" of polar ice is not the result of runaway global warming. *Daily Telegraph.*

[1512] www.cdiac.ornl.gov/epubs/ndp/ushcn/ushcn.html

[1513] www.stevegoddard.wordpress.com/data-tampering-at-ushcngiss/

[1514] www.cru.uea.ac.uk/cru/data/temperature/HadCRUT3-gl.dat

[1515] www.metoffice.gov.uk/hadobs/hadcrut4/data/versions/previous_versions.html

[1516] www.notalotofpeopleknowthat.wordpress.com/tag/temperature-adjustments/

[1517] www.data.giss.nasa.gov/gistemp/

[1518] www.wattsupwiththat.com/2015/07/24/impact-of-pause-buster-adjustment-on-giss-monthly-data/

[1519] www.dailycaller.com, 25th May, 2017: Climate scientists trying to discredit Trump's EPA chief end up proving him right.

[1520] Santer, B. D. *et al.* 2017: Causes in differences between model and satellite tropospheric warming rates. *Nature Geoscience* 10 (7), 478-485.

[1521] Pokrovsky, O. 2019: Cloud changes in the period of global warming The results of the International Satellite Project. *Izvestiya-Atmospheric and Ocean* Physics 55 (9), 1189-1197.

[1522] Svensmark, H. 2017: Increased ionization supports growth of aerosols into cloud condensation nuclei. *Nature Communications* 8, Article 2199.

[1523] www.woodfortrees.org/plot/gistemp/from:1998/to:2015/plot/gistemp/from:1998/t):2015/trend/plot/rss/from:1998/to:2015/plot/rss/from:1998/to:2015/trend

[1524] www.notalotofpeopleknowthat.wordpress.com/2015/04/09/coolin-the-past-in-holland/

[1525] www.wattsupwiththat.com/2009/12/08/thesmoking-gun-at-darwin-zero/

[1526] www.jennifermarohasy.com/2015/08/bureau-just-makes-stuff-up-deniliquin-remodelled-then-rutherglen-homogenized/

[1527] *Sunday Telegraph*, 30th August 2015.

[1528] Marohasy, J. 2016: Temperature change at Rutherglen in South-East Australia. *New Climate* https://doi.org/10.2222/nc.2016.001.

[1529] Benalla (Vic), Echuca (Vic.), Deniliquin (NSW).

[1530] Graham Lloyd, 4th September 2014: 'More time' to find Rutherglen temperature record. *The Australian.*

[1531] www.bom.gov.au/climate/change/.../ACORN-SAT-Station-Catalogue-2012-WEB.pdf

[1532] Hogan, B. 2017: Warming up in Rutherglen: A century-old weather station in Rutherglen has shown researchers how climate statistics are re-modelled to create a warming trend. *IPA Review* July 2017, 38-42.

[1533] Marohasy, J. 2016. Temperature change at Rutherglen in south-east Australia, New Climate, https://doi.org/10.22221/nc.2016.001

[1534] Deacon, E. L. 1952: Climate change in Australia since 1880. *Australian Journal of Physics* 6, 209-218.

[1535] Marohasy, J. *et al.* 2014: Modelling Australian and global temperatures. What's wrong? Bourke and Amberley as case studies. *The Sydney Papers Online,* 26.

[1536] Michael Bastasch 2017: Study finds temperature adjustments account for "nearly all of recent warming" in climate data sets. *Daily Caller* 7th July 2017.

[1537] Joe D'Aleo, 7th July 2017, *Daily Caller News Foundation.*

[1538] Quirk, T. 2017: Taking Melbourne's temperature. In: *Climate change: The facts 2017* (ed. J. Marohasy). IPA, 101-116.

[1539] www.jennifermarohasy.com, 23rd February 2019: Changes to Darwin's climate history are not logical.

[1540] www.climateconversation.org.nz/docs/Statistical%20Audit%20of%20the%20 NIWA%207-Station%20Review%20Aug%202011.pdf

[1541] www.drroyspencer, 29th January 2021

[1542] Stone, B. 2009: Land use as climate change mitigation. *Envir. Sci. Technol.* 43 (6), doi 10.1021/es902150g

[1543] Joseph D'Aleo, *Daily Caller News Foundation* 27th June 2017.

[1544] Manley, G. 1974: Central England temperatures: Monthly means 1659 to 1973. *Roy. Met. Soc.* 425, 389-405.

[1545] Parker, D. *et al.* 1992: A new daily central England temperature series 1772-1991. *Envir. Science* doi:10.1002/JOC.33790120402.

[1546] Christiansen, B. & Ljungqvist, F. C. 2012: The extra-tropical Northern Hemisphere temperature in the last two millennia: reconstructions of low frequency variability. *Clim. Past* 8: 765-786.

[1547] Ljungqvist, F. C. *et al.* 2012: Northern Hemisphere temperature patterns in the last 12 Centuries. *Clim. Past* 6: 227-249.

[1548] Jouzel, J. *et al.* 2007: Orbital and millennial Antarctic climate variability over the past 800,000 years. *Science* 317: 793-796.

[1549] Scafetta, N. 2021: Detection on non-climate biases in land surface temperature records by comparing climatic data and their model simulations. *Climate Dynamics* doi.org/10.1007/ s00382-021-05626-x.

[1550] www.oceanservice.noaa.gov/education/tutorial_currents/05conveyor2.html

[1551] www.argo.ucsd.edu/

[1552] www.ftp://ftp.cru.uea.ac.uk/

[1553] www.ftp://ftp.cru.uea.ac.uk/people/philjones/

[1554] Steve McIntyre 2009: *"Unprecedented" data purge at CRU.* Climate Audit.

[1555] Joseph D'Aleo, *Daily Caller News Foundation* 27th June 2017.

[1556] https://wattsupwiththat.com.

[1557] Watts, A. 2017: Creating a false warming signal in the US temperature record. In: *Climate change: The facts 2017* (ed. J. Marohasy). IPA, 75-91.

[1558] www.noaanews.noaa.gov/stories2009/20090319_lubchenco.html

[1559] www.motls.blogspot.com/2008/01/2007-warmest-year-on-record-coldest-in.html

[1560] www.cato.org, Andrei Illarionov, 17th December 2009, New Study: Hadley Center and CRU apparently cherry-picked Russia's climate data.

[1561] www.data.giss.nasa.gov/gistemp/stdata/

[1562] www.wattsupwiththat.com/2010/04/24/inside-the-eureka-weather-station/

[1563] www.cma.gov.cn/en2014/

[1564] www.tahmo.org/_african-climate-data/

[1565] Leeper, R. D. *et al.* 2019: Impacts of small-scale urban encroachment on air temperature observations. *Amer. Met. Soc.* 58(6), 1369-1380.

[1566] Fall, S. A. *et al.* 2011: Analysis of the impacts of station exposure on the U.S. Historical Climatology Network temperatures and temperature trends. *J Geophys Res.* doi.

org/10.1029/2010JD015146.

[1567] www.wattsupwiththat.com, 3rd May 2019, Big news—verified by NOAA- poor weather station siting.

[1568] www.drroyspencer.com, 22nd April 2021.

[1569] Seim, T. O. & Olsen, B. T. 2020: The influence of IR absorption and backscatter radiation from CO_2 on air temperature during heating in a simulated Earth/atmosphere experiment. *Atmos. Climate Sci.* 10, 168-185.

[1570] Pettenkofer method.

[1571] Beck, E. 2007: 180 years of atmospheric CO_2 gas analysis by chemical methods. *Energy and Environment* 18: 259-282.

[1572] Pales, J. C. and Keeling, C. D. 1965: The concentration of atmospheric carbon dioxide in Hawaii. *Jour.n Geophys. Res.* 70: 6053-6076.

[1573] Backastow, R. *et al.* 1985: Seasonal amplitude increase in atmospheric CO_2 concentration at Mauna Loa, Hawaii, 1959-1982. *Journ. Geophys. Res. 90*: 10529-10540.

[1574] Bischof, W. 1960: Periodical variations of the atmospheric CO_2-content in Scandinavia. *Tellus* 12: 216-226.

[1575] Bischof, W. 1962: Variations in concentration of carbon dioxide in free atmosphere. *Tellus* 14: 87-90.

[1576] Ryan, S. 1995: Quiescent outgassing of Mauna Loa Volcano 1958-1994. In: *Mauna Loa revealed: structure, composition, history and hazards* (eds Rhodes, J. M. & Lockwood, J. P.), *Amer. Geophy. Union Monogr.* 92: 92-115.

[1577] Jaworowski, Z. *et al.* 1992: Atmospheric CO_2 and global warming: a critical review; 2nd Revised Edition. *Norsk Polarinstitutt Meddelelser* 119.

[1578] Keeling, C. D. *et al.* 1989: A three-dimensional model of atmospheric CO_2 transport based on observed winds. 1: Analysis of observational data. In: *Aspects of climate variability in the Pacific and the Western Americas* (ed. Peterson, D. H.) *Amer. Geophys. Union Monogr.* 55: 165-236.

[1579] Keeling, R. F. *et al.* 1996: Global and hemispheric sinks deduced from changes in atmospheric O_2 concentration. *Nature* 381: 218-221.

[1580] Keeling, C. D. & Whorf, T. P. 2005: Atmospheric CO_2 records from sites in the SIO air sampling network. In: *Trends: A compendium of data on global change. Carbon Dioxide Analysis Center*, Oak Ridge National Laboratory, TN.

[1581] MacFarling Meure, C. *et al.* 2006: Law Dome CO_2, CH_4 and N_2O ice core records extended to 2000 years BP. *Geophys. Res. Letts* 33: L14810.

[1582] ftp://ftp.cmdl.noaa.gov/ccg/co2/in-situ

[1583] Litner, B. R., Buermann, W., Koven, C. D. and Fung, I. Y.,2006: Seasonal circulation and Mauna Loa CO_2 variability. *Journ. Geophys. Res.* 111, D13104, 10.1029/2005JD006535.

[1584] www.edberry.com, 24th November, 2015: Salby: CO_2 follows integral of temperature.

[1585] Jaworowski, Z. *et al.* 1992: Atmospheric CO_2 and global warming: a critical review; 2nd Revised Edition. *Norsk Polarinstitutt Meddelelser* 119.

[1586] Briegleb, B. P. 1992: Longwave band model for thermal radiation in climate studies. *Journ Geophys. Res.* 97: 11475-11485.

[1587] IPCC, 2001: *Climate Change 2001. The scientific basis. Contributions of working group 1 to the Third Assessment Report of the Intergovernmental Panel on Climate Change* (eds Houghton, J. T. *et al.*) Cambridge University Press.

[1588] Callender, G. S. 1938: The artificial production of carbon dioxide and its influence on temperature. *Quart. Journ. Royal Meteorol. Soc.* 66: 395-400.

[1589] Scafetta, N. & West, B. J. 2006a: Phenomenological solar signature in 400 years of reconstructed Northern Hemisphere temperature record. *Geophys. Res. Letts* 33: L17718 doi:10.1029/2006GL027142.

[1590] Scafetta, N. & West, B. L. 2006b: Phenomenological solar contribution to the 1900-2000 global surface warming. *Geophys. Res. Letts* 33: L05708 doi:10.1029/2005GL025539.

[1591] Rodhe, H. 1992: Modeling biogeochemical cycles. In: *Global biogeochemical cycles* (eds Butcher, S. S. *et al.* 55-72, Academic Press.

[1592] Houghton, J. T. *et al.* (eds) 1990: *Climate Change. The IPCC Assessment. Intergovernmental Panel on Climate Change.* Cambridge University Press.

[1593] O'Neill, B. C. *et al.* 1994: Reservoir timescales for anthropogenic CO_2 in the atmosphere. *Tellus* 46B: 378-389.

[1594] Jaworowski, Z. *et al.* 1992: Atmospheric CO_2 and global warming: a critical review; 2nd Revised Edition. *Norsk Polarinstitutt Meddelelser* 119.

[1595] Segalstad, T. V. 1996: The distribution of CO_2 between atmosphere, hydrosphere and lithosphere; minimal influence from anthropogenic CO_2 on the global "Greenhouse Effect". *The Global Warming Debate. The Report of the European Science and Environment Forum.* (ed. Emsley, J.), 41-50, Bourne Press.

[1596] Revelle, R. & Suess, H. 1957: Carbon dioxide exchange between atmosphere and ocean and the question of an increase in atmospheric CO_2 during past decades. *Tellus* 9: 18-27.

[1597] Broecker, W. S. *et al.* 1979: Fate of fossil fuel carbon dioxide and the global carbon balance. *Science* 206: 409-418.

[1598] Jaworowski, Z. *et al.* 1992: Atmospheric CO_2 and global warming: a critical review; 2nd Revised Edition. *Norsk Polarinstitutt Meddelelser* 119.

[1599] Bischof, W. 1960: Periodical variations of the atmospheric CO_2-content in Scandinavia. *Tellus* 12: 216-226.

[1600] Zhen-Shan, L. & Xian, S. 2007: Multi-scale analysis of global temperature changes and trend of a drop in temperature in the next 20 years. *Meteorol. Atmos. Phy.* 95: 115-121.

[1601] Le Quere, C. *et al.* 2003: Two decades of ocean CO_2 sink and variability. *Tellus B* 55: 649-656.

[1602] https://www.ocregister.com, 18th January 2011, Global warming baloney: Hottest year on record

[1603] https://www.barasso.senate.gov/public/index.cfm/2010/2010/3/post-b5fb5b46-a699-6acb-e43d-c14a42133346

[1604] http://scienceandpublicpolicy.org, 27th August 2010, Surface temperature records: Policy driven deception?

[1605] http://www.ncdc.noaa.gov/img/climate/research/ushcn/ts.ushcn_anom25_diffs_urb-raw_pg.gif

[1606] A new paper comparing NCDC rural and urban surface temperature data, Watts Up With That?, 26th February 2010.

[1607] Long, E. R. 2010: Contiguous U.S. temperature trends using NCDC raw and adjusted data for one-per-state rural and urban station sets. *SPPI Original Paper*, February 27, 2010.

[1608] John O'Sullivan, 4th July 2017: Fatal courtroom act ruins Michael 'hockey stick Mann

[1609] www.aeir.org, 10th August 2017: Hockey stick climate change enforcer checked hard in Canadian case.

[1610] Kenneth Schultz, 10th July 2021: Logistics and costs for Australia to achieve net zero carbon dioxide emissions by 2050. *Watts Up With That?*

[1611] www.energy.gov.au, Australian Energy Update, Commonwealth of Australia 2020 _ Guide to the Australian Energy Statistics

[1612] *The Daily Telegraph*, 14th May 2021: Net Zero will unleash monster price rises and environmental devastation.

[1613] *Financial Times*, 29th September 2021: Renewable energy output drops almost a third.

[1614] *The Guardian*, 30th September 2021: Britain's energy sector in meltdown as leading energy supplier faces collapse.

[1615] *The Guardian*, 28th September 2021: Surge in UK wholesale gas prices fuels winter energy crisis fears.

[1616] www.OilPrice.com, Are carbon taxes to blame for Europe's energy crisis?

[1617] *Daily Express*, 23rd September 2023: Russia 'to determine UK's winter fate' after Merkel 'outsmarted' by Putin on energy crisis.

[1618] *The Daily Telegraph*, 23rd September 2021: Europe wrapped around Putin's little finger.

[1619] *American Thinker*, 28th September 2021: Green energy leaves Europe in the cld – the freezing cold.

[1620] *Bloomberg,* 30th September 2021: Europe asking Russia for more coal to survive winter energy crunch.

[1621] *Daily Mail*, 23rd September 2021: Why are we facing an energy crisis when we're sitting on a gold mine? We should consider giving fracking pioneers and North Sea drillers the only 'green' light that matters: the one that says 'Go'.

[1622] Global Coal Newcastle Standard Specification.

[1623] *Reuters*, 28th September 2021: China desperate for more coal as energy crisis cripples large section of industry.

[1624] *The Daily Telegraph*, 28th September 2021: Streets go dark and lifts grind to a halt as China cuts power to meet climate demands.

[1625] *The Sydney Morning Herald*, 30th September 2021: China braces for a chilly winter as its home-grown energy crisis intensifies.

[1626] Ronald Bailey, 2nd July 2021: To stop climate change, Americans must cut energy use by 90 percent, live in 640 square feet, and fly only once every 3 years, study claims. *Reason.*

[1627] *The Daily Telegraph*, 30th September 2021: Boris should beware his Red Wall doesn't become a Yellow Vest.

[1628] *The Wall Street Journal*, 29th September 2021: Will Net Zero bring down Joe Biden?

[1629] *Forbes*, 14th August 2014: The IPPC's climate "code red" versus the real world.

[1630] www.theepochtimes.com/mkt_morningbrief/un-climate-report-reveals-the-crisis-is-about-thuth-bot-climate_3945984.html

[1631] www.cel.org, 10th August 2021: Observations concerning the newest IPCC report.

[1632] IPCC 2001, Chapter 14.2.2.2.

[1633] Voosen, P. 27th July, 2021: U.N. climate panel confronts implausibly hot forecasts of future warming. *Science.*

[1634] www.cfact.org/2021/08/17/the-un-ipcc-science-panel-opts-for-extreme-nuttiness/

[1635] Science and Environmental Policy Project, 14th August 2021.

[1636] *Vox*, 19th May, 2021: Scientists concede they haven't got a clue what will happen to clouds and temperatures as the planet warms.

[1637] Connolly, R. *et al.* 2021: How much has the Sun influenced Northern Hemisphere temperature trends? An ongoing debate. *Atmos. Ocean Phys* doi: 10.1088/1674-4527/21/6/131.

[1638] https://wattsupwiththat.com/2021/08/18/ar6-and-the-paleocene-eocene-thermal-maximum.

[1639] IPCC AR6 2921, p. 5-14.

[1640] McKitrick, R. 2021: Checking for model consistency in optimal fingerprinting: a comment. *Climate Dynamics* doi.org/10.1007/s003282-021-05914-7.

[1641] *Committee for a Constructive Tomorrow*, 19th August 2021: IPCC proposed banishment of fossil fuels would place most of the world's population at risk.

[1642] www.unifccc.int, 30th November 2015, COP21

[1643] *The Times*, 16th June 2021: Scotland's Net Zero fiasco.

[1644] *The Australian*, 17th September 2021, UN declares world is out of time on emissions.

[1645] *The Sunday Telegraph*, 28th August 2021, Alok Sharma under fire as nuclear industry claim that they have been banned from the COP26.

[1646] *Daily Mail*, 31st August 2021, Fury as travel test bypass for thousands as it emerges that delegates at Cop26 climate summit will be allowed to dodge Covid screening.

[1647] *The Hill*, 8th September 2021, John Kerry says the world is 'doomed' unless 20 nations take climate action.

[1648] China, USA, Russia, Japan, Germany, South Korea, Iran, Canada, Indonesia, Saudi Arabia, Mexico, South Africa, Brazil, Australia, Turkey, UK, Italy, Poland and France.

[1649] *Daily Mail*, 8th September 2021, Alok Sharma is accused of pandering to China as state media says he 'hailed Beijing's efforts in tackling climate change'.

[1650] *The Hindu*, 15th September 2021, India rebuffs John Kerry, won't commit to Net Zero.

[1651] *Spiked*, 9th September 2021, Tim Black: Climate alarmism is the real threat to public health.

[1652] www.quadrant.org.au/opinion/doomed-planet/2021/07/when-climateers-let-the-truth-slip-out/.

[1653] Thomas Richard, *PoliZette*, 13th July 2017.

[1654] *Clean Energy Wire*, 11th January 2021.

[1655] Luther's 95 Theses were to reform some of the theological and administrative aspects of the Catholic Church and he had no original intention to create Protestantism.

[1656] www.economicshelp.org, World Bank EN.ATM.CO2E.PC

[1657] Fight the good fight.

[1658] Erini, F. F. *et al.* 2021: Personal carbon allowances. *Nature Sustainability* doi/org 10/1038/s41893-021-00756-w

[1659] www.climatedepot.com/2016/11/03/watch-schwarzenegger-again-threatens-climate-skeptics-i-wouild-like-to-strap-their-mouth-to-the-exhaust-pipe-of-a-truck-turn-on-the-engine

[1660] www.breitbart.com/big-hollywood/2017/03/17/delingpole-climate-change-deniers-should-be-executed-gently-says-eric-idle/.

[1661] www.americanthinker.com/articles/2012/12/professor_calls_for_death_penalty_for_climate_change_deniers.html.

[1662] www.dailymail.co.uk/sciencetech/article-2566659?Are-global-warming-Nazi-People-label-sceptics-deniers-kill-MORE-people-Holocaust-claims-scientist.html

[1663] www.who.int, 31st May 2019.

[1664] www.bbc.co.uk, 28th February 2011: Study say green sector costs more jobs than it creates.

[1665] Modified from *The Life of Brian*.

[1666] Manning, S. W. *et al.* 2020: Beyond megadrought and collapse in the Northern Levant: The chronology of Tell Tayinat and two historical inflexion episodes, around 4.2ka BP, and following 3.2ka BP. *PLOS One* 29th October 2020, doi.org/1371/journal.pone.0240799.

[1667] Moran, A. 2021: No upside for electricity customers in the early closure of coal generators. *Spectator Australia* 10th March 2021.

CPSIA information can be obtained
at www.ICGtesting.com
Printed in the USA
LVHW080919040622
720508LV00015B/723